Lecture Notes in Computer Science 1948

Edited by G. Goos, J. Hartmanis and J. van Leeuwen

T0205273

Springer
Berlin
Heidelberg
New York
Barcelona
Hong Kong
London
Milan
Paris
Singapore
Tokyo

Tieniu Tan   Yuanchun Shi
Wen Gao (Eds.)

Advances in Multimodal Interfaces – ICMI 2000

Third International Conference
Beijing, China, October 14-16, 2000
Proceedings

 Springer

Series Editors

Gerhard Goos, Karlsruhe University, Germany
Juris Hartmanis, Cornell University, NY, USA
Jan van Leeuwen, Utrecht University, The Netherlands

Volume Editors

Tieniu Tan
Chinese Academy of Sciences, Institute of Automation
P.O. Box 2728, Beijing 100080, China
E-mail: tnt@nlpr.ia.ac.cn

Yuanchun Shi
Tsinghua University, Computer Department, Media Laboratory
Beijing 100084, China
E-mail: shiyc@tsinghua.edu.cn

Wen Gao
Chinese Academy of Sciences, Institute of Computing Technology
P.O. Box 2704, Beijing 100080, China
E-mail: wgao@cti.com.cn

Cataloging-in-Publication Data applied for

Die Deutsche Bibliothek - CIP-Einheitsaufnahme

Advanced in multimodal interfaces : third international conference ;
proceedings / ICMI 2000, Beijing, China, October 14 - 16, 2000.
Tieniu Tan ... (ed.). - Berlin ; Heidelberg ; New York ; Barcelona ; Hong
Kong ; London ; Milan ; Paris ; Singapore ; Tokyo : Springer, 2000
   (Lecture notes in computer science ; Vol. 1948)
   ISBN 3-540-41180-1

CR Subject Classification (1998): H.4, I.2, I.4, I.5, D.2

ISSN 0302-9743
ISBN 3-540-41180-1 Springer-Verlag Berlin Heidelberg New York

Springer-Verlag Berlin Heidelberg New York
a member of BertelsmannSpringer Science+Business Media GmbH
© Springer-Verlag Berlin Heidelberg 2000
Printed in Germany

Typesetting: Camera-ready by author
Printed on acid-free paper   SPIN 10780929   06/3142   5 4 3 2 1 0

Preface

Multimodal Interfaces represents an emerging interdisciplinary research direction and has become one of the frontiers in Computer Science. Multimodal interfaces aim at efficient, convenient and natural interaction and communication between computers (in their broadest sense) and human users. They will ultimately enable users to interact with computers using their everyday skills.

These proceedings include the papers accepted for presentation at the Third International Conference on Multimodal Interfaces (ICMI 2000) held in Beijing, China on 14–16 October 2000. The papers were selected from 172 contributions submitted worldwide. Each paper was allocated for review to three members of the Program Committee, which consisted of more than 40 leading researchers in the field. Final decisions of 38 oral papers and 48 poster papers were made based on the reviewers' comments and the desire for a balance of topics. The decision to have a single track conference led to a competitive selection process and it is very likely that some good submissions are not included in this volume. The papers collected here cover a wide range of topics such as affective and perceptual computing, interfaces for wearable and mobile computing, gestures and sign languages, face and facial expression analysis, multilingual interfaces, virtual and augmented reality, speech and handwriting, multimodal integration and application systems. They represent some of the latest progress in multimodal interfaces research.

We wish to thank all the authors for submitting their work to ICMI 2000, the members of the Program Committee for reviewing a large number of papers under a tight schedule, and the members of the Organizing Committee for making things happen as planned. We are grateful to the Chinese National 863-306 Steering Committee, the National Natural Science Foundation of China and the National Laboratory of Pattern Recognition of the CAS Institute of Automation for financial support.

August 2000

Tieniu Tan
Yuanchun Shi
Wen Gao

General Co-Chairs

Wen Gao (CAS Institute of Computing Technology, China)
A. Waibel (CMU, USA)

Program Committee Co-Chairs

Tieniu Tan (CAS Institute of Automation, China)
S. Oviatt (Oregon Grad. Institute of Science and Technology, USA)
K. Yamamoto (Gifu University, Japan)

Program Committee:

Keiichi Abe (Shizuoka University, Japan)
N. Ahuja (UIUC, USA)
S. Akamatsu (ATR, Japan)
Elisabeth Andre (DFKI, Germany)
Norm Badler (University of Pennsylvania, USA)
H. Bunke (University of Bern, Switzerland)
Ruth Campbell (University College London, UK)
P.C. Ching (Chinese University of Hong Kong, China)
Phil Cohen (OGI, USA)
James L. Crowley (I.N.P.G, France)
Trevor Darrell (MIT, USA)
Myron Flickner (IBM-Almaden, USA)
Rob Jacob (Tufts University, USA)
J. Kittler (Univ. of Surrey, UK)
H. Koshimizu (Chukyo University, Japan)
Jim Larson (Intel, USA)
S.-W. Lee (Korea University, Korea)
Bo Li (Beijing University of Aeronautics and Astronautics, China)
G. Lorette (IRISA, France)
Hanqing Lu (CAS Institute of Automation, China)
Tom MacTavish (Motorola, USA)
Dominic Massaro (University of California at Santa Cruz, USA)
Steve J. Maybank (University of Reading, UK)
Rosalind W. Picard (MIT, USA)
Steve Shafer (Microsoft Research, USA)
Paul M. Sharkey (University of Reading, UK)
Rajeev Sharma (Penn State University, USA)
Yuanchun Shi (Tsinghua University, China)
H. Shum (Microsoft Research, China)
Kentaro Toyama (Microsoft Research, USA)
Matthew Turk (Microsoft, USA)

Geoff West (Curtin University of Technology, Australia)
Bo Xu (CAS Institute of Automation, China)
Yangsheng Xu (Chinese University of Hong Kong, China)
Jie Yang (CMU, USA)
B.Z. Yuan (Northern Jiaotong University, China)
Pong Chi Yuen (Hong Kong Baptist University, China)
Shumin Zhai (IBM-Almaden, USA)

Organizing Committee:

Bo Li, Co-chair (Beijing University of Aeronautics and Astronautics, China)
Yuanchun Shi, Co-chair (Tsinghua University, China)
Mengqi Zhou, Co-chair (CIE, China)
Bing Zhou (Beijing University of Aeronautics and Astronautics, China)
Cailian Miao (CAS Institute of Automation, China)
Changhao Jiang (Tsinghua University, China)
Hao Yang (CAS Institute of Automation, China)
Wei Li (Beijing University of Aeronautics and Astronautics, China)
Yan Ma (CAS Institute of Automation, China)
Yanni Wang (CAS Institute of Automation, China)
Yuchun Fang (CAS Institute of Automation, China)
Yujie Song (CAS Institute of Automation, China)
Zhiyi Chen (CAS Institute of Automation, China)

Sponsors and Organizers:

Chinese National 863-306 Steering Committee
Natural Science Foundation of China
National Laboratory of Pattern Recognition at CAS Institute of Automation
China Computer Federation
Beijing University of Aeronautics and Astronautics
Tsinghua University
IEEE, Beijing Section
Association for Computing Machinery

International Steering Committee:

A. Waibel (CMU, USA)
Qingyun Shi (Peking University, China)
Seong-whan Lee (Korea University, Korea)
Victor Zue (MIT, USA)
Wen Gao (CAS Institute of Computing Technology, China)
Yuan Y. Tang (Hong Kong Baptist University, China)

Table of Contents

Affective and Perceptual Computing

Gesture Recognition

Face and Facial Expression Detection, Recognition and Synthesis

Multilingual Interfaces and Natural Language Understanding

Speech Processing and Speaker Detection

Object Motion, Tracking and Recognition

Handwriting Recognition

Input Devices and Its Usability

Virtual and Augmented Reality

Multimodal Interfaces for Wearable and Mobile Computing

Sign Languages and Multimodal Navigation for the Disabled

Multimodal Integration and Application Systems

Gaze and Speech in Attentive User Interfaces

Paul P. Maglio, Teenie Matlock, Christopher S. Campbell,
Shumin Zhai, and Barton A. Smith

IBM Almaden Research Center
650 Harry Rd
San Jose, CA 95120 USA

{pmaglio, tmatlock, ccampbel, zhai, basmith}@almaden.ibm.com

Abstract. The trend toward pervasive computing necessitates finding and implementing appropriate ways for users to interact with devices. We believe the future of interaction with pervasive devices lies in *attentive user interfaces*, systems that pay attention to what users do so that they can attend to what users need. Such systems track user behavior, model user interests, and anticipate user desires and actions. In addition to developing technologies that support attentive user interfaces, and applications or scenarios that use attentive user interfaces, there is the problem of evaluating the utility of the attentive approach. With this last point in mind, we observed users in an "office of the future", where information is accessed on displays via verbal commands. Based on users' verbal data and eye-gaze patterns, our results suggest people naturally address individual devices rather than the office as a whole.

1 Introduction

It is a popular belief that computer technology will soon move beyond the personal computer (PC). Computing will no longer be driven by desktop or laptop computers, but will occur across numerous "information appliances" that will be specialized for individual jobs and pervade in our everyday environment [9]. If point-and-click graphical user interfaces (GUI) have enabled wide use of PCs, what will be the paradigm for interaction with pervasive computers? One possible approach is *attentive user interfaces* (AUI), that is user interfaces to computational systems that *attend* to user actions—monitoring users through sensing mechanisms, such as computer vision and speech recognition —so that they can *attend* to user needs— anticipating users by delivering appropriate information before it is explicitly requested (see also [7]).

More precisely, attentive user interfaces (a) monitor user behavior, (b) model user goals and interests, (c) anticipate user needs, and (d) provide users with information, and (e) interact with users. User behavior might be monitored, for example, by video cameras to watch for certain sorts of user actions such eye movements [4,14] or hand gestures [1], by microphones to listen for speech or other sounds [10], or by a computer's operating system to track keystrokes, mouse input, and application use

T. Tan, Y. Shi, and W. Gao (Eds.): ICMI 2000, LNCS 1948, pp. 1-7, 2000.
© Springer-Verlag Berlin Heidelberg 2000

[4,6,7]. User goals and interests might be modeled using Bayesian networks [4], predefined knowledge structures [11], or heuristics [7]. User needs might be anticipated by modeling task demands [11]. Information might be delivered to users by speech or by text [7,10], and users might interact directly through eye gaze, gestures or speech [1,4,12,14].

Attentive user interfaces are related to perceptual user interfaces (PUI), which incorporate multimodal input, multimedia output, and human-like perceptual capabilities to create systems with natural human-computer interactions [10,13]. Whereas the emphasis of PUI is on coordinating *perception* in human and machine, the emphasis of AUI is on directing *attention* in human and machine. For a system to attend to a user, it must not only perceive the user but it must also anticipate the user. The key lies not in how it picks up information from the user or how it displays information to the user; rather, the key lies in how the user is modeled and what inferences are made about the user.

This paper reports the first of a series of studies investigating issues in AUI. Critical to AUI is multimodal input: speech, gesture, eye-gaze, and so on. How do people use speech when addressing pervasive computing devices and environments? Do people talk to devices differently than they talk to other people? Do people address each device as a separate social entity. Do they address the whole environment as a single entity? In the famous science fiction movie, *2001: A Space Odyssey*, the astronauts address HAL, the computer that runs the spaceship, directly; they do not address the individual objects and devices on board the ship. Is it really more natural to say "Open the pod-bay doors, HAL" than it is to say "Pod-bay doors, open"? In addition, eye gaze plays an important role in human-human communication; for instance, a speaker normally focuses attention on the listener by looking, and looking away or avoiding eye-contact may reflect hesitation, embarrassment, or shyness. Thus, it is natural to ask, what is the role of eye gaze when speaking in an attentive environment? Will people tend to look at individual devices when communicating with them? If so, eye gaze might provide additional information to disambiguate speech.

2   Experiment

We investigated how people naturally interact with an "office of the future". To separate conceptual issues from current technology limitations, a Wizard-of-Oz design was used to provide accurate and timely reactions to user commands and actions. The devices in the office were controlled by one of the experimenters hidden behind a wall. Users were given a script containing a set of office tasks and were told to perform them using verbal commands only. A green blinking light served as feedback that the command was understood and being executed. There was one between-subjects factor with two levels, distributed feedback and non-distributed feedback. In the distributed condition (DC), feedback in the form of the green flashing light was seen on each device. In the non-distributed condition (NC), feedback appeared in a single location on the wall representing the "office". Non-distributed feedback was meant to suggest that the office itself process the commands (kind of like talking to HAL in 2001), whereas distributed feedback was meant to suggest that individual devices process the commands (unlike talking to HAL). We were interested in whether and how people's behavior—verbal and gaze—might change with the kind of feedback provided.

2.1   Method

Thirteen volunteers were randomly placed into one of two conditions—DC or NC. In both, participants were given a script in the form of a list containing six tasks, such as get an address, dictate memo, print memo, find date from calendar, get directions, and print directions. These were to be completed using four devices available in the room: an address book, a calendar, a map, and a dictation device. Neutral language in the instructions (e.g., "You will need to make a printout of the document") was used so as not to bias participants' utterances toward giving command to individual devices (distributed) or to the room as a whole (non-distributed). As a cover story, participants were told that a wide range of voices were needed to test IBM's "office of the future" speech recognition project. This way, participants were not aware we would be looking at the nature of their verbal commands, and would feel less self-conscious about the types of language they were generating. It also ensured that participants would speak loudly and clearly. Participants were told to take their time and to repeat verbal commands if they encountered a lag or a problem in being understood. Hidden cameras on the devices recorded gaze information, and a microphone recorded verbal commands.

2.2   Attentive Office

The attentive office contained three small flat screen displays were labeled "Calendar", "Map/Directions", and "Address", and a plastic, futuristic-looking orb was labeled "Dictation". There was also a printer, and a large flat screen display without a function, but with futuristic images displayed on it (simply meant to distract user's attention). All displays were 800x600 pixel LCD flat panels. In the DC, a small black box with a green light (feedback module) was attached to the top of each screen. For the dictation device, no screen was used, so the feedback module was placed behind the futuristic orb. No traditional manual input devices (keyboards, mice, or other control buttons) were in the room. During the experiment, devices displayed the requested information immediately after the request. The information appeared in a futuristic-looking graphical display that did not look like the typical Windows desktop environment (see Figure 1).

Fig 1. Futuristic display used in the attentive office.

3   Results

The language data were coded as follows. Verbal commands were first placed into four categories based on type of request: imperative, subject noun-phrase, question, and other. An imperative request was a direct command to perform an action (e.g., "Get directions to Dr. Wenger's."). A subject noun-phrase request was a first-person statement highlighting the requestor's goal or desire (e.g., "I want directions to Dr Wenger's home"). A question was a was an interrogative statement requesting an action (e.g., "How do I get to Dr. Wenger's home?"). Finally, remaining requests were put in the other category, including fragmented requests (e.g., "Wenger").

For each type of command, utterances were then divided into one of two categories depending on whether the participant specified an addressee when giving the command. Specified addressee requests included a reference to agent or actor (e.g., "Printer, print memo" and "Calendar, when is Thanksgiving 2000?"), and non-specified addressee requests did not explicitly mention the addressee (e.g., "Print memo" and "When is Thanksgiving 2000?").

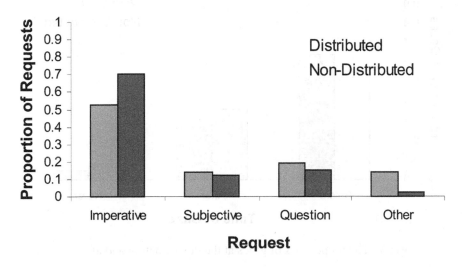

Fig 2. Proportion of utterances in the four categories.

For the verbal commands, results indicate that participants made many more imperative requests (62%) than question requests (17%), subject noun-phrase requests (13%), or other requests (8%). Figure 2 shows the proportion of requests in each category for DC and NC (differences between DC and NC were not reliable within each of the four categories). The language data also show that for all verbal commands (collapsed across category), very few of the requests were specified (< 2%). That is, only a few instances of statements such as, "Printer, give me a copy", emerged in the data. Conversely, nearly all commands featured no specific addressee, such as "Give me a copy".

Gaze patterns were coded according to if and when participants looked at the devices or the "room" (the light on the wall in NC). Namely, for each utterance, the participant either looked at the appropriate device or at the wall before (or during) speaking or after speaking. Figure 3 shows proportion of gazes occurring before/during, after, or never in relation to the verbal request. Overall, the data show more looks occurred before requests than after. In addition, participants nearly always looked at the device when making a request. As with types of requests made, gaze times did not differ for DC and NC.

4   Discussion

We explored the nature of user interaction in an attentive office setting. The pattern of results we found suggests that the DC environment was more natural for participants than was the NC environment Though most verbal requests did not address specific

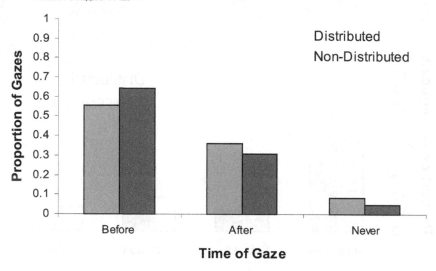

Fig 3. The proportions of gazes at the device before and after speaking.

devices, the majority of looks at devices occurred before the requests. Looking before speaking is seen in human-human communication: People often use eye gaze to establish or maintain social contact or interaction. Thus, looking before speaking may indicate that participants were specifying the recipient of the request through eye gaze, as if they default to a natural social interaction with individual devices.

Interestingly, the pattern of results—both verbal and gaze data—did not vary with experimental condition. There was no fundamental difference between the way people spoke to or directed gaze in the DC or NC environments. Our expectation prior to performing the study, however, was that in NC, centralized feedback (via a single flashing light) might bias participants not to look at individual devices, or might encourage participants to verbally specify an addressee more often. There are at least three possible explanations for why no differences arose in language and gaze patterns between the two conditions. First, multiple devices were present in both conditions. The mere presence of multiple devices in the non-distributed condition may have influenced people to interact with each device individually rather than with the single flashing light on the wall. Second, the flashing lights may not have provided a compelling a source of feedback, especially in the non-distributed condition. Third, the role of the flashing light may have been unclear in general. For instance, did it refer to an assistant, an authority, a monitor, and so on? In future studies, we plan to address this last issue by varying the script across conditions to highlight the different possible roles of the agent providing feedback.

In many voice-activated user interfaces, verbal requests typically must be spoken in a scripted and unnatural manner so that the recipient of the message can be easily determined [1]. Our data suggest that gaze information alone will disambiguate the recipient 98% of the time. For example, if one says "Wenger's home" while looking at the address book, it is clear that the address for Wenger's home is desired. By automating the attentive office to coordinate voice and gaze information in this way, a user's tacit knowledge about interpersonal communication can enable natural and efficient interactions with an attentive user interface. The preliminary study reported here marks only the first step toward understanding the ways in which people naturally interact with attentive devices, objects, and environments.

Acknowledgments

Thanks to Dave Koons, Cameron Miner, Myron Flickner, Chris Dryer, Jim Spohrer and Ted Selker for sharing ideas and providing support on this project.

References

1 Bolt, R. A. (1980). Put that there: Voice and gesture at the graphics interface. *ACM Computer Graphics 14(3)*, 262-270.

2 Coen, M. H. (1998). Design principles for intelligent environments, In *Proceedings of the Fifteenth National Conference on Artificial Intelligence (AAAI '98)*. Madison, WI.

3 Hirsh, H., Coen, M.H., Mozer, M.C., Hasha, R. & others. (1999). Room service, AI-style. *IEEE Intelligent Systems, 14(2)*, 8-19.

4 Horvitz, E. Breese, J., Heckerman, D., Hovel, D., &. Rommelse, K. (1998). The Lumiere project: Bayesian user modeling for inferring the goals and needs of software users, in *Proceedings of the Fourteenth Conference on Uncertainty in Artificial Intelligence.*

5 Jacob, R. J. K. (1993). Eye movement-based human-computer interaction techniques: Toward non-command interfaces. In Hartson, D. & Hix, (Eds.)., *Advances in Human-Computer Interaction, Vol 4*, pp. 151-180. Ablex: Norwood, NJ.

6 Linton, F., Joy, D., & Schaefer, H. (1999). Building user and expert models by long-term observation of application usage, in *Proceedings of the Seventh International Conference on User Modeling*, 129-138.

7 Maglio, P. P, Barrett, R., Campbell, C. S., Selker, T. (2000) Suitor: An attentive information system, in *Proceedings of the International Conference on Intelligent User Interfaces 2000.*

8 Maglio, P. P. & Campbell, C. S. (2000). Tradeoffs in displaying peripheral information, in *Proceedings of the Conference on Human Factors in Computing Systems (CHI 2000)*.

9 Norman, D. A. (1998). *The invisible computer*. Cambridge, MA: MIT Press.

10 Oviatt, S. & Cohen, P. (2000). Multimodal interfaces that process what comes naturally. *Communications of the ACM, 43(3)*, 45-53.

11 Selker, T. (1994). COACH: A teaching agent that learns, *Communications of the ACM, 37(1)*, 92-99.

12 Starker, I. & Bolt, R. A. A gaze-responsive self-disclosing display, in *Proceedings of the Conference on Human Factors in Computing Systems, CHI '90*, 1990, 3-9.

13 Turk, M. & Robertson, G. (2000). Perceptual user interfaces. *Communications of the ACM, 43(3)*, 33-34.

14 Zhai, S., Morimoto, C., & Ihde, S. (1999). Manual and gaze input cascaded (MAGIC) pointing, in *Proceedings of the Conference on Human Factors in Computing Systems (CHI 1999)*, 246-253.

Resolving References to Graphical Objects in Multimodal Queries by Constraint Satisfaction

Daqing He, Graeme Ritchie, John Lee

Division of Informatics, University of Edinburgh
80 South Bridge, Edinburgh EH1 1HN Scotland
{Daqing.He, G.D.Ritchie, J.Lee}@ed.ac.uk

Abstract. In natural language queries to an intelligent multimodal system, ambiguities related to referring expressions – *source ambiguities* – can occur between items in the visual display and objects in the domain being represented. A multimodal interface has to be able to resolve these ambiguities in order to provide satisfactory communication with a user. In this paper, we briefly introduce source ambiguities, and present the formalisation of a constraint satisfaction approach to interpreting singular referring expressions with source ambiguities. In our approach, source ambiguities are resolved simultaneously with other referent ambiguities, allowing flexible access to various sorts of knowledge.

1   Source Ambiguities

With the widespread use of multimodal interface, many systems integrate natural language (NL) and graphical displays in their interactions. In some systems, graphics on the screen represent entities or attributes of entities in the application domain. For example, Fig. 1 shows a system called IMIG, from [7], where icons in the DISPLAY area represent individual cars, and characteristics of the icons convey attributes of the corresponding cars. A table of how attributes of the real cars are represented is displayed in the KEY area. Each representation is called a *mapping relation*. A user who is browsing through the cars with a view to buying may use the POTENTIAL BUY area to collect the icons of cars that s/he is interested in. During the interaction, the user can ask about the cars on the screen, or perform actions (e.g. *move, remove,add*) on the icons of those cars.

Interesting ambiguities can occur during interactions with a system like this because it has no total control of what the user can enter through NL modality. An attribute used in a referring expression can be an attribute either from the display on the screen, or from the entities in the application domain. For instance, as the worst scenario, the colour attribute represented by the word "green" in (1 a) potentially can denote the colour of an icon on the screen or the colour of a car in the domain, which are different. Another example of the ambiguity is that the referent of a phrase can be either the entity in the domain or its icon on the screen. For example, the two uses of the phrase "the green car" in (1 a) and (1 d) refer to different entities: the first refers to a *car* (represented by a green icon on the screen), whereas the second refers to that green *icon*.

T. Tan, Y. Shi, and W. Gao (Eds.): ICMI 2000, LNCS 1948, pp. 8-15, 2000.

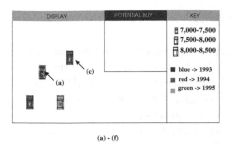 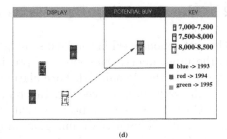

(a) - (f)   (d)

Fig. 1. Screen displays for (1)

(1) **User:** What is the insurance group of the green car?   (a)
   System: It is group 5.   (b)
      User: Move it to the potential buy area.   (c)
   System: The green car has been moved.   (d)

In the IMIG system [6], this distinction is made by storing entities and their attributes from the domain (i.e. *domain entities/attributes*) in the *world model*, and those from the screen (i.e. *screen entities/attributes*) in the *display model*. The *source* of an entity/attribute indicates whether it belongs in the domain or on the screen. The ambiguities mentioned above are hence termed *source ambiguities*.

Source ambiguities affect the interpretation of referring expressions, and hence of the input query. Also, they cannot be resolved in a simple way. A satisfactory resolution of source ambiguities seems to involve at least knowledge about the screen, about the domain and about the dialogue context.

He et al. [7] postulated a *described entity set* and an *intended referent* for each referring expression. The described entity set is an abstract construct which allows a more systematic account of the relationship between the linguistic content of the expression (roughly, its "sense") and the (intended) referent. The described entity set of a singular referring expression is either a singleton set containing a entity (domain or screen), or a two-element set containing a domain entity and a screen entity which are related to each other by the mapping relation (see above). That is, the described entity set contains the objects that might be the referent of the phrase, based purely on the descriptive content of the phrase, but with the source unresolved. The intended referent is what is called the *referent* in the literature. More about source ambiguities can be found in He et al. [5–7].

There are several relevant linguistic regularities, within and between referring expressions, related to the sources of the words [7]. An example is the *screen head noun rule*, which states that *if the head noun of a phrase unambiguously names a screen category and it does not have a classifier modifier, then the described entity set of the phrase contains only screen entities and the intended referent is a screen entity too.*

2   The Resolving Process as a CSP

The restrictions used in source disambiguation can come from many places, such as the input query, the context of previous interactions and the content of the screen and the domain [7]. Formulating these restrictions as a CSP is natural and allows flexible processing. Referent resolution has already been proposed as a CSP [4, 10], and we propose that these two processes can be viewed as an integrated CSP. Source disambiguation is necessary for finding the intended referent of a phrase; conversely, the success or failure of referent resolution provides feedback on any potential solutions to source disambiguation.

3   Related Work

Among previous work related to source ambiguities, only He et al. [6, 7] provided a systematic discussion and a set of terminology for handling source ambiguities. The CSP approach here improves upon those ideas, by being more motivated, rigorous and general.

Ben Amara et al. [2] mentioned the issue of referring to objects by using their graphical features, and indicated that, to handle this type of references, the graphical attributes on the screen should, like those in the domain, be available to referent resolution. Binot et al. [3] mentioned the referent of a phrase being a graphical icon on the screen. However, as these authors acknowledged, they did not have a solution to source ambiguities.

Andre & Rist [1] and McKeown et al. [9] both allow natural language expressions to allude to graphic attributes, so that text can refer to parts of accompanying illustrations. However, they worked on multimodal *generation* and treated graphics on the screen merely as the descriptions of the represented domain entities. They did not have a resolution mechanism for source ambiguities.

4   The Formalisation of the CSP

The formalisation of source disambiguation and referent resolution into a CSP (called *CSP-souref*) consists of two stages: the identification of the variables and the extraction of the constraints on the variables.

4.1   Sources and Referents: the Variables

There are two kinds of variables in our CSP, and all of them have a finite range. A variable ranging across *entities* (potential intended referents, for example) has a range containing the entities in the context model and the display model, together with any entities that correspond, through mapping relations, to them. A variable ranging across *sources* has the range {screen, domain}. Each value in a variable's range is a potential solution (a *candidate*) for that variable, and the constraint resolver proceeds by refining that initial set.

For the simplicity of computation, two different variables are used in CSP-souref if it is not clear that two sources/entities are conceptually identical. For example, described entities can be derived from many parts of a referring expression (i.e. adjectives, nouns and prepositional phrases). Before the exact described entities are identified, it is difficult to tell which parts give rise to the same described entities. Therefore, in CSP-souref, different parts of a referring expression contribute different variables. Once solutions are found for these variables it will be explicit which variables actually represent the same described entity.

As a simplification in this preliminary exploration of source ambiguities, we are not considering plural referring expressions. Hence, the solution for an entity variable is a single entity – there are no sets of entities to be considered as referents. Also, because an entity has exactly one source, the solution for a source variable contains only one candidate.

Our resolving process has to be general enough to cover not only phrases that have referents, such as definite noun phrases, deictic phrases and pronouns, but also phrases that do not have referents, such as indefinite phrases. Entity variables corresponding to the former type of phrase require solutions to be computed, but variables for the latter type have no such requirement. Hence this distinction must be marked on the variables to allow the resolver to recognise when there is unfinished business.

4.2   Constraints and Preferences

Not all the attributes/relations related to the variables can be used as constraints in a CSP. Rich & Knight [12] pointed out that only those that are locally computable/inferrable and easily comprehensible are suitable constraints. We distinguish between obligatory *constraints*, which must be met by any candidate, and heuristic *preferences*, which can be ignored if necessary to find a solution.

All the restrictions that occur within CSP-souref can be stated naturally as unary or binary constraints/preferences, so we have adopted the simplifying assumption that there will be no more complex forms of restrictions.

Constraints on Sources are attributes/relations involving only source variables. They can be derived from the following origins of knowledge:

1. *Domain knowledge about the sources of some particular attributes.* For example, the word "icon" in the IMIG system is always related to an entity from the screen. A source variable related to this type of word would be given a (unary) constraint stating that "the solution of variable A must be the source a", written as $must\_be(A, a)$.

2. *The screen head noun rule* (see section 1). For example, suppose variable $S1$ represents the source of the described entity related to the word "blue" in the phrase "the blue icon" and variable $S2$ represents the source of the intended referent of the same phrase. The (unary) constraints generated from the rule can be written as $must\_be(S1, screen)$ and $must\_be(S2, screen)$.

3. *Semantic preconditions of relations.* For example, the possessive relation "of" in "the top of the screen" requires the variable $S4$, which represents

the source of a described entity of "the top", to have the same value as the variable $S5$, which represents the source of the intended referent of "the screen". The (binary) constraint can be written as $same\_source(S4, S5)$.

Preferences on Sources are from the *heuristic rules* mentioned in [5,7]. For example, the words "red" and "yellow" in "delete the red car at the right of the yellow one" are preferred to have the same source because their semantic categories share a common direct super-category in the sort hierarchy, which satisfies the *same type* rule. This can be written as the binary preference $prefer\_same(S7, S8)$ where $S7$ and $S8$ represent the two source variables related to the two words respectively.

Constraints on Entities are another set of constraints involving at least one entity variable. They come from the following origins of knowledge, the first two of which were introduced by Mellish [10], while the last two are specific to CSP-souref.

1. *Local semantic information that derives from the components of a referring phrase.* In CSP-souref, the components are adjectives, nouns and prepositional phrases. For example, the noun "car" provides a constraint on its entity variable to be a car. Constraints from this origin are unary constraints.

2. *Global restrictions that derive from a sentence or the whole dialogue.* In CSP-souref, global restrictions mainly refer to the "preconditions" of operations that make the operations "meaningful". For example, the intended referent of the direct object of operation move should be movable. Constraints from this origin are unary or binary.

3. *Relations between an entity variable and a source variable representing the source of the entity.* The relations can be written as binary constraints that "the two variables represent an entity and the source of that entity".

4. *Restrictions between the described entities and the intended referent of the same phrase.* The restrictions state that the intended referent and the described entities of a referring expression are either the same entity or the entities linked with a mapping relation. They help CSP-souref to restrict the entity variables to have the same solution if the variables represent the same entity.

Preferences on Entities come from various information, including the heuristic about the relation between the salience of a dialogue entity and the likelihood of that entity being the referent. As stated in Walker [13], the more salient a dialogue entity is, the more likely it is to be the referent. Another origin of the preferences is the spatial distance between an entity and the position indicated by a pointing device. Work in the multimodality literature (such as Neal & Shapiro [11]) usually assumes that the nearer an entity to the pointed position, the more likely it is to be the entity that the user wants to point out. CSP-souref prefers the candidate nearest to the pointed position.

4.3   An Example

To help in understanding the formalisation, we use input command (2) as an example. Variables DE** and Sde** represent a described entity and its source,

Table 1. All the constraints raised from (2)

Con1	$must\_be(Sde12, domain).$	Con2	$must\_be(Sde21, screen).$
Con3	$must\_be(Sir2, screen).$	Con4	$same\_source(Sir1, Sir2).$
Con5	$has\_feature(DE11, blue).$	Con6	$has\_feature(DE12, car).$
Con7	$has\_feature(DE21, screen).$	Con8	$has\_feature(IR1, removable).$
Con9	$has\_feature(IR2, position).$	Con10	$source\_entity(Sde11, DE11).$
Con11	$source\_entity(Sde12, DE12).$	Con12	$source\_entity(Sde21, DE21).$
Con13	$source\_entity(Sir1, IR1).$	Con14	$source\_entity(Sir2, IR2).$
Con15	$same\_or\_corres(DE11, DE12).$	Con16	$same\_or\_corres(DE11, IR1).$
Con17	$same\_or\_corres(DE12, IR1).$	Con18	$same\_or\_corres(DE21, IR2).$

and IR* and Sir* represent an intended referent and its source. DE11, Sde11, DE12, Sde12, IR1 and Sir1 are identified from the phrase "the blue car", and variables DE21, Sde21, IR2 and Sir2 are from the phrase "the screen". Sde11, Sde12, Sde21, Sir1 and Sir2 range over {screen, domain}, and DE11, DE12, DE21, IR1 and IR2 range over a set containing two classes of entities (see section 4.1 above). The first are all the entities from the context and display models, and the second class are the entities related to entities in the first class through mapping relations.

(2)   "remove the blue car from the screen"

Table 1 lists all the constraints extracted from (2). Con1 to Con4 are constraints on sources where Con1 and Con2 are from the origin 1, Con3 is from the screen head noun rule and Con4 is from the semantic preconditions. The remaining constraints are constraints on entities. Con5 to Con7 come from local semantic information, Con8 and Con9 are from the global restrictions, Con10 to Con14 are from the origin 3 and Con15 to Con18 are from the origin 4.

5   Resolving CSP-souref

A binary CSP can be viewed as a constraint network, whose nodes and arcs are the variables and the constraints, respectively. We have adopted Mackworth's network consistency algorithm (AC-3) [8], which achieves node and arc consistency. This is because the algorithm seems to be the most cost-effective way of resolving CSP-souref.

We use the example (2) to explain the process of constraint satisfaction for CSP-souref. Assume the relevant knowledge bases contain:

```
THE WORLD MODEL
    car(car1),car(car2),blue(car1),have_source(car1,domain),red(car2),
    car(car3),white(car3),have_source(car2,domain),have_source(car3,domain)
THE DISPLAY MODEL
    icon(icon1), red(icon1),have_source(icon1,screen),removable(icon1),
```

Table 2. The candidate sets of variables in Table 1 during constraint satisfaction. {*} represents {car1, car2, icon3, icon1, icon2, screen1}

variable	initial candidate set	after NC	after AC-3
Sde11	{screen, domain}	{screen, domain}	{domain}
Sde12	{screen, domain}	{domain}	{domain}
Sde21	{screen, domain}	{screen}	{screen}
Sir1	{screen, domain}	{screen, domain}	{screen}
Sir2	{screen, domain}	{screen}	{screen}
DE11	{*}	{car1, icon3}	{car1}
DE12	{*}	{car1, car2}	{car1}
DE21	{*}	{screen1}	{screen1}
IR1	{*}	{icon3, icon1, icon2}	{icon1}
IR2	{*}	{screen1}	{screen1}

```
    icon(icon2),red(icon2),have_source(icon2,screen),removable(icon2),
    icon(icon3),blue(icon3),position(screen1),have_source(icon3,screen),
    removable(icon3), screen(screen1),have_source(screen1,screen).
THE MAPPING MODEL
    corres(car1, icon1),corres(car2, icon2),corres(car3, icon3)
THE CONTEXT MODEL:
    car1, car2, icon3
```

Table 2 shows the results after achieving node and arc consistency respectively. In this example, each variable has only one candidate in its candidate set when network consistency is achieved. This candidate is the solution.

However, preferences sometimes have to be used to find the solution even after network consistency has been achieved. A preference selects a candidate for one of the remaining variables that have not found their solutions. With this piece of new information, network consistency can be achieved again in a new state, which might reach the solution. However, this could also sometimes lead to an inconsistency, where backtracking is necessary to get rid of the inconsistency. In IMIG, backtracking usually starts with removing the preference just applied.

We did an evaluation involving human subjects for the usefulness of our approach, which includes examining CSP-souref using the heuristic rules, source and context information. The statistical outcome shows that the functions provide significant help in making the test dialogues free from misunderstanding [5].

6   Conclusion and Future Work

In this paper, we have presented a framework for dealing with source ambiguities. By using a constraint satisfaction method, we integrated source disambiguation with referent resolution, and provided a unified mechanism to handle various restrictions on source ambiguities or referent ambiguities. In addition, the sequence of applying the restrictions is much more flexible in our approach.

Future work lies in the following directions: 1) mapping relations are accessible during the resolution, but they are assumed not to be mentioned by the user in the dialogues. For example, the sentence "which car is represented by a blue icon?" is not considered, although it could be used in an interaction. 2) the heuristics used in CSP-souref are mainly based on our intuition and experiments on a very small number of test dialogues. This restricts the applicability of these heuristics and imposes difficulties in further developing the resolution model. It would be beneficial to have a multimodal dialogue corpus. Even just a small one that contains only the dialogues related to our research would facilitate further exploration on source ambiguities.

Acknowledgements: The first author (Daqing He) was supported by a Colin & Ethel Gordon Studentship from the University of Edinburgh.

References

1. André, E., Rist, T.: Referring to World Objects with Text and Pictures. In *Proceedings of COLING'94* Kyoto Japan (1994) 530–534.
2. Ben Amara, H., Peroche, B., Chappel, H., Wilson, M.: Graphical Interaction in a Multimodal Interface. In *Proceedings of Esprit Conferences*, Kluwer Academic Publisher, Dordrecht, Netherlands (1991) 303–321.
3. Binot, J., Debille, L., Sedlock, D., Vandecapelle, B., Chappel, H., Wilson, M.: Multimodal integration in MMI^2: Anaphora Resolution and Mode Selection. In Luczak, H., Cakir, A., Cakir, G. (eds.): *Work With Display Units–WWDU'92* Berlin, Germany (1992).
4. Haddock, N.: *Incremental Semantics and Interactive Syntactic Processing*. PhD thesis, University of Edinburgh (1988).
5. He, D.: References to Graphical Objects in Interactive Multimodal Queries. PhD thesis, University of Edinburgh (2000).
6. He, D., Ritchie, G., Lee, J.: Referring to Displays in Multimodal Interfaces. In *Referring Phenomena in a Multimedia Context and Their Computational Treatment, A workshop of ACL/EACL'97* Madrid Spain (1997) 79–82.
7. He, D., Ritchie, G., Lee, J.: Disambiguation between Visual Display and Represented Domain in Multimodal Interfaces. In *Combining AI and Graphics for the Interface of the Future, A workshop of ECAI'98* Brighton UK (1998) 17–28.
8. Mackworth, A.: Consistency in Networks of Relations. *Artificial Intelligence* **8** (1977) 99–118.
9. McKeown, K., Feiner, S., Robin, J., Seligmann, D., Tanenblatt, M.: Generating Cross-References for Multimedia Explanation. In *Proceedings of AAAI'92* San Jose USA (1992) 9–16.
10. Mellish, C.: *Computer Interpretation of Natural Language Descriptions*. Ellis Horwood series in Artificial Intelligence. Ellis Horwood, 1985.
11. Neal, J., Shapiro, S.: Intelligent Multimedia Interface Technology. In Sullivan, J., Tyler, S. (eds.) *Intelligent User Interfaces* ACM Press New York (1991) 11–44.
12. Rich, E., Knight, K. *Artificial Intelligence*. McGraw-Hill, New York, 2 Ed (1991).
13. Walker, M.: Centering, Anaphora Resolution, and Discourse Structure. In Walker, M., Joshi, A., Prince, E. (eds.): *Centering Theory in Discourse*. Oxford University Press (1997).

Human-Robot Interface Based on Speech Understanding Assisted by Vision

Shengshien Chong[1], Yoshinori Kuno[1,2], Nobutaka Shimada[1] and Yoshiaki Shirai[1]

[1]Department of Computer-Controlled Mechanical Systems, Osaka University, Japan
[2]Department of Information and Computer Sciencies, Saitama University, Japan

Abstract. Speech recognition provides a natural and familiar interface for human beings to pass on information. For this, it is likely to be used as the human interface in service robots. However, in order for the robot to move in accordance to what the user tells it, there is a need to look at information other than those obtained from speech input. First, we look at the widely discussed problem in natural language processing of abbreviated communication of common context between parties. In addition to this, another problem exists for a robot, and that is the lack of information linking symbols in a robot's world to things in a real world. Here, we propose a method of using image processing to make up for the information lacking in language processing that makes it insufficient to carry out the action. And when image processing fails, the robot will ask the user directly and use his/her answer to help it in achieving its task. We confirm our theories by performing experiments on both simulation and real robot and test their reliability.

Keywords : human-robot interface; speech understanding; vision-based interface; service robot; face recognition

1. Introduction

As the number of senior citizens increases, more research efforts have been aimed at developing service robots to be used in the welfare service. However, these developments depend very much on the technology of human interface. It should allow even handicapped persons to be able to give commands to the robot in a simple and natural way. The demand for user-friendliness leads naturally to a dialogue controlled speech interface, which enables everyone to communicate easily with the robot. This is not only needed for convenience but also for lowering the inhibition threshold for using it, which still might be a problem for widespread usage. Not only do we want the robot to understand robotic commands but also we want the robot to be able to understand human-like commands and be more flexible in its language understanding.

For strongly handicapped persons, unable to use keyboards or touch screens, speech understanding and dialogue is one of the main preliminaries. Moreover, we have seen an improvement in the quality of the speech recognition technology in recent years, and we foresee that this technology will be widely used in the coming future as the technology further improves. It is also possible to operate the computer using voice nowadays. However, there is a need to memorize those commands and

T. Tan, Y. Shi, and W. Gao (Eds.): ICMI 2000, LNCS 1948, pp. 16-23, 2000.

say them with correct pronunciation. A good service robot should be able to reduce the burden of having to memorize these commands. Moreover, human beings tend to use different words to say the same thing at different times, and there is a need for it to respond to this. And also, human beings tend to miss out certain information in a natural conversation. The human utterances provide incomplete information (where is the user), (where is the corner) and (what is a corner). References occurred to already spoken concepts (corner is –at the red signboard). To do so, it has to use both its speech and vision capabilities. Processed images need to be analyzed and fused with the corresponding part of the spoken utterances. The robot needs to understands human being's intentions from these processes.

We consider a case of a service robot that is able to bring something that the user asks it. Here, we consider one of the robot's human interfaces and that is the ability to move the robot using voice. Human beings use many different kinds of words alone just to move the robot. In addition to overcoming this, we use other information to overcome the ambiguity in human being's utterances and aim to decipher the correct meanings of human being's commands.

One of the ambiguities is the missing of information. Human beings tend to miss out information that they think the other person knows while talking. For example, consider the case when a person says "some more" to a robot. If the robot was moving forward by the previous command, this "some more" means, "go some more to the front." We propose a method of using past commands to the robot and the robot's own movements to solve the problem of such ambiguities and thus, to understand the intended instruction.

One can see from current research efforts that learning and vision play key roles in developing robots, which are aimed to aid and work together with humans. At the workshop with the 7th ICCV, many methods involving the understanding of both speech and image were introduced [1], [2]. However, these speech and image understanding processes work independently with only the results incorporated together to get the final result [1]. Here, we focused on the ambiguity of speech. We initialize image processing to get visual information in order to remove such ambiguities.

If the robot still fails to understand the user's intention after using the above method, it will perform text-to-speech operation to form questions to ask the user for additional information. For this part, we use the same method [3] proposed earlier to ask the user for additional information.

In addition, we include emotions in our system. Okada has done such research [2] using information of emotions. In this paper, we only use the speed of speech and the types of words that the user adopts to determine the urgency of tasks.

We use the above information to analyze the user's speech input. We implement our system on our intelligent wheelchair [4] proposed earlier and perform experiments.

2.   System Configuration

Our system as shown in Figure 1, consists of a speech module, a visual module, a central processing module and an action module.

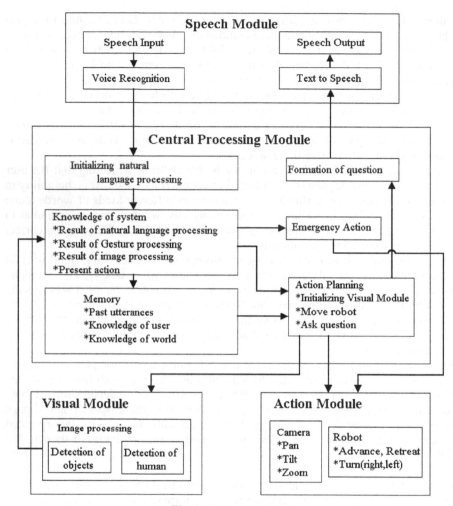

Fig. 1. System configuration

The speech module is made up of voice recognition and text-to-speech. We use ViaVoice Millennium Pro software (Japanese version) by IBM to do voice recognition, and ProTalker97 software also by IBM to do text-to-speech. The result of voice recognition is sent to the central module. When there is a command from the central processing module for it to output sound, it will perform text-to-speech and output the sound.

The visual module will perform image processing when there is a command from the central processing module. Presently, we have equipped it with the ability to detect human beings and the ability to detect objects based on color detection. The approximate distance between the robot and the object is calculated using the size of the target detected.

The central processing module is the center of the system. It uses various information and knowledge to analyze the meanings of recognized voice input. It will send a command to the visual module and uses the result for analysis of meanings when it thinks visual information is needed. And it will send command to the speech module to make sentences when it thinks it needs to ask the user for additional information. It will send command to the action module when action is to be carried out by the robot.

The action module waits for the command from the central processing module to carry out the action intended for the robot and camera.

3.   Central Processing Module

This is an important module as it monitors all the operations going on within the system. It constantly checks the activities of the other modules, receives their information, and plans whatever actions or processing necessary.

In the language processing part, various information and knowledge are being used together with the recognized result from the speech module to find out the meanings of verbal natural human voice input. And it carries out effective communication with the user in order to obtain additional information when it is not able to analyze the utterances correctly. We do so in the following ways.

3.1   Basic Operation

As our goal is to get to our destination, we do not need a vast vocabulary. Here, we define the various words recognized into to the following categories: objects, actions, directions, adjectives, adverbs, emergency words, numerals, colors, names of people and others.

First we listen for the end of an utterance. The utterance (speech input) is converted into text. Next, we compare our existing vocabulary to the text and search for the words that were defined and we group them into the above categories. After this we rearrange the words back to the correct spoken order, then we analyze its significance in a robot's movement. Next, we analyze what actions should be taken and what additional information is required. Here, emergency words refer to words or short phrases that are being used to stop the robot in an emergency or when the robot did something that is not intended for it. Examples are words like "stop, no, danger, wrong". When these words are recognized, the system stops whatever operations it is doing and performs an independent operation.

Here, we make an important assumption. We assume that humans always make decisions from their point of view. For instance, when the user is facing the robot, the robot's left is the user's right, and thus if the user wants the robot to move in the direction of his/her right, all he/she has to do is to just say "go to the right" like what he/she normally does and the robot will know that it is supposed to move to its left. To do so, the robot obtains information on its positional relation with the user by vision.

After categorizing the words, if the system contains enough information to start the robot's action, the system sends the command to the action module for the action to be carried out. However, if there exists any ambiguities in the command due to lack of information, the system does the following to work out an appropriate solution by using other information.

3.2   Context Information

The ambiguity of commands caused by abbreviated utterances is solved using context information. Here, context information refers to the history of actions that had been carried out so far and the present state of the robot.

The history of actions is used in the following way. Just a while ago the robot was moving forward, and now the user says, "A little bit more." The robot analyzes this as meaning to go forward a little bit more. However, if the user said, "Go faster," previously, then the "a little bit more" means go a little bit faster.

The present state of the robot is used in the same way. If the user says "some more" while the robot is turning to the left now, the robot analyzes it as meaning turning some more to the left. If the user says "some more" while the robot is moving at a high velocity now, the robot analyzes it as meaning increasing the velocity to move faster. If the user says "wrong" while the robot is moving forward now, the robot analyzes it as meaning that it should move backward instead of forward.

3.3   Visual Information

The system knows the user's command from the results of the basic operation, but some important information is lacking to actually activate the action. When the central processing module knows that it can make up for it with its vision, it initializes the visual module to look for additional information. As our objective this time is to move the robot using voice input, the main information lacking is mostly related to directions. Visual processing is activated when the meanings of places and directions are uncertain and when the location of persons or target objects is unknown. We have three main objectives here for our visual module, namely, detection of human beings, detection of objects specified by the user and calculation of the distance between the target and the robot.

The followings are our examples.

Example 1

In a case when the user said, "Come over here."

As the word "here" signifies the location of the user, the visual module searches for a human being and will move in that direction if it is located. The following explains more in detail. First, the basic operation realizes that the move action is designated for the robot. However, in order to move, it needs to know the direction to move, and in this case it is not specified. The central processing module checks whether it is possible to decide which direction to move. In this case, the word "here" contains the meaning of "the person who gives the command" and thus, the command to start the program to search for human beings is given. Next, the visual module searches for human beings. The direction of the detected person is the direction where the robot should move and the command is sent to the action module. If no other command is given, the robot will continue to follow the user. In such a situation, when "come here" is being said again, analysis is possible even without new visual processing.

Example 2

In a case when the user said, "Go to Mr. A."

The robot does not know the location of Mr. A but it knows it can look for the human being on its own and thus its vision is initialized to search for the human being and goes towards him. By doing so, the robot does not need to ask for directions.

3.4. Emotional Information

We use the following two methods in determining the feelings of the user.

1. The time lap between speech input.
   If the time lap is large, it means that the user is not in a hurry to go to the destination and vice versa.

2. The words that the user uses.
   If the user uses words that are more polite such as please, it means that he/she is not in a hurry. If he/she uses words that are more direct, it means that he/she is in a hurry.

3.5. Getting Information through Communication

When the robot still fails to determine the proper action to be carried out despite using the above information, it activates the text-to-speech to ask questions in order to obtain new information. For example, when object or user detection fails consecutively for several times, the robot will say that it does not know the location of the target and will ask for the location of the target.

4. Experiment

We have installed our proposed system onto our intelligent wheelchair [4]. By treating the wheelchair as a mobile robot, we send commands to the robot while we are not sitting on it.

We performed experiments in various cases. The following is an example. The robot is told to go to Mr. A, but it does not know where Mr. A is, and asks the user for information. There is a wall there and the user thinks that Mr. A might be behind it. The user wants the robot to go and look at the other side of the wall to see if it can find Mr. A there. Thus, he directs the robot there and the robot succeeds in finding Mr. A. We ask a person to wear a red jacket and use color detection here to search for the target.

User	:	"Go to Mr. A."
Robot	:	(Initiates the search for Mr. A but it results in failure.)
Robot	:	(Initiates question asking.)
Robot	:	"Where is Mr. A?"
User	:	(Thinks that Mr. A might be behind the wall.)
User	:	"I will lead you to him. Go forward."

Robot   :   (Goes forward.)
Robot   :   (Found Mr. A.)
Robot   :   "I have found Mr. A, shall I go towards him on my own? "
User   :   "Yes."
Robot   :   (Goes towards the target on its own after calculating the approximate
          distance between the target and the robot.)

Figure 2 shows our experiments.

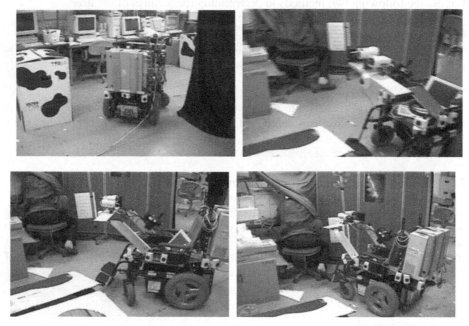

Fig. 2. In the process of "go to Mr. A".

In our experiments, we succeeded in using vision to assist the speech input. In addition, we also used emotional information as described in section 3.4.

5. Conclusion

We have developed a system that allows us to direct the robot by voice input using a variety of different words. It can work even when we miss out some information in our utterances as vision is used to assist the speech input. Experimental results show that this approach is promising for human-robot interface.

We are still in the preliminary stages here in determining the possibility of our system and there are still some problems in the hardware part and in transmitting the voice command to the robot. Also, the image processing used in this system is still a simple one. Our system can be expanded and modified to use in many different areas like hospitals, offices, schools or even shopping complexes where there are many human activities. In the future, we will consider an actual task of a service robot and implement the required image processing functions, improve the hardware part and make the service robot a reality.

Acknowledgements

This work was supported in part by the Ministry of Education, Science, Sports and Culture under the Grant-in-Aid for Scientific Research (09555080, 12650249) and Ikeuchi CREST project.

References

[1] S. Wachsmuth and G. Sagerer, "Connecting concepts from vision and speech processing," Proc. Workshop on Integration of Speech and Image Understanding, pp.1-19, 1999.
[2] N. Okada, "Towards affective integration of vision, behavior and speech processing," Proc. Workshop on Integration of Speech and Image Understanding, pp.49-77, 1999.
[3] T. Takahashi, S. Nakanishi, Y. Kuno, and Y. Shirai, "Human-robot interface by verbal and nonverbal communication," Proc. 1998 IEEE/RSJ International Conference on Intelligent Robots and Systems, pp.924-929, 1998.
[4] Y. Kuno, S. Nakanishi, T. Murashima, N. Shimada, and Y.Shirai, "Intelligent wheelchair based on the integration of human and environment observations," Proc. 1999 IEEE International Conference on Information Intelligence and Systems, pp.342-349, 1999.

Designing Multi-sensory Models for Finding Patterns in Stock Market Data

Keith Nesbitt[1]

[1] Department of Computer Science and Software Engineering,
University of Newcastle,University Drive, Callaghan, NSW. 2308. Australia
knesbitt@cs.newcastle.edu.au

Abstract. The rapid increase in available information has lead to many attempts to automatically locate patterns in large, abstract, multi-attributed information spaces. These techniques are often called 'Data Mining' and have met with varying degrees of success. An alternative approach to automatic pattern detection is to keep the user in the 'exploration loop'. A domain expert is often better able to search data for relationships. Furthermore, it is now possible to construct user interfaces that provide multi-sensory interactions. For example, interfaces can be designed which utilise 3D visual spaces and also provide auditory and haptic feedback. These multi-sensory interfaces may assist in the interpretation of abstract information spaces by providing models that map different attributes of data to different senses. While this approach has the potential to assist in exploring these large information spaces what is unclear is how to choose the best models to define mappings between the abstract information and the human sensory channels. This paper describes some simple guidelines based on real world analogies for designing these models. These principles are applied to the problem of finding new patterns in stock market data.

1 Introduction

A problem facing many areas of industry is the rapid increase in the amount of data and how to comprehend this information. This data often contains many variables, far more than humans can readily comprehend. 'Data-mining' has been proposed in many areas to uncover unknown patterns, rules or relationships in this data. Two broad approaches exist, one uses automated techniques and the other relies on keeping the user central to the task and providing appropriate models the user can explore. The success of human-based, data-mining relies on effective models being developed for the exploration of large, multivariate and abstract data sets.

Virtual Environment technology claims to increase the number of variables we can mentally manage. With multi-sensory interfaces we can potentially perceive and assimilate multivariate information more effectively. The hope is that mapping

T. Tan, Y. Shi, and W. Gao (Eds.): ICMI 2000, LNCS 1948, pp. 24-31, 2000.

different attributes of the data to different senses such as the visual, auditory and haptic (touch) domains will allow large data sets to be better understood.

Multi-sensory interaction is a natural way for us to operate in the real world. Despite this, designing multi-sensory models is complicated due to the many perceptual issues that arise for both an individual sense and when multiple senses interact. Further difficulties exist when we are trying to design models for exploring abstract data. Abstract data from a domain such as the stock market has no real-world model or analog about which we can structure the information in a way that is intuitive to the user. It is desirable to avoid some complex mapping between the display artifacts and data parameters. When the user wishes to explore this data and search for new patterns it is important they are not distracted from their 'data-mining' task. The goal is to choose mappings from the abstract data attributes to the different senses so it is intuitive and provides an effective model for exploration.

MODALITY		
 Visual	 Auditory	 Haptic
System Displays OUTPUT to User as:		
Colour, Lighting	Speech	Force Feedback
Transparency	Pitch	Vibration
Shape (2D, 3D)	Timbre	Texture
Patterns, Texture	Rhythm	Tactile Display
Structure (2D, 3D)	"Earcons"	Body motion
Movement	Melody (music)	Temperature
Spatial relations	Direction (2D,3D)	Pain
Text, Icons	Movement	Chemical
User provides INPUT from:		
Eye-gaze direction	Sound	Keyboard
Head position &	Speech	Mouse (2D, 3D)
orientation		Props - control dials
		Joystick (2D, 3D),
		Wands, Phantom ™
		Body Position
		Gestures

Fig. 1. Possible sensory artifacts for user interaction [1]

This paper describes some simple guidelines for analyzing an abstract data domain and subsequently designing and implement multi-sensory models. The focus is to design models for human-based exploration of large abstract data spaces. By a model I mean a mapping between data attributes and the artifacts of the multi-sensory display. This approach is described in the context of developing applications for finding new patterns in stock market data.

2   Designing Multi-sensory Models of Abstract Data

A multi-sensory model is defined as a mapping between the visual, auditory and haptic artifacts of the display (Fig. 1) and the parameters that characterize the abstract data. Display of abstract data visually is well understood in many areas, but many challenges still exist in displaying information effectively to the auditory and haptic domains [2,3]. The basic over-riding principle I use to design the models is that these mappings be are based on real world analogies or metaphors.

Many of the more successful applications within Virtual Environments involve exploration of 3-D real world models. These include prototyping and simulations but also data interpretation tasks in the medical diagnostic and petroleum exploration domain [4,5]. The use of analogies has been successfully used for design of auditory displays [6]. The natural use of haptics for displaying structural surfaces of real world models has also demonstrated its effectiveness [7].

Finally, trying to understand and measure the perceptual capabilities of individual senses is difficult. The problem becomes even more difficult when the interaction of multiple senses must be considered. However, we interact intuitively with the world in a multi-sensory way all the time. In effect we have already learned many ways of interacting with multi-sensory models. The problem then, becomes finding appropriate ways of expressing the abstract data in a real-world context.

		DISPLAY SENSE		
		 Visual	 Auditory	 Haptic
Metaphor **Types**	SPATIAL	1. Spatial Visual Metaphors	2. Spatial Auditory Metaphors	3. Spatial Haptic Metaphors
	TEMPORAL	4. Temporal Visual Metaphors	5. Temporal Auditory Metaphors	6. Temporal Haptic Metaphors
	SIGHT	7. Sight Metaphors		
	SOUND		8. Sound Metaphors	
	TOUCH			9. Touch Metaphors

Fig. 2. Classification of metaphors used for domain analysis

3   Guidelines for Choosing Mappings

The first step in designing the model is to analyze the abstract data for possible multi-sensory interactions. To do this I consider the data attributes in terms of five types of natural metaphors or mappings: spatial, temporal, sight, sound and touch (Fig. 2).

Metaphor Category	Display Sense	GUIDELINE FOR USAGE
1.Spatial Visual Metaphors	 👁	Use 'visual-spatial-metaphors' where compare and contrast of data or relationship over space or where the overall data structure is critical.
2.Temporal Visual Metaphors	 👁	Use 'visual-temporal-metaphors' where the steps of change or transition are important and a change in spatial relationships needs to be understood.
3. Sight Metaphors	 👁	Use 'sight-metaphors' as the first choice for displaying quantitative information or the value of data attributes. More than one visual metaphor may be combined.
4.Spatial Auditory Metaphors	 👂	Use 'auditory-spatial-metaphors' to draw attention to specific structure as a navigation aid. Especially where the eyes may be oriented in another direction.
5.Temporal Auditory Metaphors	 👂	Use 'auditory-temporal-metaphors' to display time varying data. Especially where a sudden change to constant information needs to be detected or to compress the time scale in a time series.
6. Sound Metaphors	 👂	Use 'sound-metaphors' for displaying complementary quantitative attributes, where the visual sense is likely to be overloaded or a different resolution of data is required.
7.Spatial Haptic Metaphors	 ✋	Use 'haptic-spatial-metaphors' to describe ancillary structural data.
8. Temporal Haptic Metaphors	 ✋	Use 'haptic-temporal-metaphors' as navigation or selection aids.
9. Touch Metaphors	 ✋	Use 'touch-metaphors' as a final choice for displaying complementary data attributes, where the visual and auditory senses are likely to be overloaded.

Fig. 3. Guidelines for choosing the most appropriate mappings

'Spatial Metaphors' relate to scale, location and structure and can be described for the visual, auditory and haptic senses. They concern the way pictures, sounds and forces are organized in space. 'Temporal Metaphors' are concerned with how data changes with time and can also be applied to all senses. They concern how we perceive changes to pictures, sounds and forces over time. 'Sight Metaphors', 'Sound Metaphors' and 'Touch Metaphors' are all concerned with direct mappings between a sensory artifact and some quantitative information. For example, a particular color, sound volume or surface hardness represents a particular data value. A more detailed description of these metaphors can be found in previous work [8].

Initial analysis of any domain using the 5 types of metaphors will reveal a number of *potential* mappings between the application data and the artifacts of the different sensory displays. These mappings will be categorized into one of the nine groups:

1. 'Spatial-Visual-Metaphors' are mappings to 2-D or 3-D visual structures or positions in this space.
2. 'Spatial-Auditory-Metaphors' are mappings to 2-D or 3-D auditory structures or positions in this space.
3. 'Spatial-Haptic-Metaphors' are mappings to 2-D or 3-D haptic structures or positions in this space.
4. 'Temporal-Visual-Metaphors' are mappings that convey information by the way the visual artifacts change over time.
5. 'Temporal-Auditory-Metaphors' are mappings that convey information by the way the auditory artifacts change over time.
6. 'Temporal-Haptic-Metaphors' are mappings that convey information by the way the haptic artifacts change over time.
7. 'Sight-metaphors' are quantitative mappings to the visual artifacts of the display.
8. 'Sound-metaphors' are quantitative mappings to the auditory artifacts of the display.
9. 'Touch-metaphors' are quantitative mappings to the haptic artifacts of the display.

The next step in the process is to reduce the number of *potential* mappings and decide on the most appropriate use of the senses. To do this I follow a series of guidelines (Fig. 3). These guidelines are based on user feedback from previous development experience [7] and by considering the way we use our senses within the real world. It implies a natural hierarchy of the senses, with the visual dominating sound and haptic interactions at the lowest level.

4 Finding Patterns in Stock Market Data

The stock market data I model is from the domain of 'technical analysis'. 'Technical analysis' is defined as "the study of behavior of market participants, as reflected in price, volume and open interest for a financial market, in order to identify stages in the development of price trends"[10]. This field of study has grown out of Dow Theory and studies patterns in the data itself in an attempt to understand the balance of supply and demand. This is in direct contrast to more traditional 'fundamental analysis', as it does not attempt to understand the underlying factors that determine the value of an instrument in a financial market. It considers that all such information is discounted by data attributes such as price and volume. The data is abstract, multivariate and large and can be analyzed over multiple time scales from minutes to months depending on the investment strategy of the trader.

Metaphor Category	Display Sense	Example mapping from Technical Analysis domain.
1.Spatial Visual Metaphors		It is traditional to use an abstract 2D space representing time and price and to use this to analyse price levels. In 3-D space this model can be augmented by volume. This allows points in the visual space to relate time, price and volume.
2.Temporal Visual Metaphors		The 'volatility' of the market is a quantity that changes over time. The rate of change of price may be as important as an actual price. This can be modeled for example as a visual object that flickers very quickly in volatile markets and slowly in markets that are constant.
3. Sight Metaphors		'Candlestick charting' is a 2-colour code for visual representation of the relationship between open and close price for a trading period. Black for when close price < open price and white when open price > close price.
4.Spatial Auditory Metaphors		In a 3D environment it is possible to surround the user by a number of distinct sound sources, where the placement of the sound source identifies the financial instrument. This allows auditory monitoring of multiple trading instruments simultaneously.
5.Temporal Auditory Metaphors		A traditional use of sound in the stock market is the noise from a ticker-tape which displays the rate of trading. A rapid 'tick' indicating greater market action.
6. Sound Metaphors		Pitch could be mapped to price level. A higher pitch indicating higher prices. Sound volume may be mapped to trading volume.
7.Spatial Haptic Metaphors		Understanding levels of support or resistance in price can be directly mapped as haptic surfaces that the user can feel and push through. The amount of force required to penetrate these surfaces indicating the strength of support or resistance at that level.
8. Temporal Haptic Metaphors		'Momentum' of price is derivative of price and is important for understanding how price level may be changing. Force can be used to model the way price momentum changes with time.
9. Touch Metaphors		Degree of vibration may be used to display the trade volume for a time period.

Fig. 4. Examples of each metaphor category from the domain of 'Technical Analysis' [9].

Example mappings from the application domain of technical analysis to each of the metaphors can be described. (Fig. 4). Using the guidelines outlined above I have modeled the domain of technical analysis and designed a number of multi-sensory models for specific technical analysis tasks. A 3-D auditory and haptic charting model for detecting patterns in price and volume data is an extension of traditional charting techniques. A moving landscape designed to find trading opportunities in the bidding prices from buyers and the asking prices from sellers is a new application idea. Finally, a model for finding temporal ('trailing' or 'leading') relationships between several financial instruments is also being developed.

5 Conclusions and Further Work

Like many new emerging computer technologies much hype and speculation has surrounded the value and application of Virtual Environments. Realistically, everyday use of these environments for applications such as technical analysis is not likely in the short term. High cost, many usability issues and the lack of commercial software make it infeasible for rapid adoption of these environments. A shorter-term possibility is the use of such environments to investigate the discovery of new relationships in abstract data, such as that provided by the stock market. In such cases the potential reward may offset the risks against success.

Metaphors that provide totally new ways of seeing, hearing and touching financial data may reveal patterns that have not before been understood. This may in turn create new and unique trading opportunities. Many systems have been developed which use algorithmic or heuristic rules for trading the market based on price trends. Once new patterns or rules are discovered the opportunity then exists to incorporate them into such automatic trading systems.

Further work needs to be done in developing and testing these models and the guidelines I have proposed. What has been presented is work in progress. However, early feedback from users confirms that the use of these guidelines has resulted in intuitive and easy to understand displays.

References

1. Stuart, R. The Design of Virtual Environments. McGraw-Hill, New York. 1996. ISBN 0-07-063299-5
2. Kramer, G. et al. Sonification Report: Status of the Field and Research Agenda. International Conference of Auditory Display 1997.
3. Srinivasan, M. A. and Basdogan, C., Haptics in Virtual Environments: Taxonomy, Research Status, and Challenges. Computer & Graphics, Vol.21, No. 4, pp. 393-404. (1997)

4. Durlach, NI, Mavor AS. Virtual Reality. Scientific and Technological Challenges. National Academy Press, Washington, DC. (1996)
5. Froehlich, B., Barrass, S., Zehner, B., Plate J. and Goebel, M. Exploring GeoScience Data in Virtual Environments. Proc IEEE Visualization, 1999.
6. Barrass, S. (1996) EarBenders: Using Stories About Listening to Design Auditory Displays, Proceedings of the First Asia-Pacific Conference on Human Computer Interaction AP-CHI'96, Information Technology Institute, Singapore.
7. Nesbitt, K.V., Gallimore, R.J., Orenstein, B.J. Investigating the Application of Virtual Environment Technology for use in the Petroleum Exploration Industry. Proceedings of 23rd Australasian Computer Science Conference. ACSC 2000. Australian Computer Science Communications, Vol. 22, No. 1. pp181-188.
8. Nesbitt, K. V. A Classification of Multi-sensory Metaphors for Understanding Abstract Data in a Virtual Environment. Proceedings of IV 2000, London. 2000.
9. Nesbitt, K. V. and Orenstein B.J. Multisensory Metaphors and Virtual Environments applied to Technical Analysis of Financial Markets. Proceedings of the Advanced Investment Technology, 1999. pp 195-205. ISBN: 0733100171.
10. Technical Analysis: Course Notes from Securities Institute of Australia course in Technical Analysis (E114), 1999. http://www.securities.edu.au

Audio-visual Segmentation and "The Cocktail Party Effect"

Trevor Darrell,* John W. Fisher III, Paul Viola
MIT AI Lab
William Freeman
Mitsubishi Electric Research Lab

Abstract

Audio-based interfaces usually suffer when noise or other acoustic sources are present in the environment. For robust audio recognition, a single source must first be isolated. Existing solutions to this problem generally require special microphone configurations, and often assume prior knowledge of the spurious sources. We have developed new algorithms for segmenting streams of audio-visual information into their constituent sources by exploiting the mutual information present between audio and visual tracks. Automatic face recognition and image motion analysis methods are used to generate visual features for a particular user; empirically these features have high mutual information with audio recorded from that user. We show how audio utterances from several speakers recorded with a single microphone can be separated into constituent streams; we also show how the method can help reduce the effect of noise in automatic speech recognition.

1 Introduction

Interfaces to computer systems generally are tethered to users, e.g., via a keyboard and mouse in the case of personal computers, through a touch-screen when dealing with automatic tellers or kiosks, or with a headset

*contact: MIT AI Lab Room NE43-829, 545 Technology Square, Cambridge MA 02139 USA. Phone:617 253 8966, Fax:617 253 5060, Email: trevor@ai.mit.edu

T. Tan, Y. Shi, and W. Gao (Eds.): ICMI 2000, LNCS 1948, pp. 32-40, 2000.
© Springer-Verlag Berlin Heidelberg 2000

microphone or telephone when using automatic speech recognition systems. In contrast, humans interact at a distance and are remarkably adept at understanding the utterance of remote speakers, even when other noise sources or speakers are present. The "cocktail party effect"–the ability to focus in on a meaningful sub-stream of audio-visual information–is an important and poorly understood aspect of perception [1].

In this paper we show how multimodal segmentation can be used to solve a version of the cocktail party problem, separating the speech of multiple speakers recorded with a single microphone and video camera. Our technique is based on an analysis of joint audio-visual statistics, and can identify portions of the audio signal that correspond to a particular region of the video signal, and vice-versa. Automatic face recognition is used to identify locations of speakers in the video, and mutual information analysis finds the portions of the audio signal that are likely to have come from that image region. We can thus attenuate the energy in the audio signal due to noise sources or other speakers, to aid automatic speech recognition and teleconferencing applications.

2   Source Separation

Most approaches to the problem of observing and listening to a speaker in a noisy and crowded room rely on active methods, either physically steering a narrow field microphone or adjusting the delay parameters of a beam-forming array [9, 10, 4].

These approaches are valuable, but require special sensors and sophisticated calibration techniques. We have developed passive methods which work on broadly tuned, monaural audio signals and which exploit time-synchronized video information. Our approach works using readily available PC teleconferencing camera and microphone components, as well as on video from broadcast and archival sources.

We base our method on the statistical analysis of signals with multiple independent sources. Prior statistical approaches to source separation often took audio-only approaches, and were successful when multiple microphones were available and the number of sources were known. The "blind source separation" problem has been studied extensively in the machine learning literature, and has been shown to be solvable using the Independent Components Analysis technique [2, 11, 13] and related methods. However these methods required knowledge of the number of sources and microphones, and could not solve the general source separation problem with a single microphone (as humans do).

3   Multimodal Mutual Information

Mutual Information Analysis is a powerful technique that has been shown to be able to accurately register signals despite a wide range of visual sensor types (e.g., intensity, range) [14]. Here we apply the technique to multi-modal data types, including both audio and visual channels, and show how it can identify pixels in a video sequence which are moving in synchrony with an audio source.

Classically, the mutual information between two random vectors can be written as

$$I(X, Y) = H(X) + H(Y) - H(X, Y)$$

where $H(X)$ is the entropy of vector X and $H(X, Y)$ is the joint entropy of X and Y. In the case where X and Y are normally distributed with mean u_X and covariance Σ_Y, and jointly distributed with mean u_{XY} and covariance Σ_{XY}, then this is simply

$$I(X, Y) = \frac{1}{2} log(2\pi e)^n |\Sigma_X| + \frac{1}{2} log(2\pi e)^n |\Sigma_X| - \frac{1}{2} log(2\pi e)^{n+m} |\Sigma_{XY}|$$

$$= \frac{1}{2} log \frac{|\Sigma_X||\Sigma_Y|}{|\Sigma_{XY}|}$$

where m, n are the length of X and Y. This formalism has been applied with a scalar but time-varying X representing the audio source, and a multi-dimensional time-varying Y representing the video source [8].

The Gaussian assumption in unrealistic in many environments; a more general method is to use non-parametric density models. In order to make the problem tractable high dimensional audio and video measurements are projected to low dimensional subspaces. The parameters of the sub-space are learned by maximizing the mutual information between the derived features [6].

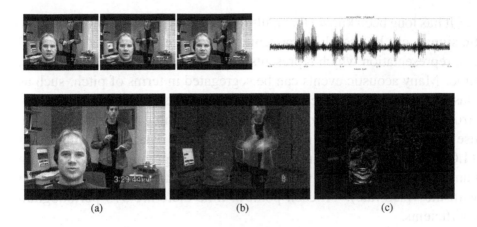

Figure 1: Top: Video (left) and associated audio (right) signals. Bottom: Video frame (left), pixel-wise standard deviation image (middle), and mutual information image (right).

These methods have been used to identify which locations in a video signal correspond to a single audio source [8, 6]. For example, Figure 1(top) shows example joint video and audio signals. Figure 1(bottom) shows a frame from the video, the pixels identified as high variance from video information alone, and the pixels identified has having high mutual information with the audio source. One can see that analysis of image variance or motion alone fails to distinguish the pixels moving in synchrony with the audio source. Analyzing the the joint mutual information over both video and audio signals approach easily identifies the pixels moving in the mouth region, corresponding to the speech sound source.

4   Spectral Audio Representation

We are interested in the inverse problem: we wish to enhance or attenuate audio components based on a particular video region. Unfortunately, an instantaneous scalar audio model makes it difficult in general to divide the audio signal into constituent sources. To overcome this problem, we extend the model to use a multi-dimensional time-frequency sound representation, which makes it easier to segment different audio sources.

It has long been known that multidimensional representations of acoustic signals can be useful for recognition and segmentation. Most typical is a representation which represents the signal in terms of frequency vs. time. Many acoustic events can be segregated in terms of pitch, such as classical musical instruments and human vowel sounds. While the cepstrum, spectrogram, and correlogram are all possible representations, we use a periodogram-based representation. Audio data is first sampled at 11.025 KHz, and then transformed into periodogram coefficients using hamming windows of 5.4 ms duration sampled at 30 Hz (commensurate with the video rate). At each point in time there are 513 periodogram coefficients.

We use a non-parametric density estimation algorithm, applied to multi-dimensional, time-varying audio and image features. Specifically, let $v_i \sim V \in \Re^{N_v}$ and $a_i \sim A \in \Re^{N_a}$ be video and audio measurements, respectively, taken at time i. Let $f_v : \Re^{N_v} \mapsto \Re^{M_v}$ and $f_a : \Re^{N_a} \mapsto \Re^{M_a}$ be mappings parameterized by the vectors α_v and α_a, respectively. In our experiments f_v and f_a are single-layer perceptrons and $M_v = M_a = 1$. The adaptation method extends to any differentiable mapping and output dimensionality [5]. During adaptation the parameters vectors α_v and α_a (the perceptron weights) are chosen such that

$$\{\hat{\alpha}_v, \hat{\alpha}_a\} = \arg \max_{\alpha_v, \alpha_a} I(f_v(V, \alpha_v), f_a(A, \alpha_a)) \tag{1}$$

into two scalar values (one for video and one for audio).

This process is illustrated in figure 2 in which video frames and sequences of periodogram coefficients are projected to scalar values. A clear advantage of learning a projection is that rather than requiring pixels of the video frames or spectral coefficients to be inspected *individually* the projection summarizes the entire set efficiently.

In [7] we first demonstrated how this framework could be utilized to segment the audio information based on raw pixel data in a user specified window of a video sequence. To create a fully automatic system for human-computer interface applications, we add pre-processing steps which transform the pixel data into a more suitable representation.

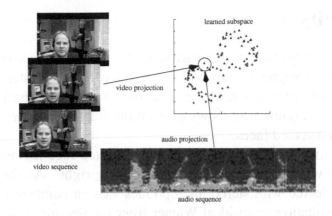

Figure 2: Fusion Figure: Projection to Subspace

5   Automatic Face and Motion Detection

To make the mutual information technique useful for human-computer interface applications, we use face detection and motion estimation as preprocessing steps. These provide an analysis of the video data to extract visual features that will correspond to the acoustic energy of an utterance in real-world conditions.

A face detection module provides the location of pixels in a video stream which belong to an individual face. We restrict the adaptation algorithm to only consider these pixels, and thus only find components of the audio stream which have high mutual information to that individual's face. Our implementation used the CMU facedetector library, which is based on a neural-network algorithm trained with both positive and negative examples of face images [12]

Image motion features are estimated and used as the video input to the mutual information analysis. Computing mutual information based on flow rather than raw pixel intensity change can help in cases where there is contrast normalized motion, such as with random dot patterns. We used the well-known robust optic flow implementation detailed in [3], which combines outlier rejection, multi-resolution processing, segmentation, and regularization. In practice we set the parameters of this algorithm to strongly regularize the motion estimates, since precise motion boundary localization was not important for this task.

6 Results

We first tested our system on an image sequence with two speakers recorded with a single microphone (the speakers were recorded with stereo microphones so as to obtain a reference, but the experiments used a single audio source). Figure 3(a) shows an example frame from the video sequence with detected faces.

By selecting data from one of the two detected face regions we can enhance the voice of the speaker on the left or right. As the original data was collected with stereo microphones we can compare our result to an approximation to an ideal Wiener filter (neglecting cross channel leakage). Since the speakers are male and female, the signals have better spectral separation and the Wiener filter can separate them. For the male speaker the Wiener filter improves the SNR by 10.43 dB, while for the female speaker the improvement is 10.5 dB. Our technique achieves a 9.2 dB SNR gain for the male speaker, and a 5.6 dB SNR gain for the female speaker.

It is not clear why performance is not as good for the female speaker, but figures 3(b) and (c) are provided by way of partial explanation. Having recovered the audio in the user-assisted fashion described we used the recovered audio signal for video attribution (pixel-based) of the entire scene. Figures 3(b) and (c) are the images of the resulting α_v when using the male (b) and female (c) recovered voice signals. The attribution of the male speaker in (b) appears to be clearer than that of (c). This may be an indication that the video cues were not as detectable for the female speaker as they were for the male in this experiment.

Our second test evaluated the ability of our method to improve accuracy in speech recognition where other noise sources were present. A single user spoke into a handheld videocamera with built-in microphone, at a distance of approx. 4 feet. A second noise source was synthetically added to the audio stream at varying SNR levels (8-13db). Recognition was performed using a commercially available transcription package; the system was trained as specified without added noise.

Preliminary results from this test are encouraging; in recognition tests involving a combination of digits and spoken phrases, we obtained approximately a 33% reduction in error rate. At higher noise levels (8db), the observed error rate was 55% unfiltered, and 38% with our technique

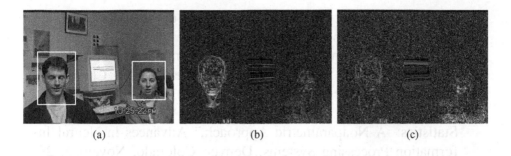

(a)   (b)   (c)

Figure 3: User assisted audio enhancement: (a) example image, with detected face regions, (b) image of α_v for region 1, (c) image of α_v region 2

applied as a preprocessing step. At lower noise levels (13db), the observed error rate was 3322% after our technique was applied. Averaged over all SNR levels, the unfiltered error rate was 49rates are unnaturally high, since the system was not trained with the added noise source. Nonetheless we believe the relative error rate improvement is a promising sign.

We hope that this technique can aid in the eventual goal of enabling audio interface without an attached microphone for casual users. Our lab is exploring the use of these techniques in intelligent environments, which should support hands-free browsing of information resources via pointing gestures and speech commands.

References

[1] B. Arons. A Review of The Cocktail Party Effect. Journal of the American Voice I/O Society 12, 35-50, 1992.

[2] A. Bell and T. Sejnowski. An information maximisation approach to blind separation and blind deconvolution, Neural Computation, 7, 6, 1129-1159, 1995.

[3] M. Black and P. Anandan, A framework for the robust estimation of optical flow, Fourth International Conf. on Computer Vision, ICCV-93, Berlin, Germany, May, 1993, pp. 231-236

[4] M. A. Casey, W. G. Gardner, and S. Basu "Vision Steered Beamforming and Transaural Rendering for the Artificial Life Interactive

Video Environment (ALIVE)", Proceedings of the 99th Convention of the Aud. Eng. Soc. (AES), 1995.

[5] J. Fisher and J. Principe. Unsupervised learning for nonlinear synthetic discriminant functions. In D. Casasent and T. Chao, eds., SPIE Optical Pattern Recognition VII, vol 2752, p 2-13, 1996.

[6] J. W. Fisher III, A. T. Ihler, and P. Viola, "Learning Informative Statistics: A Nonparametric Approach," Advances in Neural Information Processing Systems, Denver, Colorado, November 29-December 4, 1999.

[7] J. W. Fisher III, T. Darrell, W. Freeman, and P. Viola, Learning Joint Statistical Models for Audio-Visual Fusion and Segregation, in review.

[8] J. Hershey and J. Movellan. Using audio-visual synchrony to locate sounds. In T. K. L. S. A. Solla and K.-R. Muller, editors, *Proceedings of 1999 Conference on Advances in Neural Information Processing Systems 12*, 1999.

[9] Q. Lin, E. Jan, and J. Flanagan. "Microphone Arrays and Speaker Identification." IEEE Transactions on Speech and Audio Processing,Vol. 2, No. 4, pp. 622-629, 1994.

[10] T. Nakamura, S. Shikano, and K. Nara. "An Effect of Adaptive Beamforming on Hands-free Speech Recognition Based on 3-D Viterbi Search", ICSLP'98 Proceedings Australian Speech Science and Technology Association, Incorporated (ASSTA), Volume 2, p. 381, 1998.

[11] B. Pearlmutter and L. Parra. "A context-sensitive generalization of ICA", Proc. ICONIP '96, Japan, 1996.

[12] H. Rowley, S. Baluja, and T. Kanade, Neural Network-Based Face Detection, Proc. IEEE Conf. Computer Vision and Pattern Recognition, CVPR-96, pp. 203-207,. IEEE Computer Society Press. 1996.

[13] P. Smaragdis, "Blind Separation of Convolved Mixtures in the Frequency Domain." International Workshop on Independence & Artificial Neural Networks University of La Laguna, Tenerife, Spain, February 9 - 10, 1998.

[14] P. Viola, N. Schraudolph, and T. Sejnowski. Empirical entropy manipulation for real-world problems. In *Proceedings of 1996 Conference on Advances in Neural Information Processing Systems 8*, pages 851–7, 1996.

Visual Recognition of Emotional States

Karl Schwerdt*, Daniela Hall**, and James L. Crowley***

Project PRIMA, Lab. GRAVIR - IMAG
INRIA Rhône-Alpes
655, ave. de l'Europe
38330 Montbonnot St. Martin, France

Abstract. Recognizing and interpreting a human's facial expressions and thereby his mood are an important challenge for computer vision. In this paper, we will show that trajectories in eigenspace can be used to automate the recognition of facial expressions. Prerequisite is the exact knowledge of position and size of the face within a sequence of video images. Precision and stability are two properties deciding if a tracking algorithm is suitable for subsequent recognition tasks. We will describe in this paper a robust face tracking algorithm that can sufficiently normalize a video stream to a face allowing for facial expression recognition based on eigenspace techniques. The presented face tracking algorithm is mainly based on using probability maps generated from color histograms.
Keywords: Computer Vision, Visual Appearance, Face Tracking, Recognition of Emotional State

1 Introduction

Recognition and identification of visual and aural cues have met the interest of researchers since the late 80's, when the first smart rooms were built. Making computers, or computer-supported environments see and hear is a first step [1, 2] in creating intelligent environments. If, however, computers are to give more than automatic responses or reflexes to what they perceive, they have to have *some kind* of awareness about their environment. This implies that computers need to have the means to *recognize* what they see. Eventually, people should be able to talk to their counterpart, to make gestures that are recognized, to give and receive visual cues.

There are currently two approaches in computer vision of how to automatically extract information from an image or a stream of video images. One is called model based. That is, an *a priori* assumption about the image content is made, and a model of that content is created. This might be a (talking) head, a hand, or even a set of models, for instance, of cars and people. The extracted image content is then mapped onto those models.

The second approach to extract information from video images, is to project the video data into an orthonormal space, for instance, an eigenspace. The image content

* Karl.Schwerdt@inria.fr
** Daniela.Hall@inria.fr
*** Jim.Crowley@inria.fr

T. Tan, Y. Shi, and W. Gao (Eds.): ICMI 2000, LNCS 1948, pp. 41-48, 2000.

is thus encoded as it *appears* to an observer. Objects and their movements and deformations can then be recognized as trajectories in that space. This is made possible by pre-processing the video data using *a priori* assumptions about the video content, e.g., tracking a talking head, which is a standard situation in video telephony.

In this paper we will show, that it is generally possible to recognize facial expressions as trajectories in eigenspace, given that the input video stream is well normalized to the face [3]. We will also show that we have a face tracking algorithm [4], which has originally been designed to enable efficient video compression based on orthonormal basis coding and has the required precision to make possible facial expression recognition by eigenspace trajectories.

Section 2 will concisely present the face tracking algorithm, whereas section 3 discusses the algorithm for the recognition of the facial expression. In both sections there is a performance evaluation for each algorithm. Section 4, finally, draws some conclusions and gives an outlook on further work.

2   Robust Tracking of Faces Using Color

Detecting pixels with the color of skin provides a reliable method for detecting and tracking faces. The statistics of the color of skin can be recovered from a sample of a known face region and then used in successive images to detect skin colored regions. Swain and Ballard have shown how a histogram of color vectors can be back-projected to detect the pixels which belong to an object [5]. Schiele and Waibel showed that for face detection, color RGB triples can be divided by the luminance to remove the effects of relative illumination direction [6]. We give here only a very short introduction of the face tracker. The reader is referred to [4] for a thorough discussion and further references.

2.1   Probability of Skin

The reflectance function of human skin may be modeled as a sum of a Lambertian and a specular reflectance function. In most cases the Lambertian component dominates. For a Lambertian surface, the luminance of reflected light varies as a function of the cosine of the angle between the surface normal and illumination. Because the face is a highly curved surface, the observed luminance of a face exhibits strong variations. These variations may be removed by dividing the three components of a color pixel, (R, G, B) by the luminance. This gives an luminance-normalized color or chrominance vector, with two components, (r, g).

$$r = \frac{R}{R + G + B} \qquad g = \frac{G}{R + G + B}$$

The luminance-normalized pixels from a region of an image known to contain skin can be used to define a two dimensional histogram, $h_{skin}(r, g)$, of skin color. The effects of digitizing noise can be minimized by smoothing this histogram with a small filter. A second histogram of the same dimensions, $h_{total}(r, g)$, can be made from all of the pixels of the same image. This second histogram should also be smoothed by the same filter. These two histograms make it possible to obtain the probability that a given pixel has skin color (see [4] for a derivation of equation 1):

$$p(skin|r,g) \approx \frac{h_{skin}(r,g)}{h_{total}(r,g)} \tag{1}$$

In order to detect a skin color region we must group skin pixels into a region. Let $P_{skin}(i,j)$ represent the probability map of skin for each color pixel $(r(i,j), g(i,j))$ of an image (see figure 1) at position (i,j) (see figure 2).

$$P_{skin}(i,j) = p(skin|r(i,j), g(i,j)) \tag{2}$$

The center of gravity or first moments, μ, of the probability map gives the position of the skin colored region, and the second moments, $C(i,j)$, gives a measure for the spatial extent of the skin colored region.

$$\mu = \begin{bmatrix} \mu_i \\ \mu_j \end{bmatrix} \qquad C(i,j) = \begin{bmatrix} \sigma_i^2 & \sigma_{ij} \\ \sigma_{ji} & \sigma_j^2 \end{bmatrix} \tag{3}$$

Unfortunately, skin color pixels in any other part of the image will contribute to these two moments. This effect can be minimized by weighting the probability image with a Gaussian function placed at the location where the face is expected. The initial estimate of the covariance of this Gaussian should be the size of the expected face. Once initialized, the covariance is estimated recursively from the previous image.

Fig. 1. Raw color input image (176 x 144 pixels).

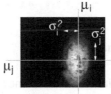

Fig. 2. Probability images for the Center of Gravity Algorithm. The bigger the graylevel pixel value, the higher the probability of skin color

For each new image, a two dimensional Gaussian function, $g(i,j;\mu,C)$, using the mean and covariance from the previous image is multiplied with the probability map as shown in equation 4 to give new estimates for the mean and covariance. The coefficients of equation 3 are calculated in the following manner:

$$
\begin{aligned}
\mu_i &= \frac{1}{S} \sum_{i,j} P_{skin}(i,j) \cdot i \cdot g(i,j,\mu,C), \\
\mu_j &= \frac{1}{S} \sum_{i,j} P_{skin}(i,j) \cdot j \cdot g(i,j,\mu,C), \\
\sigma_i^2 &= \frac{1}{S} \sum_{i,j} P_{skin}(i,j) \cdot (i-\mu_i)^2 \cdot g(i,j,\mu,C), \\
\sigma_j^2 &= \frac{1}{S} \sum_{i,j} P_{skin}(i,j) \cdot (j-\mu_j)^2 \cdot g(i,j,\mu,C), \text{ and} \\
\sigma_{ji} = \sigma_{ij} &= \frac{1}{S} \sum_{i,j} P_{skin}(i,j) \cdot (i-\mu_i)(j-\mu_j) \cdot g(i,j,\mu,C),
\end{aligned}
\tag{4}
$$

where $S = \sum_{i,j} P_{skin}(i,j) \cdot g(i,j,\boldsymbol{\mu},\mathbf{C})$. The effect of multiplying new images with the Gaussian function is that other objects of the same color in the image (hands, arms, or another face) do not disturb the estimated position of the region being tracked.

2.2   Behavior Discussion

The use of a Gaussian weighting function for new input data poses two problems: 1) Even if we assume the distribution function of the tracked object – even if it does not move – to be approximately the same for subsequent images, the combined *pdf* of weighting and distribution function will shrink with each cycle, if no measure of compensation is taken. 2) The object being tracked moves above a certain speed such that the *pdf* of weighting and distribution function do not overlap. Combining the results of compensation for Gaussian function overlapping and object motion compensation, we get as the covariance matrix of the Gaussian weighting function for new incoming data [4]:

$$\mathbf{C}'(i,j) = \begin{bmatrix} 2(\sigma_i^2 + \Delta\mu_i^2) & \sigma_{ij} \\ \sigma_{ij} & 2(\sigma_j^2 + \Delta\mu_j^2) \end{bmatrix}, \tag{5}$$

where σ_i^2, σ_{ij}, and σ_j^2 are those from equation 3, and $\Delta\mu_i$ and $\Delta\mu_j$ are the differences between the coordinates of the center of the tracked object in the current and in the previous image.

2.3   Performance Evaluation

Our robust tracking algorithm carries a somewhat higher computational cost than, e.g., connected components of a thresholded image. This is illustrated with the computing times shown in figure 3. This figure shows the execution time for a 176x144 pixels sized image on a SGI 02 workstation for the robust tracking or center of gravity (RA) algorithm, a connected components (CCO) algorithm, and connected components algorithm assisted by a zeroth order Kalman filter. Average execution times are around 25 milliseconds per image for the connected components and 70 milliseconds for the robust algorithm.

Jitter is the number of pixels that the estimated position moves when the target is stationary. Jitter is the result of interference with illumination, electrical noise, shot noise, and digitizer noise. Algorithms which employ a threshold are especially sensitive to such noise. Table 1 illustrates the reduction in jitter for the robust tracker when compared to connected components.

Figure 4 compares the precision of tracking an object moving in the horizontal direction. All three trackers were applied to the same image sequence. The output of the color tracker using the connected components algorithm is shown with and without Kalman filter. The Kalman filter eliminates position jitter but reduces precision of global position estimation.

	RA	CCO w/o KF	CCO w/ KF
Jitter Energy	29	308	151

RA : Robust Algorithm based on center of gravity
CCO : Connected Components Algorithm
KF : Kalman Filter

Table 1. Jitter energy measured for a stationary object by the robust estimator, and by connected components with and without a Kalman Filter

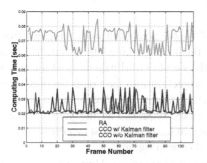

Fig. 3. Computing time per image for the robust estimator (RA), Connected Components Algorithm without (CCO w/o Kalman) and with (CCO w/ Kalman) assistance from a Kalman filter recursive estimator.

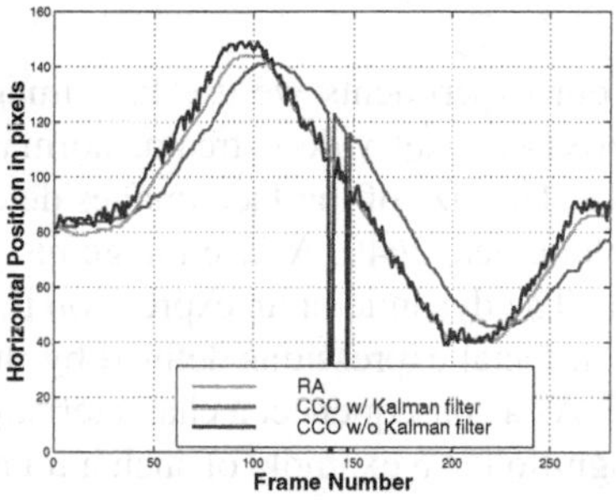

Fig. 4. Comparing tracking precision of a moving object. The dips correspond to tracking failures of the color tracker using a threshold to decide if a pixel has skin color or not. The threshold was determined empirically beforehand, but can not be adapted to varying lighting conditions during tracking.

3   Representing Emotional State in Eigenspace

3.1   Appearance Representation in Eigenspace

The use of principal components analysis for face recognition was introduced by Sirovich and Kirby [7], and explored by Turk and Pentland [8]. The technique has since been applied to many difficult computer vision problems [9, 10]. The principal difficulty is that the images must be normalized in position and orientation in order for the method to be reliably used. Our initial goal has therefore been to investigate whether it is possible to adapt this method to recognize facial expressions.

Let us consider an image as a vector of (pixel) values. Calculating the principal components of a large set of face images (vectors) then yields a set of orthogonal images (eigenvectors). These images give an optimal description of the variance between a set of training images, where each image of the set defines a dimension in this basis and is referred to as an *eigenface*. Figure 5 shows mean and first few eigenimages of sample of face images. Similar images tend to have similar projections into an eigenspace.

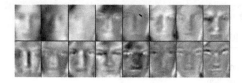

Anger   Raising   Disgust   Happiness   Surprise
         Eyebrows

Fig. 5. Mean and first few eigenimages ordered by decreasing eigenvalues.

Fig. 6. Facial expressions treated in the experiment.

3.2 Facial Expressions

For our experiments we use face images of 44 by 60 pixels cut out of images from a 176x144 pixel video stream, normalized by the face tracking algorithm discussed above. The size of the face images determines the number of dimensions of the eigenproblem, here 2640. A face image results after projection into face space in one single point. The dynamic of an expression is an important parameter for recognition. For this reason facial expressions defined by the entire image sequence during performance is used. As a consequence, a characteristic trajectory in face space must be considered for recognition. An example of such a trajectory is shown in figure 7.

We use histograms for the recognition of facial expressions. However, the use of histograms is limited to four or five dimensions for memory space reasons. On the other hand, the first four eigenvectors are not sufficient for a distinction of the five different expressions plus a neutral state. Figure 6 shows the five expressions. A technique for the selection of appropriate eigenface dimensions must be found in order to reduce the dimensions.

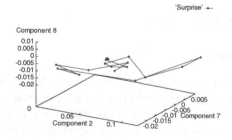

Fig. 7. Example of a trajectory cycle, representing *surprise* expression on the principle components 2, 7, and 8.

As can be seen in figure 5, eigenvectors with big eigenvalues represent the coarse differences between the images. Detailed movements of single features like eye brows are captured by eigenvectors with smaller eigenvalues, and it is those eigenvectors which are important for the distinction of facial expressions. Therefore, the first twelve eigenvectors are considered to be possible candidates for histogram generation. Eigenvectors from this group not contributing to the distinction of the expressions are removed. The recognition task is done after creating a histogram from the remaining eigenvectors.

In our experiment, four dimensional histograms of the 2nd, 7th, 8th and 11th eigen-vectors are built. These are the eeigenvectors that divide the facial expression classes best. They can be found by analyzing the distribution of the training samples.

We experimented with cell numbers of 16 and 25 cells per dimension and Gaussian filter of size 3. Table 2 shows the recognition results.

Type	Success	Reliability	Type	Success	Reliability
Anger	100%	61.5%	Anger	75.0%	100%
Brows	66.7%	57.1%	Brows	44.4%	44.4%
Disgust	100%	100%	Disgust	100%	66.7%
Happy	100%	50.0%	Happy	100%	63.6%
Surprise	76.9%	83.3%	Surprise	53.8%	70.0%
Experiment 3a: 16 cells			Experiment 3b: 24 cells		

Table 2. Experimental results of facial expression recognition with four dimensional histogram

An important parameter of the histogramms are the number of cells. Is the number too low, the histogramm is too coarse to reflect the distribution of the training samples. Many false positives would be expected. Is the number of cells too high, the training samples are described very precisely. The histogram does not tolerate noise in the ob-served trajectories. This can be overcome by including a very large number of training samples. In our example only few training samples are available. The experiments have shown, that histogramms with 16 cells result in good recognition. This is convenient because the memory space requirements are lower.

The reason for the somewhat inferior results in the spontaneous expressions like raising eye brows and surprise is that only a smaller absolute number of training images is available due to the shorter duration of spontaneous expressions. A larger number of training sequences would improve the results.

Our technique is successful for the recognition of expressions spanning a larger set of images, like anger, disgust, and happiness. All test expressions are correctly detected. The reliability is computed by:

$$\text{reliability} = \frac{(\text{number of detections} - \text{incorrect detections})}{(\text{number of detections})} \quad (6)$$

The reason for the decreased reliability can be found in the small number of training images and the large number of dimensions of the histogram, both resulting in many empty histogram cells.

As expected, a precise normalization of face position in the images is crucial in this experiment. Translations of only a few pixels can cause important changes in the eigenspace projection of the image, thus creating discontinuities in the trajectory rep-resenting the facial expression. Such discontinuities break up projection into the his-togram. A systematic evaluation of recognition under controlled noise conditions re-mains to be done.

4   Conclusions

Eigenspace techniques provide a way to identify and recognize facial gestures, given that we have a face tracker that can reliably normalize a video input stream to a face. In order to do this, various techniques from computer vision have been used to create a fast and robust face tracking system, mainly based on color detection.

Moreover, face tracking generally enhances the usability of a video communication system by allowing the user to freely move in front of the camera while communicating. It is however crucial that the face-tracking system be stable and accurate in order to provide the best results for facial expression detection, since normalization errors can introduce large jumps in eigenspace, making trajectory matching unreliable. We showed in this paper that our face tracking algorithm has these properties.

For the recognition of facial expressions very good results are obtained under the condition of a perfect normalization. The most reliable eigenvectors for face expression recognition are not necessarily the eigenvectors with the largest eigenvalues. Eigenvalues are caused by scatter in the sample population, and this scatter can be due to sources such as lighting and normalization, which are independent of face expressions.

Acknowledgement

This work has been sponsored by the EC DG XII Human Capital and Mobility Network SMART II.

References

1. J. Crowley, J. Coutaz, and F. Bérard, "Things that see," *Communications of the ACM*, vol. 43, pp. 54–64, March 2000.
2. A. Pentland, "Perceptual intelligence," *Communications of the ACM*, vol. 43, pp. 35–44, March 2000.
3. D. Hall, J. Martin, J. Crowley, and R. Dillmann, "Statistical recognition of parameter trajectories for hand gestures and face expressions," in *IAR*, (Mulhouse, France), 1998.
4. K. Schwerdt and J. Crowley, "Robust face tracking using color," in *Proc. of the Fourth IEEE Int. Conf. on Automatic Face and Gesture Recognition*, (Grenoble, France), pp. 90–95, March 2000.
5. M. Swain and D. Ballard, "Color indexing," *Int. Journal of Computer Vision*, vol. 7, no. 1, pp. 11–32, 1991.
6. B. Schiele and A. Waibel, "Gaze tracking based on face color," in *Proc. of the Int. Workshop on Automatic Face and Gesture Recognition*, (Zurich, Switzerland), pp. 344–349, June 1995.
7. I. Sirovich and M. Kirby, "Low–dimensional procedure for the characterization of human faces," *Journal of Optical Society of America*, vol. 4, pp. 519–524, March 1987.
8. M. Turk and A. Pentland, "Eigenfaces for recognition," *Journal of Cognitive Neuroscience*, vol. 3, pp. 71–86, March 1991.
9. M. Black and A. Jepson, "Eigen tracking: Robust matching and tracking of articulated objects using a view-based representation," in *Proc. of the 4th European Conf. on Computer Vision*, vol. 1065 of *Lecture Notes in Computer Science*, pp. 329–342, Springer, April 1996. in Volume I.
10. V. Colin de Verdière and J. Crowley, "Visual recognition using local appearance," in *Proc. of the 5th European Conf. on Computer Vision*, vol. 1406 of *Lecture Notes in Computer Science*, (Freiburg, Germany), pp. 640–654, Springer, June 1998. in Volume I.

An Experimental Study of Input Modes for Multimodal Human-Computer Interaction

Xiangshi REN[1], Gao ZHANG[2] and Guozhong DAI[3]

[1] Kochi University of Technology, Kochi 782-8502, Japan
[2] Microsoft Research, Beijing 100080, China
[3] Chinese Academy of Sciences, P.O. Box 8718, Beijing 100080, China

Abstract. Two experimental evaluations were conducted to compare interaction modes on a CAD system and a map system respectively. For the CAD system, the results show that, in terms of total manipulation time (drawing and modification time) and subjective preferences, the "pen + speech + mouse" combination was the best of the seven interaction modes tested. On the map system, the results show that the "pen + speech" combination mode is the best of fourteen interaction modes tested. The experiments also provide information on how users adapt to each interaction mode and the ease with which they are able to use these modes.
Keywords: multimodal interface, mode combination, interface evaluation, CAD systems, map systems, pen-based input, interactive efficiency.

1 Introduction

Multimodal interfaces have long been considered as alternatives and as potentially superior. Many studies of multimodal user interfaces have been reported for tasks such as text processing [10], map-based tasks [8], and in the communication environment [6]. Studies on speech and keyboard input [2, 4], mouse and speech input [1], speech and gesture input [5] have also been conducted. These studies did not compare all reasonable combination modes, such as uni-modal and tri-modal combinations.

Researchers generally believe that a multimodal interface is more effective than a unimodal interface, e.g. Hauptmann et al. (1989)[5] who observed a surprising uniformity and simplicity in the user's gestures and speech, and Oviatt et al. (1997) [8] who reported that users overwhelmingly preferred to interact multi-modally rather than single-modally. However, little quantitative research and evaluation has been reported on multimodal combinations which would certify that one combined input mode is more natural and efficient than another in particular environments, e.g. a map system, a CAD system etc.. Suhm et al. (1999) [9] asked the question "which modality do users prefer?", however, the authors only answer with reference to single modes. They did not state which is the most effective mode, nor did they report on combined modes.

This study evaluates the differences in modalities and their combinations through usability testing on a CAD system and a map system respectively. We

T. Tan, Y. Shi, and W. Gao (Eds.): ICMI 2000, LNCS 1948, pp. 49-56, 2000.

look at how interaction modes are adapted to different applications. We are interested in what is the most effective modality that users prefer in a given application. We also seek to provide information on how users choose different interaction modes when they work on an application.

2 Experiment One: a CAD System

A CAD system is a complex interactive system, with which users can draw graphic objects with a mouse, choose and drag objects and tools, select colors and other properties from menus or dialogue boxes, and manage their plans through the file system. We consider that a CAD system is an application-oriented system, and the study of the usability of its multimodal interface ought to be based on an applied CAD system. Furthermore, we give consideration to all manipulations including drawing, location, modification and property selection (thickness/color etc.). Driven by this, we set up a multimodal user interface environment on an AutoCAD system, where users could use pen, speech, and/or mouse to draw an engineering plan.

2.1 Method

Participants Twenty-four subjects (20 male, 4 female, all right handed; 14 students, 10 workers) were tested for the experiment. Their ages ranged from twenty to thirty five years. Ten of them had had previous experience with AutoCAD systems, the other had no experience, but they could all use the mouse and keyboard proficiently. Only five of them had had previous experience with the pen used in the experiment.

Apparatus The hardware used in the experiment was a pen-input tablet (WA-COM), a stylus pen, a microphone, and a personal computer (P166, IBM Corp.). The software used in the experiment was Windows 95, AutoCAD 12.0 for Windows, a drawing recognition system we developed and a speech recognition system (CREATIVE Corp.).

Design We did not use a keyboard in the experiment because the task was only drawing. There are seven possible interface combinations for a mouse, a pen, and speech: mouse, pen, speech used individually, mouse + pen, mouse + speech, pen + speech, pen + speech + mouse. Obviously, speech-only cannot accomplish the drawing tasks efficiently.

In order to simplify the experiment, we performed a preliminary experiment to compare the difference between the use of the mouse and the use of the pen in the CAD system. The result showed the pen was suitable for drawing the outline of the plan. The pen's efficiency and subject preference rating were better than the mouse's but the pen was not as accurate as the mouse. We inferred from this result that the pen + speech mode was better than the mouse + speech mode for outlining and we therefore omitted tests for the mouse + speech mode. Furthermore, the frequent change between mouse and pen takes a lot of time and is not convenient. We therefore assigned the pen to drawing tasks and the mouse to modification tasks and we made speech simultaneously available to both pen and mouse operations.

Thus, in order to investigate the differences between different input modes and their combined use, each subject tested four modes: the mouse, pen, pen + speech, and pen + speech + mouse modes. The use of pen and mouse in the pen + mouse + speech mode interface eliminated frequent changing between mouse and pen. The mouse was used as a supplemental device to the pen at the modification stage to ensure accuracy.

Task and Procedure First the experiment was explained to each of the subjects, who were each given 30 minutes to learn how to use the pen and the speech input equipment.

The CAD system chose one of the four interface modes randomly and showed the corresponding instruction information on the title bar of the AutoCAD system. After receiving the mode information, a dialogue box with three buttons appeared: "beginning to draw", "beginning to modify", and "finishing drawing". The subject chose "beginning to draw" to begin the test.

The AutoCAD system was opened and a sample plan appeared on the screen. This plan was selected as the test object and appeared as a sample, which could not be altered by the users. In the test, the subject tried his/her best to match the sample plan. In order to establish the drawing time and modification time, the subject was not allowed to modify the drawing before he/she had finished all of the drawing. After finishing all of the drawing, the subject could choose "beginning to modify" to begin the modification stage. After finishing all the modifications, the subject could choose "finishing drawing" to finish the current test.

The subject was asked to do six tests for each interface mode. The first was a practice and the results of the other five were recorded as formal tests. Each subject had to test all four modes. Whenever they finished a test, they were allowed to have a rest.

Data for each interface mode was recorded automatically as follows: (1) The time taken to draw the plan: This is the time lapsed from the selection of the "beginning to draw" button to the selection of the "beginning to modify" button. (2) The time taken to modify the plan: The time lapsed between selection of the "beginning to modify" button and the selection of the "finishing drawing" button. (3) Accuracy of drawing: At the beginning of each test, the system provided each subject with a background paper. During the test, the subjects were told to trace the background drawing. When the drawing was finished, the system calculated the matching percentage between the background paper and drawing paper. We reckoned the matching percentage to be the degree of accuracy. (4) Subject preference: The subjects were questioned about their preferences after they finished testing each interface mode. They were asked to

rank (on a scale of 1-10) the mode just tested according to their satisfaction
with the mode and their desire to use that mode.

2.2   Results

We performed an ANOVA (analysis of variance) with repeated measures on the
within-subject factors on the interface modes used, with drawing time, modification time, total time (drawing time + modification time), accuracy, and subjective preference as dependent measures. Accuracy was calculated according to
the matching rate between the sample plan and the user's drawing plan.

Pen + speech + mouse mode The pen + speech mode was faster than the
other three in drawing time (mean = 9.9 minutes), $F(3,92) = 185.97$, $p < 0.0001$,
however, the pen + speech + mouse mode was faster than the other three in
modification time (mean = 5.0 minutes), $F(3,92) = 145.06$, $p < 0.0001$, in total
time (mean = 15.0 minutes), $F(3,92) = 35.44$, $p < 0.0001$.

The mouse-based interface was the most accurate (mean accuracy = 91.6%),
$F(3,92) = 136.88$, $p < 0.0001$.

The pen + speech + mouse mode also had the highest satisfaction rating
(mean = 7.8), $F(3,92) = 7.18$, $p < 0.0001$.

Location and modification issues The result shows that the pen-based interface was slower than the mouse interface for modification, $F(1,46) = 240.47$,
$p < 0.0001$. The pen-based interface was less accurate than the mouse interface,
$F(1,46) = 515.8$, $p < 0.0001$. The subjective rating results also show that the
pen-based interface was not as satisfactory as previously thought.

We noted that users took a lot of time and energy learning to use the pen
with the complex menus, especially the 19 novices (out of 24 subjects) who had
no experience with the pen. However, the pen + speech mode was faster than
the pen-based mode in total time, $F(1,46) = 91.46$, $p < 0.0001$.

Regarding accuracy, a significant difference was found between the pen +
speech mode and the pen-based mode, $F(1,46) = 86.14$, $p < 0.0001$, the results
show that the combination of pen + speech was more accurate than the pen on
its own. Subjective preferences show that the pen + speech mode had higher
ratings (mean = 7.29) than the pen-only mode (man = 6.54), $F(1,46) = 8.41$, $p
< 0.005$.

2.3   Discussion

We presented an evaluation experiment based on the applied CAD system. The
experimental results show that the pen + speech + mouse mode was the best of
the seven interaction modes on the CAD system. In particular the pen + speech
+ mouse mode was faster than the pen + mouse mode in total drawing and
modification time, $F(1,46) = 96.77$, $p < 0.0001$. We do not only show that the

multimodal interface is better than the unimodal interface for CAD systems but we also show the results of comparisons between combined modes.

The results also show that a proper combination of input modes can improve the interactive efficiency and user satisfaction of CAD systems. The pen was suitable for drawing the outline of the plan. The mouse was useful for accurate modification procedures. Speech was suitable for inputting the descriptive properties of graphic objects and selecting menus.

Many studies suggest the use of new interactive modes such as the pen or speech to replace the traditional mouse and keyboard. However, based on our experiment, the mouse is still useful for modification and location procedures because location and modification with the pen may not be accurate enough. We recommend the pen for outlining, initial location and layout procedures and the mouse for modification because it is more accurate. We should pay more attention to location and modification technology. In the meantime, we suggest that the traditional mouse continue to be used for modification procedures in CAD systems.

3   Experiment Two: a map system

Map systems are usually used in public places. They require a more convenient user interface. Some multimodal interactive methods, such as spoken natural language, gesture, handwriting, etc. have been introduced into map systems to produce more natural and intuitive user interfaces [7, 3].

We set up a prototype multimodal map system where users can use pen, speech, handwriting, pen-based gestures, as well as mouse and keyboard modes (typing, selecting, and dragging), to accomplish a number of trip plan tasks, e.g. to get the necessary information to plan their travel routes. For this environment, we designed an experiment to investigate which is the best of the different combination modes.

3.1   Method

Participants Twenty-four subjects (12 male, 12 female, all right handed; 12 students, 12 workers) were tested for the experiment. Their ages ranged from twenty to thirty-five. None of them had had any experience in using this kind of trip plan system.

Apparatus The hardware was the same as used in Experiment One (see Sect. 2.1). The software was Windows 95, a drawing recognition system (developed by us) and a speech recognition system (CREATIVE Corp.).

Design We used the keyboard (as well as the mouse, pen and speech) because we considered that the keyboard was useful for information retrieval. Thus the possible combinations for keyboard, mouse, pen and speech are: mouse, keyboard, speech, pen used individually, mouse + keyboard, mouse + speech, mouse + pen, keyboard + speech, keyboard + pen, speech + pen, m + k + s, m + k + p, k + s + p, s + p + m.

We designed the experiment in two steps. Step One was to exclude modes and combinations seldom-used by the subjects. The mean success rate was 1/14 = 7%, we treated those combinations with success rates above 7% as useful combination modes. Step Two was to more accurately compare the differences between useful multi-modal combinations using combination modes above 7%.

Task and Procedure There were four classes of task in the map system: distance calculation; object location; filtering; information retrieval. All these tasks could be accomplished by multiple modal combination modes.

In Step One, each of the subjects had 30 minutes to learn how to use the input equipment (mouse, keyboard, pen, and speech) to accomplish trip plan tasks. After they were familiar with the experimental environment, the experiment began. The subjects were asked to accomplish four tasks for each mode combination selected. The system allowed 10 minutes for each class of task. All possible mode combinations were given randomly to the subjects to perform. The subjects were asked to perform the task as soon as possible by use of the appointed mode combination. If the subjects accomplished the appointed task in a given time (20 seconds), an automatic program running in the background recorded the performance time and procedure. If the task took longer than twenty seconds the testing system assumed that the user had failed to perform this task.

In Step Two, the subjects were asked to accomplish 24 tasks for each mode combination selected. The tasks were randomly assigned in test of each combination. Data for each mode combination was recorded automatically as follows. The program running in the background automatically recorded the performance time. This was the time lapsed from the beginning of the first task to the end of the last task. The subjects were questioned about their preferences after they finished testing each mode combination. They were asked to rank (on a scale of 1-10) the mode just tested according to their satisfaction and their desire to use it.

3.2   Results

An ANOVA (analysis of variance) was conducted on the results of the experiment. The manipulation efficiency and subjective preference differences of these multimodal interactive modes were compared.

Useful combination modes From Step One, we analyzed useful combination modes for accomplishing the trip plan tasks. The statistical results showed that in single-modality mode, the success rate for the mouse was 1%, for the keyboard it was 10%, speech 15% and pen 12%.

In bi-modality mode, the success rate for mouse + speech was 13%,, keyboard + speech 11% and speech + pen 18%. The success rate for mouse + keyboard was 5%, mouse + pen 3%, and keyboard + pen 5%. The tri-modality modes were seldom used with success rates of less than 3%.

We therefore chose the following six modes for further testing in Step Two: keyboard, speech, pen, mouse + speech, keyboard + speech, pen + speech.

Pen + speech mode Significant differences in the six modes were found in mean manipulation time, $F(5,138) = 105.6$, $p < 0.0001$. The results revealed the pen + speech combination was faster than the other five modes in total time (mean manipulation time of pen + speech was 6.8 minutes). On the other hand, the pen-only interface was the slowest among the six modes with mean manipulation time of 9.1 minutes.

There was also a significant difference in the six modes for subject preferences, $F(5,138) = 105.6$, $p < 0.0001$. The pen + speech mode had the highest satisfaction rating (mean = 8.5). The speech-only mode had the lowest satisfaction rating (mean = 6.3).

Based on the analyses, the pen + speech combination was the best of the six interaction modes.

3.3   Discussion

Regarding individual modes, the success rate for the mouse was 1%, the keyboard was 10%, the speech was 15%, the pen was 12%. This reveals that the mouse (though it is accurate) is not suitable for trip plan tasks. In combination modes (pairs), speech plays an important supplemental role in trip plan tasks (mouse + speech 13%, keyboard + speech 11%, pen + speech 18%). All tri-modality modes rated less than 3%. Tri-modality modes were seldom used.

Overall, the pen + speech combination was the best of the fourteen interaction modes.

On the other hand, in Experiment One, we show that the mouse is still useful for modification and location procedures in CAD systems however, map systems do not need a mouse because these kinds of systems do not call for accurate fine tuning.

This experiment shows that modes used in pairs were better than tri-modality modes. The results also show that more modes may not necessarily be better than less modes, e.g. the pen + speech + mouse mode was better than pen + speech mode on CAD systems, however, the pen + speech mode was better than the tri-modality modes on map systems. The optimal number of combined modes for each environment should be investigated.

4   Conclusions

First, the pen + speech + mouse mode was the best of seven interaction modes tested on CAD systems. The pen + speech mode was the best of fourteen interaction modes tested on map systems. Second, we also show that more combination modes may not necessarily be better than less modes. Third, our tests for the first time give statistical support to the view that the mouse is still useful for accurate modification and location procedures especially in multimodal interfaces for CAD systems. Finally, we contributed to the body of information about how users adapt to each interaction mode, and the ease with which they are able to use them. These tests included keyboard, mouse, pen, and speech. Other interaction modes such as gaze input and touch, should also be tested in combination modes.

Acknowledgements

This study is supported by the key project of National Natural Science Foundation of China (No.69433020) and the key project of China 863 Advanced Technology Plan (No. 863-306-ZD-11-5)

References

1. Bekker, M.M., Nes, F.L.van, and Juola, J.F., A comparison of mouse and speech input control of a text-annotation system, Behaviour & Information Technology, **14**, 1 (1995), 14-22.
2. Damper, R.I., and Wood, S.D., Speech versus keying in command and control applications, Int. J. of Human-Computer Studies, **42** (1995), 289-305.
3. Cheyer, A. and Julia, L., Multimodal maps: an agent-based approach; SRI International, 1996, http:// www.ai.sri.com/~ cheyer/papers /mmap/ mmap.html.
4. Fukui, M., Shibazaki, Y., Sasaki, K., and Takebayashi, Y., Multimodal personal information provider using natural language and emotion understanding form speech and keyboard input, The special interest group notes of Information Processing Society of Japan, **64**, 8 (1996), 43-48.
5. Hauptmann A.G., Speech and gestures for graphic Image manipulation, In Proc. of the CHI'89 Conference on Human Factors in Computing Systems, ACM, New York (1989), 241-245.
6. Ohashi, T., Yamanouchi, T., Matsunaga, A., and Ejima, T., Multimodal interface with speech and motion of stick: CoSMoS, Symbiosis of Human and Artifact, Elsevier B.V. (1995), 207-212.
7. Oviatt. S., Toward empirically-based design of multimodal dialogue system, In Proc. of AAAI 1994-IM4S, Stanford (1994), 30-36.
8. Oviatt, S., DeAngeli, A., and Kuhn, K., Integration and synchronization of input modes during multimodal human-computer interaction, In Proc. of the CHI'97 Conference on Human Factors in Computing Systems, ACM, New York (1997), 415-422.
9. Suhm B., Myers B., and Waibel A., Model-based and empirical evaluation of multimodal interactive error correction, In Proc. of the CHI'99 Conference on Human Factors in Computing Systems, ACM, New York (1999), 584-591.
10. Whittaker, S., Hyland, P., and Wiley, M., Flochat: handwritten notes provide access to recorded conversations, In Proc. of the CHI'94 Conference on Human Factors in Computing Systems, ACM, New York (1994), 271-277.

This article was processed using the LaTeX macro package with LLNCS style

Emotion Expression Function in Multimodal Presentation

Yuan Zong[1], Hiroshi Dohi[2], Helmut Prendinger[2], and Mitsuru Ishizuka[2]

[1] IBM Japan Systems Engineering Co., Ltd.,
1-1, Nakase, Mihama-ku, Chiba-shi, Chiba 261-8522, Japan
Tel: 81-43-297-6055, Fax: 81-43-297-4836
yzong@jp.ibm.com
[2] Department of Information and Communication Engineering,
School of Engineering, University of Tokyo,
7-3-1, Hongo, Bunkyo-ku, Tokyo 113-8656, Japan
Tel: 81-3-5841-6755, Fax: 81-3-5841-8570
{dohi,helmut,ishizuka}@miv.t.u-tokyo.ac.jp

Abstract. With the increase of multimedia content on the WWW, multimodal presentations using interactive lifelike agents become an attractive style to deliver information. However, for many people it is not easy to write multimodal presentations. This is because of the complexity of describing various behaviors of character agents based on a particular character system with individual (often low-level) description languages. In order to overcome this complexity and to allow many people to write attractive multimodal presentations easily, MPML (Multimodal Presentation Markup Language) has been developed to provide a medium-level description language commonly applicable to many character systems. In this paper, we present a new emotion function attached to MPML. With this function, we are able to express emotion-rich behaviors of character agents in MPML. Some multimodal presentation content is produced in the new version of MPML to show the effectiveness of the new emotion expression function.

1 Introduction

An interface is a necessary part of human-computer interaction. The ideal user interface would let us perform our tasks without being aware of the interface as the intermediary. The longevity and ubiquity of the now two-decade old graphical user interface should not mislead us into thinking that it is an ideal interface.

Among many possible Post-GUI interfaces, a multimodal interface is supposed to be the most promising one. A multimodal interface uses the character agent as the middle layer between user and computer, interacting with user and controlling the device. The character agent recognizes the user's command and runs a task as the user requests. After the task is completed, the character reports the result by using verbal output or actions. By employing an character agent, the user can get informa-

T. Tan, Y. Shi, and W. Gao (Eds.): ICMI 2000, LNCS 1948, pp. 57-64, 2000.
© Springer-Verlag Berlin Heidelberg 2000

tion from many information channels (e.g., speech with intonation, emotion, actions and so forth).

One important implementation of a multimodal interface is multimodal presentation, which is an attractive way to present research work or products. With the development of the multimedia technology, presentation technology evolved. Centuries ago, people used text to appeal the audience. Because text only conveys information through a single channel, it was not a very effective presentation method. Recently, people use various presentation tools to make presentation (e.g., OHP, PowerPoint, and so forth).

Fig. 1. Current Presentation   **Fig. 2.** Presentation with Lifelike Agent

As shown in Fig.1, multimodal presentation allows conveying different information through different channels, such as images, movies, text, and presenters' speech. Because this presentation method conveys information through different channels, it is more effective and became the most popular presentation method at present. However, its disadvantage is that the presenter has to be at the meeting hall, thereby restricting the presentation to a certain time and place.

The solution is Multimodal presentation. It is a new presentation method to make presentation without the restriction of time and place. Fig. 2 illustrates this kind of presentation. The character agents make the presentation instead of a human presenter. You can download the presentation content from the WWW, then ask character agent make the presentation according the content.

However, this attractive presentation method did not replace the current popular PowerPoint presentation tools yet. The reason is that it is too difficult to write the multimodal presentation content. There are many character agents, and different script language specifications are defined to control different character agents, respectively. Most of these script languages require rather low-level programming skills.

In order to overcome the complexity of describing various behaviors of character agents, and to write attractive presentation content easily, we developed MPML (Multimodal Presentation Markup Language).

2 MPML 1.0

The goal of MPML (Multimodal Presentation Markup Language) is to enable every-one to write attractive multimodal presentation easily [9]. Current multimodal presen-tation content is mostly written for a particular character system. In many cases, one has to program a detailed description to control the particular agent system [8].

We envision that people can write multimodal presentations easily, just as people can write homepage easily using HTML. So MPML is designed to write multimodal presentation content independent of specific character agents.

Some features of MPML Version 1.0 are:

* *Independent of the Character agent system.*
* *Easy to describe*, i.e., anyone who understands HTML should be able to learn MPML in short time.
* *Media synchronization supported*, because MPML conforms to SMIL.
* *Easy to control character*
* *Interactive presentation supported*

3 Emotion and MPML 2.0e

As the interface layer between computer and user, a character agent should not only have communication abilities, but also personality traits which lets users feel affec-tion. If a character agent, which has the face and body, but can only perform ma-chine-like reactions, the audience will soon feel bored when communicating with the character agent [7,2]. Considering personality and social behavior of the character agent, we focus on emotion expression functions [4,6].

Emotion can be expressed as *joy, sadness, anger, surprise, hate, fear* and so forth. There is no generally accepted classification system of emotions yet. So we focus on research about emotions in cognitive psychology. In 1988, Andrew Ortony, Gerald Clore, and Allan Colins published a book called *The Cognitive Structure of Emotion*, in which they provide a detailed analysis of emotions [5]. Their analysis became well known as the OCC model.

According to the OCC model, all emotions can be divided into terms according to the emotion-eliciting situation. Emotion-eliciting situations can be divided roughly into 3 types. The first type of emotion-eliciting situation is *consequences of events*. The second type of emotion-eliciting situation is *actions of agents*. The third type of emotion-eliciting situation is *aspect of objects*. According to the classification of emotion-eliciting situations, all emotions can be divided into three classes, six groups and twenty-two types of emotion (Fig.3).

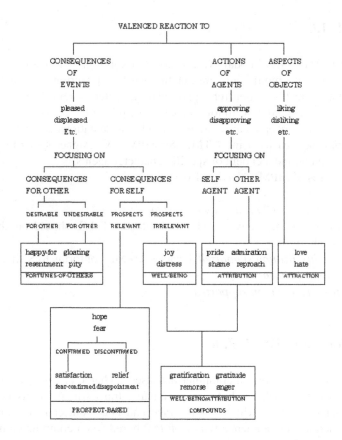

Fig. 3. The Cognitive Structure of Emotion

In MPML Version 2.0e, we provide an emotion expression function to control agents' emotion more conveniently. The content provider can specify the twenty-two types of emotion defined in OCC emotion model, and accordingly modify the action performed by character agent. The character agent expresses the emotion by performing different actions and changing speech parameters (pitch, volume, speed, and emphasis of certain words). For example, when the emotion type is specified as "pride", the character agent would wave his hands, then speak loudly with the emphasis at the beginning of the sentence.

Except for the emotion expression functions, some new functions are added in Version 2.0e:

- *Page:* Every presentation is divided into individual pages. Content providers may describe content page by page.
- *Fast-forward:* The audience can request to go to the next or previous page when watching the presentation.
- *Presentation-macro:* Some templates are prepared for particular presentation purposes.

Fig. 4 illustrates the tag structure for MPML Version 2.0e.

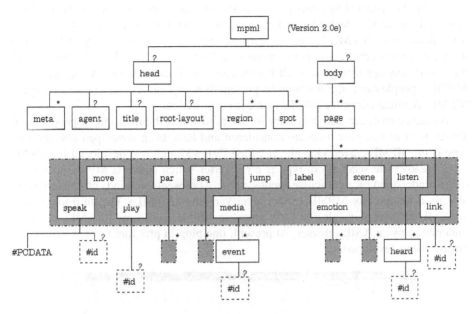

Fig. 4. Tag Structure of MPML

The below is a sample for MPML script:

```
<mpml>
 <head>
  <title> MPML Presentation </title>
  <agent id="PD" character="peedy">
 </head>
 <body>
  <page id="first" ref="self_intro.html">
    <emotion type="pride">
      <speak>
       My name is Zong Yuan,
       I am from Tokyo University.
      </speak>
    </emotion>
  </page>
 </body>
</mpml>
```

According to the above script, the character agent called "peedy" would give a self-introduction with the "pride" emotion activated.

4   Tools for MPML

In order to be accepted by many people, authoring tools and audience tools should be provided for MPML. As for authoring tools, two types are conceivable. One is a plain text editor. Since MPML is easy to learn and write, it should be easy to be written with a plain text editor. Another authoring tool might be a visual editor. Just as people use Homepage Builder to built homepages - with the help of a visual editor for MPML - people can script multimodal presentation content without the knowledge of MPML. A visual editor for MPML is under construction.

Audience tools are also necessary for users to watch the multimodal presentation. Three types of audience tools are considered and have been developed already. One type is the MPML player. One player called "ViewMpml" was developed for MPML 2.0e already. The second tool type is a converter which converts MPML to a script that is understood by a particular agent system. At present two kinds of converters are already developed for MPML 1.0 (an older version of MPML). The third tool type is an XML browser with plug-in [10]. Because MPML conforms to XML, it can be understood by an XML browser. At present, one plug-in program written in XSL has already been developed.

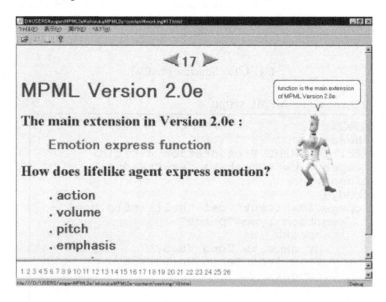

Fig. 5. MPML Player (ViewMpml)

Fig. 5 displays ViewMpml, a MPML player developed for MPML Version 2.0e. It supports all tags defined in MPML Version2.0e's specification. It is free and can be download from the following site:

http://www.miv.t.u-tokyo.ac.jp/MPML/en/2.0e/

Moreover, a movie file (1.4 Mbytes) for a 15 seconds multimodal presentation is provided at the next site:

http://www.miv.t.u-tokyo.ac.jp/MPML/en/2.0e/movies/mpmlmovies.mpg

5 Conclusions

The goal of MPML is to enable many people to publish multimodal presentation content easily. In MPML Version 2.0e, we keep the features of Version 1.0 and applied some new functions to MPML. The effectiveness of using character agents for presentation relies on the so-called "persona effect", which says that the mere presence of an animated character makes presentations more enjoyable and effective [3]. One of our main goals was that presentations can be run anytime and anywhere. In particular, presentations should be run client-side in a web-browser (Microsoft Internet Explorer, 5.0 or higher). This restriction ruled out other possibilities, such as running pre-recorded video-clips, since they have long loading times and are expensive to produce. However, we are aware that experiments suggest video-recordings of real people to be the most effective presentation method (except for human presentation performance, of course).

The main improvement of Version 2.0e is an emotion expression function, which integrates the emotions identified in the OCC model to MPML. The mapping from the emotions to the character agent' s behavior (action and speech) is done by common sense (intuition) rather than according to empirical investigation. However, we can change the emotion parameters easily by changing the text setting files. A prime candidate would be the work on "basic emotions" [1], which identifies a set of emotions that have distinctive signals (e.g., distinctive facial expressions or distinctive speech).

The currently available character agents were not designed for emotion expression. Therefore we started developing customized 3D character agents to express emotion more freely and naturally. Another idea is to let the character agent reason about the emotion-eliciting situation.

References

1. Ekman, P.: An Argument for Basic Emotions. Cognition and Emotion, 6, 3-4, 1992, 169-200
2. Elson, M.: The Evolution of Digital Characters. Computer Graphics World, Vol. 22, No. 9, Sept. 1999, 23-24
3. Lester, J.C, Converse, S.A., Stone, B.A., Kahler, S.E.: Animated Pedagogical Agents and Problem-solving Effectiveness: A Large-scale Empirical Evaluation. Artificial Intelligence in Education, IOS Press: Amsterdam, 1999, 23-30

4. Nagao, K., Takeuchi, A.: Speech Dialogue with Facial Displays. Multimodal Human-Computer Conversation. 32nd Annual Conference of the Association of Computational Linguistics. 1994, 102-109.
5. Ortony, A., Clore, G.L., Collins, A.: The Cognitive Structure of Emotions. Cambridge Univ. Press, 1988.
6. Proceedings Workshop on Recognition, Analysis, and Tracking of Faces and Gestures in Real-Time Systems. IEEE Computer Society Press. Los Alamitos, CA, 1999.
7. Thomas, F., Johnson, O.: Disney Animation: The Illusion of Life. Abbeville Press, New York, 1981.
8. http://msdn.microsoft.com/workshop/imedia/agent/
9. http://www.miv.t.u-tokyo.ac.jp/MPML
10. http://www.w3.org/TR/REC-xml/

A Sound MagicBoard

Christophe Le Gal, Ali Erdem Ozcan, Karl Schwerdt, and James L. Crowley

Projet PRIMA, INRIA Rhône-Alpes,
655 ave. de l'Europe,
38330 Montbonnot St. Martin, France

Abstract. Vision and audio can be used in a complementary way to efficiently create augmented reality tools. This paper describes how sound detection can be used to enhance the efficiency of a computer vision augmented reality tool such as the MagicBoard. A MagicBoard is an ordinary white board with which a user can handle both real and virtual information, allowing, e.g., for a copy&paste operation on a physical drawing on the board. The electronic part of the MagicBoard consists of a video–projector and the concurrent use of several detectors such as a camera and microphones. In our system, sound is used to support a purely vision–based finger tracker, especially during the most critical phase, i.e., when trying to detect and localize a finger tap as a click on the board. The relative phase of a signal caused by a finger tap on the board, detected by several microphones is used to estimate the position of the finger tap. In the probable case of several possible estimates, the estimates are eventually verified by the visual tracker. The resulting system is surprisingly precise and robust.
Keywords: Computer vision, Sound tracking, Multi-modal tracking, Augmented reality

1 Introduction

Intelligent office environments, as described by Coen [1, 2], are designed to reduce the work load devoted to tasks such as handling e–mail, phone or other office equipment, while providing new tools for collaboration and communication. The work described in this paper is part of an effort to develop an intelligent office assistant [3, 4]. This paper focuses on the MagicBoard, which is one of the Interaction tools provided in this environment.

The MagicBoard (see figure 1) is an augmented reality tool similar to the DigitalDesk [5]. The magic board combines a normal white board on which a user can write with a normal marker, with virtual information projected by a video projector. The user's actions are observed by a camera and detected by Microphones. The system employs a vision-based tracker to estimate the position of the users hands.

In order to obtain a robust and fast tracking, we combine two different techniques, based on two different modes. One is vision–based, the other uses sound. We present here briefly both techniques and explain how they interact to provide precise and reliable results.

Section 2 discusses the MagicBoard and its vision tools in more details, whereas section 3 concerns the new sound processing techniques employed. Measurements and results are discussed in section 4, and section 5 draws some conclusions.

T. Tan, Y. Shi, and W. Gao (Eds.): ICMI 2000, LNCS 1948, pp. 65-71, 2000.

Fig. 1. The MagicBoard allows for the combination of digital information and physical information on the board. The user can run augmented commands, for example, copy&paste-ing of physical information using a certain gesture. An electronic calculator and a tool-bar are projected onto the white board

2   Vision Based Augmented Reality

2.1   MagicBoard

The MagicBoard is an augmented reality tool, which allows an easy interaction between the user and the office intelligence. The board integrates the following devices (see figure 2) :

- A simple **White-board**.
- A **video–projector** displays virtual drawing in addition to the physical drawings. It can also be used to add a projected menu to the board.
- A **camera** is used by the system to analyze the content of the board.
- Four **Microphones** are used for the recognition of spoken commands, and, as we will see in section 3, to enforce the detection of the "click" position.

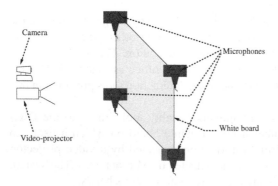

Fig. 2. The MagicBoard is a normal white board augmented by a camera and a video–projector. Several microphones are used for speech detection and for the localization of the clicks

    The collaboration of these device allows augmented operation to be performed on the board, such as copy&paste of physical or numerical drawings, or automatic analysis of a drawn chart. The camera is also used to track the user's hand, which allows a more "computer–style" interaction, such as drag&drop, buttons, or gesture interaction. More details on the MagicBoard can be found in [6].

A variety of techniques have been demonstrated for finger tracking including correlation, template matching, color, contour description, image differencing, and model based approaches. The description of a robust method and some references can be found in [7]. Each of those techniques is strong under specific conditions.

Kendon [8] stated that gestures are composed of three phases: *preparation*, *stroke* and *retraction*. *Preparation* and *retraction* phases do not have to be analyzed very precisely, whereas the effective gesture (*stroke*) needs a higher level of precision. This is especially true in the case of selection or clicking gestures. During the preparation phase a fast feedback, i.e. about 10 ms delay, is necessary for a comfortable use. During the stroke, a more accurate localization is needed, but speed is no longer crucial; a 100 ms delay between the click action and the system reaction is acceptable.

The vision–based system provides information about for the first phase, since the sound detector is not able to detect a hand movement through air. Consequently, the vision–based tracker needs to be *fast* and *reliable*. A lack of accuracy, however, is acceptable at this step, for during the stroke phase, the sound source detector will be used for an accurate positioning. We therefore decided to use a simple but fast skin–color detector as the vision tracker.

2.2 Skin Color Detection

For the initialization, the system projects an outline of a hand on the board and the user is requested to put his hand on this image. The detector uses the pixels of this area to create a histogram of skin color h_{hand}. Color is represented the chrominance pair (r, g) [9], where

$$r = \frac{R}{R + G + B}, \text{ and } g = \frac{G}{R + G + B}$$

The constructed histogram $h_{hand}(r, g)$ shows the number of occurrences of each pair (r, g) of the hand image, i.e., This histogram provides an approximation of the a posteriori probability $p(r, g|\text{skin})$. A second histogram $h_c(r, g)$ of the same size is constructed from the pixels from the entire image. This histogram provides an approximation of the a posteriori probabilities for all possible chrominance values $p(r, g)$. Under the assumption that $p(\text{skin})$ is constant, the Bayes rules gives the probability for a given pixel to be a skin–color pixel

$$p(\text{skin}|r, g) = \frac{p(r, g|\text{skin}) \cdot p(\text{skin})}{p(r, g)}$$
$$= \frac{h_{hand}(r, g)}{h_c(r, g)} = h_{rel}(r, g) \qquad (1)$$

where h_{rel} is a third histogram generated from h_{hand} and h_c. See [10] for a derivation and more details about color tracking.

A threshold is applied in order to obtain a binary image of skin-color. Since this image is noisy, it is filtered using a technique inspired from classical mathematical morphology operators

$$I_{filtered}(x, y) = \begin{cases} 1 & \text{if} \quad \displaystyle\sum_{j=x-w}^{x+w} \sum_{i=y-h}^{y+h} I(j, i) > \sigma \\ 0 & \text{else} \end{cases} \tag{2}$$

This technique is simple enough to be easily performed within the 20 ms between two consecutive frames. A Kalman filter is used to enforce the robustness of the tracker. The conjunction of the filtering technique and the Kalman filtering provides results very reliable but not very accurate.

The techniques in this section provide a robust but not very accurate finger tracker based on color. The following section will introduce a simple, fast, and precise sound localization method based on sound energy to quickly detect position and clicks (finger taps on the board).

3   Sound Tracking

Several microphones are installed along the sides of the MagicBoard in order to estimate the position where a finger taps the board (see figure 2). The delay between the reception of a signal by two different microphones is proportional to the difference of the distance between the microphones and the impact point.

Sound detection is used only during stroke phase. When a finger taps on the board, a sound wave propagates in the board, the same way a seismic wave does. The position of the impact point can be computed by analyzing the signals received at several points on the board.

When an impact occurs, a visible peak in the sound energy curve can be detected. The lag in the detection of a peak by two microphones yields a first approximation of the Interaural Arrival-Time Difference. This first approximation is not sufficiently accurate. The signal begins before the maximum of the peak, and a simple threshold can detect peaks with an error of one or two periods because of the gain difference and high frequency noise.

We obtain more accurate results for the difference by computing the least square of the difference of the energy received by two microphones. Energy provides a representation for sound that gives the best results. This is barely surprising since the human auditive system uses also sound energy [11] for localization. Given two signals, x_1 and x_2, received by two microphones, we compute

$$\varphi \text{ such that } \sum_{t=-W}^{W} (x_1^2(t) - x_2^2(t - \varphi))^2 \text{ is minimum,} \tag{3}$$

which roughly corresponds to the classical IPD *(Interaural Phase Difference)* [12].

The time window for this operation should be chosen large enough to provide distinguishable minima. On the other hand, a too large window would lead to the confusion of two impacts in the case of a double-click, and would imply an augmentation of the reaction time. Experiments showed that a good time window size is about $W = 50ms$.

4 Results

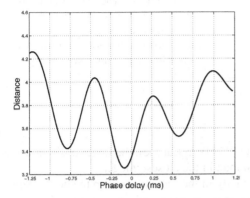

Fig. 3. Square difference between two received signals for delay from -1.25 ms to 1.25 ms. The minimum is reached for the effective delay between the reception of the signal by the two microphones

Figure 3 shows the variation of the square difference between to signals with a phase difference. In this example the absolute minimum is clearly visible. Other local minima are also present because of the periodic nature of sound waves. In some cases it is not very clear which local minima should be considered for the real phase difference. Figure 4 shows the results of the detector. As it can be seen, the sound tracker is very precise; almost all the detection are done in a one centimeter large window centered at the impact point. Of course some errors are observed. These correspond to the selection of the wrong minimum.

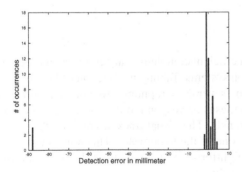

Fig. 4. Evaluation of the sound detector. The x axis is the distance between the impact point and the detected point in millimeters. Each bar shows the number of occurrence of the error. The graph shows that the errors are very low, but outliers do occur

Figure 5 shows the same results for the vision–based detector. This tracker is less precise because of several reasons :

– **Resolution.** When the camera "sees" the whole board a finger has a size of only a few pixels,

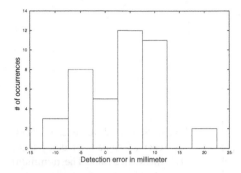

Fig. 5. Evaluation of the vision–based tracker. The axis meaning is the same as in fig 4. The graph shows that the vision–based detector is not very accurate but always provides a good first approximation

- **Delay.** Because of the acquisition rate (25 Hz in Europe), the finger position is not exactly the same at the impact instant than in the closest image,
- **Projection.** The whole hand is not exactly in the same plane than the board. Some errors when projecting the hand image in the board coordinate are inevitable,
- **Method.** The method chosen is intrinsically not very precise. It is quite reliable, especially since the Kalman filter forbids too big errors, but this also means that the answer of the tracker may only be an extrapolated response.

Each local minimum provides an hypothesis for an impact position. The combination of these hypotheses (discrete distribution of probability) with the information provided by the vision–based tracker (continuous distribution of probability) permits the selection of the correct local minima, thus providing the precise location of the impact. In addition hypothesis generated by the sound locator can also be used with a prediction–verification algorithm, by using for example a hand detector to verify the presence of a hand at the predicted position.

5   Conclusion

This paper describes how sound can be an efficient complementary component for computer vision tools when building interaction systems. Taking the difference of the phases of signals of the same event, but detected by several microphones, we obtain hypotheses for the position of that event. These hypotheses correspond to the local minima of the curve of the difference of the detected signals. The visual tracker, knowing the rough position of the hand, can quickly validate one of the hypotheses. This multi-modal approach demonstrates that, even with straightforward techniques, as described in this paper, fast and precise information can be achieved.

References

1. M.H. Coen, "Design principals for intelligent environments," in *Proceeding American Association for Artificial Intelligence 1998 Spring Symposium on Intelligent Environments*, Stanford, CA, USA, Mar. 1998.

2. M. H. Coen, "Building brains for rooms: Designing distributed software agents," in *Ninth Conference on Innovative Applications of Artificial Intelligence. (IAAI97)*, Providence, R.I., 1997.

3. C. Le Gal, J. Martin, and G. Durand, *"SmartOffice*: An intelligent and interactive environment," in *Proc. of 1st International Workshop on Managing Interaction in Smart Environments*, P. Nixon, G. Lacey, and S. Dobson, Eds., Dublin, Dec. 1999, pp. 104–113, Springer-Verlag.

4. Monica Team, "MONICA : Office Network with Intelligent Computer Assistant," http://www-prima.inrialpes.fr/MONICA.

5. P. Wellner, "Interacting with paper on this digitaldesk," *CACM*, vol. 36, no. 7, pp. 86–95, 1993.

6. D. Hall, C. Le Gal, J. Martin, O. Chomat, T. Kapuscinski, and J. L. Crowley, "Magicboard: A contribution to an intelligent office environment," in *Proc. of the International Symposium on Intelligent Robotic Systems*, 1999.

7. D. Hall and J. L. Crowley, "Tracking fingers and hands with a rigid contour model in an augmented reality," in *MANSE'99*, 1999, submission.

8. A. Kendon, "The biological foundations of gestures : Motor and semiotic aspects," in *Current Issues in the Study of Gesture*, Nespoulous, Perron, and Lecours, Eds. Lawrence Erlbaum Associates, Hillsday, N.J., 1986.

9. Bernt Schiele and A. Waibel, "Gaze tracking based on face color," *International Workshop on Automatic Face- and Gesture-Recognition*, June 1995.

10. K. Schwerdt and J. L. Crowley, "Robust face tracking using color," in *Proc. of 4th International Conference on Automatic Face and Gesture Recognition*, Grenoble, France, 2000, pp. 90–95.

11. J.F. Lamb, C.G. Ingram, I.A. Johnston, and R.M. Pitman, *Essentials of Physiology, 2nd edition*, Blackwell Scientific Publication, Oxford, 1986.

12. Keith Dana Martin, "A computational model of spatial hearing," M.S. thesis, Cornell University, 1995.

A Head Gesture Recognition Algorithm

Jinshan Tang   and   Ryohei Nakatsu

ATR Media Integration & Communications Research Laboratories
2-2 Hikaridai, Seika-cho, Soraku-gun, Kyoto 619-02 Japan
{stang,nakatsu}@mic.atr.co.jp

Abstract. *In this paper, we present a head gesture recognition algorithm that is based on the tracking technique developed by Kanade, Luca, and Tomasi (KaLuTo) and a one-class-in-one neural network algorithm. In our method, human skin colors are used to detect the head from complex backgrounds and the KaLuTo algorithm is applied to detect and track feature points on the head. Feature points tracked in successive frames are used to construct a feature vector as the input to the one-class-in-one neural network, which is employed to classify the head gestures into different classes. This method can offer some robustness to different background conditions. Experimental results prove the effectiveness of this method*

1 Introduction

In the past several years, gesture recognition has drawn a lot of attention because of its broad range of applicability to more natural user interface designs and to the understanding of human intention. Many researchers have come to focus on this area and as a result many papers have been published. Some of the famous work includes the sign language recognition, the surveying of human activities and so on. In gesture recognition, head gesture recognition is an important research area. Especially in the surveying of human activities, head gestures are more important than other gestures, such as hand gestures. For example, people are used to nodding to agree with other people, shaking head to disagree with other people and so on.

In this paper, we propose a head gesture recognition algorithm. Our algorithm is suitable for complex backgrounds. It first uses a head detector to segment the head from a complex background so that the recognition algorithm can be applied to different background conditions. Then, feature points such as local extremum or saddle points of luminance distributions are tracked over successive frames by using the *KaLuTo* algorithm. Next, a feature vector for an image sequence to be used as the input of the neural network is extracted from the coordinates of the feature points. Finally, the vector is fed to a neural network, which has been trained in advance, and the gestures are classified into different classes.

The paper is organized as follows. Section 2 describes the implementation of our head gesture recognition algorithm. Experimental results and the conclusion are presented in section 3 and 4, respectively .

T. Tan, Y. Shi, and W. Gao (Eds.): ICMI 2000, LNCS 1948, pp. 72-80, 2000.
© Springer-Verlag Berlin Heidelberg 2000

2 System Implementation

An overview of our head gesture recognition algorithm is illustrated in Figure 1. The input is the video sequence to be recognized. The head gesture recognition processing consists of three steps: 1) the detection of the face from complex backgrounds; (2) the tracking of face feature points by the *KaLuTo* algorithm and feature vector construction; and (3) the classification of all head gestures with the one-in-one neural network that has been trained in advance.

2.1 Head Location

In order to recognize head gestures from complex backgrounds, the first thing for us to do is to locate the head in the complex backgrounds. In our algorithm, the location of the head is realized by the detection of the face. The face detection adopted in this paper is skin-color-based method.

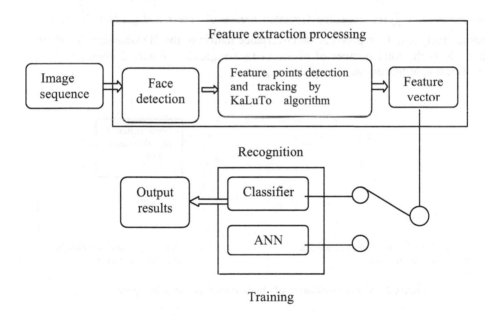

Figure 1. System architecture of our head gesture recognition algorithm

In skin color information based face detection, color space selection is a very important factor, because the distribution of human skin color depends on the color spaces. In order to eliminate the influence of light condition, r-g color space is used instead of the original R,G,B color space in our paper. It can be obtained by the following processing:

$$r = \frac{R}{R+G+B} \quad (1) \qquad g = \frac{G}{R+G+B} \quad (2) \qquad b = \frac{B}{R+G+B} \quad (3)$$

where r,g,b are normalized color values of R, G, and B, respectively.

From (1)(2)(3), we know that r+b+g=1. Accordingly, the color image pixels can be expressed by using any two of the three color components r, g, and b. It has been proven that in the r-g color space, the skin color distributions of different persons under different lighting conditions have similar Gaussian distributions. Therefore, the face color distribution can be represented by the 2D Gaussian distribution $G\,(m\,,V^{2})$ with

$$m = (\bar{r}, \bar{g}) \quad (4) \qquad \qquad \bar{r} = \tfrac{1}{N}\sum_{i=1}^{N} r_i \quad (5)$$

$$\bar{g} = \tfrac{1}{N}\sum_{i=1}^{N} g_i \quad (6) \qquad \qquad V = \begin{pmatrix} \sigma_{rr} & \sigma_{rg} \\ \sigma_{gr} & \sigma_{gg} \end{pmatrix} \quad (7)$$

where $\bar{r}$ and $\bar{g}$ represent the Gaussian means of the r and g color distributions respectively and V represents the covariance matrix of the 2D Gaussian distribution and N is the total number of pixels in face regions. Figure 2 shows the color distribution of human face in r-g color space.

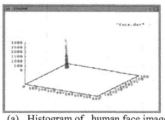

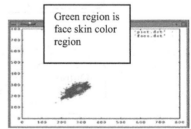

(a) Histogram of human face image   (b)   human face skin color distribution
     in r-g color space   in a complex background

Figure 2. Color distribution of human face in r-g color space

After we obtain the model of the skin color off-line, we can use it to detect the face. The detection of the face is implemented by measuring the possibility of each pixel belonging to the face region. This can be realized by calculating

$$d(x,y) = (I(x,y) - m)V^{-1}(I(x,y) - m) \quad (8)$$

where I(x,y) is the color intensity vector at (x,y) in the r-g color space.

A threshold (D) is set in advance to decide whether a pixel belongs to the face region or not. If d(x,y) <D, then the pixel belongs to the face region.

In order to save time in detecting the face in the next frame, we use a prediction algorithm to predict the face area in the next frame and the face detection in the next frame is limited to the prediction area. Experimental results of face detection are shown in Figure 3(a)-(b).

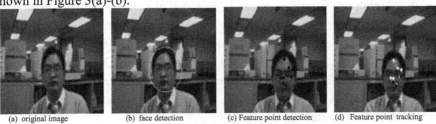

| (a) original image | (b) face detection | (c) Feature point detection | (d) Feature point tracking |

Figure 3. Feature point detection and tracking in face region

2.2   Construction of the Feature Vector of an Image Sequence
2.2.1 KaLuTo Feature Points Tracking Algorithm

The KaLuTo tracking algorithm consists of two steps. The first step is to select feature points from the local extremum or saddle points of luminance distributions. The second step is to track the feature points selected in the previous frame.

In the first step, for each pixel, the following matrix is defined [1].

$$Z = \begin{pmatrix} g_x^2 & g_x g_y \\ g_x g_y & g_y^2 \end{pmatrix} \qquad (9)$$

where g_x, g_y denote the average of the gradient in the x and y directions, respectively, over a square-shaped window centered at the pixel. If the matrix is nonsingular, then the point is considered as a local extremum or saddle point of the luminance distributions. Those points with a larger second eigenvalue or higher trackability are selected to be feature points.

In the second step, those feature points selected in the previous frame are tracked in the present frame by solving the following equation.

$$Z_d = \iint_W [I(x) - J(x)] \begin{bmatrix} g_x \\ g_y \end{bmatrix} dx \qquad (10)$$

where $I(x)$ and $J(x)$ denote the luminance

Examples of feature point selection and tracking are shown in Figure 3 (c) and (d), respectively. In the face, the black points represent the feature points selected in the previous frame, and the white points represent the tracking result of feature points in the present frame.

2.2.2 Construction of a Feature Vector as the Input for a Neural Network

The feature vector for a video sequence to be used as the input of a neural network is extracted from the coordinates of the feature points.

Let the length of a sequence be N, and let the (t-1)th and the t-th be two successive frames. Then, for each frame, for example, the t-th (t>1) frame, we select eight pairs of points in the face area denoted by $(x_{t1}, y_{t,1}), (x'_{t,1}, y'_{t,1})$, $(x_{t,2}, y_{t,2}), (x'_{t,2}, y'_{t,2}), ..., (x_{t,8}, y_{t,8}), (x'_{t,8}, y'_{t,8})$ (t=2,3,....,N-1) where $(x_{t,i}, y_{t,i})$ (i=1,2,..,8) are the coordinates of the feature points selected in the (t-1)th frame and $(x'_{t,i}, y'_{t,i})$ (i=1,2,..,8) are the coordinates of the feature points tracked in the t-th frame.

After we obtain the coordinates of the eight pairs of feature points in each frame from the second frame to the last frame, we can use the information extracted from them to construct the feature vector of an image sequence. The steps for the construction of such a feature vector can be described as follows.

First, use the method developed in [2] to estimate the rotation parameters of the head motion for two successive frames. For example, for the (t-1)th and the t-th frame, by the method in [2], we can obtain three angles θ_t, ϕ_t, ρ_t, which represent different head motion rotations. θ_t is the rotation around the optical axis, ϕ_t is the angle of the axis of rotation Φ parallel to the image plane from the x axis of the coordinate attached to the object, and the angle ρ_t is the angle around the axis Φ. For more information on the estimation method, refer to [2].

Second, some other information is added to strengthen the robustness of features. This information includes the centers of the head in motion, $C_{t,x}, C_{t,y}$ (t=2,...,N), the directions of the head in motion, $E_{t,x}, E_{t,y}$, and the motion energy distributions in the x and y directions D_t (t=2,...,N) for the two successive frames. Their values can be calculated as follows:

$$E_{t,x} = \begin{cases} 1 & if \quad C_{t,x} > 0 \\ 0 & if \quad C_{t,x} = 0 \\ -1 & if \quad C_{t,x} < 0 \end{cases} \quad (11) \qquad E_{t,y} = \begin{cases} 1 & if \quad C_{t,y} > 0 \\ 0 & if \quad C_{t,y} = 0 \\ -1 & if \quad C_{t,y} < 0 \end{cases}$$

(12)

$$D_t = \begin{cases} 1 & if \quad D_{t,x} > D_{t,y} \\ 0 & if \quad D_{t,x} = D_{t,y} \\ -1 & if \quad D_{t,x} < D_{t,y} \end{cases} \qquad (13)$$

$$C_{t,x} = \frac{\sum_{i=1}^{8} x_{t,i}}{8} \quad (14) \qquad C_{t,y} = \frac{\sum_{i=1}^{8} y_{t,i}}{8} \quad (15)$$

where

$$D_{t,x} = \frac{\sum_{i=1}^{8} |x_{t,i}|}{8} \quad (16) \qquad\qquad D_{t,y} = \frac{\sum_{i=1}^{8} |y_{t,i}|}{8} \quad (17)$$

Third, the parameters obtained by equations (11) to (17) in each frame are used to construct the feature vector as follows:

$$V^T = (\theta_2, \phi_2, \rho_2, ..., \theta_N, \phi_N, \rho_N, C_{2,x}, C_{2,y}, ..., C_{N,x}, C_{N,y}, E_{2,x}, E_{2,y}, ..., E_{N,x}, E_{N,y}, D_2, ...D_N) \quad (18)$$

In the above, N is assumed to be fixed, but most of the time, this assumption is incorrect. In order to process a varying N in our algorithm, we adopt the following method. First, we fix N, and then we compare it with the number of frames. If the frame count of the sequence is more than N, we use the first N frames of the sequence to construct the feature vector. If the frame count (suppose it is M) is less than N, we pad the frames M+1,M+2,...,N with frame M. That is, the frames from M to N are assumed to be still.

2.3   Head Gesture Classification with One-Class-in-One Neural Network

Neural networks have been successfully applied to many classification problems. The conventional Multilayer perceptron uses an all-class-in-one-network structure and so the number of hidden units is large, so many researchers use a one-class-in-one-network structure [3]. The structure can be described in Figure 4. In this kind of neural network, each subnet is trained individually and one subnet is used for one special gesture. The output of every subnet is 0 or 1. Before training, each gesture is allocated to one subnet. If the input feature vector is from the gesture class which the subset is allocated to, then the output of this subnet is set to be 1, otherwise it is set to be 0. The advantage of a one-class-in-one-network in classification is that it is suitable for distributed computing. Therefore, a greater speedup can be anticipated [3].

For each subnet, a pi-sigma neural network with a modified learning algorithm is proposed. The output of a pi-sigma neuron is given by

$$y = \sigma\left(\prod_{j=1}^{K} h_j \right) \quad (19)$$

where h_j is the output signal of the j-th summing unit:

$$h_j = \sum_{k=1}^{N} w_{kj} x_k \quad (20)$$

Here x_k is the k-th element of the input vector, K is the order of the Pi-sigma neuron, and w_{kj} is the element of the weight matrix. The learning is performed using an LMS-type approach:

$$\Delta w_l = \eta(t^p - y^p) \bullet (1 - y^p) \bullet (y^p) \bullet \left(\prod_{j \neq l} h_j^p \right) \bullet X^p \quad (21)$$

where t^p and y^p denote the desired and actual neuron outputs for the p-th input vector, respectively, and η is the learning rate.

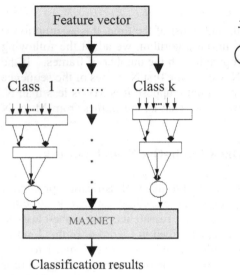

Figure 4. Structure of neural network   **Figure 5.** Second order pi-sigma
              head gesture classification   neural network

3 Experimental Results

The proposed algorithm was applied to recognize different head gestures [5], as shown in Figure 6. The image sequence of each gesture was captured by a CCD camera attached to an SGI workstation, from only one person. The size of each image was 320×240 and the frame rate was 30 frames/second. 60 image sequences were used for each gesture in our experiment, i.e., 30 for training the neural network and 30 for testing the ability of the recognition algorithm.   A second-order Pi-sigma neural network was designed and was expected to respond with the maximum value for a correctly classified gesture; otherwise, the minimum value. In the training, the learning rate was set to be 0.0000001 and the number of iterations was set to be 5000. The recognition results are shown in table 1. From table 1, we know that the effectiveness is good.

Table 1   Head gesture recognition experimental results

Classes	Nod (Yes)	Shake head (No)	Look left	Look right	Look up

Recognition rate	90.4%	89.1%	85.6%	86.5%	84.3%
Classes	Look down	Circle to left	Circle to right	Maybe	Level moving
Recognition rate	87.6%	92.1%	93.34%	95.7%	87.8%

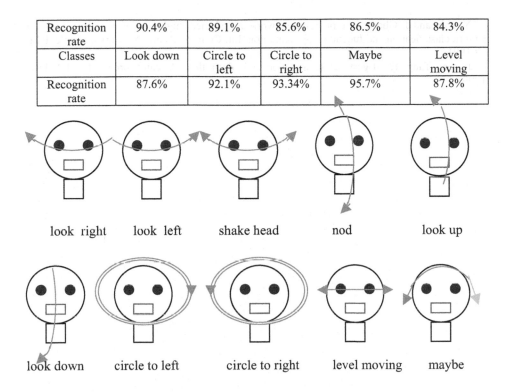

look right   look left   shake head   nod   look up

look down   circle to left   circle to right   level moving   maybe

Figure 6. Recognizable head gestures in our experiment

4 Conclusion

In this paper, we proposed a new algorithm to recognize head gestures. Skin colors are used to detect the face from complex backgrounds and the Kanade-Luca-Tomasi algorithm is applied to detect and track feature points to be used to construct a feature vector as the input to a one-class-in-one neural network. Several head gestures are used in our experiment to test the algorithm, and the results prove that the algorithm is effective.

References

1 Jianbo Shi and Carlo Tomasi: Good features to track. IEEE conference on Computer Vision and Pattern Recognition, Seattle, June 1994.
2 Takahiro Otsuka and Jun Ohya: Real-time estimation of head motion using weak perspective epipolar geometry. Proc. of Fourth IEEE Workshop on Applications of Computer Vision, pp. 220-225, Oct. 1998.
3 S.H. Lin,S.Y. Kung, and L.J. Lin: Face Recognition/Detection by Probabilistic Decision-Based Neural Network. IEEE Trans. Neural Network, Vol. 8, Jan. 1997.

4 Y. shin and J. Ghosh: The pi-sigma network: an efficient higher-order neural network for pattern classification and function approximation. Int. Conf. on Neural Networks, Seattle, I:12-18,1991.
5. Jinshan Tang: Head gesture classification and definition. Research proposal, ATR, Dec. 1999.

Gesture Recognition for a Walking Person by Using a Sequence of Spatially Reduced Range Image

T. Nishimura[1], S. Nozaki[2], K. Makimura[2], S. Endo[2], H. Yabe[1], and R. Oka[1]

[1]Tsukuba Research Center, Real World Computing Partnership, 1-6-1, Takezono, Tsukuba, 305-0032 Japan, Fax:+81-298-53-1640, Tel;+81-298-53-1686
nishi@rwcp.or.jp
[2]Mediadrive Corp.

Abstract. In order to obtain flexibility of gesture recognition system, a sequence of range image has been used for both erasing background noise and distinguishing gestures of which categories are depending on depth features. The essential processing was to spatially divide the space in front of the person into, for example, 3 by 3 by 3 voxels and calculate the ratio of hand area in each voxels. This paper newly introduces a tracking method for a walking person who is also moving.

1 Introduction

Methods to recognize the motion of bodies and hands of humans are important in order to construct a flexible man-machine interface [1]. Especially those systems using only images are useful because it is not necessary to wear contact type sensors such as data gloves etc. [11].

In this research field, many researches have been proposed methods that treat vision as a dynamic process and acquire visual features from time-varying images [1-4]. Yamato et al. [6] suggested gesture recognition of the swinging action of tennis players using Hidden Markov Model. Although this technique enables learning of a gesture model, it requires many data to make a model. Darrell, et al. [7] described the motion of a palm as a series of views and compared it with models using Dynamic Time Warping. However, this method requires a monotonous. Ishii et al. [8] measured amount of movements from the 3-dimensional positions of skin colors of the hands and the face of a performer by processing color images and using stereo matching technique.

On the other hand, we have realized a real-time recognition system using spotting method [11]. The spotting recognition defined as time sequence patterns of input are automatically segmented in the recognition process simultaneously. Consequently, a gesture interface without letting the system know either the start or the end of gestures is feasible. We realized this spotting system by using the Continuous DP (Dynamic Programming) [10] that we proposed in the field of voice recognition.

T. Tan, Y. Shi, and W. Gao (Eds.): ICMI 2000, LNCS 1948, pp. 81-87, 2000.

The Continuous DP recognizes gestures continuously, since it produces the result of recognition frame-wisely, that is, simultaneously with each frame of an input image sequence.

We have also proposed appearance-based features under the following assumptions.

Assumption 1: There is one gesture performer in constant position and orientation without any change of the backgrounds.

Assumption 2: Only large motion gestures are supposed to be recognized.

We call this feature "low resolution feature (LRF)". This LRF can get motion information from time-varying images efficiently without any 3D model matching } The LRF is extracted from a time difference binary image as shown in Fig. 1. Dividing the image into 3 by 3, the ratios of the changed pixels are calculated. Therefore we adopted a LRF with nine dimensions. The feature is robust to a little change of motion as is often inevitable for people and to the change of cloths and backgrounds with those when models are taken.

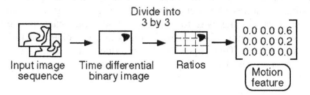

Fig. 1. **Low resolution feature using time-differential binary image.**

But the LRF has two disadvantages: influenced by the temporal change of background, difficult to distinguish depth-changing motion. In order to solve these problems, range images have been used for gesture recognition in the past works. [8][9] But the approach to use low-resolution feature of range images has never been considered. Therefore we proposed reduced range feature by dividing range image into depth in addition to plane directions and calculating the ratio in the each voxels. The feature is independent form the background temporal change and the gestures that differ only in the depth direction can be distinguished.

This paper newly introduces a tracking method for a walking person who is also moving hands.

2   Spatially Reduced Range Feature

In order to overcome the disadvantages of LRF as shown in the previous section, we propose spatially reduced range feature (RRF). This feature is calculated in the following. Firstly we assume that the space of a gesture performer is segmented in the range image. Then, as shown in Fig. 2, feature vector is extracted from the segmented space dividing the space into N_2 by N_2 ($N_2 = 3$ in this figure) in the image and N_l ($N_l = 3$ in this figure) in the depth direction. And the number of pixels in each divided space is calculated and the feature vector of N_2 by N_2 by N_l dimensions is calculated.

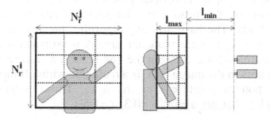

Fig. 2. **Segmented voxels for reduced range feature.**

The experiment on recognition of eight types of gesture using the conventional low-resolution features [11] resulted in the recognition rate of appropriately 80%, with the feature vectors in 3 by 3 ($N_2 = 3$) dimensions and the size of the input images as small as 12 by 12. Since low-resolution features using range images are qualitatively the same as the conventional method in the image plane, features are hereafter extracted assuming $N_2 = 3$. Therefore, the RRF vectors are in $9 \cdot N_l$ dimensions.

3   Human Tracking Algorithm

We propose a new human tracking algorithm in order to realize gesture recognition for walking person who is also moving hands.

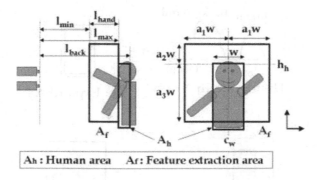

Ah : Human area   Af : Feature extraction area

Fig. 3. **Human model and feature extraction area.**

3.1   Human Model

We suppose the human stands upright and the camera captures his or her whole upper body. And a cube (width: w, horizontal position of center of gravity : c_w, vertical position of the top of the head : h_h, the length from the cameras : l_{max}, l_{min}) approximates the body. The basic tracking algorithm mentioned in the next section estimates human area A_h. Feature extraction area A_f is estimated from A_h as shown in Fig. 3. The minimum distance between the hand and the camera l_{min} is calculated as $l_{max} - l_{hand}$. And the area in the image of A_f is a rectangle with left top point $(c_w-a_1w, h_h + a_2w)$, with right bottom point $(c_w+a_1w, h_h -a_2w)$. Those parameters are decided experimentally as $l_{hand}=0.4$(m), $a_1=1.0$, $a_2=0.3$, $a_3=1.5$.

3.2   Basic Tracking Algorithm

In this section, a basic tracking algorithm independent of hands motion is proposed. First, we suppose the human stands nearest from the camera without anything within l_1(m) around him or her. The basic tracking algorithm has 5 steps as shown in Fig. 4.
Step1 Get the whole range image.
Step2 Calculate the histogram in search area As. Find the nearest peak 1 peak larger than a threshold thr_d. Then $l_{max}=l_{peak}-l_f$, $l_{back} = l_{peak} - l_b$. Hear l_f and l_b is defined as the distance from the peak to the front and back of the human.
Step3 Make a binary image converting pixels within human body (from l_{max} to l_{back}) to 1 and pixels without the body to 0. Project the binary image horizontally and

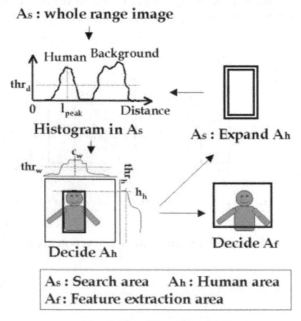

As : whole range image

Histogram in As   As : Expand Ah

Decide Ah   Decide Af

As : Search area   Ah : Human area
Af : Feature extraction area

Fig. 4. **Tracking algorithm.**

decide the head's vertical position h_h by threshold thr_h. Next project the binary image vertically and decide the center of gravity c_w. By a threshold thr_w, position of left shoulder s_l, that of right shoulder s_r and human width w are decided. But we change the shoulder positions as $s_l = c_w - w/2$, $s_r = c_w + w/2$ in order to reduce the influence of noise.

Step4 Extract the human area A_f by the parameters decided in the previous step.

Step5 In order to cope with shift of human area, expand the human area by p pixel, and make it the search area. Go back to Step2.

Thresholds thr_d , thr_h and thr_w are set r_d, r_h and r_w times of the maximum value in each histograms. Experimentally, we decided $r_d = 0.2$, $r_h= 0.2$, $r_w= 0.4$, p=5, $l_f = 0.15$(m), $l_b=0.2$(m), $l_i = 0.6$(m).

4   Experiments for Walking Person

The experiment apparatus was a commercially available PC (Pentium III, 500 MHz CPU) and a range image generator (Triclops from Pointgrey Corp.). Triclops has three cameras, which obtains the disparity and the distance by calculating the minimum value of the sum of SAD (Sum of Absolute Difference) in images both with the horizontal pair cameras and vertical pair cameras. In addition, each image from the camera is low-pass filtered before SAD calculation and the camera alignment is corrected. The size of a single image is digitized 120 by 160($N_r^i =120$, $N_r^j =160$) and that of the window in which the SAD is calculated is 7 by 7. The subject stood approx. 1 m from the camera and another person always moved in the background. The resolution in the depth direction was approx. 8 cm.

Gesture models were selected for imaginary object manipulation and mobile robot instruction s follows:
(1) Larger, (2) Smaller, (3) Forward, (4) OK, (5) No, (6) Backward, (7) Left, (8) Right, (9) Bye, (10) Rotate, (11) Go, (12) Stop, (13)-(21) Push $k-l(k,l=1,2,3)$, (22)-(30) Take $k-l$, 30 types in total (Numbers (1)-(30) will be used as gesture numbers hereafter.) Gestures (13)-(21) are those pressing imaginary buttons on a shelf that exists in front of the subject. Gestures (22)-(30) are those taking some object from this shelf. This shelf was assumed to correspond to the divided domains in 3 by 3 for feature extraction. The standard patterns were created by manually extracting only the gesture portions from image sequences that captured gestures (1)-(30). The frame length T of the standard patterns used in this experiment ranged from 7 to 13. While the subject performed each action at a normal speed, range images were sampled at 7 frames/s.

Then, sequences of input images were recoded with the subject performing gestures (1)-(30) in this order, repeating 10 times. To make natural fluctuations in gestures, approx. 5-minute intervals were inserted between the recordings. In this experiment, the recognition rate was defined as the number of correct answers divided by the total number of gestures 30 ×10.

The experimental method was same as Section 3. But human walked inside the slanted line area about 0.1(m/s) as shown in Fig. 5. Three walking patters are tested as shown in Fig. 6 with the division number of reduced range feature $N_l=2$.

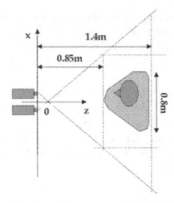

Fig. 5. **Experimental setting. Human walks inside the shadowed area.**

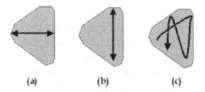

(a)   (b)   (c)

Fig. 6. **Walking pattern.**

Fig. 7. **Tracking images. Top: raw images. Middle: Range images showing the human area. Bottom: Range images showing hands area.**

The recognition rates were 75.2 % for Fig. 6(a), 81.8 % for Fig. 6(b), 78.5 % for Fig. 6(c). When human walks in depth direction, the recognition rates fell down. This is considered to be caused by the low resolution in depth direction. But the results shows the effectiveness of our method because the recognition rate was more than 80 % even when the random walk.

Fig. 7 shows tracking images during gesture 'larger'. The top images are raw images. The middle images are range images showing the human area. The bottom images are range images showing hands area. These images show the robustness of the tracking algorithm when hands are hiding the body.

5   Conclusions

The proposed human tracking algorithm enables gesture recognition of walking person. The experimental results showed the effectiveness realizing about 80% recognition rate for walking person. In future work, we integrate the proposed system into mobile robot or virtual reality system.

References

1. V.I. Pavlovic, R.Sharma, T.S. Huang: "Visual Interpretation of Hand Gestures for Human-Computer Interaction: A Review, " IEEE Trans. Pattern Analysis and Machine Intelligence, vol. 19, no. 7, pp.677-695, 1997.
2. H.H. Baker and R.C. Bolles: ``Generalizing Epipolar-Plane Image Analysis on The Spatio-temporal Surface", Proc. CVPR, pp. 2-9, 1988.
3. R.C. Bolles, H.H. Baker and D.H. Marimont: ``Epipolar-Plane Image Analysis: An Approach to Determining Structure from Motion", International Journal of Computer Vision, 1,pp. 7-55, 1987.
4. A.P. Pentland : ``Visually Guided Graphics", International AI Symposium 92 Nagoya Proceedings, pp.37-44, 1922.
5. M.A. Turk, A.P. Pentland : ``Face Recognition Using Eigenfaces", Proc. CVPR, pp.586-590, 1991.
6. J. Yamato, J. Ohya, K. Ishii: ``Recognizing Human Action in Time-Sequential Images Using Hidden Markov Model", Proc. CVPR, pp.379-385, 1992
7. T. J. Darell and A. P. Pentland: ``Space-Time Gestures", Proc.IJCAI'93 Looking at People Workshop(Aug. 1993)
8. II. Ishii, K. Mochizuki and F. Kishino: ``A Motion Recognition Method from Stereo Images for Human Image Synthesis", The Trans. of the EIC, J76-D-II,8,pp.1805-1812,(1993-08)
9. Kazunori UMEDA, Isao FURUSAWA and Shinya TANAKA: "Recognition of Hand Gestures Using Range Images," Proc. 1998 IEEE/RSJ Int. Conf. on Intelligent Robots and Systems, pp.1727-1732, 1998.10.
10. R. Oka: "Continuous Word Recognition with Continuous DP," Report of the Acoustic Society of Japan, S78, 20, 1978. [in Japanese]
11. T. Nishimura, T. Mukai, R. Oka: "Spotting Recognition of Human Gestures performed by People from a Single Time-Varying Image," IROS'97, vol. 2, pp.967-972, 1997-9.
12. T. Nishimura, S. Nozaki and R. Oka: `` Spotting Recognition of Gestures by Using a Sequence of Spatially Reduced Range Image," ACCV2000, vol.II, pp.937-942 (2000-1)

Virtual Mouse----Inputting Device by Hand Gesture Tracking and Recognition[+]

Changbo HU, Lichen LIANG, Songde MA, Hanqing LU

National Laboratory of Pattern Recognition, Institute of Automation,
Chinese Academy of Sciences, P.O. Box 2728, Beijing 100080, China
cbhu,lcliang,masd,hqlu@nlpr.ia.ac.cn

Abstract. In this paper, we develop a system to track and recognize hand motion in nearly real time. An important application of this system is to simulate mouse as a visual inputting device. Tracking approach is based on Condensation algorithm, and active shape model. Our contribution is combining multi-modal templates to increase the tracking performance. Weighting value is given to the sampling ratio of Condensation by applying the prior property of the templates. The recognition approach is based on HMM. Experiments show our system is very promising to work as an auxiliary inputting device.

1 Introduction

Hand gesture based human-computer interfaces have been proposed in many virtual reality systems and in computer aided design tools. In these systems, however, the user must wear special physical devices such as gloves and magnetic device. Comparatively, vision based system is a naturer way. It is very attractive to utilize hand gesture as a kind of "mouse" using only visual information. But it is in fact an inherently difficult task although it is very easy for human being. One obvious difficulty is that hand is complex and highly flexible structure. Tracking and

---

[+] This work is funded by research grants from the NSFC (No.69805005) and the 973 research project (G1998030500).

T. Tan, Y. Shi, and W. Gao (Eds.): ICMI 2000, LNCS 1948, pp. 88-95, 2000.

recognizing hand motion is the basic techniques needed for this task. Several attempts to recognize hand gesture can be referred in [1,2,3,4].

Generally, gesture researches divide the recognition process into two stages. First, some low-dimensional feature vector is extracted from an image sequence. Most classical method is to segment the object out from image. And the information is obtained to describe the object. Second, recognition is preformed directly or indirectly on these observation data. However it is highly desirable to develop systems where recognition feeds back into the motion feature extraction because the motion style is tightly related to the activity. A great potential advantage of the multi-model approach is that recognition and feature extraction are preformed jointly, and so the form of the expected gesture can be used to guide feature search, potentially making system more efficient and robust.

In this paper we apply active shape model in Condensation framework to deal with hand shape tracking and apply multi-modal templates to use prior knowledge to increase system performance. This extended Condensation algorithm can achieve higher accuracy with lower size of sample-set. Standard Condensation algorithm is used to produces an approximation of an entire probability distribution of likely object position and pose, represented as a weighted sample-set. The weighted mean of this sample-set is used as a starting approximation of ASM. The output ASM is a refined estimation of the object position and pose, which is expected to have a high accuracy. This refined estimation is added to the sample-set with a relatively high weight. The advantages of this method are that: since the Condensation is just to provide a coarse estimation, the size of samples-set can be reduced, so increasing the computational speed. ASM produces a refined estimation, and adding this result to the sample set will improve the sampling of next frame. HMM net is used to train and recognize the hand motion continuously. Although performance of HMM largely depend on training, HMM can be performed in a real time. In general, the system does work well, in near real-time (10 frames/sec on PII 400without special hardware), and it can copes with cluttered backgrounds.

2   Related Works

Active Shape Models (ASM) proposed by Cootes[5] is a successful method to track deformable objects. It can get a high accuracy and can cope with clutter. But its tracking performance greatly depends on a good starting approximation, so the object movement must be not too large, that limits its application. Random sampling filters [6,7] were introduced to address the need to represent multiple hypotheses when tracking. The Condensation algorithm [7] based on factored sampling has been applied to the problem of visual tracking in clutter. It has the striking property: despite its use of random sampling which is often thought to be computationally inefficient, the Condensation algorithm runs in near real time. This is because tracking over time maintains relatively tight distributions for shape at successive time-steps, and particularly so given the availability of accurate, learned models of shape and motion. The Condensation algorithm has a natural mechanism to trade off speed and robustness. Increasing the sample set size N can lower the tracking speed, but obtain a higher accuracy.

3   Our Approaches

3.1   Multi Hand Templates and PCA Representation

Assuming one hand model is described by a vector x_e, a training set of these vectors is assembled for a particular model class, in our case the hand in its various different poses. The training set is aligned (using translation, rotation and scaling) and the mean shape calculated by finding the average vector. To represent the deviation within the shape of the training set, Principle Component Analysis (PCA) is preformed on the deviation of the example vectors from the mean. In order to do this the covariance matrix S of the deviation is calculated:

$$S = \frac{1}{E} \sum_{e=1}^{E} (x_e - \overline{x})(x_e - \overline{x})^T \qquad (1)$$

The t unit eigenvectors of S corresponding to the t largest eigenvalues supply the variation modes; t will generally be much smaller than N, thus giving a very compact model. It is this dimensional reduction in the model that will enable simple gesture recognition. A deformed shape x is generated by adding weighted combinations of v_j to the mean shape:

$$x = \bar{x} + \sum_{j=1}^{t} b_j v_j \qquad (2)$$

where b_j is the weighting for the jth variation vector.

In our case, we set $t=5$. The hand model uses a mean shape and the first five modes of variation (the five eigenvectors that correspond to the largest five eigenvalues). The model hand shape is described solely in terms of these vectors and the mean shape. The model can be projected onto an image by specifying its location, in terms of scale s, rotation θ, x-translation t_x, and y-translation t_y, and its pose in terms of the variation vector weights b_j $i=1,..5$.

3.2   Active Shape Model for Hand Contour Representation

Active Shape Models (ASM) [5] were originally designed as a method for exactly locating a feature within a still image, given a good initial guess. A contour, which is roughly the shape of the feature to be located, is placed on the image, close to the feature. The contour is attracted to nearby edge in the image and can be made to move towards these edges, deforming (within constraints) to exactly fit the feature. The process is iterative, with the contour moving in very small steps.

The ASM uses a Point Distribution Model (PDM) to describe the shape of the deformable contour. The model is described by a vector $x_e = (x_1, y_1, \ldots\ldots x_N, y_N)$, representing a set of points specifying the outline of an object.

Given that the hand model's current projection onto the image is specified by s, θ, t_x, t_y and b_i $i=1,..5$, we use the ASM algorithm [5] to approach a new position. The algorithm is used iteratively in order to converge on a stable solution.

3.3   Condensation Tracker

A full description and derivation of the Condensation algorithm is given in [7]. We describe here our improvement of the Condensation algorithm.

Given time-step t, Condensation algorithm is activated which produces an approximation of an entire probability distribution of likely object position and pose, represented as a weighted sample-set. The weighted mean of this sample-set is used as a starting approximation of ASM. The output of ASM is an estimation of the object position and pose. This estimation is added to the sample-set.

The advantages of this method are that: since the Condensation is just to provide a coarse estimation, the number of samples can be reduced, so increasing the tracking speed. ASM provides a fine estimation, and this result can guide the Condensation algorithm in next time-step. Following gives a synopsis of the algorithm.

From the "old" sample set $\{s_{t-1}^{(n)}, \pi_{t-1}^{(n)}, n = 1,...N\}$ at timstep t-1, construct a "new" sample set $\{s_t^{(n)}, \pi_t^{(n)}, n = 1,...N\}$ for time t.

Construct the nth of N-1 new samples as follows:

1. Select a sample $s_t'^{(n)} = (x_t'^{(n)}, i)$ as follows:

   (a) Generate a random number j with probability proportional to $\pi_{t-1}^{(j)}$. This is done efficiently by binary subdivision using cumulative probabilities.

   (b) Set $s_t'^{(n)} = s_{t-1}$

2. Predict by sampling from $p(X_t \mid X_{t-1} = s_t'^{(n)})$ to choose each $s_t^{(n)}$. In our case, the new sample value may be generated as

   $$s_t^{(n)} = s_t'^{(n)} + Bw_t^{(n)}$$ where $w_t^{(n)}$ is a vector of standard normal random

   variates, and BB^T is the process noise covariance.

3. Measure and weight the new position in terms of the measured features Z_t:
   $$\pi_t^{(n)} = p(Z_t \mid X_t = s_t^{(n)})$$
   then normalize so that $\sum_n \pi_t^{(n)} = 1$

4. Once the N-1 samples have been constructed: estimate the mean of the sample-set as

   $$E(X_t) = \sum_{n=1}^{N-1} \pi_t^{(n)} s_t^{(n)}$$

5. Activate ASM algorithm with initialization being $E(X_t)$, and get a refined estimation $\hat{X}_t$

6. Add $\hat{X}_t$ with its weight $\pi_t^{(N)}$ to the sample set at time t.

This algorithm is also used iteratively until convergence or reaching time limit.

3.3   Apply Template Knowledge to Weight Sampling

In order to increase the quality and speed of the tracker, a prior weight related to template knowledge is multiplied to sampling weight at different position. Assume the knowledge of the template is vector x. and x satisfy the Gaussian distribution

$$p(x) = \frac{1}{2\pi} \left| \Sigma \right|^{-1/2} \exp\left[-\frac{1}{2}(x-\mu)^T \Sigma^{-1}(x-\mu) \right] \qquad (3)$$

where $\mu = \frac{1}{n}\sum_{k=1}^{n} x_k$, $\Sigma = \frac{1}{n}\sum_{k=1}^{n}(x_k - \mu)(x_k - \mu)^T$ is the mean and variance

learned beforehand from examples.

When the $X_t = s_t^{(n)}$ is sampled, a number of points $a_i, i = 1.N$ are randomly selected inside the shape, for simplification we take the prior weight of this sampling as the mid-value of $p(a_i), i = 1..N$.

Using skin color and texture of the template in this way can improve the tracking performance dramatically.

3.4   Mouse Action Recognition Using HMM

Three actions are currently defined to the virtual mouse: left button down, right button down. and mouse moving. According our hand tracker, observation vector $(t_x, t_y, \theta, s, b_1, b_2, b_3, b_4, b_5)$ is concatenated to a time series. In HMM we use only vector $(b_1, b_2, b_3, b_4, b_5)$ to train and recognize mouse action. (t_x, t_y) is used to compute the position of the mouse. We use three-state, first order discrete HMM to perform this task. The topological net is as figure 1.

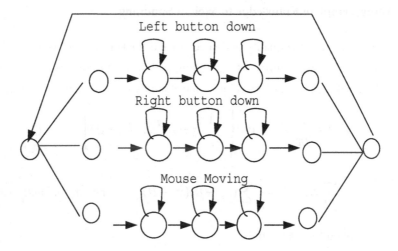

Fig. 1. . HMM network for continuous mouse action recognition

4   Experiments

The system run on a PII450 PC machine, about 10 frames can be tracked using our tracker and after about 2 frame delay the mouse's action is determined. The function of system is like this: the application shows a window, displaying the video images being seen by a camera. When a hand moving on the desk in the view of the camera, the hand was located and tracked until it moved out of view. Another window shows an mouse icon moving according the position; the color and pattern of the icon are changing with the action of the mouse.

The hand model includes 51 points. When motion is detected, the initial sampling is concentrated around the motion region. The size of sample-set for initial localization is 1500. Once the hand is located, the size of sample-set is reduced to 200. Three of the hand models are shown in Figure 2.

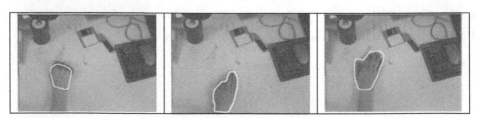

Fig. 2. Three ASM hand models

5   Conclusion

A system for hand tracking and gesture recognition has been constructed. A new method which extends the Condensation algorithm by introducing Active Shape Models and multi-modal templates, is used to fulfill this task. The system works in near real time, and virtual mouse position and action are interestingly controlled by hand.

Reference

1.   Rehg, J. and Kanade, T. Visual Tracking of high dof articulated structures: An application to human hand tracking. In Eklundh, *Proc. 3rd ECCV*, 35-46, Springer-Verlag, 1994.
2.   Freeman, W.T. and Roth, M. Orientation histograms for hand gesture recognition. In Bichsel, *Intl. Workshop on automatic face and gesture recognition*, Zurich. 1996.
3.   Cohen, C. J., Conway, L., and Koditschek, D. .Dynamical system representation, generation, and recognition of basic oscillatory motion gestures. In *Proc. 2nd Int. Conf. on Automatic Face and Gesture Recognition*, 151-156, 1996.
4.   Yaccob, Y. and Black, M. Parameterized modeling and recognition of activities. In *Proc. 6th Int. Conf. on Computer Vision*, 6, 120-127, 1998.
5.   Cootes, T., Hill, A., Taylor, C. and Haslam, J. The use of active shape model for locating structures in medical images. *J. Image and Vision Computing*, 12,6, 355-366, 1994.
6.   M. Isard and A. Blake. Visual tracking by stochastic propagation of conditional density. In *Proc. 4th ECCV*, 343-356, Cambridge, England, Apr. 1996.
7.   G. Kitagawa. Monte Carlo filter and smoother for non-Gaussian nonlinear state space models. *J. of Computational and Graphical Statistics*, 5(1):1-25, 1996.
8.   U.Grenander, Y. Chow, and D.M. Keenan. *HANDS. A Pattern Theoretical Study of Biological Shapes*. Springer-Verlag. New York, 1991.
9.   B.D. Ripley. *Stochastic simulation*. New York: Wiley, 1987.

A Self-reliance Smoothness
of the Vision-Based Recognition of Hand Gestures

Xiao-juan Wu, Zhen Zhang, and Li-fang Peng

Department of Electronic Engineering, Shandong University
Ji'nan (250100), Shandong, China

Shchen@public.sdu.edu.cn, zzhen@sina.com, plf@public.sdu.edu.cn

Abstract. The hand gestures are a natural and intuitive mode for human-computer interaction. The vision-based recognition of hand gestures is a necessary method for future human-computer interaction. On the other hand, the edge detection in the course of the hand recognition is also a key technique. In this paper based on the classical self-reliance smoothness operator, we propose a new method of midpoint threshold (the averaged grads method). According to this method, the optimal threshold and edge image of hand gestures can be got by using the first or each information of grads of reference image.

1   Introduction

With the rapid development of computer technology, the research of human-computer interaction that is adaptive to human custom has been more and more regarded as a very important problem. Because hand gestures are move intuitive than other interaction methods, they have been the favor of more and more people. So in a vision-based system of hand detection, we should firstly detach the hand signal from a great deal of video signals, and then operate the detection of them.

The edge detection of a hand is to find out the edges image in which include plenty of points with various scales and intensity. At the same time, in order to detect the edge of what we need, we should get rid of a lot of useless and redundant details and discontinuous points. In a hand gesture, there are two features of direction and amplitude on edges. The scale of pray being parallel to edge varies stably, while the sale of pray being vertical to edge varies sharply, and this variation maybe is jumping or sloping. In the edges, the first derivative has maximum value, while the second derivative is zero. So the edge points are the points whose first derivative value is great while whose second derivative is zero. Therefore, it is a powerful method that applies

T. Tan, Y. Shi, and W. Gao (Eds.): ICMI 2000, LNCS 1948, pp. 96-102, 2000.
© Springer-Verlag Berlin Heidelberg 2000

the maximum of grads and the zero points of second derivation to detect edges. However, in realities, images are not always clear and reliable. They are often contaminated by noise and a lot of useless background. So in realities, we often need a smoothness function to fit with the pray of points surrounding around the edge points and then calculate the first and second derivative of the curve.

Because of the defects presented above, this paper introduced the method of self-reliance smoothness filter to images. Its basic thought is to use a small averaged template iteratively convolute with the signal to be smoothed. This course has the feature of distributing from various directions. The template function is the points' grads function, which varies at each time of iteration. After many times of iteration, the result is that the image is composed by a lot of areas with average intensity, between which there have very fine jumping edges. Therefore, there have two obvious merits of Self-reliance Smoothness Filter (SSF). The first one is to sharpen the edges and to make them as the edges between coherent areas, the second is to smooth the internal section of each area. Smoothness is slow and stepwise, while the edge sharpness only needs few steps of iteration. If processing the signal after SSF by simple differential operation, we can realize edge detection well with both the maximum values of first derivative and the zero values of second derivative.

2   The Basic Principle of Self-reliance Smoothness Filter and the Obtain of the Most Optimized Threshold

SSF applies the computation to discontinuous features of each points of original to estimate the adding coefficient of the template, such as equation (1)

$$w^{(t)}(x) = f(d^{(t)}(x)) \qquad (1)$$

In the equation, $d^{(t)}(x)$ is the measurement to discontinuous degree of signal; $f(d^{(t)}(x))$ is a monotone sub-function, $f(0) = 1$, with the increase of $d^{(t)}(x)$, $f(d^{(t)}(x)) \rightarrow 0$. Here signal's grads is the estimation value of $d^{(t)}(x)$. So get

$$w^{(t)}(x) = f(S'^{(t)}(x)) = \exp(-\frac{|S'^{(t)}(x)|^2}{2k^2}) \qquad (2)$$

$S'^{(t)}(x)$ is the first derivative of $S^{(t)}(x)$, k is the scale coefficient, which is the threshold when performing SSF. It decided the grads' amplitude to be cut during the course of SSF. If considering the planar image signal, is defined as the amplitude of grads of (3X3 window)

$$(\frac{\partial f^{(t)}(x,y)}{\partial x}, \frac{\partial f^{(t)}(x,y)}{\partial y})^T = (G_x, G_y)^T \qquad (3)$$

So the value reflected the continuous measurement is presented by equation 2.1.4

$$w^{(t)}(x,y) = f(d^{(t)}(x,y)) = \exp(-\frac{(d^{(t)}(x,y))^2}{2k^2}) \qquad (4)$$

$$d^{(t)}(x,y) = \sqrt{G_x^2 + G_y^2}$$

Then the smoothed signal is defined as function 2.1.5

$$f^{(t+1)}(x,y) = \frac{1}{N^{(t)}} \sum_{i=-1}^{+1} \sum_{j=-1}^{+1} f^{(t)}(x+i,y+j)w^{(t)}(x+i,y+j) \qquad (5)$$

$$N^{(t)} = \sum_{i=-1}^{+1} \sum_{j=-1}^{+1} w^{(t)}(x+i,y+j)$$

k from equation (4) is the scale coefficient. If the grads of signal is great enough (>k), the amplitude of it will increase with the increase of times of iteration, and then get the goal of sharpening edges. If the grads of points is too small (<k), the uneven section of signal will be smoothed. The detail operation of SSF can be got from document [5].

From the opinion of statistic, in an image there have a lot of edges that we need and a lot of edges we should abandon. Generally we need the edges that vary sharply, but don't consider a lot of detail edges.

When an image is smoothed by SSF, there has a very key problem to define the coefficient k. Because k presents the intensity and deepness of smoothness and sharpness, the definition of k is very important. Here we proposed a method obtaining the most optimized coefficient k, the averaged grads method, to self-adaptively satisfy coefficient estimation of different images. There have two methods of threshold process introduced below.

The process consulting the first scale of grads.

According to the equations (5), (6) and (7),

$$G_x(x,y) = \frac{1}{2}[f(x+1,y) - f(x-1,y)] \qquad (6)$$

$$G_y(x,y) = \frac{1}{2}[f(x,y+1) - f(x,y-1)] \qquad (7)$$

we get the first derivative (first grads) of image. x and y is the dimension of image.

$$k'(x,y) = \sqrt{Gx^2 + Gy^2} \qquad (8)$$

Then we get the average of k

$$\bar{k} = [\sum_x \sum_y k'(x,y)]/(xy) \qquad (9)$$

To get the sharper edges, we process it by multiplying a coefficient to it.

$$k = 1.5 * \bar{k} \qquad (10)$$

The process consulting each scale of grads.

All scales of grads are averaged and added with coefficient, such as

$$\bar{k}_i = (\sum_x \sum_y k'(x,y)_i)/(xy) \qquad (11)$$

$$k = (\sum_i \bar{k}_i)/n \qquad (12)$$

In the equations, i is the label of one certain scale, n is the total scales. One problem should be noticed that before obtaining the grads of a scale, we should do iteration according to the last scale's grads.

3   Simulation

In order to justify the validity of the most optimized threshold process method we propose above, in the simulation, a real hand picture (190X218) that is 256 scaled pray image with 10 percents noise added was processed. Through the comparison between the most optimized coefficient and the coefficient that were not processed by any methods, we got the results below.

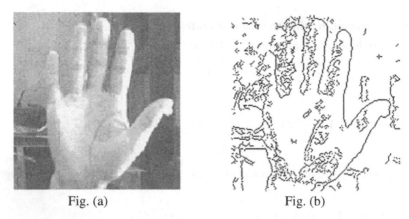

Fig. (a)   Fig. (b)

Figure (a) was the original image; Figure (b) presented the image that was not processed by SSF but directly detected by Canny operator.

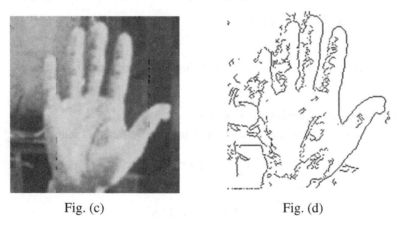

Fig. (c)   Fig. (d)

Figure (c) was the image after being processed by SSF, with the coefficient k (8) defined before; Figure (d) was the detected image by Canny operator.

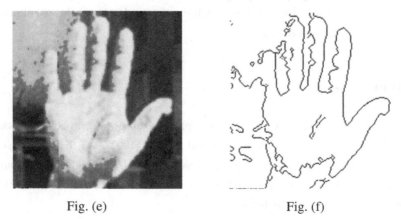

Fig. (e)   Fig. (f)

Figure (e) was the image after SSF, with the coefficient k (18.9311) of resulting first scale of grads; Figure (f) is its detection image.

Figure (g) was the image that has increased the contrast of pray from image (e); Figure (h) was its detection image.

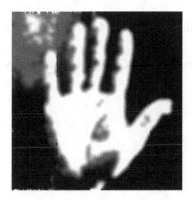

Fig. (g)

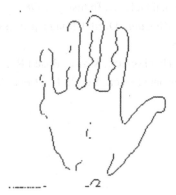

Fig. (h)

4   Conclusion

From the results above, we can get the conclusion below:

The image after SSF can get much better effect in edge detection than the original image; Edge detection using the most optimized coefficient is much better than the edge detection without any threshold process, and it can basically find out the hopeful edges; Though they have the similar effect in edge detection, the process consulting the first grads information needs less amount of computation than the process consulting each scale of grads information. So in the realities, the former improves the effect of real-time operation, which is helpful to realize the real-time hand recognition in the single-vidicon-based complicated background.

References

1. H.S, Hou and H. C. Andrews, Cubic splines for image interpolation and digital filtering, IEEE Trans. Acoustic Speech Signal Process. 26(6), 508-517(1978)
2. F. Mokhtarian and A. Mackworth, A theory of multiscale, curvature-based shape representation for planar curves, IEEE Trans. Pattern Anal. Mach. Intell. 14(8), 789-805(1992)

3. Fridtjof and G. Medioni, Structural indexing: efficient 2-D Object recognition, IEEE Trans. Pattern Anal. Mach. Intell. 14(12), 1198-1204(1992)
4. Jintae Lee, Tosiyasu L Kunii, Model-based Analysis of Hand Posture, Computer Graphics and Applications. 15(5), 339-348(1995)
5. Zheng Nanning, Computer vision and pattern recognition, Publishing house of National Defence Industry, 1998
6. Zhu Zhigang, Digital image process, Publishing house of Tsinghua University, 1998
7. Ren Haibing ect., Vision-Based Recognition of Hand Gestures: A Survey, ACTA Electronic Sinica, Vol.28 No.2 Feb. 2000

Hand Shape Extraction and Understanding
by Virtue of Multiple Cues Fusion Technology

Lin Xueyin, Zhang Xiaoping, Ren Haibing

Department of Computer Science and Technology, Tsinghua University, Beijing 100084
Fax: 010-62771138 Tel:010-62778682
Email: zxxppp@263.net

Abstract.   In order that information included in the hand shape can be extracted for gesture communication aiming to the new technology of Human Computer Interaction, hand shape should be reliably extracted based on image sequences. In this paper a hand shape extraction approach is proposed .By using multiple cues such as motion and color information, embedded in image sequences, a set of complicated hand shapes which compromise a small dictionary of hand postures, can be reliably extracted within a rather sophisticated environment, In this paper the proposed shape extraction strategy is addressed and preliminary results are indicated to prove its effectiveness

Keywords.   Hand posture extraction, multiple information fusion

1.   Introduction

Recently gesture understanding has been one of the hot research topics in computer vision and computer graphics area. Since hand gesture is expected to be one of the modal of human computer interactive, research has been conducted almost worldwide. Though successful result have been reported in recently, most of them are mainly concerned the body motion or arm motion .In that cases hand is usually modeled as a blob moving around and its trajectory is captured so that the information embedded in the blob trajectory be extracted.

In practice, however, hand gesture consists of not only arm (and body) movement but hand shape as well. Hand shape plays a very important role in human communication. Therefore, hand shape analysis should be one of the most important issues in gesture understanding.

Hand Shape extraction and understanding turn out to be very difficult for several reasons. First, hand consists of palm and fingers. The joints of fingers can be manipulated to form complicated articulated movement with tens degrees of freedom. Second the self-occlusion of fingers makes the analysis complicated. Finally hand usually moves around within a rather complicated environment. As a result, the background in the image is usually sophisticated. Realized by such hardness, most researches use data glove to sense the joint angles instead of hand shape analysis based on visual information.

T. Tan, Y. Shi, and W. Gao (Eds.): ICMI 2000, LNCS 1948, pp. 103-110, 2000.

Recently, some research efforts have turn to retrieve the 3D posture of the human hand by the using vision-based methods. For example, Wu and Huang [1] decouple global hand motion and local finger motion separately, the global hand motion is formulated as a least median of squares (LMS) problem and the local finger motion is resolved by an inverse kinematics problem. Shimada et al. [2] aimed to estimate the joint angles of figures and 3D shape of hand model, from a monocular image sequence without depth data. Chua, Guan and Ho [3] try to estimate3D hand posture by using the constraints based on a hand model with 18 degrees of freedom. Chong and Tsai [4] use light spots projected by laser beams on the palm to estimate the plane equation of the lack of the hand first, and then the joint angles of the finger are generated by an inverse kinematical technique. Imagawa et al. [5] have suggested an appearance-based hand shape recognition strategy in which global and local feature of hand gesture has been considered in a unified framework. They suggest a two-step strategy. At first, a candidate set of words represented by the hand gesture is formed by using detected global features and they narrows the words down to one by using local features

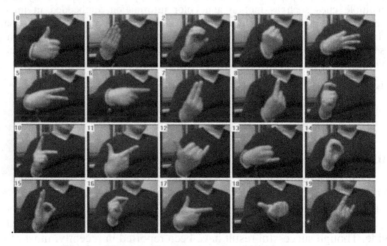

Fig. 1. The set of hand gestures used in this paper

In this paper an appearance-based hand shape analysis strategy is proposed. The aim of the proposed method is to extract and analysis the hand shape robustly with a vision-based methodology. A set of twenty different kinds of hand gestures has been selected as examples to check the effectiveness of our method with each of them represents a word (Fig. 1). In order to extract hand shape reliably, multiple cues, such as motion and color, are used in a harmonic way. The motion boundaries are extracted first, by using a set of trinary images. After that a dominant motion extraction procedure is used to filter out the hand boundary. Besides, the color formation is also used to help extract the hand region completely. The preliminary experimental results indicate that hand shape can be extracted in a rather complicated environment. The paper is organized as follows, in the second section the outline of the proposed algorithm is explained and then each component procedure is discussed in detail. Some preliminary experiment results are shown in the section 3, and followed by a conclusion in the last.

2.   Hand Shape Analysis Method Based on Multiple Cues Fusion

The algorithm we use consists of three main procedures, and its flow chart is shown in Fig. 2. The first procedure --- **trinary image analysis** procedure is used to extract the entire motion boundary, so that the effect of stationary background will be discarded completely. The second procedure is called dominant motion analysis procedure, in which the Dominant motion of the hand will be extracted by means of a motion clustering strategy, Color analysis procedure is the third one, in which the color attribution is used to distinguish hand boundary from the other and the hand region will be filled out as well.

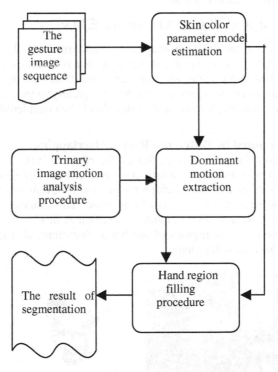

Fig. 2.   The flow chart of our algorithm

2.1   Trinary Image Analysis Procedure

As mentioned before, our hand shape extraction algorithm is mainly based on motion analysis, which uses a trinary image analysis strategy [6,7]. In our method, however, the strategy used is slight different from theirs. The principle of this method is simple and can be explained as follows.

Suppose that a dark block is moving in front of a bright back ground, form left to right. Therefore a set of trinary images I_{k-1} , I_k , I_{k+1} can be captured. Since all the trianary images are captured while block is moving, the differences between each pair of images can be expressed as two difference regions in D_k and D_{k+1} respectively. Since both the difference images are relative to the same image I_k, the corresponding difference regions in both difference images share a common edge. As a result, the block boundary in I_k can be extracted by using the intersection operation between D_k and D_{k+1} as:

$$B_k(u,v) = D_k(u,v) \ \wedge \ D_{k+1}(u,v) \tag{1}$$

Where u, v is the pixel coordinate in image. Since our aim is to extract boundary. We use the following equation instead.

$$B_k(u,v)=D_k(u,v) \wedge D_{k+1}(u,v) \wedge E_k(u,v) \tag{2}$$

Where $E_k(u, v)$ expresses the edge image extracted form image I_k.

From the principle mentioned, it can be seen that the motion boundary can be reliably distinguished from stationary boundary, and therefore the entire boundary related to the stationary background can be completely wiped out. Though the principle is simple, some implementation issues should be considered.

2.1.1 Noise Edge Caused by Difference Region Overlapping

From the equation (1), it can be seen that all the edges of $E_k(u,v)$ will remain when the very pixel (u,v) is located in the intersection area of both difference images. If the intersection area is related to the overlapping of two irrelevant difference regions in D_k and D_{k+1}, all the edge inside will remain as noise. This situation usually happens while the movement is large. For example, the region A in Fig. 3 is the intersection area relative to two different regions of the hand. Therefore, all the irrelevant edge lines included in this area will remain.

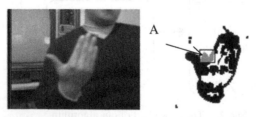

Fig. 3.   An instance of extra region caused by irrelevant difference regions

Besides, difference region overlapping might be caused by motion with two opposite direction, as shown in the figure 4. In that case, noise edge will remain either.

From the analysis mentioned above the movement between a pair of image should be kept with a reasonable amount. If the distance between two opposite boundary of the object is W, then the width of inter-frame difference region should be smaller that half of W. Besides, the bi-directional motion should be avoided. In practice, however, it is not a serious problem, and using a post-processing procedure can eliminate the noise edge.

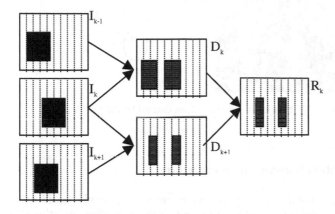

Fig. 4.   The error caused by to-and-fro motion

2.1.2 Motion Direction Sensitivity Problem

From the Fig. 5, it can be seen that if the boundary segment is parallel to the direction of inter-frame motion, this segment will not be detected by the motion analysis method, and is called the motion direction sensitivity problem.

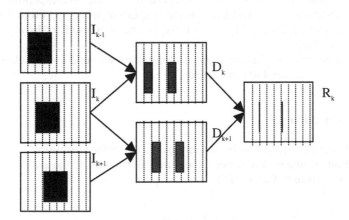

Fig. 5. The illustration of the motion direction sensitivity problem

For example, the edge segment in the A region of the Fig. 6, is lost caused by this problem. In our method the lost edge segment is recovered by an edge tracking procedure.

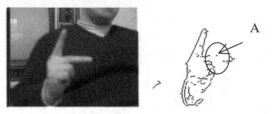

Fig. 6. An instance of missing edges caused by the problem

2.2 Dominant Motion Determination Procedure

It is noted that our hand boundary detection method is mainly based on motion information. Any moving edge will be filter out in the procedure mentioned above, But it is unavoidable that person's body will move more or less while he (or she) gestures, so the hand boundary should be filter out from the others by means of some strategy. In this algorithm, two kinds of information are used to do it. One is based on motion evaluation, and will be discussed in this section. The other exploits color information, and will be addressed in the next section.

Based on the observation that hand motion is usually the dominant motion while person gestures. As a result, the inter-frame motion relative to hand boundary will be the largest in image sequence. Under this consideration, a cluster algorithm in a motion displacement space is developed to find the cluster relative to hand moving. In the implementation of the algorithm the trinary image analysis procedure is performed in succession, and the inter-frame displacement of the corresponding motion boundary is calculated, and cumulated in the displacement space. From the largest cluster in this space, a threshold value of displacement is used to distinguish hand motion from the others.

From our preliminary experiments, most of the motion boundary not relative to the hand motion can be successfully discarded.

2.3 Skin Color Filtering

In order that hand region can be reliably extracted color attribution is used to fill out hand region in image. From our experience skin color of the hand can usually expressed as a mixture of Gauss in the (r, g, b) space:

$$p_i(C) = \frac{1}{(2\pi)^{3/2} \prod_{j=1}^{3} \delta_{ij}} \exp\left\{ -\frac{1}{2} \sum_{j=1}^{3} \frac{(c_j - m_{ij})^2}{\delta_{ij}^2} \right\} \tag{3}$$

$$P(C) = \max\,(pi(C)) \qquad 0 < i < 4$$

Where C is vector composed by the r, g and b component of a pixel, and P(C) expresses the probability of the color C is the color of skin.

In order to establish the probability model, two strategies have been adopted in our method. One is an off-line training in a training phase, the other is performed in an online manner. In the training phase, person will be asked to show a couple of different hand shape. By collect the color attribution of both the hand region and background region, the parameters of the mixtures of Gauss are estimated. In the on-line phase while hand boundary has been detected by the proceeding procedures, the color attribution in the small neighborhood is collected   and the parameters of the probability model are re-estimated so that the model be rectified time to time.

3.   Experiment Result

The hand shape extraction algorithm has been conducted in our preliminary experiment. In our experiment, person is allowed to wear colorful cloth and move his (her) body in a natural wary while gesturing in a general condition. The experimental process is shown in the Fig. 7 with some interim results.

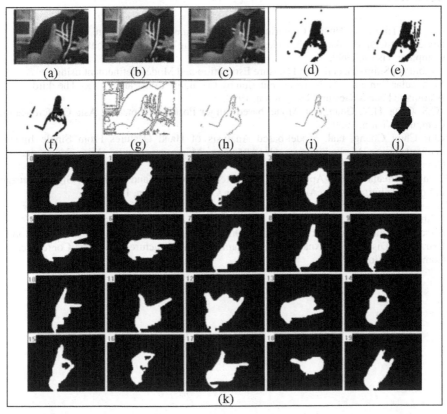

Fig. 7. Some experiment results

In the fig. 7, images (a), (b) and (c) are the source image. (d) is the difference image between (a) and (b), and (e) is the difference image between (b) and (c). (f) is the intersection of (d) and (e). (g) is the edge image extracted from image (b). (h) is the intersection of (g) and (f), and (i) which is corresponding to (a) is gained by the same method with (h). After the motion amount analysis, we get (i), and through fill out (i) by the color attribute, we get the final result image (j), and Image (k) is a set of results corresponding to the images shown in Fig. 1.

4. Conclusion

In this paper, we have presented about the methods about how to extract the hand from the background. It is obvious that hand shape extraction is just the first step of our gesture recognition system. Now we are working hard to develop the method of shape description and recognition.

References

1. Ying Wu, Thomas S. Huang. Capturing Articulated Human Hand Motion: A Divide-and-Conquer Approach in Gesture Recognition, International Conference on Computer Vision, 1999.
2.   Nobutaka Shimada, etal. Hand Gesture Estimation and Model Refinement using Monocular Camera – Ambiguity Limitation by Inequality Constraints, Proc. The third Automatic Face & Gesture Recognition, 1999.
3.   C.S. Chua, H.Y. Guan, etal. Model-based Finger Posture Estimation, Asia Conference on Computer Vision, 1999.
4. Chin-Chun Chang, etal. Model-based Analysis of Hand Gestures From Single Images Without Using Marked Gloves Or Attaching Marks on Hands, Asia Conference on Computer Vision, 1999.
5. Kazuyuki Imagawa, etal. Appearance-based Recognition of Hand Shapes for Sign Language in Low Resolution Image, Aisa Conference on Computer Vision, 1999.
6. Ross Cutler and Matthew Turk, View-based Interpretation of Real-time Optical Flow for Gesture Recognition, Proc. The third Automatic Face & Gesture Recognition, 1999.
7. Bisser Raytchev etal, User-Independent Online Gesture Recognition by Relative-Motion Extraction, Discriminant Analysis and Dynamic Buffer Structures, The Asia Conference on Computer Vision, 1999.

Head Pose Estimation for Video-Conferencing with Multiple Cameras and Microphones

Ce Wang and Michael Brandstein

Harvard University
Division of Engineering and Applied Sciences
340 Maxwell Dworkin Hall
33 Oxford Street
Cambridge, MA 02138, USA
wangc,msb@hrl.harvard.edu,
WWW home page: http://himmel.hrl.harvard.edu

Abstract. *An automatic video-conferencing system is proposed which employs acoustic source localization, video face tracking and pose estimation. The audio portion of the system provides the initial localization of the talkers and the video component tracks the talkers by utilizing source motion, contour geometry, color data, and simple facial features. Decisions involving which camera to use are based on an estimate of the head's orientation. This head pose estimation is achieved using a very general head model which employs hairline features and a learned network classification procedure powered by Support Vector Machines. The procedure is capable of accurately evaluating head orientations over a complete 360 degree interval. By relying on a facial criterion that is easily extracted from video images acquired across a range of lighting and zooming conditions, the estimator is designed to be effective in practical situations such as those encountered in video conferencing or surveillance scenarios.*

KEYWORDS: Pose Estimation, Face Detection, SVM, Video-Conferencing

T. Tan, Y. Shi, and W. Gao (Eds.): ICMI 2000, LNCS 1948, pp. 111-118, 2000.
© Springer-Verlag Berlin Heidelberg 2000

1   Introduction

A popular area of multi-media research is automatic video conferencing, where a set of computer-controlled cameras capture the images of one or more individuals, adjusting for orientation, range, and source motion. In [1, 2] we proposed a hybrid face tracking system which made use of an acoustic-based localization algorithm for initial camera steering followed by a motion-based technique to develop body contours and detect facial regions. The result was a computationally efficient multi-source tracker. This paper extends these ideas by adding head pose estimation based on Support Vector Machine learning. The head pose estimation algorithm is used to decide which camera is best suited to capture a scene. This method is successful for estimating facial orientations, and does not require very high quality images or detailed facial features.

2   Video-Conferencing System Setup and Face Detection

The components of the hybrid tracking system are shown in Figure 1. The system consists of video (face detection and tracking, pose estimation, shot selection) and audio (talker localization, speech enhancement) components. The video system uses a set of cameras with three degrees of freedom: pan, tilt, and zoom. The audio system consists of a set of four 4-element microphone arrays and is capable of detecting voice activity and localizing the positions of talkers in the acoustic environment through time-delay estimation and an ensuing triangulation procedure. The location estimate derived from the microphone array data is used as a coarse position measurement, providing an initial camera pointing vector and determining the region of active talkers. The visual portion of the system is then activated to refine the location estimate, frame the talker, estimate the poses of the targets, select the camera with the best view, and track any subsequent motion. The tracking information is fed to the speech enhancement procedure which then extracts the acoustic signal associated with the desired talker(s), attenuating unwanted sources, noise, and reverberation.

The methods used for face detection and tracking were first presented in [1, 2]. Multi-talker face detection is accomplished via a hierarchical procedure employing four features derived from the video stream: (1) source motion, (2) contour geometry, (3) color statistics, and (4) facial analysis. Source motion and contour geometry identify potential face regions while color and facial analysis are used as validation criteria. Because these latter criteria are used only for confirmation purposes rather than as the primary means of face detection, both the color test and facial analysis can be quite general and robust to varying conditions. At each stage of the algorithm, the appropriate features are estimated from the image data with only a limited amount of computation.

Figure 2 displays a sample image from the face detection process. The large boxes delineate the segmented regions derived from the motion information. The ovals are the result of the geometrical analysis which generates a two-sided contour boundary for each region and then estimates a neckline location. For

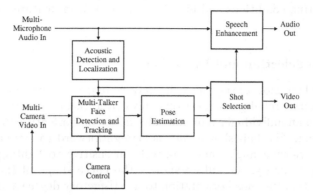

Fig. 1. Video-Conferencing System Setup

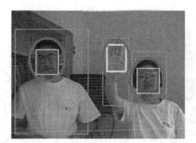

Fig. 2. Face Tracking Example

this example each of the three candidates (two faces and a hand) were found to possess a headlike geometry. The color confirmation test is then applied within each of the ovals. The detected skin color regions are indicated by a light box. To this point, the hand is still misclassified as a face candidate due to its similar shape and color. After searching for eye regions, the hand is discarded from the set of possible face regions, as indicated by the darker boxes within the lighter boxes.

3   Head Pose Estimation

Given the tracking data provided by the face detection procedure, we now address the problem of automatically estimating an individual's head orientation. This information is useful for determining shot relevance and deciding which camera best captures the scene of interest. A number of methods for facial orientation or head pose estimation have been proposed in the literature [3–5]. The general approach involves estimating the positions of specific facial features in the image and fitting this data to a head model. Many of these methods still require manually selecting feature points and assume near-frontal views and high-quality images. For our applications, such conditions are not usually satisfied. Poor lighting conditions and hidden or unclear facial features are common problems.

3.1 Feature Selection and Detection

In situations where most of the fine facial details, (e.g. eyes, lips, and mouth) are difficult to estimate reliably, we focus on a more overt facial property, namely the border between an individual's hair and skin, as the major feature for estimating head orientation. The hair-skin border is straightforward to estimate across a wide range of viewing angles and relatively insensitive to lighting conditions and poor image quality. As will be shown, the curve shape of the hairline is sufficient for estimating head orientation to a satisfactory degree and is a robust and consistent statistic across a variety of people and hairstyles. Figure 3(A) illustrates the hairline contour derived from three people in different orientations.

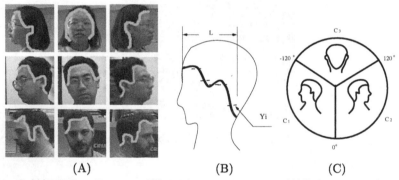

(A)   (B)   (C)

Fig. 3. (A)Hairline Patterns (B)Hairlines as a Feature (C)Subset Classification

Estimation of the hairline contour is performed in conjunction with the head tracker described previously. Within the head regions, hairlines are generated by clustering brightness, color and connectivity into hair and skin regions. Having the pre-segmented head region available significantly simplifies and improves this process. Once estimated, the hairline is parameterized into a simple feature vector. Referring to Figure 3(B), denote the extent of the hairline along the horizontal axis as L. Along L, the hairline is evenly segmented into 6 pieces. The average vertical position of the contour segment for each piece, Y_k, is normalized to form the 6-point feature vector y, whose elements are given by $y_k = [Y_k - (Y_3 + Y_4)/2]/L, (k = 1, \ldots, 6)$. The purpose of this normalization is to make the feature vector invariant to the head size and relative frame position.

3.2 Orientation Estimation by Support Vector Machine Regression

The hairline contours shown in Figure 3 display specific patterns for various facial orientations. The problem of orientation estimation becomes one of finding the mapping function $f : y \to \theta, \theta \in [-\pi, +\pi]$, where θ is the facial orientation. Given a training data set consisting of N vector-orientation pairs $\{y_i, \theta_i\}$, the mapping function is specified from local linear projections associated with specific training data subsets. Specifically, the training samples are first partitioned into three classes C_1, C_2 and C_3, based upon their coarse orientation angles (Figure 3(C)). These classes roughly correspond to right, left, and rear facing head orientations. For each class and its corresponding subset of training vector indexes, S_i, the local orientation estimate is found from $\hat{\theta} = w_i^T y + \beta_i$ which is the mapping function for linear Support Vector Machine regression. The mapping function is obtained by solving the following optimization problem:

$$\min_{w_i, \xi, \xi^*} \left[\frac{1}{2}||w_i||^2 + K_i \left(\sum_{j \in S_i} \xi_j + \sum_{j \in S_i} \xi_j^* \right) \right]$$

Subject to the constrains:

$$\theta_j - w_i^T y_j - \beta_i \leq \epsilon_i + \xi_j \quad j \in S_i$$
$$w_i^T y_j + \beta_i - \theta_j \leq \epsilon_i + \xi_j^* \quad j \in S_i$$
$$\xi_j \geq 0 \qquad j \in S_i$$
$$\xi_j^* \geq 0 \qquad j \in S_i$$

where $||w_i||^{-1}$ is relevant to the lower bound on the distance between the points y_j and the hyperplane (w_i, β_i), and is thus called margin. ϵ_i and K_i are predefined parameters. ϵ_i determines the deviation tolerance of the training data from regression. And K_i controls the trade off between margin and the empirical risk which is proportional to $\sum \xi_j + \sum \xi_j^*$. The idea underlying SVM for regression is to minimize the empirical risk as well as enlarging the margin at the same time. Details of finding the optimal (w_i, β_i) are referred to [6]. This solution gives the training of local mapping functions.

Figure 4 illustrates the whole process. In practice, an observed feature vector y is first classified into one of the three subsets using a Bayesian Classifier, i.e. finding the class which maximizes the *a posteriori* probability given by: $p(C_j|y) = p(y|C_j)p(C_j)$, $j = 1, 2, 3$. The *a priori* term $p(C_j)$ is calculated from the ratio of the subset cardinality, $|S_j|$, to the total number of training pairs, i.e.

$$p(C_j) = \frac{|S_j|}{N}$$

and class-conditional density is found from

$$p(y|C_j) = \frac{1}{|S_j|} \sum_{i \in S_j} \Phi(y, y_i, \Sigma).$$

Without loss of generality, assume C_1 is the class which maximizes $p(C_j|y)$ and $C_k, k \in \{2, 3\}$ is the class which is second most likely. If the likelihood associated with Class 1 is significantly greater than that of the C_k then the pose

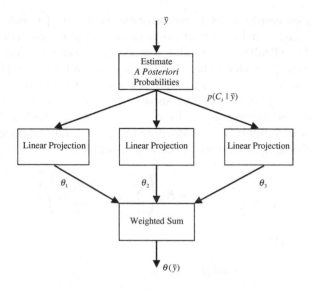

Fig. 4. Pose Estimation Outline

orientation is estimated directly from the class 1 projection. Specifically, given a predefined threshold δ, if

$$|p(C_1|\boldsymbol{y}) - p(C_k|\boldsymbol{y})| > \delta$$

then the estimated orientation is :

$$\theta(\boldsymbol{y}) = \hat{\theta}_1 = \boldsymbol{w}_1^T \boldsymbol{y} + \beta_1$$

Otherwise the orientation angle is found using a weighted sum of the individual subset orientation angles according to:

$$\theta(\boldsymbol{y}) = (\boldsymbol{w}_1^T \boldsymbol{y} + \beta_1) \left(\frac{p(C_1|\boldsymbol{y})}{p(C_1|\boldsymbol{y}) + p(C_k|\boldsymbol{y})} \right)$$
$$+ (\boldsymbol{w}_k^T \boldsymbol{y} + \beta_k) \left(\frac{p(C_k|\boldsymbol{y})}{p(C_1|\boldsymbol{y}) + p(C_k|\boldsymbol{y})} \right)$$

Note that the number of rough classes into which the continuous range of orientations are segmented is subject to the balance between the efficiency of the local prediction and the class separability. Increasing the number of classes improves prediction accuracy of the linear projection, but at the cost of decreased subset classification performance. The use of three subset classes was found to provide a reasonable compromise between these effects.

3.3   Performance of Facial Orientation Estimation

Supervised training for the proposed scheme was achieved using a total of 60 hand labeled vector-orientation pairs derived from video sequences of the three

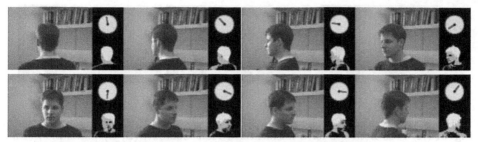

Fig. 5. Examples of Face Orientation Estimation

individuals shown in Figure 3 across the full range of potential orientation angles. The individuals were selected to provide a variety of hairstyles. Figure 5 illustrates the pose estimation results achieved for sample images of the video stream for a fourth subject. In each picture a single image frame is shown along with a clock indicating the estimated facial orientation (6 o'clock corresponds to fully forward-facing, 12 o'clock is fully rear-facing) and a plot illustrating the segmented head region along with the detected hair and skin blobs used to extract the feature vector.

The pose estimation results are very accurate across the complete range of possible orientation angles despite the presence of object shadows and a complex image background. This illustrates the appropriateness of the extracted image features and the effectiveness of the proposed pose learning and estimation procedures. Our experiments indicate that the hairline contour can be reliably estimated from facial images encompassing a range of lighting conditions and source-camera distances using only relative intensity statistics, color disparity, and region connectedness. In situations where the skin or hair regions are partially misidentified, the pose estimator still provides reasonable results. Figure 6(A) illustrates an example where the head has been incorrectly segmented from the background and the hair blob takes on a peculiar shape. In this case the orientation is correctly estimated.

Besides accurately estimating orientation angles appropriate for their specific class, the local linear projections extrapolate well to pose angles outside their defined range. In Figure 6(B)(C) a pair of images which would fall in the Class 2 angle region are estimated entirely from the projection associated with Class 1, i.e. $\theta(y) = \hat{\theta}_1$. In both instances the estimated pose angles are very reasonable.

4   Conclusion

This paper has presented our work with multi-talker face tracking and pose estimation. Video clips of the results demonstrated here as well as the overall tracking system are available from the webpage: http://himmel.hrl.harvard.edu. The pose estimation procedure introduced is capable of accurately evaluating head orientations over a complete 360° interval. The algorithms are designed to be simple enough for real-time applications and, by relying on facial criteria that are easily extracted from images across a range of lighting and zooming

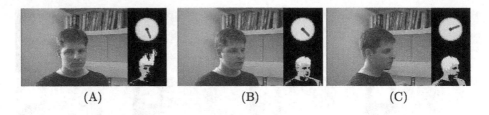

(A)   (B)   (C)

Fig. 6. (A) Pose Estimate for a Misidentified Hair Contour; (B, C) Estimation of a Class 2 Pose Angle Using a Class 1 Projection

conditions, to be effective in practical environments such as those encountered in video conferencing or surveillance scenarios. A number of avenues are available for improving the capability and functionality of the proposed method. An obvious point is that the algorithm relies on a criterion which many individuals may not possess, namely a hairline. To address this issue we are extending the pose estimation scheme to include additional facial features such as the rough locations of the eye and mouth regions. Ideally, the estimator will incorporate a number of facial cues and will take into consideration their reliability when calculating a final result. Finally, in this work, only the pan rotation of the head is considered. The learning-based method employed here may be adapted to evaluate other orientation features such as tilt.

References

1. Wang, C., Brandstein, M. S.: A Hybrid Real-Time Face Tracking System. ICASSP'98, vol.6, pp.3737-3741, Seattle, Washington, USA, May 12-15, 1998.
2. Wang, C., Brandstein, M. S.: Multi-Source Face Tracking with Audio and Visual Data. Proceedings of IEEE 3rd Workshop on Multimedia Signal Processing, pp.169-174, Copenhagen, Denmark, September 13-15, 1999.
3. Lopez, R., Huang, T. S.: Head Pose Computation for Very Low Bit-Rate Video Coding. 6th International Conference on Computer Analysis of Images and Patterns, pp.440-447, Springer-Verlag Berlin Heidelberg, 1995.
4. Kruger, N., Potzsch, M., Malsburg, C.: Determination of Face Position and Pose with a Learned Representation Based on Labeled Graphs. Image and Vision Computing, vol.15, no.8, pp.665-673, August, 1997.
5. Shimizu, I., Zhang, Z., Akamatsu, S., Deguchi, K.: Head Pose Determination from One Image Using a Generic Model. 3rd IEEE International Conference On Automatic Face and Gesture Recognition, pp.100-105, Nara, Japan, April, 1998.
6. Vapnik, V. N.: Statistical Learning Theory. Wiley, 1998.

A General Framework for Face Detection

Ai Haizhou   Liang Luhong   Xu Guangyou

Computer Science and Technology Department, Tsinghua University, State Key Laboratory
of Intelligent Technology and Systems, Beijing 100084, PR China
ahz@mail.tsinghua.edu.cn

Abstract.   In this paper a general framework for face detection is presented
which taken into accounts both color and gray level images. For color images,
skin color segmentation is used as the first stage to reduce search space into a
few gray level regions possibly containing faces. And then in general for gray
level images, techniques of template matching based on average face for
searching face candidates and neural network classification for face verification
are integrated for face detection. A qualitative 3D model of skin color in HSI
space is used for skin color segmentation. Two types of templates:
eyes-in-whole and face itself, are used one by one in template matching for
searching face candidates. Two three-layer-perceptrons (MLPs) are used
independently in the template matching procedure to verify each face candidate
to produce two list of detected faces which are arbitrated to exclude most of the
false alarms. Experiment results demonstrate the feasibility of this approach.

Keywords: Face detection, skin color segmentation, template matching, MLP

1. Introduction

Face detection problem, originated in face recognition research as its first step to
locate faces in the images, is now regarded as rather an independent problem which
may have important applications in many different fields including security access
control, visual surveillance, content-based information retrieval, advanced human and
computer interaction. Therefor, face detection problem has recently been intensively
researched both in computer vision and pattern recognition society and also in
multimedia society.

T. Tan, Y. Shi, and W. Gao (Eds.): ICMI 2000, LNCS 1948, pp. 119-126, 2000.
© Springer-Verlag Berlin Heidelberg 2000

There have been many different approaches to the problem of face detection which include Dai and Nakano [1] based on color features, Yang and Huang [2] based on heuristic mosaic rules, Moghaddam and Pentland [3-4] based on eigenfaces, Sung and Poggio [5] based on classification, Rowley, Baluja and Kanade [6] based on neural networks, Osuna, Freund and Girosi [7] based on SVM (support vector machines) , Schneiderman and Kanade [8] based on probabilistic modeling, Garcia and Tziritas [9] based on color and wavelet packet analysis. Our approach presented in this paper is based on template matching of an average face and neural network verification for gray level images, and in general with a skin color segmentation procedure at the first step for color images.

2.   Overview of Face Detection Framework

As illustrated in Fig.1, the proposed face detection algorithm consists of mainly two parts of which one is skin color segmentation for color images and the other is template-based face detection in gray level images.

Skin color segmentation groups pixels of skin color into rectangle regions in which faces are to be searched. According to the fact that in particular image usually face color is relatively different with non-body background color and that color in face area is rather uniform, a fast segmentation algorithm based on local color uniformity is developed.

In segmented regions or the gray level image, techniques of template matching based on average face for searching face candidates and neural network classification for face verification are integrated for face detection. Two types of templates: eyes-in-whole and face itself, are used one by one in template matching for searching face candidates. And two three-layer-perceptrons (MLPs) are used independently in the template matching procedure to verify each face candidate to produce two lists of detected faces which are arbitrated to exclude most of the false alarms.

The face detection framework is shown in Fig.2 with the principles based on machine learning methods by training MLP over samples collected through template matching and   MLP verification by bootstrap techniques. Compared with ANN-based face detection [6], this method has the advantage of constrained non-face sample collection that guarantees greatly reduced training space and much lower training costs in computation power.

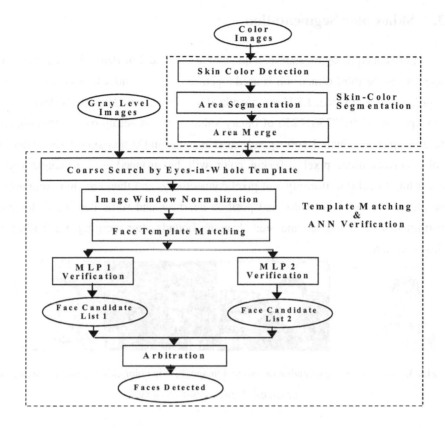

Fig. 1.   Flow chart of Face Detection Algorithm

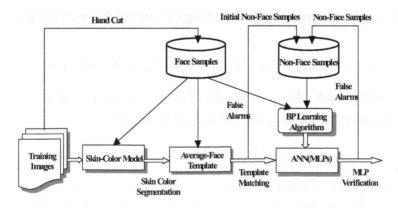

Fig. 2.   Face Detection Framework

3.   Skin Color Segmentation

For skin color segmentation, a qualitative 3D model in HSI color space in which intensity as the third dimension is used together with hue and saturation chrominance coordinates. In practice, I,S,H are quantized separately into 16,256,256 levels, and a lookup table of $16 \times 256 \times 256$ of binary value 0 or 1 is set up over a training color face samples in which 1 corresponds to skin color and 0 non-skin color. After skin color classification, pixels of skin color will be grouped into rectangle regions according to color uniformity and pixel connectivity, and then rectangle regions will be merged according to color and position nearness and some heuristic rules about permitted scale difference and area changing bounds after merging. Fig.3 illustrates this procedure.

Fig. 3.   Skin color segmentation (a, original image; b, skin color classification; c, skin color grouping; d, after region merging)

4.   Template Generation and Image Window Normalization

Given an intensity image window $D[W][H]$ with average intensity $\overline{\mu}$ and squared difference $\overline{\sigma}$. The transformation to a normalized gray distribution of the same average (μ_0) and squared difference (σ_0) is as follows:

$$\hat{D}[i][j] = \frac{\sigma_0}{\overline{\sigma}}(D[i][j] - \overline{\mu}) + \mu_0 \tag{1}$$

An average face is generated from a set of 40 mug-shot photographs containing frontal upright faces. For face detection with different poses (rotation in image plane), six groups of templates corresponding to the upright ones are generated via affine transformations with rotation and stretch, shown in Fig 4. They account for a pose variation from -30° to 30° with a 10° step that cover most of face poses contained in normal images.

The affine transformation with stretch ratio t and rotation θ that maps image point (x, y) to the point (x', y') by which other groups of templates are easily generated is as follows:

$$\begin{pmatrix} x' \\ y' \end{pmatrix} = \begin{pmatrix} \cos\theta & -t \cdot \sin\theta \\ \sin\theta & t \cdot \cos\theta \end{pmatrix} \begin{pmatrix} x \\ y \end{pmatrix}, \qquad 0 \le \theta < 2\pi \tag{2}$$

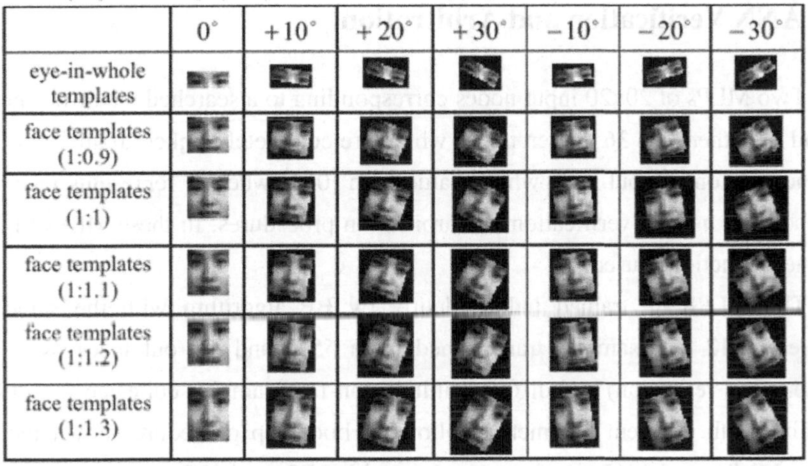

	$0°$	$+10°$	$+20°$	$+30°$	$-10°$	$-20°$	$-30°$
eye-in-whole templates							
face templates (1:0.9)							
face templates (1:1)							
face templates (1:1.1)							
face templates (1:1.2)							
face templates (1:1.3)							

Fig.4 Templates of various poses

5.   Template Matching

Given a template $T[M][N]$ with the intensity average μ_T and the squared difference σ_T, an image region $R[M][N]$ with μ_R and σ_R, the correlation coefficient $r(T, R)$ is as follows:

$$r(T,R) = \frac{\sum_{i=0}^{M-1}\sum_{j=0}^{N-1}(T[i][j]-\mu_T)(R[i][j]-\mu_R)}{M \cdot N \cdot \sigma_T \cdot \sigma_R} \tag{3}$$

The matching algorithm is composed of the following procedures:

1. Given an image, initialize a face candidate list;
2. At each point of the image, firstly, eyes-in-whole templates of various poses are matched, of which top m matches over an eyes-in-whole threshold are further

matched by corresponding groups of face templates, selecting the maximum one from those over a face threshold and passed MLP verification;

3. Compare the maximum one with those in the face candidate list, if there is no overlapping region found, put it in the list, otherwise the bigger will replace the smaller;

4. Subsampling the image by a ratio 1.2, repeat from step 2, until a specified size is reached;

6.   ANN Verification and Arbitration

Two MLPs of 20×20 input nodes corresponding to a searched window, one with 32 and the other with 36 hidden nodes which are completely linked to all input nodes and the only one output node whose value is in [0, 1] where 1 represents face and 0 non-face, are used in verification and arbitration procedures. In these MLPs Sigmoid nonlinear function is used.

The MLPs are trained independently by BP algorithm with the same face samples (3312 face samples transformed from 552 hand cut out faces by slightly rotations and reflection) and different initial non-face samples collected by template matching with different parameters following bootstrap procedure shown in Fig.2. Finally MLP1 used 4065 non-face samples and MLP2 used 3274 ones.

As shown in Fig.1, the two MLPs are used independently in the searching procedure for verification that results in two lists of detected faces among which arbitration procedure is processed to exclude most of false alarms and output those left as the final faces detected.

7.   Experimental Results

Experimental results are listed in Table 1. Some processed images are given in Fig.5-7.

Fig. 5.   Single rotated face detection of MIT set (114 faces out of 144 faces are detected (79.2%), 1 rejected and 29 failed)

Table 1. Results on some test sets

Test set	Description	Correctly detected	Missed faces	False alarms	Correct Rate
1	Gray level images (123 images with 398 faces)	342 /398	56	132	85.9%
2	Gray level images (20 images with 132 faces selected from MIT test set [5][6]: 23 images with 149 faces by excluding 3 line drawings)	103 /132	29	78	78.0%
3	Color images from image grabber (110 images with 122 faces)	116 /122	6	14	95.1%
4	Color images of scanned pictures (111 images with 375 faces)	310 /375	65	119	82.7%

Fig.6.   Face detection on gray level images (the middle and the right are from CMU web side)

Fig. 7.   Face detection on color images

8.   Summary

This paper has demonstrated the effectiveness of detecting faces by using integrated method of skin color segmentation, template matching and ANN verification techniques. Compared with direct neural network based methods which are extremely computationally expensive in both training and testing procedures, this method is simple and yet effective.

In this face detection framework, templates derived from average face play a very important role. We have a strong belief in that average face model will play a very important role in future face detection researches as well as in face synthesis. Systematic analysis still needs to be done about average face and its application in face detection.

At present this face detection algorithm is designed for frontal faces with rotations in image plane and small variations off the image plane. Profile detection is not yet considered.

References

1. Y. Dai and Y. Nakano, Face-Texture Model Based on SGLD and its Application in Face Detection in a COLOR Scene, *Pattern Recognition,* Vol.29, No.6, pp1007-1016, 1996.

2. G. Z. Yang and T. S. Huang, Human Face Detection in a Complex Background, *Pattern Recognition,* Vol.27, No.1, pp53-63, 1994.

3. B. Moghaddam and A. Pentland, Probabilistic Visual Learning for Object Representation, *IEEE Trans. Pattern Analysis and Machine Intelligence,* Vol.19, No.7, pp.696-710, 1997.

4. B. Moghaddam and A. Pentland, Beyond Linear Eigenspaces: Bayesian Matching for Face Recognition, pp.230-243, in Face Recognition from Theory to Applications, edited by H. Wechsler et al., Springer 1998.

5. K. Sung and T. Poggio, Example-Based Learning for View-Based Human Face Detection, *IEEE Trans. Pattern Analysis and Machine Intelligence,* Vol. 20 No.1, pp39-51, 1998.

6. H. A. Rowley, S. Baluja and T. Kanade, Neural Network-Based Face Detection, *IEEE Trans. Pattern Analysis and Machine Intelligence,* Vol. 20 No. 1, pp23-38, 1998.

7. E. Osuna, R. Freund and F. Girosi, Training Support Vector Machines: an application to face detection, in *Proc. of CVPR*, Puerto Rico, 1997.

8. H. Schneiderman and T. Kanade, Probabilistic Modeling of Local Appearance and Spatial Relationships for Object, in Proc. *CVPR*, 1998.

9. C. Garcia, G. Tziritas, Face Detection Using Quantized Skin Color Regions Merging and Wavelet Packet Analysis, IEEE Trans. Multimedia, 1999, 1(3): 264~277

Glasses Detection for Face Recognition Using Bayes Rules

Zhong Jing, Robert Mariani, Jiankang Wu

RWCP Multi-modal Functions KRDL Lab
Heng Mui Keng Terrace, Kent Ridge
Singapore 119613
{jzhong, rmariani, jiankang}@krdl.org.sg

Abstract. In this paper, we propose a method to detect, extract and to remove the glasses from a face image. The aim is to achieve a face recognition system robust to glasses presence. The extraction is realised using Bayes rules that incorporate the features around each pixel, and the prior knowledge on glasses features that were learnt and stored in a database. Glasses removal is achieved with an adaptive median filter conducted in the points classified as glasses. The experiments indicate the performance of this method is better than that of Deformable Contour, and are satisfying.

1 Introduction

Robust face recognition system requires recognising faces correctly under different environments, such as face scale, orientation, lighting, hairstyle and wearing glasses. However glasses affects the performance of face recognition system. Previous work [1] has investigated the detection of glasses existence, however the extraction of glasses in face image is rarely addressed in literatures. Originally, we used a Deformable Contour [2] based method. Since it is only based on edge information, the result would be altered by edges of nose, eyebrows, eyes, wrinkles, etc. To solve this drawback, in this paper, we propose a new glasses point extraction method, which combines some visual features to classify glasses points and non-glasses points. Based on this method, a glasses removal approach was developed to remove the glasses points from face image in order to achieve a glasses invariant face recognition system. In our approach, Bayes rule is used to incorporate the features around one pixel extracted from face image and the prior-knowledge on glasses properties stored in a database. The prior-knowledge is represented by conditional probabilistic distribution, which is obtained by training over face image database and quantized into a feature vector table. The feature vectors describe the properties around the pixels, such as texture, intensity and direction of edge etc. After the extraction of glasses points, an adaptive median filter is applied in each pixel classified as glasses point. The pixel value is substituted by result of median filter. The method is outlined in Figure 1.

T. Tan, Y. Shi, and W. Gao (Eds.): ICMI 2000, LNCS 1948, pp. 127–134, 2000.

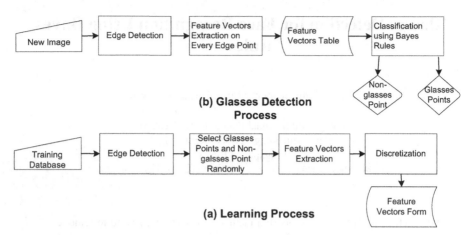

Fig. 1. Outline of our system

2   Glasses Point Detection Using Bayes Rule

The visual features of glasses points at the glasses frame differ drastically due to the diversity of glasses material, colours, etc. This fact indicates that it is difficult to have a unique model representing the visual features of all glasses categories. Thus, we choose a supervised learning scheme, under which the visual features of glasses are learnt and stored into a feature vector table, and in the glasses detection phase, the learnt knowledge will be recalled to determine whether the input pixel is a glasses point or not. The process can be realised by the Bayes rule [4].

2.1   Overview of Bayes Rule

The possibility of whether a pixel in face image is classified as glasses point or not can be represented as the posterior probability given the visual features at the given pixels:

$$P(Glasses \mid features) \qquad for\ glasses \qquad (1)$$
$$P(\overline{Glasses} \mid features) \qquad for\ non-glasses$$

We have the following decision rule to determine the pixel point:

If $P(Glasses \mid features) > P(\overline{Glasses} \mid features)$ then the pixel point is glasses   (2)
point, otherwise it is not a glasses point.

However, in practice, it is not applicable to directly obtain the posterior probability distribution $P(Glasses \mid features)$. Instead, we have to compute this posterior probability function from the prior probability and the conditional probability distribution (CPD) within glasses class and non-glasses class. This can be realised by Bayes formula:

$$\begin{cases} P(Glasses \mid features) = \dfrac{P(features \mid Glasses)P(Glasses)}{P(features)} \\[2mm] P(\overline{Glasses} \mid features) = \dfrac{P(features \mid \overline{Glasses})P(\overline{Glasses})}{P(features)} \end{cases} \quad (3)$$

The conditional probability distribution $P(features \mid Glasses)$ can be obtained by learning from database of face images wearing glasses. However, the prior probability $P(Glasses)$ and $P(\overline{Glasses})$ is unknown. For solving this problem, combining the decision rule (1) and the Bayes formula (2), we have the Bayes decision rule for glasses point detection:

$$\frac{P(features \mid Glasses)}{P(features \mid \overline{Glasses})} \underset{\overline{Glasses}}{\overset{\overset{Glasses}{>}}{<}} \lambda = \frac{P(\overline{Glasses})}{P(Glasses)} \quad (4)$$

If the left side of Equation 3 is larger than λ, we consider the pixel is glasses point, otherwise it's not. The uncertainty of prior probability is handled by the constant λ, which is used to adjust the sensitivity of glasses point detection, and its value is determined empirically.

2.2 Learning the Conditional Probability Distribution

For the conditional probability distribution (CPD), it is necessary to define its form. In many applications of Bayes rule, it is usually assumed that the CPD has the form of Gaussian function. Unfortunately, for glasses detection, this method is not applicable, because the features of glasses points differ too much for variable types of glasses so that the Gaussian assumption is hardly accurate. Therefore, we instead use a non-parameter estimation approach.

Before training, a set of face images wearing glasses is selected from face database. Sobel edge detector is performed on the input image to produce an edge map. Only those pixels that are marked as edge would be selected for learning and glasses detection. Glasses points and non-glasses points are labeled manually within these images, fifty points for either left or right part of glasses frame in each image. And the nine neighbors of these labeled points would be added as well. This produces $50 \times 9 \times 2$ glasses points and non-glasses points respectively in one image. The feature vectors are extracted at those labeled pixels to constitute a training set. We assume the training set has M training samples, $M/2$ for glasses and $M/2$ for non-glasses. A feature vector quantization method is employed by k-mean clustering to separate the training set into N non-overlapped clusters. The mass centroid of each cluster is computed as the Representative Feature Vector (RFV) of the cluster. The process is illustrated as Figure 2. The result of quantization is a table, in which each entry represents a cluster with its RFV. To compute the CPD of the feature vector **f** within glasses or non-glasses, first find the nearest RFV in the table, which represent the cluster i:

$$\|\mathbf{f}, RFV_i\| < \|\mathbf{f}, RFV_j\|, \forall i \neq j, \ i, j = 1,2,\ldots,N \tag{5}$$

and the CPD is computed by:

$$\hat{p}(feature \mid Glasses) \approx \frac{N_{ig}}{M}, \quad \hat{p}(features \mid \overline{Glasses}) \approx \frac{N_{i\overline{g}}}{M} \tag{6}$$

where N_{ig} is the number of glasses points within ith cluster, $N_{i\overline{g}}$ is the number of non-glasses points within ith cluster, and M is the size of sample set.

2.3   Glasses Detection Process

Given a new face image, Sobel Filter is employed to obtain an edge map. For each edge pixel of the edge image, a feature vector is extracted. The nearest RFV of this feature vector is computed, and the CPD of both glasses and non-glasses of this feature vector are acquired by searching RFV Table. Finally, both CPD are fed into formula (3) to output a classification result. In our experiments, λ is set to 1.0.

3.   Extraction of Feature Vectors

Feature vector is extracted at those pixels marked as edge in the edge map, which has been computed using Sobel edge detector. Three kinds of visual features are involved

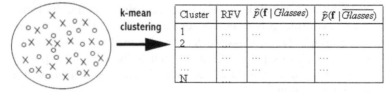

Training Set
x : Glasses feature vector
o: Non-glasses feature vector

RFV Table

Fig. 2. Training process

in glasses point detection. They are texture features, shape features using moment, and geometrical features that include the edge orientation and intensity, the distance between the point and eye, as well as the Polar Co-ordination.

Texture features are used to identify the glasses points from the eyebrows, since in some cases both of them might be very similar in shape and grey scale distribution, but different from each other in texture. We extract texture features in the original grey scale image. The computation is based on Fast Discrete Fourier Transform [5] within an 8x8 region around the pixel. Texture features are derived from the power spectrum of original image after FFT. The Discrete Fourier Transform is:

$$X(u,v) = \frac{1}{NM} \sum_{x=0}^{N-1} \sum_{y=0}^{M-1} I(x,y) e^{-j2\pi(\frac{xu}{N}+\frac{yv}{M})} \tag{7}$$

where $I(x,y)$ is original image, and N, M is 8. The power spectrum is an 8x8 block:

$$P(u,v) = \sqrt{X(u,v)X(u,v)^*} \tag{8}$$

The texture features is:

$$T_x = \frac{P(0,1)}{P(0,0)}, T_y = \frac{P(1,0)}{P(0,0)} \tag{9}$$

The shape features are used to identify the glasses points from the nose and other confusing points around the glasses. For this purpose, the general invariant moment [6] is employed, which is invariant to translation, rotation, scale and contrast. The moment features are computed on the original image. Given a digital image $I(x,y)$ with the size of MxN, the $(p+q)th$ order geometrical moments are:

$$m_{pq} = \sum_x \sum_y x^p y^q I(x,y) \tag{10}$$

And the central moments are:

$$\mu_{pq} = \sum_x \sum_y (x-\bar{x})^p (y-\bar{y})^q I(x,y), \bar{x} = \frac{m_{10}}{m_{00}}, \bar{y} = \frac{m_{01}}{m_{00}} \tag{11}$$

The first five Hu's invariant moments are:

$$\phi_1 = \eta_{20} + \eta_{02}, \phi_2 = (\eta_{20} - \eta_{02})^2 + 4\eta_{11}^2,$$
$$\phi_3 = (\eta_{30} - 3\eta_{12})^2 + (3\eta_{21} + \eta_{03})^2, \phi_4 = (\eta_{30} + \eta_{12})^2 + (\eta_{21} + \eta_{03})^2, \tag{12}$$
$$\phi_5 = (\eta_{30} - \eta_{12})(\eta_{30} - \eta_{12})[(\eta_{30} + \eta_{12})^2 - 3(\eta_{21} + \eta_{03})^2]$$
$$+ (3\eta_{21} - \eta_{03})(\eta_{21} + \eta_{03})[(3(\eta_{30} + \eta_{12})^2 - (\eta_{21} + \eta_{03})^2]$$
$$\eta_{pq} = \mu_{pq} / \mu_{00}^\tau, \tau = \frac{p+q}{2}, p+q = 2,3,...$$

General Invariant moment is based on Hu's moments but is invariant to scale and contrast further. The first four general invariant moments are used:

$$\beta(1) = \sqrt{\phi(2)}/\phi(1), \beta(2) = \phi(3) \cdot \mu_{00}/\phi(2) \cdot \phi(1) \tag{13}$$
$$\beta(3) = \phi(4)/\phi(3), \beta(4) = \sqrt{\phi(5)}/\phi(4)$$

As illustrated in Figure. 3. The geometrical features extracted on the edge map consist of the edge orientation and intensity (e, φ), the distance l between the edge pixel and the eye centre, and the direction θ of the line that links the edge pixel and the eye centre. The geometrical features are used to distinguish the edge points belonging to the glasses frames and the edge points of the eyes and wrinkles.

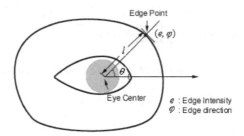

Fig. 3. Geometrical features

Finally, we have a feature vector with the dimension of 10, where 2 features for texture, 4 features for moment and 4 features for geometrical property. In our experiments, these features are discriminative to identify the glasses points.

4. Determination of Glasses Presence and Removal of Glasses

After glasses points classification, the number of pixels that are classified as glasses is taken as the criteria to determine whether the glasses exists or not. If the number of glasses points is greater than a threshold, we conclude that glasses exist in the face.

For every pixel in the face images wearing glasses that is classified as glasses points, the median filter is applied in a rectangular window centred at the given pixel. The rectangular window is adaptively resized to enclose as many non-glasses points as the median filter needs. The median value of the grey level of the non-glasses pixels within this region is computed and substituted to the central pixel of this region. This operation is performed for every glasses point. The result image is a face image without glasses.

5. Experiments and Results

One hundred face images wearing glasses are collected from a face image database. The image size is 200x150. The glasses varies in colour, size and shape. For training, fifty face images are randomly selected from this face database. The remaining images are used to test the performance of our method.

Before training process, we manually labeled the glasses points in glasses frames. Fifty glasses points are labeled both in left part and right part of glasses frame. These points with their neighboring points (9 points) are used in the training step to produce the RVF table for glasses class. For non-glasses points, similar process is undertaken to produce non-glasses RVF table.

The remaining fifty face images are used to test the performance of the approach as well to compare the performance with Deformable Contour method. Figure 4 shows the results of our algorithms for the glasses detection and removal, which are original images, glasses points extraction based on Bayes rules, and glasses removal from the images, respectively. After glasses points detection based on Bayes Rules, a simple linear interpolation algorithm is employed to obtain the contour of glasses frame. The comparison results of average errors of glasses contour are shown in Table 1. The average error is computed by:

$$E = \frac{1}{M} \sum_{i=1}^{M} \left\| \mathbf{g}_i - \mathbf{t}_i \right\|$$ (14)

where, $\mathbf{t}_i$ is the ith point (x_{t_i}, y_{t_i}) of glasses chain obtained by manual labeling, and $\mathbf{g}_i$ is the ith glasses point (x_{g_i}, y_{g_i}) computed from glasses detection results corresponding to point $\mathbf{t}_i$. M is the number of the glasses chain points, which we set it to 20. The comparison result of glasses false detection is shown in Table 2.

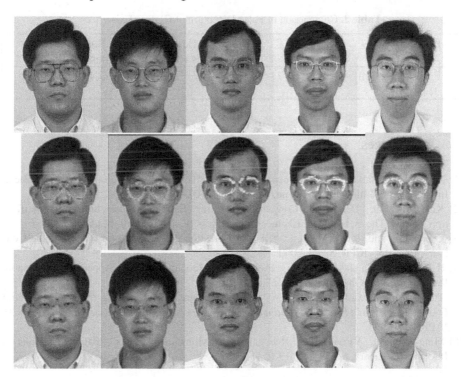

Fig. 4. Results of our algorithms of glasses detection and removal.

6. Conclusion

This paper describes a method for detecting glasses and an approach to remove the glasses points in face images. The experiments we conducted within a face image database showed that our method is effective and has a better performance than that of Deformable Contour based method. The visual effect of glasses removal algorithm also indicates the method can be applied in face recognition sensitive to glasses presence.

Table 1. The comparison of Deformable Contour method and Linear Interpolation based on Glasses Detection using Bayes Rule

Algorithm	Average error of Left part glasses	Average error of Right part glasses
Based on Deformable Contour	27.3297	28.9375
Linear Interpolation based on Glasses Detection using Bayes Rule	11.0963	19.627

Table 2. Comparison of False Detection of Glasses Existence

Detection Algorithm	Based on Edge Of Glasses Ridge [1]	Based on Bayes Rules
False Detection Rate	2/419	0/419

7. References

[1] X. Jiang, M. Binkert, B. Achermann, H. Bunke, "Towards Detection of Glasses in Facial Images", Proc. 14th ICPR, Brisbane, 1998, 1071 - 1073
[2] Kok Fung Lai, "Deformable Contours: Modelling, Extraction, Detection and Classification", Ph. D. Thesis, Electrical Engineering, University of Wisconsin-Madison, 1994.
[3] Zhong Jing and Robert Mariani, "Glasses Detection and Extraction by Deformable Contour", Proc. 15th, ICPR, Spain, 2000.
[4] H. Schneiderman and T. Kanade, "Probabilistic Modeling of Local Apperance and Spatial Relationships for Object Recognition". CVPR98.
[5] Tao-I. Hsu, A.D.Calway, and R. Wilson. In "Texture Analysis Using the Multiresolution Fourier Transform", *Proc 8th Scandinavian Conference on Image Analysis*, pages 823--830. IAPR, July 1993
[6] Ming-Keui Hu, "Visual Pattern Recognition by Moment Invariants", IRE Transactions on Information Theory, pp179-187, 1962.

Real-Time Face Tracking under Partial Occlusion and Illumination Change[+]

Zhihong Zeng and Songde Ma

National Laboratory of Pattern Recognition
Institute of Automation, Chinese Academy of Sciences
{zhzeng,masd}@nlpr.ia.ac.cn

Abstract. In this paper, we present an approach which tracks human faces robustly in real-time applications by taking advantage of both region matching and active contour model. Region matching with motion prediction robustly locates the approximate position of the target , then active contour model detects the local variation of the target's boundary which is insensitive to illumination changes, and results from active contour model guides updating the template for successive tracking. In this case, the system can tolerate changes in both pose and illumination. To reduce the influence of local error due to partial occlusion and weak edge strength, we use *a priori* knowledge of head shape to re-initialize the curve of the object every a few frames. To realize real-time tracking, we adopt region matching with adaptively matching density and modify greedy algorithm to be more effective in its implementation. The proposed technique is applied to track the head of the person who is doing Taiji exercise in live video sequences. The system demonstrates promising performance , and the tracking time per frame is about 40ms on Pentium II 400MHZ PC.

Key word: real-time tracking , region matching , active contour model

1 Introduction

This paper addresses the problem of tracking human heads in real time applications. The real-time face tracking technique has attracted a lot of attention from researchers in computer vision because it plays an important role in man-computer interface, low bit rate video teleconferencing[14], etc.

What makes tracking difficult under real-time constraints is the potential variability in the images of an object over time. This variability arises from three principle sources: variation in target pose or target deformation, variation in illumination, and partial or full occlusion of the target.

During the last years, a great variety of face tracking algorithms have been proposed [4][9][10][12][13]. In [10], the authors surveyed the work in face tracking which had appeared in the literature prior to May, 1998. We note that, in order to

---

[+] This work was supported by the 973 Research Project (G1998030500)

T. Tan, Y. Shi, and W. Gao (Eds.): ICMI 2000, LNCS 1948, pp. 135–142, 2000.

track human faces in a complex and changing environment, multiple visual cues and models are integrated in many works [9][12]. In [12], intensity change, color, motion continuity, shape, and contrast range are adaptively integrated in the tracking system. [9] describes a face tracking system that harnesses Bayesian modality fusion, a technique for integrating the analyses of multiple visual tracking algorithms with a probabilistic framework. However, with increasing of multiple cues and modalities, computational cost will consequentially increase. That means that it will become difficult to realize real-time tracking.

Our approach is to capitalize on the strengths of region matching and active contour model (snake), while minimizing the effect of their weaknesses. Region matching can robustly tracking the targets in many situations due to using the whole region information rather than local one, but suffers from a variety of limitation. It implicitly assumes that the underlying object is translating parallel to the image plane and is being viewed orthographically. These assumptions are often violated in reality. Furthermore, it can not address the problem of illumination changes. In contrast, active contour models (snake)[1][2][3][6][7][8] which use the edge strength is insensitive to illumination and are able to fit objects more closely and have proven to be effective in detecting shapes subject to local deformation. Thus, snake is an effective complement to region matching. We can use the result from snake to update the template of interest. In this way, region matching can stand changes of both pose and illumination.

There are some ideas of this paper similar to those in [11][5] , but [11][5] didn't mention the tracking speed in their experiments. Maybe they are due to high computational cost. We instead use an effective scheme to track changes of head profiles in real-time applications. To reduce the computational cost, we adopt region matching with adaptively matching density and modify greedy algorithm presented by Williams and Shah[7] to be more effective in its implementation. In our experiments, the tracking time per frame of the system is about 40ms on Pentium II 400MHZ PC.

To be robust during the tracking process, tracking technique must have ability to reduce the influence of partial error due to some disturbances and recover from mistakes, or else the error is subsequently propagated through the entire sequence. In our system, we can assume the human head shape is roughly a rectangle or ellipse. during tracking, the system can re-initialize the curve of the objects by the rectangle or ellipse whose parameters are estimated through robust estimation every a few frames.

The rest of the paper is organized as follows. In section 2, we describe the first stage of our tracking algorithm: region matching with motion prediction. Section 3 presents the second stage: greedy algorithm with our improvement. Section 4 describes how to deal with partial error of tracking resulted from occlusion and weak edge. The experimental results of our approach are presented in section 5. We conclude the paper in section 6 with our discussion .

2 Region Matching with Motion Prediction

As we already know, region matching is robust in many applications because of making use of whole region information instead of local one. Given the first frame of image sequences, we define the template of interest object in the image to get the intensity distribution of the object and original coordinate. In the successive frames, we search for optimal matching with the template at the neighboring area to locate the tracked object by minimizing the following equation:

$$D(x) = \arg\min_{x} \sum_{x' \in R} [I(x, x') - T(x')]^2 \qquad (1)$$

R: Area of the template

x : the coordinate of the center of region candidate in the image

$I(x')$: the intensity value of the region candidate of current frame

$T(x')$: the intensity value of the template

$$x \in R^2, x' \in R^2$$

According to Kalman filtering theory[16], we can predict the position of the moving objects in next frame by using the information of the position in the last and current frames. The equations of position prediction of the moving object are as follows:

$$x^*_{i|i-1} = x_{i-1} + v^*_{i-1} \Delta t \qquad (2)$$

$$v^*_{i-1} = v_{i-2} + (x_{i-1} - x^*_{i-1|i-2}) / \Delta t$$

$$v_o = 0$$

$x^*_{i|i-1}$: position of moving object in ith frame predicted from i-1th   frame

$x^*_{i-1|i-2}$: position of moving object in i-1th frame predicted from i-2th frame

x_{i-1} : actual position of object in i-1th frame

v_{i-2} : actual velocity of object in i-2th frame

v^*_{i-1} : velocity of moving object in i-1th frame predicted from i-2th frame

Δt : time interval between two successive frames

With the help of motion prediction, the system can track well moving objects even when they move fast in the image as long as the acceleration is within the user-defined range .

The disadvantages of region matching include: 1) it need updating the template when the surfaces of targets have obvious changes due to transformation, deformation or changes in illumination, and small error of automatic template updating is often accumulated with time and that conduces to the failure of the system. 2)the computational cost is high when the area of the template is large. To reduce the computational cost of large region matching, we adaptively change sampling rate of matching between the current frame and the template. In this way, tradeoffs between matching size and matching density are made. That means that we sacrifice some degree of precision of tracking results in region matching to save computation cost. That is reasonable in our method because the goal of region matching is coarse estimation of tracked ob-ject's position to provide subsequent active contour model with initial curve. As for the first disadvantage of the region matching, we use active contour model to capture local variation of boundary of targets to eliminate error of updating templates. In other words, the result from active contour model guides updating template of region. Thus, the kind of region matching can stand changes in both pose and illumination.

3 Greedy Algorithm with Our Improvement

After getting the approximate location of the tracked object in the above stage, we apply active contour model(snake) to get the boundary of it.

To reduce the computational cost in real time applications, we adopt the greedy algorithm presented by Williams and Shah[7] with a complexity of O(nm) where n is the number of control points and m is the size of the neighborhood in which a point can move during a single iteration.

The energy functional which will be minimized is

$$E = \int (\alpha E_{cont} + \beta E_{curv} + \gamma E_{image}) ds \tag{3}$$

Both the first term about continuity and second term about curvature correspond to the internal energy of the snake. The last term measures some image quantity such as edge strength. The values of α, β and γ are taken as 1.0, 1.0 and 1.2 respectively. These are chosen so that the image quantity will has more importance than both of the other terms in determining where points on the contour move.

This method is iterative. At each iteration, points in the neighborhood of the current point are examined and the value of the energy functional is computed at each of them. Then, one of the points in the neighborhood, giving the smallest energy value, is chosen as the new location of the current point(Fig.1).

Determining the first term, E_{cont}, of equation (3) presents some difficulties. If we use $| v(i)-v(i-1)|$, as did Kass [1]and Amini[2], the contour tends to shrink while minimizing the distance between the points. It also contributes to the problem of points bunching up on strong portions of the contour. A term encouraging even spacing will reflect the desired behavior of the contours. In this case, the original goal, first-order continuity, is still satisfied. So the algorithm uses the difference between d, the average distance between points, and $| v(i)-v(i-1)|$, the distance between the current points: $d - | v(i)-v(i-1)|$. By this formula, points having distances near the average will minimum values. The values are normalized by dividing by the largest value in the neighborhood to which the point may move, thus lying in the range [0,1]. At the end of each iteration, a new value of d is computed.

In equation (3), the second term, E_{curv} is about curvature. Since the continuity term(the first term in (3)) causes the points to be relatively evenly spaced, $|v(i-1)-2v(i)+v(i+1)|$is a reasonable estimate of curvature. This formulation has also given good results in the work of Williams[7]. In [7], the curvature term is normalized by dividing by the largest value in the neighborhood, giving a value in [0,1].

We find that it is better that given the curvature(curv) at a point and maximum(max) and minimum(min) in each neighborhood, normalized curvature term is computed as (curv-min)/(max-min). In this case, the curvature term is more consistent with the other two terms of energy functional than the original term [7] and experimental results have proven that the performance of our modified formula is better.

The third term, *Eimage*, is the image force, which is the gradient magnitude. Gradient magnitude is computed as an eight-bit integer with values 0-255. Given the magnitude (mag) at a point and the maximum (max) and minimum (min) magnitude in each neighborhood, normalized edge strength term is computed as (min- mag)/(max-min). This term is negative, so points with large gradients will have small values. If the magnitude of the gradient at a point is high, it means that the point is probably on an edge of the image. If (max-min) < 5 then min is given the value (max-5). This prevents large differences in the value of this term from occurring in areas where the gradient magnitude is nearly uniform.

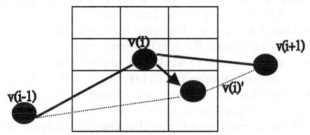

Fig. 1. New Location of a point in an iteration

4 Reducing the Influence of Error Resulted from Occlusion and Weak Edge

We note that in some environment, occlusion and weak strength of boundary would lead the snake to converge to wrong curve. That would result in losing target in a long tracking process due to accumulated error. Fortunately, in most case, we can assume that most part of target contour points from snake convergence are correct, and we could have *a priori* knowledge about the shape of the target. Therefore, we can reduce the influence of this kind of error under these assumptions. In other words, we can re-initialize the curve points for snake using the rough shape of the object whose parameters are estimated through robust estimation.

In our head tracking process, the head has clear boundary compared with the background, but weak edge exists between head and neck. So after a few frames, the lower part of the head contour shrinks toward head interior. In some cases, the head is partially occluded by the hand so that part of head contour converges to wrong edge. To solve this problem, we can surely assume that the head profile is approximately a ellipse or rectangle and we can get the ratio between the width and length of the rectangle or semminor and semimajor axes of the ellipse in the first frame. Thus, to prevent the error of snake from being accumulated, we can estimate the parameters of the ellipse or rectangle every a few frames by using robust estimation, then use these parameters and the approximately constant ratio to re-initialize curve points for snake in the next frame.

5 Experimental Results

The proposed algorithm works as follows: Given the first frame, user defines the rectangle of template including the target . Then system is activated, tracking the moving object. The tracking time per frame of our system is about 40ms on Pentium II 400MHZ PC. In our experiments, the acceleration limit is set to be ±20 pixels in both X and Y axes, that is, whenever the change of the object's velocity from frame to frame is within the range of ±20 pixels in both axes, it can be well tracked.

We successfully track the face of the person who is doing Taiji exercise (Fig.2). In Fig.2(a), user defined the template of face in the first frame. Then our system is automatically tracking the face of the person whose pose is changing due to different directions of the face relative to the camera. We note that some errors of lower part of the head profile occurs in some frames due to weak edge strength between the head and neck in Fig.2(d) and occlusion of hands in Fig.2(e)-(f). As we expect, the system can reduce the errors after re-initialization of curve points by using the rough shape of the object whose parameters are estimated through robust estimation.

To test the robustness of the approach to different illumination, we apply the approach to track the person's face when the table lamp as a light source is turned on then off (Fig.3).

These promising experimental results show that our proposed framework is practical in real-time applications.

6 Conclusion

In this paper, we presented an approach of integration of region matching and active contour model(snake). Region matching with motion prediction can robustly locate the approximate position of the target , then snake can detect the local variation of the target's boundary and results from snake can guide updating the template. In this case, the system can stand changes in pose and illumination. To reduce the computational cost, we adopt region matching with adaptively matching density and modify greedy algorithm to be more effective in its implementation. To be robust in tracking process, we propose a solution to reduce the influence of partial error of tracking resulted from occlusion and weak edge.

Even though experimental results have proven the effectiveness and robustness of our algorithm, we are in the process of testing and improving this approach in more complicated applications. In addition, we don't take into consideration the recovery method which can recover to track the target after the target is lost totally. Apparently, the recovery method can use the technique of face detection to find the lost face in a scene.

Acknowledges

We are very grateful to Professor Baogang Hu and Dr. Yuan Tian for helpful discussion, Dr. Baihua Xiao and Dr. Changbo Hu for making the image sequences.

References

1. M.Kass, A.Witkin, and D.Terzopoulos. "Snake: Active Contour Models".IJCV,1:321-331,1988
2. A.A.Amini, T.E.Weymouth, and R.C.Jain. "Using dynamic programming for solving variational problems in vision". IEEE PAMI, 12: 855-867, September, 1990
3. K.F.Lai and R.T.Chin. "Deformable Contour: modeling and extraction". IEEE PAMI,17:1084-1090,1995.
4. K.Schwerdt and J.L.Crowley. "Robust face tracking using color" . Proc. IEEE. Int. Conf. on Automatic Face and Gesture Recognition, 90-95, 2000
5. N.Paragios and R.Deriche. "Unifying Boundary and Region-based information for Geodesic Active Tracking". CVPR, 300-305, 1999
6. V.Caselles, R.Kimmel, and G.Sapiro. "Geodesic Active Contours".IJCV,1:61-79, 1997
7. D.J.Williams and M.Shah. "A fast algorithm for active contours and curvature estimation".CVGIP: Image Understanding, 55:14-26,1992
8. F.Leymaric and M.D.Levine. "Tracking deformable objects in the plane using an active contour model". IEEE PAMI, 15:617-634, June 1993
9. K.Toyama and E. Horvitz. "Bayesian modality fusion: probabilistic Integration of multiple vision algorithms for head tracking" . ACCV, 1014-1015, 2000
10.K. Toyama. "Prolegomena for robust face tracking". Microsoft Research Technical Report, MSR-TR-98-65, November 13,1998
11.B. Bascle and R. Deriche. "Region tracking through image sequences". INRIA Technical Report, RR-2439, December, 1994
12.J.Triesch and C.V.D.Malsburg . "Self-organized integration of adaptive visual cues for Face tracking". Proc. IEEE. Int. Conf. on Automatic Face and Gesture Recognition, 102-107, 2000
13.G.D.Hager and P.N.Belhumeur. "Efficient region tracking with parametric models of geometry and illumination". IEEE PAMI, 20:1025-1039, October 1998
14.J. Daugman. "Face and gesture recognition: overview". IEEE PAMI, 19:675-676, July, 1997
15.Songde Ma and ZhengYou Zhang. Computer Vision. Science Press . 1998

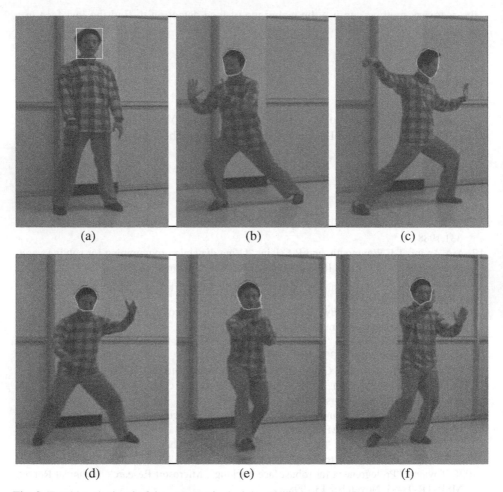

Fig. 2. Tracking the head of the person who is doing Taiji exercise

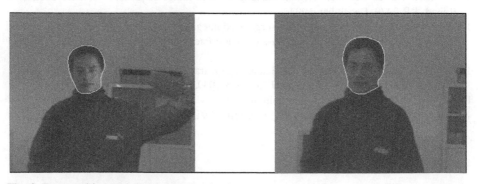

Fig. 3. Face tracking under different illumination

Eye-State Action Unit Detection by Gabor Wavelets

Ying-li Tian [1]   Takeo Kanade[1] and Jeffrey F. Cohn[1,2]

[1] Robotics Institute, Carnegie Mellon University, Pittsburgh, PA 15213
[2] Department of Psychology, University of Pittsburgh, Pittsburgh, PA 15260
Email: {yltian, tk}@cs.cmu.edu   jeffcohn@pitt.edu
http://www.cs.cmu.edu/~face

Abstract Eyes play important roles in emotion and paralinguistic communica-tions. Detection of eye state is necessary for applications such as driver awareness systems. In this paper, we develop an automatic system to detect eye-state ac-tion units (AU) based on Facial Action Coding System (FACS) by use of Gabor wavelets in a nearly frontal-viewed image sequence. Three eye-state AU (AU 41, AU42, and AU43) are detected. After tracking the eye corners in the whole sequence, the eye appearance information is extracted at three points of each eye (i.e., inner corner, outer corner, and the point between the inner corner and the outer corner) as a set of multi-scale and multi-orientation Gabor coefficients. Then, the normalized Gabor coefficients are fed into a neural-network-based eye-state AU detector. An average recognition rate of 83% is obtained for 112 images from 17 image sequences of 12 subjects.

1. Introduction

Facial Action Coding System (FACS) action unit recognition attracts attention for facial expression analysis[1, 5, 6, 16, 14]. Eyes play important roles in emotion and paralinguistic communications. Detection of eye state (i.e. whether the eye is open or closed) is also necessary for applications such as driver awareness systems. Although many methods exist for eye feature extraction and eye tracking, detecting qualitative changes of eye states is relatively undeveloped [2, 4, 7, 9, 18, 19]. In our facial expression analysis system, we developed a dual-state eye model for eye tracking[15]. In that paper, two eye states are detected by geometry feature information of the iris. However, when the eye is narrowly-opened or the iris is difficult to detect, the eye state may be wrongly identified as closed. We believe that the eye appearance information will help to solve this difficulty and increase the number of AU that can be recognized in the eye region.

Recently, Gabor wavelet has been applied to image analysis, face recognition, facial expression analysis [3, 5, 10, 13, 17, 20]. This research suggests that Gabor wavelet is a promising tool to extract facial appearance information.

In this paper, we develop a facial appearance information based eye-state AU de-tection system to detect AU 41 (upper-lid droop), AU 42 (slit), and AU 43 (closed). Figure 1 depicts the overview of the eye-state AU detection system. First, the face position is detected and the initial positions of the eye corners are given in the first

T. Tan, Y. Shi, and W. Gao (Eds.): ICMI 2000, LNCS 1948, pp. 143-150, 2000.
© Springer-Verlag Berlin Heidelberg 2000

frame of the nearly frontal face image sequence. The eye corners then are tracked in the image sequence. Next, a set of multi-scale and multi-orientation Gabor coefficients of three eye points are calculated for each eye. Finally, the normalized Gabor coefficients are fed into a neural-network-based detector to classify three states of the eye.

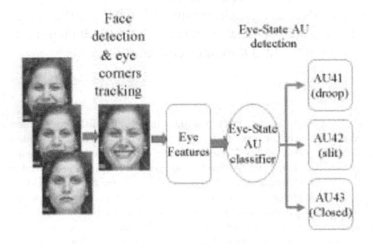

Fig. 1. Eye state detection system.

2. Eye-State AUs

In FACS, there are nine eye-state AUs i.e. AU5, AU6, AU7, AU41, AU42, AU43, AU44, AU45, and AU46. We recognized AU5(eye wide), AU6 (infra-orbital raise), and AU7 (lower-lid raise) in previous work by feature-based information[11, 14]. In this paper, we recognize AU41, AU42, and AU43 by appearance information. The examples of these AUs are shown in Table 1. We classify these AU into three eye states: open (AU41), very narrow (AU42), and closed (AU43). The closed eye is defined as closure of the eyelid brought about by total relaxation of the levator palpebrae superioris muscle, which controls the motion of the upper eyelid. The closed eye may also involve weak contraction of the orbicularis oculi pars palpebralis muscle, a sphincter muscle that surrounds the eye orbit. The very narrow eye is defined as the eyelids appearing as narrowed as possible without being closed. Their appearance resembles a slit, the sclera is not visible, and the pupil may be difficult to distinguish. Relaxation of the levator palpebrae superioris is not quite complete. The open eye is defined as a barely detectable drooping of the upper eyelid or small to moderate drooping of the upper eyelid. See paper [8] for complete list of FACS action units.

3. Localizing Eye Points

To extract information about change of eye appearance, the eye position first must be localized. Three points for each eye are used. As shown in Figure 2, these are the inner and outer corners, and the mid-point between them. At each point, multi-scale and multi-orientation Gabor wavelet coefficients are calculated.

Table 1. Eye states and corresponding FACS action units

Open	Very narrow	Closed
AU41	AU42	AU43/45/46

Upper-lid is slightly lowered.	Eyes are barely. opened.	Eyes are completely closed.

Fig. 2. Three points for each eye are used to detect eye states.

Inner corner: We found that the inner corners of the eyes are the most stable features in a face and are relatively insensitive to deformation by facial expression. We assume the initial location of the inner corner of the eye is given in the first frame. The inner corners of the eyes then are automatically tracked in the subsequent image sequence using a modified version of the Lucas-Kanade tracking algorithm [12], which estimates feature-point movement efficiently with sub-pixel accuracy.

We assume that intensity values of any given region (feature window size) do not change but merely shift from one position to another. Consider an intensity feature template $I_t(x)$ over a $n \times n$ region R in the reference image at time t. We wish to find the translation d of this region in the following frame $I_{t+1}(x + d)$ at time $t + 1$, by minimizing a cost function E defined as:

$$E = \sum_{x \in R}[I_{t+1}(x + d) - I_t(x)]^2. \tag{1}$$

The minimization for finding the translation d can be calculated in iterations (See paper[15] for details).

Outer corner and mid-point: Because the outer corners of the eyes are difficult to detect and less stable than the inner corners, we assume they are collinear with the inner corners. The width of the eye is obtained from the first frame. If there is not large head motion, the width of the eye will not change much. The approximate positions of the outer corners of eyes are calculated by the position of the inner corners and the eye widths.

After obtaining the inner and outer corners of the eyes in each frame, the middle points are easy to calculate from the position of the inner- and outer corners of the eyes.

4. Eye Appearance Information

We use Gabor wavelet to extract the information about change of eye appearance as a set of multi-scale and multi-orientation coefficients. The response image can be written as a correlation of the input image $I(x)$, with

$$a_\mathbf{k}(\mathbf{x_0}) = \int I(x) p_\mathbf{k}(\mathbf{x} - \mathbf{x_0}) dx, \tag{2}$$

where the Gabor filter $p_\mathbf{k}(\mathbf{x})$ can be formulated [3]:

$$p_\mathbf{k}(\mathbf{x}) = \frac{k^2}{\sigma^2} exp(-\frac{k^2}{2\sigma^2} x^2) \left(exp(i\mathbf{kx}) - exp(-\frac{\sigma^2}{2}) \right) \tag{3}$$

where $\mathbf{k}$ is the characteristic wave vector.

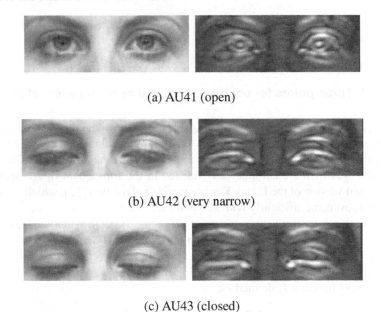

(a) AU41 (open)

(b) AU42 (very narrow)

(c) AU43 (closed)

Fig. 3. Gabor images for different states of the eyes when the spatial frequency=$\frac{\pi}{4}$ in horizontal orientation.

In our system, we use $\sigma = \pi$ and three spatial frequencies with wavenumbers $k_i = (\frac{\pi}{2}, \frac{\pi}{4}, \frac{\pi}{8})$ and six orientations from 0 to π differing in $\pi/6$. Only the magnitudes are used because they vary slowly with the position while the phases are very sensitive. Therefore, for each point of the eye, we have 18 Gabor wavelet coefficients. Figure 3 shows the examples of different eye state and the corresponding Gabor filter responses for the second spatial frequency ($k_i = (\frac{\pi}{4})$) and horizontal orientation. The Gabor coefficients appear highly sensitive to eye states even when the images of eyes are very dark.

5. Eye State Detection

5.1. Image Databases

We have been developing a large-scale database for promoting quantitative study of facial expression analysis [8]. The database currently contains a recording of the facial behavior of 210 adults who are 18 to 50 years old; 69% female and 31% male; and 81% Euro-American, 13% Africa-American, and 6% other groups. Subjects sat directly in front of the camera and performed a series of facial expressions that included single AUs and AU combinations. To date, 1,917 image sequences of 182 subjects have been FACS coded for either the entire sequence or target action units. Approximately fifteen percent of the 1,917 sequences were coded by a second certified FACS coder to validate the accuracy of the coding.

In this investigation, we focus on AU41, AU42, and AU43. We selected 33 sequences from 21 subjects for training and 17 sequences from 12 subjects for testing. The data distribution of training and test data sets for eye states is shown in Table 2.

Table 2. Data distribution of training and test data sets.

Data Set	Eye states included			
	Open	Narrow	Closed	Total
Train	92	75	74	241
Test	56	40	16	112

To assess how reliably trained observers could make these distinctions, two research assistants with expertise in FACS independently coded image sequences totaling 139 frames. Inter-observer agreement between them averaged 89%. More specifically, inter-observer agreement was 94% for AU41, 84% for AU42, and 77% for AU43. For FACS coders, the distinction between very narrow (AU 42) and closed (AU 43) was more difficult.

5.2. Neural network-based eye state detector

As shown in Figure 4, we use a three-layer neural network with one hidden layer to detect eye states. The inputs to the network are the Gabor coefficients of the eye feature points. The outputs are the three states of the eyes.

In our system, the inputs of the neural network are normalized to have approximately zero mean and equal variance.

5.3. Experimental Evaluations

We conducted three experiments to evaluate the performance of our system. The first is detection of three states of the eye by using three feature points of the eye. The second is the investigation of the importance of each feature points to eye state detection. Finally, we study the significance of image scales.

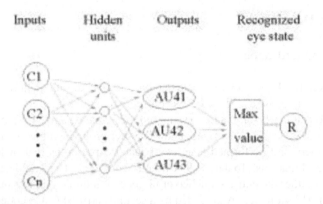

Fig. 4. Neural network-based detector for three states of the eye. The inputs are the Gabor coefficients of the eye feature points, and the output is one label out of the three states of the eyes.

Results of eye state detection: Table 3 shows the detection results for three eye states when we use three feature points of the eye and three different spatial frequencies of Gabor wavelet. The average recognition rate is 83%. More specifically, 93% for AU41, 70% for AU42, and 81% for AU43. These are comparable to the reliability of different human coders.

Compared to expression analysis, three eye states are unnecessary for driver awareness systems. Very narrow eye and closed eye can be combined into one class in driver awareness systems. In that case, the accuracy of detection increases to 93%.

Table 3. Detection results by using three feature points of the eye. The numbers in bold means can be combined into one class in driver awareness systems.

	Recognized eye states		
	Open	Narrow	Closed
Open	52	4	0
Narrow	4	**28**	**8**
Closed	0	**3**	**13**
Recognition rate of three states: **83%**			
Recognition rate of two states: **93%**			

Importance of eye feature points: We also carried out experiments on detection of the three eye states by using one point (the inner corner) of the eye and two points (the inner corner and the middle point) of the eye. The recognition rates for using different points of the eye are list in Table 4. The recognition rate of 81.3% for two points is close to that (83%) for three points. When only the inner corner of the eye is used, the recognition rate decreases to 66%. When only the outer corner of the eye is used, the recognition rate decreases to 38%. The inner corner and middle point carry more useful information than the outer corner for eye state detection.

Table 4. Detection results for three eye states by using different feature points of the eye. We found that the inner corner and middle point carry more useful information than the outer corner for eye state detection.

Used eye feature points				
1 point		2 points		3 points
Inner corner	Outer corner	Outer & middle	Inner & middle	Inner, outer, & middle
66%	38%	61.2%	81.3%	83%

Significance of different image scales: To investigate the effects of the different spatial frequencies, we evaluated the experiments by using two of the spatial frequencies (i.e., wavenumber $k_i = (\frac{\pi}{2}, \frac{\pi}{4}, \frac{\pi}{8})$). Table 5 shows the resulting comparisons. An 80% recognition rate is achieved when we use $k_i = (\frac{\pi}{4}, \frac{\pi}{8})$. It is higher than the recognition rate 74% when we use the higher spatial frequencies (i.e., $k_i = (\frac{\pi}{2}, \frac{\pi}{4})$).

Table 5. Detection results for three eye states by using different spatial frequencies.

Spatial frequencies		
$k = (\frac{\pi}{2}, \frac{\pi}{4})$	$k = (\frac{\pi}{4}, \frac{\pi}{8})$	$k = (\frac{\pi}{2}, \frac{\pi}{4}, \frac{\pi}{8})$
74%	80%	83%

6. Conclusion

In this paper, we developed an appearance-based system to detect eye-state AUs: AU41, AU42, and AU43. After localizing three feature points for each eye, a set of multi-scale and multi-orientation Gabor coefficients is extracted. The Gabor coefficients are fed to a neural-network-based detector to learn the correlations between the Gabor coefficient patterns and specific eye states. A recognition rate of 83% was obtained for 112 images from 17 image sequences of 12 subjects. This is comparable to the agreement between different human coders. We have found that the inner corner of the eye contains more useful information than the outer corner of the eye and the lower spatial frequencies contribute more than the higher spatial frequencies.

Acknowledgements

The authors would like to thank Bethany Peters for processing the images, Lala Ambadar and Karen Schmidt for coding the eye states and calculating the reliability between different coders. This work is supported by NIMH grant R01 MH51435.

References

[1] M. Bartlett, J. Hager, P.Ekman, and T. Sejnowski. Measuring facial expressions by computer image analysis. *Psychophysiology*, 36:253–264, 1999.
[2] G. Chow and X. Li. Towards a system for automatic facial feature detection. *Pattern Recognition*, 26(12):1739–1755, 1993.
[3] J. Daugmen. Complete discrete 2-d gabor transforms by neutral networks for image analysis and compression. *IEEE Transaction on Acoustic, Speech and Signal Processing*, 36(7):1169–1179, July 1988.
[4] J. Deng and F. Lai. Region-based template deformation and masking for eye-feature extraction and description. *Pattern Recognition*, 30(3):403–419, 1997.
[5] G. Donato, M. S. Bartlett, J. C. Hager, P. Ekman, and T. J. Sejnowski. Classifying facial actions. *IEEE Transaction on Pattern Analysis and Machine Intelligence*, 21(10):974–989, October 1999.
[6] B. Fasel and J. Luttin. Recognition of asymmetric facial action unit activities and intensities. In *Proceedings of International Conference of Pattern Recognition*, 2000.
[7] L. Huang and C. W. Chen. Human facial feature extraction for face interpretation and recognition. *Pattern Recognition*, 25(12):1435–1444, 1992.
[8] T. Kanade, J. Cohn, and Y. Tian. Comprehensive database for facial expression analysis. In *Proceedings of International Conference on Face and Gesture Recognition*, March, 2000.
[9] K. Lam and H. Yan. Locating and extracting the eye in human face images. *Pattern Recognition*, 29(5):771–779, 1996.
[10] T. Lee. Image representation using 2d gabor wavelets. *IEEE Transaction on Pattern Analysis and Machine Intelligence*, 18(10):959–971, Octobor 1996.
[11] J.-J. J. Lien, T. Kanade, J. F. Chon, and C. C. Li. Detection, tracking, and classification of action units in facial expression. *Journal of Robotics and Autonomous System*, in press.
[12] B. Lucas and T. Kanade. An interative image registration technique with an application in stereo vision. In *The 7th International Joint Conference on Artificial Intelligence*, pages 674–679, 1981.
[13] M. Lyons, S. Akamasku, M. Kamachi, and J. Gyoba. Coding facial expressions with gabor wavelets. In *Proceedings of International Conference on Face and Gesture Recognition*, 1998.
[14] Y. Tian, T. Kanade, and J. Cohn. Recognizing upper face actions for facial expression analysis. In *Proc. Of CVPR'2000*, 2000.
[15] Y. Tian, T. Kanade, and J. Cohn. Dual-state parametric eye tracking. In *Proceedings of International Conference on Face and Gesture Recognition*, March, 2000.
[16] Y. Tian, T. Kanade, and J. Cohn. Recognizing lower face actions for facial expression analysis. In *Proceedings of International Conference on Face and Gesture Recognition*, March, 2000.
[17] L. Wiskott, J. M. Fellous, N. Kruger, and C. von der Malsburg. Face recognition by elastic bunch graph matching. *IEEE Transaction on Pattern Analysis and Machine Intelligence*, 19(7):775–779, July 1997.
[18] X. Xie, R. Sudhakar, and H. Zhuang. On improving eye feature extraction using deformable templates. *Pattern Recognition*, 27(6):791–799, 1994.
[19] A. Yuille, P. Haallinan, and D. S. Cohen. Feature extraction from faces using deformable templates. *International Journal of Computer Vision,*, 8(2):99–111, 1992.
[20] Z. Zhang. Feature-based facial expression recognition: Sensitivity analysis and experiments with a multi-layer perceptron. *International Journal of Pattern Recognition and Artificial Intelligence*, 13(6):893–911, 1999.

Web-PICASSO: Internet Implementation of Facial Caricature System PICASSO

Takayuki Fujiwara+, Masafumi Tominaga+, Kazuhito Murakami* and Hiroyasu
Koshimizu+

+ SCCS, Chukyo University, 101 Tokodachi, Kaizu-cho, Toyota, 470-0393 Japan
{fuji, tomy}@koshi-lab.sccs.chukyo-u.ac.jp, hiroyasu@sccs.chukyo-u.ac.jp
http://www.koshi-lab.sccs.chukyo-u.ac.jp/

* Faculty of I.S.T. Aichi Pref. University, Japan
murakami@ist.aichi-pu.ac.jp

Abstract. The expansion of the site for the facial caricature presentation and evaluation could be opened toward the world through the Internet door. In order to show the works of caricatures and to get the evaluations for them, it is indispensable to have the open channel which can break the limited scope of the community. This is the reason why we proposed the Internet implementation of facial caricature system Web-PICASSO. And, in order to expand the database of this system at the same time, a mechanism to collect the face data was also implemented. Web-PICASSO was located at the Internet home page addressed by http://www.koshi-lab.sccs.chukyo-u.ac.jp/~fuji/pica2/. Web-PICASSO system was described by Java in order to cope with the various client environments. First this paper presents the basis for generating facial caricature, and the outline and experiments of Web-PICASSO are explained to show the possibilities for collecting the evaluations and questionnaires from visitors. In the future these collected evaluations will be utilized to polish the caricature due to the visitor's individual KANSEI, and furthermore the mechanism to get the visitor's face image through Web-PICASSO environment can be constructed in the caricaturing process.

1   Introduction

We are now on the project to develop a facial caricaturing system PICASSO for providing the better multimedia interface on the computer network environment. The site for the caricature presentations and evaluation could be expanded to everywhere and every time through the Internet. This site has the mechanism which could provide the visitor's face image data to expand the database of PICASSO system.

In order to show the caricature works and to get the evaluations for them, it is indispensable to have the open channel which can break the limited scope of the

T. Tan, Y. Shi, and W. Gao (Eds.): ICMI 2000, LNCS 1948, pp. 151–159, 2000.
© Springer-Verlag Berlin Heidelberg 2000

community. This is the reason why we proposed the Internet implementation of facial caricature system Web-PICASSO.

In facial caricaturing, though the caricaturist, model and gallery should be basically collaborating to generate the caricature, the relation between caricaturist and gallery is not fatally taken into account. Web-PICASSO is one of the attempts to provide the new channel between them.

Web-PICASSO was located at the Internet home page addressed by

http://www.koshi-lab.sccs.chukyo-u.ac.jp/~fuji/pica2/.

Web-PICASSO system was described by Java in order to cope with the various client environments. First this paper presents the basis for generating facial caricature, and the outline and experiments of Web-PICASSO are explained to show the possibilities for collecting the evaluations and questionnaires from visitors. In this paper, first the principle of the caricature generation is investigated, and the implementation of Web-PICASSO is presented precisely. Experiments and some investigations are also presented to show the Web-PICASSO potential.

2   The Principle of Facial Caricaturing

2.1   PICASSO System ant Its Inherent Subjects

PICASSO system is a computer system to generate the facial caricature. [4] In order to realize this, human interface between the model and PICASSO for providing face tracking should be prepared, the facial parts should be recognized to get the line drawings, the features characterizing the model should be extracted in advance, and these features should be deformed in such a way that the features of the face hit the visual KANSEI of the respective gallery.[5] The fundamentals of PICASSO are introduced as follows:

$$Q = P + b * (P - S) \qquad (1)$$

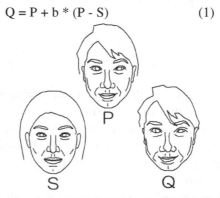

Fig.1. The fundamental principle PICASSO system (Provided by Magazine House)

The most important aspect of this principle is the relationship between the input face P and the mean face S in attribute. The examples of the attributes of the person P are age, sex, nationality, race, and so on. The problem is that it is basically impossible to adjust the set of attributes between P and the mean face S. Therefore the individuality features extracted by (P-S) are strongly affected by the differences between P and S in attribute. In other words, since the caricature Q will be altered in accordance with the different mean face S, it is expected to introduce the method to control the selection of the mean face S especially by basing on the visual KANSEI of the gallery[2][3][6].

2.2   Methodology to Control Mean Face Hypothesis via Web Environment

PICASSO system on the Web network can provide an open channel to everyone connected to the network.   This means that the visitor of Web-PICASSO can select the mean face S in consistent way to his visual KANSEI to improve the facial caricature Q.

A preliminary experiment to show this fact is given in Fig.2.   In Fig.2, the caricatures Q and Q' of The Prince of Wales were generated based on the different mean faces S and S', respectively. Since the mean faces S and S' were introduced from Japanese students and European people, the caricature Q would be more impressive especially to Japanese people who used to be familiar with Japanese faces. For example, the impression for the distance between the eyes and eyebrows of P is exaggerated in Q more strongly than the case of Q'. This fact shows that the PICASSO system must have a channel to choose the mean face and that the Internet would easily provide an open environment to PICASSO system.

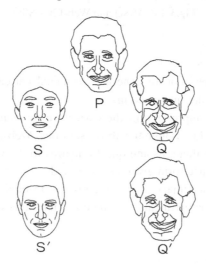

Fig.2. The possibility of the mean face hypothesis development (Provided by Tokyo Shinbun)

3.   Outline of Web-PICASSO System

3.1   Implementation

Web-PICASSO system was implemented on the WWW Internet Home Page of our Research laboratory: http://www.koshi-lab.sccs.chukyo-u.ac.jp/~fuji/pica2/ Web-PICASSO was described by Java, because this language does not depend on the computer environments of the visitors. Fig. 3 shows the relationship between PICASSO and Web-PICASSO, and Fig.4 shows the Java applet of this system.

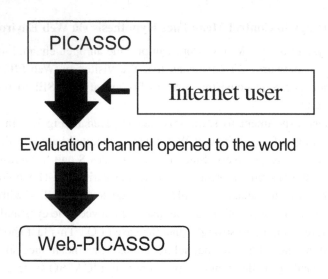

Fig.3. PICASSO and Web-PICASSO

In Web-PICASSO, the visitors
(1) can enjoy the operation to generate the facial caricatures by selecting the model whom they are interested in,
(2) can improve the impression of the caricature by changing the exaggeration rate. Simultaneously, through the above sessions, Web-PICASSO system
(3) can collect the evaluations and questionnaires to the works from the visitors by using the interface form given in Fig.5, and
(4) can acquire the face image data of the visitors through the Internet.
The details of these experiments and the considerations will be presented in the following sections.

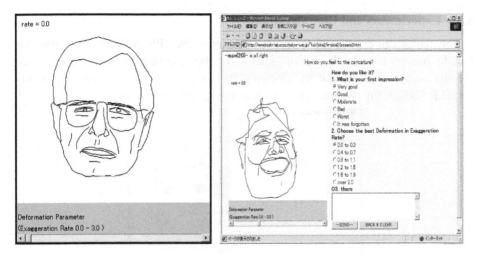

Fig.4. The Java applet of PICASSO   **Fig.5.** The example of the questionnaire form

3.2   Face Data Acquisition

As shown in Fig.6 Web-PICASSO presents an user interface to the visitors to ask them to provide their face image data through the Internet.   The provided face image data are stored in Web-PICASSO system together with the additional information on the age, race, sex, and so on. For example the face image data can be aquired through the page shown in Fig.6. This mechanism to acquire the face image data can enforce the input channel of the PICASSO system.

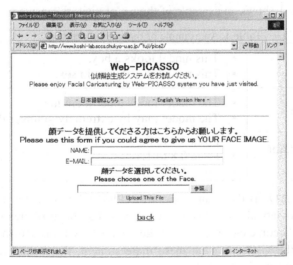

Fig.6. The front page of Web-PICASSO and the form for acquiring face image

3.3   Experiments and Considerations

The visitor can experience facial caricaturing in Web-PICASSO by selecting the face he prefers to from the list of facial icons, by selecting the mean face among the list of mean faces, and by changing the exaggeration rate according to his visual KANSEI. Fig. 7 shows the GUI which is given to the visitor.

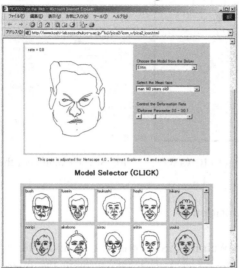

Fig.7 GUI for facial caricaturing in Web-PICASSO

In this experiment, the visitor is expected to answer the questions shown in Fig.5, and the answers are recorded and are taken into account to analyze the relation between the visitor and the model he selected from the viewpoints of the attributes of the face. Table 1 shows an example of the set of visitors' answers.

Table 1. An example of the visitor' answers

Sex	<ul><li>This model looks a boy.</li><li>This model is different in sex.</li></ul>
Age	<ul><li>The model looks older them me.</li><li>The model may by extremely younger than me.</li></ul>
Exaggeration rate	<ul><li>From 0.0 to 0.3.</li><li>From 0.8 to 1.1.</li><li>From 1.6 to 1.9.</li></ul>

In Web-PICASSO, as the channel from the PICASSO system to the visitor, the mean face is automatically selected in such a way that the attributes vector A(S) of the mean face S become closer to that of the visitor P, A(P) as shown in eq.(2). For example, when the visitor is a Japanese male student, the mean face made from Japanese male students is selected.

$$\min_j \; (\; A(P) \; - \; A(S_j) \;) \qquad\qquad (2)$$
, where $A(P) = (a_1, a_2, \; ..., \; a_k, ...)$ (P)
    is the attribute vector of the model P

Successively in Web-PICASSO, as the reverse channel from the PICASSO to the visitor, the visitor are requested to control the parameters of the facial caricaturing such as the selection of the mean face and the exaggeration rate.

From these bilateral procedure, the facial caricature will be improved in accordance with the respective visitor' visual KANSEI.   Fig. 8 shows an example of the successive change in the generated facial caricature.

(a) By using the 60th mean face   (b) By using the 40th mean face   (c) By using the 20th mean face

Fig.8. An example of the successive change in the generated caricatures

4   The Feedback Mechanism between PICASSO and Visitor for Caricature Improvement

In this chapter, we introduce a feedback mechanism to improve the facial caricature based on the interactive procedure between PICASSO and the respective visitor.

4.1   Feedback Procedure

The preliminary experiment given in the previous section can maintain the effectivity of the following feedback procedure to improve the facial caricature by refining the mean face adaptively to the interaction between the PICASSO system and the visitor. Fig. 9 shows the flow of this procedure.

```
Step 1   S_j'   :=   S_j   (initial value)
Step 2   E = min_j ( A(P) - A(S_j') )
Step 3   Terminate  if E < E_0.
Step 4   Revise the attribute vector A(S_j'')
         of the revised mean face S'' successively
         defined by the narrower set of face samples.
```

Step 4   S$_j$' := S$_j$'' , and go to Step 2.

In this procedure, the sample size of the faces for defining the revised mean face S'' is gradually reduced smaller than that of the S' in such a way that the value E in the procedure becomes gradually smaller.

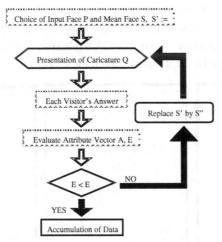

Fig.9. The Feedback Flow of Web-PICASSO

4.2   Considerations

The attribute vectors A(P) and A(S) were specified experimentally and some experiments are executed based on this feedback procedures. Fig. 10 shows an example of the process to improve the caricature. It was clarified that process would be easily controlled according to the quality of the selected mean face S.

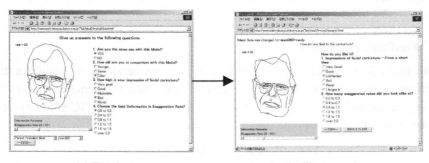

(a) The first page   (b) The second page
(The age of the mean face was increased for the better evaluation.)

Fig.10. An example of experiment by the feedback mechanism

5   Conclusion

This paper proposed a new facial caricaturing system PICASSO installed on the WWW Internet environment. The purpose of this system Web- PICASSO is to open the door to acquire the evaluations and the face image data from the world.

Basing on the preliminary experiments, we proposed a feedback mechanism to generate facial caricature basing on the interactive process between the system PICASSO and the visitor. Some experimental results demonstrated the effectively of the proposed mechanism.

As the coming subjects, it is expected to put the GUI be more friendly to the visitors, and to propose the more precise method to analyze the questionaires.

6   Acknowledgements

We would like to express many thanks for giving facial caricatures and helpful discussion to Mr. Suzuki of SKEN , Mr. You Nakagawa of Tokyo Shinbun , Mr. Kawamura of Yomiuri Shinbun and Ms. Kishibe of Magazine House. This paper was partially supported by the Ministry of Education, High-Tech Research Center Promotion, and IMS HUTOP Research Promotion.

7   References

[1] H. Koshimizu: "Computer Facial Caricaturing", Trans. The Institute of Image Information and Television Engineers, Vol.51, No.8, pp.1140-1146 (1997.8)

[2] K. Murakami, M. Takai and H. "Koshimizu; Based on a Model of Visual Illusion", 3rd Asian Conference on Computer, Proc. pp, 663-670, Hong Kong, January

[3] T, Fujiwara, M, Tominaga, K, Murakami and H, Koshimizu; "The caricature computer PICASSO inserted in the Internet", Forum faces, Proc. p-55, Tokyo, August

[4] S. E. Brennan: "Caricature Generation", Degree of Master of Science in Vision Studies as MIT (Sep.1982)

[5] M. Kaneko and O. Hasegawa: "Research Trend in Processing of Facial Images and Its Applications", Annual Meeting of Electronics/Information/System, IES. Japan (Aug.1999)

[6] M. Murakami, M. Tominaga and H. Koshimizu: "An Interactive Facial Caricaturing System Based on the Gaze Direction of Gallery", Proc.ICPR2000, (Sep.2000; Barcelona)

[7] Brennan, S.E.: "Caricature Generation", Degree of Master of Science in Visual Study (MIT) (Dec.1982)

[8]   For   example: "Deformed   Caricaturing   Gallery",
http://www.asahi-net.or.jp/kq2y-hmgc/index.htm

Recognition of Human Faces
Based on Fast Computation of
Circular Harmonic Components

Osei Adjei[1] and Alfred Vella[2]

[1] Faculty of Science, Technology and Design
Dept. of Computing and Information Systems
University of Luton
Park Square, Luton, England, LU1 3JU
osei.adjei@luton.ac.uk
[2] 194 Buckingham Road, Bletchley
Milton Keynes, England, MK3 5JB
alfred@thevellas.freeserve.co.uk

Abstract. This paper discusses facial recognition as applied to the classification of two-dimensional images and proposes a new architecture that allows a fast derivation of compact and invariant features from the Radon space. Our approach has been inspired by the review of feature-based non-connectionist and connectionist models of facial recognition. In the feature based non-connectionist model, a large part of the computational effort is focused on the extraction of facial features or the geometrical encoding of the face and the measurement of statistical parameters to describe their relationship. The connectionist model focuses on two-dimensional intensity values of the facial image allowing the geometrical encoding to be measured implicitly. The connectionist model is thus susceptible to variations in lighting conditions, spatial position and orientation of the images and can result in a poor detection of faces. An additional bottleneck of the connectionist model is the large feature vector size applied to its input that can cause non-convergence problems during training. The Radon transform is a generic transformation that is capable of representing shapes and it is used to compute harmonic components from which compact and invariant features can be derived. It is shown in this paper that these features when applied to a connectionist model result in a system that is capable of achieving high recognition rates and at high significance levels.

1 Introduction

Face recognition has received much attention in the last two decades due to the expected commercial benefits, particularly in the surveillance and security industries. The evidence of this is demonstrated in several papers on face recognition to date that have appeared in the literature. The survey by Samal et al. discussed automatic face identification systems that assume feature-based representation of the faces [6]. Feature extraction models identified in the paper

T. Tan, Y. Shi, and W. Gao (Eds.): ICMI 2000, LNCS 1948, pp. 160-167, 2000.
© Springer-Verlag Berlin Heidelberg 2000

included those that represented faces in terms of distances, angles, and areas between elementary features such as the eyes, nose, and chin, or in terms of mathematical functions such as moment invariants or auto-correlation, extracted from facial views or from the silhouette profiles. For such systems, facial recognition is achieved by computing the similarity between a face and pre-stored data held in a database. Many feature extraction methods used in facial recognition systems focus on finding individual features and measuring statistical parameters to describe their relationship; however, it is not an easy task to select features that capture all the information associated with a given face.

More recently, work on facial recognition has been directed towards the use of connectionist machines [8]. Connectionist approaches to pattern recognition are defined as those using computational algorithms that can be carried out in parallel and use distributed or non-localised mechanisms of memory storage. The main advantage of the connectionist architecture is that the information is directly derived from the statistical nature of the faces and hence the difficult problem of selecting individual features is avoided. A connectionist model is constructed by using a single layer network to map from an input space to an output space of the same dimension. Such a system possesses auto-associative memory and can be used to store and retrieve facial images. A facial image may be represented by a normalised column vector constructed by concatenating the rows of pixel values of the digitised face. Such a network can be constructed by feeding each component of the input vector to neurons connecting all other neurons with bi-directional synaptic weights between them. A set of faces is stored in the network by successfully applying each face vector and updating the weights according to a simple Hebbian learning rule. In this case, the weight matrix W between the two layers is updated using the outer product formula. Thus, assuming initial weights to be zero, weights are updated according to the formula:

$$W(n+1) = W(n) + xx^T. \tag{1}$$

This simplifies to:

$$W = \sum_i x_i x_i^T. \tag{2}$$

The matrix multiplication of x_i and W can be used to retrieve the i-th face stored in the association using the formula:

$$\nu_i = W x_i \tag{3}$$

where ν_i represents the response of the memory or the estimation of the i-th face by the memory. The cosine of the angle between ν_i and x_i is a measure of the quality of facial image reconstructed. A cosine value of unity is indicative of a perfect reconstruction.

In general, if the input vectors stored in the memory are close and not orthogonal, some cross-talk or noise will cause the desired response to distort. Recognition may be improved by using the delta rule to update the weights.

Thus, where α is a small learning constant, the weight update becomes:

$$W(n+1) = W(n) + \alpha(x_i - W(n)x_i)x_i^T. \tag{4}$$

The Delta rule is prone to long training times or failures due to the likely presence of a local minimum in the error function. This work addresses these disadvantages by using a new architecture, which combines the normalisation of input images with data reduction. Our technique allows facial features to be extracted from harmonic components. The new architecture comprises of the following stages:

1. An image acquisition stage that is used to capture facial images. No special lighting conditions are provided except for the standard overhead lights.

2. A pre-processing stage that is primarily used to improve the quality of the acquired image by removing noisy artefacts derived from the image acquisition stage. Additional functions of the section include edge detection and computation of translation and scale invariance.

3. A transformation stage that computes rotation invariant features by exploiting the theory of Radon transform to compute harmonic components of facial images. The technique has the advantage that any noise introduced into the recognition system is reduced by the integrative properties associated with the computation of the Radon transform.

4. A classification stage based on the back propagation paradigm.

Our approach has resulted in a facial recognition system that is able to recognise human faces in real time, but at modest computational costs.

2   The Fourier Slice Theorem

The Fourier Slice Theorem is an important property of the Radon transform [2] and it is the basis for the extraction of rotation invariant features in this work. The theorem states that if $F(u,v)$ represents the Fourier transform of a two-dimensional function $f(x,y)$ then the Radon transform $R(p,\theta)$ of the function $f(x,y)$ at an angle θ is equivalent to the inverse Fourier transform of points along a slice through the centre of the $F(u,v)$ plane, and making an angle θ to the u-axis. In operator notation, it is given by:

$$F_2 f(x,y) = F_1 R_2 f(x,y) \tag{5}$$

where F_2 is the two-dimensional Fourier transform, F_1 is the one dimensional Fourier transform and R_2 is the two-dimensional Radon transform. There is an implicit assumption that the Fourier transform and the Radon transform are related by an interpolation between the rectangular and polar co-ordinates as indicated by the following equations:

$$F_2 f(x,y) = F(u,v) \tag{6}$$

$$F_1 R(p,\theta) = F_r(w,\theta) \tag{7}$$

where $w = (u^2 + v^2)^{1/2}$ and $\theta = tan^{-1}(\frac{v}{u})$. Equation (7) indicates that the 1-D Fourier transform of the projection of an object is the same as the 2-D Fourier transform of that object at the same angle as the projection. Hence, computing the Fourier transform of an object's projection results in its harmonic components.

3   Invariant Feature Extraction

The pre-processing stage not only was responsible for noise reduction but also for edge detecting facial images and normalising them with respect to scaling and translation. Rotation invariance was accomplished by extracting features from images transformed into the Radon space using the image rotation and projection technique described in [1].

3.1   Translation Invariance

The simplest way to normalise an object of interest in an image with respect to its position is to move the object's centroid to the centre of the image matrix. The co-ordinates of the object's centroid (x_c, y_c) within the existing image matrix is computed as:

$$x_c = \frac{M_{10}}{M_{00}}, \tag{8}$$

$$y_c = \frac{M_{01}}{M_{00}} \tag{9}$$

where M_{00} is the zero-th order moment. M_{10} and M_{01} are first order moments. The axes are then re-labelled to centre the object at point (x_c, y_c), to produce the transformed image. The original image function $f(x, y)$ becomes a new image function:

$$g(x, y) = f(x + x_c, y + y_c). \tag{10}$$

3.2   Scale Invariance

Scale invariance [5] was achieved by scaling the image up or down such that its zero-th moment is set equal to a predetermined value β. This was achieved by transforming the original image function $f(x, y)$ to a new function $f(\frac{x}{a}, \frac{y}{a})$ where

$$a = (\frac{\beta}{M_{00}})^{1/2}. \tag{11}$$

Note that since β can assume any value, equation (11) can be used for both reduction and enlargement. Thus, by keeping β constant for all images, the same size images are applied to the feature extraction stage.

3.3 Rotation Invariance

The theory of the Radon transform shows that the rotation of an object by an angle θ_0 in the spatial domain causes a linear shift of θ_0 in the θ-axis of the p-θ Radon space. The relationship between p and θ can be formulated simply as a mapping from the spatial domain to the Radon space and given by:

$$f(r, \theta + \theta_0) \Rightarrow R(p, \theta + \theta_0). \tag{12}$$

Equation (12) demonstrates that a linear shift in θ does not cause any changes in the magnitude of the projection data. This property together with the Fourier slice theorem holds the key to the computation of rotation invariant features in this work. From the Fourier slice theorem, each sample of 1-D power spectrum represents a sample of the 2-D power spectrum at the appropriate angle and radius. To obtain the 2-D power spectrum of the object, the square modulus $u(n)$ of the 1-D Fourier transform of the projection is computed as:

$$u(n) = [|a(n)|^2 + |b(n)|^2]^{\frac{1}{2}} \tag{13}$$

where $a(n)$ and $b(n)$ are the real and imaginary parts of the Fourier transform and n is an integer. Rotation invariant features can be obtained from the corresponding elements of the arrays containing the results of the Fourier transform of the projections. To prove this, consider an object represented in polar form as $f_p(r, \theta)$, then it may be composed into circular harmonic components in the manner of references [4],[7] as:

$$f_p(r, \theta) = \sum_{M=0}^{\infty} f_M(r) e^{jM\theta} \tag{14}$$

where

$$f_M(r) = \frac{1}{2\pi} \int_0^{2\pi} f_p(r, \theta) e^{-jM\theta} \, d\theta. \tag{15}$$

The same object rotated by angle θ_0 can be expressed as:

$$f_p(r, \theta + \theta_0) = \sum_{M=0}^{\infty} f_M(r) e^{jM(\theta + \theta_0)}. \tag{16}$$

Equation (16) reduces to equation (17) after combining with equation (14):

$$f_p(r, \theta + \theta_0) = f_p(r, \theta) e^{jM\theta_0}. \tag{17}$$

The magnitude of the exponential part of equation (17) is equal to one, hence the magnitude of Fourier coefficients is not affected by rotation if the object is expressed in polar form. We conclude that rotation affects only the phase angle of the Fourier coefficients, therefore rotation invariant features can be derived from the power spectrum. Since translation and scale are firstly normalised, harmonic components computed in the transformation stage constitute features invariant to translation, scale and rotation. An additional advantage of the technique is that, features extracted are compact provided only components below the Nyquist frequency are considered.

4 Classification

The design of the classification stage can be formulated by exploring neural networks' strength in non-linear, adaptive and parallel processing. Mathematically, classification may be formulated as a two-phase operation. The first is the training phase that may be stated as: Given a system's inputs x_i with corresponding target symbols t_i, find a weight vector W_i such that the system's output $z_i = t_i$. The second is the retrieving phase that may be stated as: Given a system's inputs x_i and weight vectors W_i, determine the symbol t_i. The back propagation paradigm fulfils this formulation. Hence, the design of the classification stage was divided into two sections; training mode and the retrieving mode.

4.1 Determination of the Network Structure

The network structure was based on a three-layer architecture. The number of harmonic components computed from the transformation stage and the number of class labels in the output layer fixed both populations of the input and output layer neurons respectively. To determine the population of hidden layer neurons the relationship that connects the size of the input vector I to the number of separable regions M (i.e. classes) required and the number of hidden layer neurons J. The relationship is given in reference [10] as:

$$M(J, I) = \sum_{k=0}^{I} \binom{I}{k} \tag{18}$$

where for $\binom{I}{k} = 0$ for $J < k$). Thus for a large input vector where $I \leq J$:

$$M = \binom{I}{0} + \binom{I}{1} + \binom{I}{2} + \cdots + \binom{I}{I} = 2^J \tag{19}$$

$$J = log_2 M. \tag{20}$$

Although the above formula strictly applies only to linearly separable regions, it can be used as a guideline to compute the required number of neurons in the case of non-linearly separable problems.

5 Experimental Study

The network's ability to classify facial images is demonstrated only by the percentage of faces that network is able to recognise under noisy and noiseless conditions. Thus, the recognition rate R_c for our experiments was defined as:

$$R_s = \frac{R_c}{R_t} . 100 \tag{21}$$

where R_c is the number of faces recognised and R_t is the total number of faces to be classified. It is interesting to note also that, in all experiments concerning noise tests, the system was trained with noiseless facial images but tested with noisy ones. The controlled noise injected into the test samples was characterised by a probability distribution function [9]. The noise distribution was uncorrelated Gaussian with a zero mean that was derived from the Rayleigh distribution. The noise source used for our experiments was the additive type but the variance was used to control the amount of noise.

6   Hypothesis Tests

A method of ensuring that the performance of the proposed system is acceptable is to compare the expected frequency distribution of the categories against those observed. The observed frequency distribution is measured from the proposed system's output by applying test samples to the input and counting the number of times a particular category occurs. The chi-squared test is thus used to determine whether a set of output categories of the proposed system when compared with an expected distribution, yields a variance from probability or a pre-defined expectation than would be expected by chance alone. The chi-square distribution [3] is defined as:

$$\chi^2 = \sum_i^n (O_i - E_i)^2 / E_i \tag{22}$$

where E_i is the expected frequency of occurrence of particular class, O_i is the frequency of observations for that class and n is the number of categories. A hypothesis test is conducted by first stating a null hypothesis and computing a test statistic using equation (22). The observed value of the test statistic is compared to the critical value and a decision is made whether to reject the null hypothesis or not. For this work, the null hypothesis is that the frequency of the observed classes is the same as the frequency of the expected or theoretical classes. To prove the null hypothesis, test samples are applied to the proposed system. The frequency of occurrence of the output patterns is recorded and the chi-squared values computed. For each chi-squared value computed the critical chi-squared value is deduced from a χ^2-table. The tests show that the hypothesis of no difference between the expected and observed distributions are sustained at the significance levels as indicated in Table 1 for test samples applied at the input.

6.1   Discussion of Results

Results of hypothesis tests conducted on 10 classes of facial images demonstrate that within the constraints of the experiments, the observed distribution of classes is similar to those of the expected. These results show that high significance levels between 0.78.25-0.95.5 are achievable if 30-70 percent of the available samples are used for training the network.

Table 1. Results of hypothesis tests showing percentage of training samples (T_s) against recognition rate (R_s), chi-squared value (χ^2), critcal chi-squared value $(\chi^2_{c.v})$ and significance (Sig).

T_s	R_s	χ^2	$\chi^2_{c.v}$	Sig
20	47.60	46.85	27.877	0.001
30	78.25	23.43	27.877	0.001
50	91.23	5.48	6.393	0.700
70	94.50	3.25	3.325	0.950

7   Conclusion

Results from our experiments have confirmed that high recognition rates can be achieved from very poor quality images and thus demonstrating the robustness of our feature extraction technique. Features generated are global and compact allowing easy feature extraction and a faster system training. This makes it possible to adapt the new technique to resolve a number of other classification problems in different applications. The new technique is simple to implement and does not rely on any other feature extraction method for support.

References

1. Adjei, O., Mrozek, Z.: Robot Vision: Real Time Identification of Simulated Machine Parts Using Features Modelled From the Radon Space. Fourth Beijing International Conference on Systems Simulation and Scientific Computing, China, (1999).
2. Deans, S.R: The Radon Transform and Some of its Applications. Wiley-Interscience, New York (1983).
3. Eberhart, R.C., Dobbins, R.W., Hutton, L.V.: "Performance Metrics" Neural Network PC Tools (ed. Eberhart and Dobbins). Academic Press Inc, New York (1990).
4. Hsu, Yuan-Neng, Arsenault, H.H., April, G.: Rotation Invariant Digital Pattern Recognition Using Circular Harmonic Components. Applied Optics (1982).
5. Khotanzad, A., Lu, J.: Shape and Texture Recognition by a Neural Network. Artificial Neural Networks and Statistical Pattern Recognition (ed. Sethi, I.K and Jain, A.K), North-Holland, Amsterdam (1991).
6. Samal, A., Iyenga, P.A: Automatic Recognition and Analysis of Human Faces Facial Expressions: A survey. Pattern Recognition (1992) Vol. 25, No. 1, 65-77.
7. Ravichandran, G., Casasent, D.: Advanced In-phase Rotation Invariant Correlation Filters. IEEE Trans. on Pattern Anal. and Mach. Intell. (1994) Vol 16, 415-420.
8. Valentin, D., Abdi, H., O'Toole, A.J., Cotrell, G.W: Connectionist Model of Face Processing: A survey. Pattern Recognition (1994) Vol. 27, No.9.
9. Weeks, A.R., Myler, H.R., O'Toole, A.J., Weenas, H.G.: Computer Generated Noise Images for the Evaluation of Image Processing Algorithms. Optical Engineering (1993) Vol. 32, No. 5.
10. Zurada, J.M.: Artificial Neural Systems. West Publishing, New York (1992)

Multiple Faces Tracking and Zooming System

Mitsunori Ohya[1], Hitoshi Hongo[2], Kunihito Kato[1],
Caihua Wang[2], and Kazuhiko Yamamoto[1]

[1]Gifu University, Faculty of Engineering
1-1, Yanagido, Gifu Japan 501-1193
{ohya, kkato, yamamoto}@yam.info.gifu-u.ac.jp

[2]HOIP, Softopia Japan and JST
4-1-7, Kagano, Ogaki, Gifu Japan 503-0006
{hongo, c-wang}@softopia.pref.gifu.jp

Abstract. We propose a multi-camera system that can track multiple human faces and hands as well as focus on them for face and hand-sign recognition. Our current system consists of four video cameras. Two cameras are fixed and used as a stereo camera to estimate position. The stereo camera detects faces and hands using a standard skin color method we proposed. The distances of targets are then estimated. To track multiple targets, we evaluated position and size of targets in consecutive frames. The other two cameras perform target tracking. Our system selects a target for recognition by using size and motion information sequentially. If size of the selected target is too small for recognition, tracking cameras acquire its zoomed image. Using our system, we experimented on multiple target tracking.

1 Introduction

In order to make computers interact with humans in the form of human communication, it is important to focus on human intentions as indicated by facial expression, eye direction and so on. In practice, using a vision-based system should allow us to solve many problems such as occlusion, getting image resolution necessary to achieve a particular aim, and dealing with multiple faces and hands.

In recent years, various methods based on computer vision technologies to improve man-machine interfaces and security systems have been developed [1][2][3]. The research projects achieved good results. However, on some face and hand recognition systems and eye and mouth detection systems, the results depend on the image size. We already proposed a multiple camera system that can track multiple targets independent of their position [4]. This system had one fixed camera and two tracking cameras. However, since we tracked multiple skin regions using skin region's center of gravity and size, sometimes we could not track them when people overlapped. In this paper, we propose a multiple camera system that can track multiple skin regions using spatial position and size. In order to estimate the position, our system uses stereo vision. Our system can control the pan, tilt angles and zoom ratio of the cameras to acquire an appropriately

T. Tan, Y. Shi, and W. Gao (Eds.): ICMI 2000, LNCS 1948, pp. 168-175, 2000.

sized image. In this paper, we describe our system configuration, and algorithm for detecting and tracking multiple skin regions. Experiments showed that our system could track multiple skin regions.

2 Skin Region Detection System

Our system has four video cameras and acquires images that have appropriate sizes for recognition. In this section, we give an overview of our system.

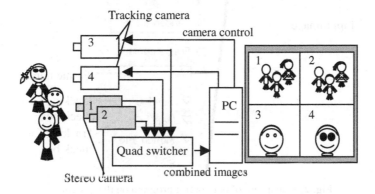

Fig. 1. System configuration

2.1 Overview of the System

Figure 1 shows the configuration of our system. Our previous system used the EWS (O2, SGI) [4]. Now, we have implemented our method on a standard PC (Windows98, PentiumIII 700MHz). Our new system uses video cameras (EVI-G20, Sony) whose pan, tilt angles and zoom ratio can be controlled by the computer. Two of the cameras are set in parallel as a stereo camera. The stereo camera detects skin faces and hands and measures their respective distances from stereo camera. The other two cameras are used for tracking and zooming in on the detected regions. All of images acquired by stereo camera and tracking cameras are combined using a quad switcher (YS-Q430, Sony), and the combined image is input into the PC. Figure 2 shows a snapshot of our system control program. We set the Tracking Camera 1 check box and Tracking Camera 2 check box to use these cameras.

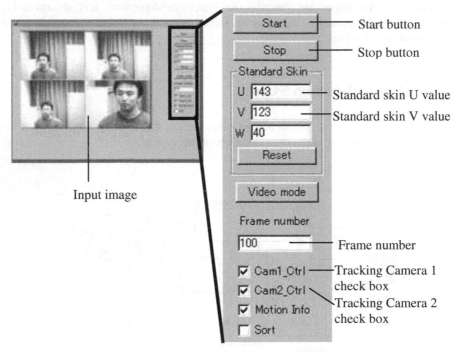

Fig. 2. A snapshot of our control program on the system

In this section, we explain our system flow chart. First, our system detects skin regions (for Asian subjects) using color information from input stereo camera image. Then, our system estimates skin region's spatial position using stereo view, and it tracks skin regions using their position and size. If the target does not have an appropriate size for recognition, tracking cameras zoom in on the target, and then our system detects skin regions using color information from the inputted tracking camera. If detected region is still not an appropriate size for recognition, tracking cameras track same region until it has an appropriate size.

2.2 Skin Color Detection

Various methods of face detection using color information have been developed [5][6]. Typically, such methods used hue and chromatic values in HSV color system. However, these methods do not work well if face objects are not bright enough. LUV color system consists of lightness (L) and color values (U and V). In general, LUV color system is as capable of representing the color distance between any two colors as humans. We have already proposed a method of face detection using U and V values in LUV color system [4]. Our method to detect skin color regions is described as follows.

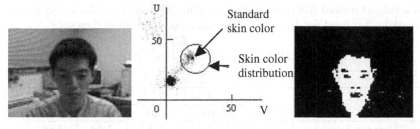

(a) Original image   (b) Color distribution   (c) Detected skin color region

Fig. 3. Face detection

First, a two-dimensional UV values histogram is made from the previous frame. From the two-dimensional histogram, we determine the standard skin color that denotes maximum number of pixels within the range of skin colors. Second, the UV value of each pixel in input image is converted to the color deviation from the standard skin color. Figure 3 (a) is the original image, and Figure 3 (b) shows standard skin color and color distribution of original image. Finally, we make a histogram of color distance from above results and then extract the skin region by discriminant analysis. Figure 3 (c) shows the results of the detected region for the facial image in Figure 3 (a).

2.3 Stereo Matching

To estimate spatial position of detected regions, we use stereo vision. Our system consists of a parallel stereo camera setup, where the disparity of two cameras is 100 mm.

As mentioned above, skin color regions are detected by stereo camera. Furthermore, we perform object matching on detected regions from left and right cameras. If multiple candidates exist for matching, we need to resolve the combination of regions. Since parallel stereo has optical axes in parallel, the epipolar line becomes horizontal. Regarding the relationship and disparity of stereo camera and the distance between humans and our system, we assume that targets cannot change their location within stereo images because the distance between two cameras is small. Therefore, we use positions of detected region's centers of gravity to match up targets as follows:

    (1) Regions with a same epipolar lines
    (2) Regions of a same size
    (3) Regions on a same half-line, with no switching of right and left

First, to examine how precisely our system measures spatial position, we experimentally estimated the spatial position of a square marker and subject's face. We took 50 images respectively, in which the marker was in front of stereo camera at distances of 1, 1.5, 2, 2.5, and 3 m for data set 1. The marker was detected by subtracting the background. We took 50 images respectively, in which a subject stood for data set 2 in the same experimental conditions as data set 1. We took 50 images respectively, in which a subject moved from side to side for data set 3, and in which a subject stood and shook his/her head for data set 4 in the same environment as data set 2.

Table 1. Stereo matching results (m)

Distance from the camera (m)		1	1.5	2	2.5	3
Data set 1	average	0.94	1.45	2.04	2.58	3.08
mark	stdv	0.00	0.00	0.02	0.03	0.05
Data set 2	average	1.04	1.57	2.03	2.65	3.08
face (stay)	stdv	0.02	0.03	0.11	0.14	0.21
Data set 3	average	1.02	1.56	2.04	2.56	3.23
face (move)	stdv	0.03	0.08	0.12	0.23	0.36
Data set 4	average	1.00	1.56	2.06	2.63	3.40
face (shake his/her head)	stdv	0.02	0.07	0.11	0.20	0.38

Table 1 shows average distance from stereo camera and standard deviation (stdv) for data sets 1, 2, 3 and 4 respectively. Our system could accurately estimate distance of the mark when the mark was within 3 m of the camera. Our system also could correctly estimate distance of a subject within 2 m. When a subject moved and shook his/her head, face region's center of gravity moved, but our system could estimates distance of detected face.

2.4 Tracking Camera Control

Tracking cameras are used to acquire zoomed image. We set left camera as the base point, and it bases the tracking cameras on. Our system directs tracking cameras based on target detected by left camera. The zoom rate of the camera is determined by size of target. We categorized zooming rates into five levels. In practice, tracking camera system exhibits some delay. To reduce the delay in camera control, our system has two tracking cameras to track independent targets at the same time. In addition, to collect target images efficiently, our system controls tracking cameras as follows:

(1) Decide the tracking priority of the detected targets based on motion and target size.
(2) Select the camera whose angle is aligned most closely with the target.
(3) Direct the selected camera to the target.

In this work, we based the tracking priority on motion and size. Detected target images are acquired in order. After acquiring the target image, its priority is set to the lowest level. Figure 4 shows an example image in which our system detected two faces and zoomed in on them and an example image in which our system detected a people's face and hand and zoomed in on them.

Fig. 4. Example of images taken by our system

3   Multiple-target Tracking

In this section, we explain our method that tracks multiple targets and show the experimental results.

3.1   Multiple-skin Regions Tracking Method

Multiple skin regions are tracked over continuous frames by finding the closest match for each region in next frame based on the consistency of the features such as position and visual features. If multiple candidates exist for matching, all combinations of skin regions are tested. Then we select the minimum value of E_t in equation (1) as the best match, where P_t is skin region's position, S_t is size of skin region at time t, and w_p and w_s are the weight coefficients of the corresponding features.

$$E_t = w_p \sum \sqrt{(P_t - P_{t-1})^2} + w_s \sum \sqrt{(S_t - S_{t-1})^2} \qquad (1)$$

To reduce the computational cost, we exclude any combination of regions whose sum of the distances is greater than a certain threshold. The threshold was determined by the previous experiment.

3.2   Multiple-target Tracking Experiment 1

Using our system, we experimented on tracking multiple faces [4]. As shown in Table 2, we took image data under five situations, each having a different number of subjects and incidents of overlapping. We gathered three data sets with different subjects in each situation. Each data set consists of 60 frames. The sampling rate was 5 frames/sec. Each captured image was digitized to a matrix of 640 x 480 pixels with 24-bit color. Table 3 shows the accuracy rate of multiple-face tracking.

From the results, our system achieved an accuracy rate of 100% under situations 1, 2 and 3. However, in the case of situations 4 and 5, the accuracy rate decreased because the images of two subjects overlapped each other.

Table 2. Experimental conditions

Situation	1	2	3	4	5
Number of subject	1	2	3	2	3
Number overlapping	0	0	0	1	2

Table 3. Accuracy rate in multiple-face tracking experiment (%)

Situation	1	2	3	4	5
Data set 1	100	100	100	96.7	92.2
Data set 2	100	100	100	95.8	95.0
Data set 3	100	100	100	95.8	95.0

3.3 Multiple-target Tracking Experiment 2

Using our system, we experimented on tracking a subject's face and right hand. We took image data of the subject holding his right arm straight out in front of him, moving it out 90 degrees to the right, then up 90 degrees, then back to the starting point. During the experiment, he kept his right arm straight. Each data set consists of 40 frames.

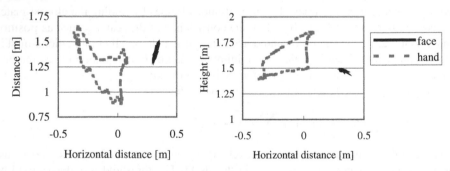

Fig. 5. Tracking results

3.4 Experimental Results and Discussion

Figure 5 shows tracking results of subject's face and right hand. He stood at a position of 1.5 m from the camera and 0.3 m left of center. His height is about 1.7 m. The length from his hand to his shoulder is about 0.6 m. We were able to examine the movable range of his hand.

4   Conclusions

We have proposed a system that can track multiple skin regions using multiple cameras to focus on detected region. We are able to acquire zoomed images. Therefore, these images may be use for face and hand-sign recognition and facial component detection. Our current system consists of a stereo camera and two tracking cameras. Stereo camera detects skin regions using the standard skin color method and estimates their spatial position. Tracking cameras zoom in on images to obtain a appropriate size of the regions.

Experiments showed that our system could estimate skin region's spatial position using standard skin color method and track multiple regions using position and size of them. Therefore we could acquire zoomed images.

Our current system takes approximately 0.4 sec per frame to complete all processes, including stereo matching and tracking camera control. We need to improve the speed. We can do this by using hardware to convert RGB to LUV color and process the Gaussian filters. Our next task will be to improve accuracy rate of the detected region's position.

References

1. S.Morishima, T.Ishikawa and D.Terzopoulos: "Facial Muscle Parameter Decision from 2D Frontal Image", Proc. ICPR'98, pp.160-162 (1998)
2. Jin Liu: "Determination of Point of Fixation in a Head–Fixed Coordinate System", Proc. ICPR'98, pp.501-504 (1998)
3. T.Shigemura, M.Murayama, H.Hongo, K.Kato and K.Yamamoto: "Estimating the Face Direction for the Human Interface", Proc. Vision Interface'98, pp.339-345 (1998)
4. M.Ohya, H.Hongo, K.Kato and K.Yamamoto: "Face Detection System by Using Color and Motion Information" Proc. ACCV2000 pp.717-722 (2000)
5. K.Sobottka and I.Pitas: "Extraction of Facial Regions and Features Using Color and Shape Information", Proc. IAPR 96, Vol.3, pp.421-425 (1996)
6. G.Xu and T.Sugimoto: "A Software-Based System for Realtime Face Detection and Tracking Using Pan-Tilt-Zoom Controllable Camera", Proc. ICPR'98 , pp.1194-1197 (1998)

Face Warping Using a Single Image

Yan Bai, Denis Cardon, and Robert Mariani

RWCP *, KRDL Multi-Modal Function Laboratory
Kent Ridge Digital Laboratory
21, Heng Mui Keng Terrace, Kent Ridge, Singapore 119597
{baiyan, denis, rmariani}@krdl.org.sg

Abstract. Given an image of a face and the co-ordinates of the eyeballs' centre, we provide a method which creates the corresponding frontal face and gives a qualitative and quantitative measure of the horizontal tilt of the face, using the natural symmetry of the human head. By multiscale dynamic programming, we obtain the optical flow (2D warp) which transforms the input face into its mirror image, this mirror image being a good approximation of the same head that has undergone a 3D rotation. The frontal face is obtained by applying half of the magnitude of the optical flow, and any pose is obtained by applying a fraction of this optical flow. The nose axis, which gives a quantitative and qualitative tilt estimation, is obtained by the reverse optical flow applied on the symmetrical axis of the generated frontal face. Experimental results on 500 images and on the Surrey database confirm the performance of this system.

1 Introduction

In many face recognition or facial expression recognition systems the face is assumed to be in a frontal view, like a mug-shot photo [1]. An important challenge for face recognition is to recognise face under varying poses [2]. There are two streams of research in this field. The first one tries to transform off-frontal views into frontal views by image warping [3, 4] and applies well-known frontal face recognition algorithms. The second one tries to identify the pose of the face in order to use a specific recogniser trained for this pose [5].

Systems for 3D pose estimation using a single image has been proposed combining skin and hair regions information [6], and 2D face pose discrimination between the left, right and frontal pose has been proposed using the support vector machine [5]. The first one provides quantitative and qualitative measure of the head pose, while the second one provides qualitative face pose estimation.

In this paper, we propose a robust method that transforms a tilted face into its frontal version and gives a qualitative and quantitative measure of the horizontal tilt of the head. Potential applications of this system are frontal face recognition system using the generated frontal face, automatic frontal face selection, pose-dependent face database indexing and face recognition using the measure of the face pose. . .

---

* Real World Computing Partnership

T. Tan, Y. Shi, and W. Gao (Eds.): ICMI 2000, LNCS 1948, pp. 176-183, 2000.

In the second section, we give our general methodology; section 3 details the line by line optical flow computation and its limitation, section 4 describes the use of multi-band approach computation and its limitations, then section 5 shows how to use face shape a priori knowledge to constrain the warp through supervised learning, finally section 6 gives the results and performances of this algorithm, before to conclude and to give some future perspectives to this work.

2   General Methodology

Given a face image and the co-ordinates of the two eyeballs' centres, horizontally aligned, our aim is both to create a frontal view and to obtain a qualitative and quantitative measure of the horizontal tilt of the head. The synthesis of the frontal face is done by computing the optical flow between the face and its mirror image, this mirror image being a good approximation of the same head that has undergone a 3D rotation. By construction, if we apply half of the optical flow to the face, we obtain the frontal face, and by computing the reverse optical flow, we obtain the nose axis.

In our system, the input faces are normalised in scale, rotation and translation, by using an automatic eyes detection and setting the two eyes at fixed co-ordinates. If this operation is successful, that is if the normalised image has the two eyes horizontally aligned, then the estimation of the optical flow between the input image and its mirrored version is done line by line. However, in order to be robust to the mislocation of the eyes and to larger horizontal tilt, we have extended this line by line process into a multiple overlapping bands algorithm, which produces frontal faces. Still, the overlapping band algorithm is not robust enough to large tilt, so we designed a constrained warping which deals with the reverse problem, that is, given a face and a set of optical flow, to try to find the correct one.

Since the two eyeballs have been horizontally aligned, the symmetrical plane of the head is in vertical position. Thus the optical flow between the input image and the mirror image will be considered as horizontal. That is, for every pixel (i, j) in the first image, we have to compute the translation vector $\overrightarrow{t} = (t_i, t_j)$ to be applied to that pixel in order to obtain the mirror image.

The method used to compute the optical flow is dynamic programming. We will first explain the computation algorithm for one line and then explain in the next section how we take into account the computation for the whole image using multiscale approach.

3   Line by Line Optical Flow Estimation

Let $L = \{P_1, \dots, P_N\}$, and $M = \{Q_1, \dots, Q_N\}$ the ith line of the input image and mirrored image respectively composed of N pixels P_i and Q_j. We compute by dynamic programming, the optimal operation that transforms L into M, by maximising the grey level correlation and minimising the overall stretching.

Let $H_i = \{(P_i, Q) : Q \in M : |x(P_i) - x(Q)| < \delta\}$, the set of corresponding pixels hypothesis (P_i, Q_j), associated to the pixel P_i, where $x(P_i)$ is the abscissa of the pixel P_i in the line L and δ is a fixed threshold. We associate to the set $H = \{H_1, \ldots, H_N\}$ a directed and weighted graph $G = (H, T)$, where $T = \{T_1, \ldots, T_{N-1}\}$ is the set of the transitions between two following hypothesis, where $T_i = \{H_i \times H_{i+1}\}$. Here the transitions incorporate the geometrical compatibility between two hypothesis.

The weight associated to a hypothesis $H_{ij} = (P_i, Q_j)$ depends on the color correlation between the pixels P_i and Q_i

$$weight(H_{ij}) = \lambda |gl(P_i) - gl(Q_j)| . \tag{1}$$

where λ is a constant factor, and $gl(p)$ is the grey level of the pixel p.

The geometrical cost associated to the transition $t(H_{ij}, H_{i+1,k})$ is proportional to the amount of stretching or compression generated by the application of these two hypothesis

$$\begin{aligned} weight(t(H_{ij}, H_{i+1,k})) &= |x(P_i) - x(Q_j) - x(P_{i+1}) + x(Q_k)| \\ &= \mu| - 1 + x(Q_k) - x(Q_j)| . \end{aligned} \tag{2}$$

We reject $(weight = \infty)$ all the transitions for which $x(Q_k) - x(Q_j) < 0$, that is that two pixels can't cross over. This restriction doesn't exactly reflect reality since that when the head is very tilted, the nose may occlude some part of the face and some pixel are in fact "behind" the nose pixels. However, this restriction does not reduce the quality of the result and it prevents the algorithm from exploring many possibilities that would not be possible.

Using a shortest path algorithm we compute the best set of hypothesis which represents the local best flow for the line L between the input image and the mirrored image. Repeating this process for each line of the pictures allow us to obtain the optical flow, expressed as a matrix O of horizontal translation vectors

$$O(i, j) = x(Q_{ik}) - x(P_{ij}) = k - j . \tag{3}$$

where i is the line index, j is the column index in the first image, P_{ij} is the pixel of the input image situated at the coordinates (i, j) and Q_{ik} is the corresponding point in the mirror image situated at the coordinates (i, k), $x(P_{ij}) = j$ denoting now the column index of the image.

The parameter λ, which balances the contribution of the color correlation and the geometrical constraints, has been calculated through supervised learning. We obtained the best result for $\lambda = 15$. The parameter δ regulates the size of the hypothesis set H, and is arbitrarily fixed to half width of the image.

This line by line optical flow estimation only give us the local minima: the best match for the whole image is not the best match of its individual lines. With the following images, we illustrate this fact: input image, the warped frontal face and the mirror image. For most of the lines, the algorithm find the right match. But for a few line, we can figure out that the local match is not the correct one.

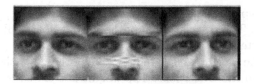

Fig. 1. Result of the line by line optical flow showing the problems of local minima

4   The multiple band approach

For solving the problem illustrated previously, we have developed a multiple bands algorithm. It simply integrates optical flow constraints across several adjacent lines, to encode the fact that two adjacent lines should be roughly subjected to the same warp. As the line by line process is already satisfying, the introduction of this multiple band process corrects the local errors.

To do so, we propose an iterative algorithm which starts with large bands and large translation hypothesis (δ). At each iteration, we reduce the width of the bands and the allowed translations, and we repeat this process until the band is a few pixel wide and the translation hypothesis are no more than a few pixels in magnitude. In order to strengthen the continuity of the optical flow constraints, the bands overlap, and equal contribution from the two bands are attributed to the mutual area.

We fixed the overlapping region equal to half the width of the band. So for a $N \times M$ pixels image and a band width of γ, the set of band will be

$$\{[1 \ldots \gamma], [\tfrac{1}{2}\gamma \ldots \tfrac{3}{2}\gamma], \ldots\} \,. \tag{4}$$

The algorithm to compute the warping using bands is the same as the one used before. We just change the set of hypothesis. We compute the optical flow $O^k = O$ after k iterations. For the first iteration, we don't make any assumption on the optical flow, so we have $O^0 = 0$.

Formally, given a band B of width γ, the average optical flow OB estimated at the iteration k for the band B is given by

$$OB^k(j) = \frac{1}{\gamma} \sum_{i=B_{start}}^{B_{start}+\gamma} O^{k-1}(i, j) \,. \tag{5}$$

This average optical flow, combined with the maximum of authorized translation δ^k allow us to compute a refined optical flow to be applied to the band B, by defining the new set of allowed translations for all the pixels in the band B

$$H(j) = \{OB^k(j) - \delta^k, \ldots, OB^k(j), \ldots, OB^k + \delta^k\} = \{H^1(j), \ldots, H^{2\delta+1}\} \,. \tag{6}$$

Here, for convenience of notations, we have described the set of hypothesis as a set of translation vectors instead of couple of corresponding points. Therefore,

$H(j)$ denotes the hypothesis set associated to all pixels $\{P_{B_{start},j}, \ldots, P_{B_{start}+\gamma,j}\}$ of the input image. The translation hypothesis that don't fall into the range of the image are rejected.

We modified the weight associated to a hypothesis, without changing the weight associated to the transitions. The weight associated to the hypothesis is generalized to the band as

$$weight(H^h(j)) = \frac{1}{N} \sum_{i=B_{start}}^{B_{start}+\gamma} |gl(P_{i,j}) - gl(P_{i,j} + tr(H^h(j)))| . \qquad (7)$$

where $gl(p)$ is the gray level of the pixel p and $tr(H)$ is the value of the translation hypothesis H.

The dynamic programming gives a new approximation of the optical flow for this band. We do this computation for all the set of band that we defined earlier. In the overlapping region, we just take the average of the warping of the overlapping bands. Through those computations, we obtain a new approximation of the overall optical flow. In our experiments, we fixed 4 iterations with the bandwidth γ and the allowed translations δ to $(10, 50), (5, 17), (3, 11), (1, 7)$ for 60×60 pixels images.

Here are some result of this method. The problems observed with the first algorithm are not observed any more. However for very tilted head, the most tilted part of the face is so small that it cannot be well match and the result warp is not satisfying. That is why we have implemented constrained warping.

Fig. 2. Results on surrey database. Left: Original images; Right: Warped images

5   Supervised Learning and Constrained Warping

In the preceding sections, we only made the assumption that the object to warp has a vertical symmetry plane. Even if those assumption should be enough to compute the optical flow, it appears that for very horizontally tilted faces (more than 25 degrees), the algorithm cannot match the pixels correctly. Then instead of computing the algorithm and then deducing its position and the corresponding optical flow, we took the inverse problem: We suppose that the head is in a certain position, we apply a pre-registered optical flow corresponding to this position, and then, we check whether the result looks like a frontal face or not, and repeat this process for a whole set of supposed positions.

For learning and testing, we use the KRDL database of tilted faces. It contains the pictures of 19 persons including men and women, Asians and Europeans, with 25 different positions for each person (up, down, right, left...). Fig. 3 give a sample of the 25 images for one person in the database. The tilt of the head varies from about −35 degrees to 35 degrees in horizontal tilt and −20 degrees to 20 degrees in vertical tilt (0 degree being the frontal face). We registered satisfying optical flows estimated from a subset of this database using the multiple band algorithm. Given an input image, we apply the different registered optical flows (our training set) to this image and select automatically the optical flow O^* that creates the most frontal face. Then, in order to take into account the little differences between faces, we have to improve the best optical flow O^* that has been automatically selected. For this purpose, we apply the multiple bands algorithm to this input image and its mirror image by using O^* as the initial optical flow (i.e. $O^0 = O^*$) and improve O^* by the iterative process, but with very small translations and band width.

The choice of the correct optical flow OF is based on three measures of quality: First the right-left correlation algorithm on the generated frontal face, and the two other correlations $corr(OF_1(II), MI)$ and $corr(OF_{\frac{1}{2}}(II), OF_{\frac{1}{2}}^{-1}(MI))$, with $OF_m(I)$ being the result of the application of the magnitude m of the optical flow OF on the input image I, $OF^{-1}(I)$ being the reverse function of $OF(I)$, e.g. $OF^{-1}(OF(I)) = I$, $corr(I_1, I_2)$ being the correlation between image I_1 and image I_2, II being the input image and MI being the mirror image.

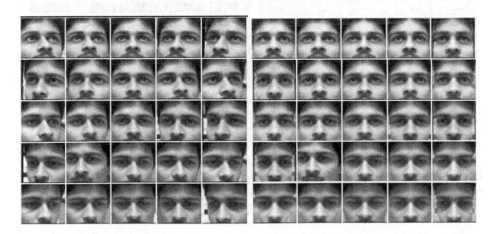

Fig. 3. Sample of KRDL database of tilted faces and of the corresponding warped faces. Please note that one of the image was not correctly normalized and has been rejected by the system (input image returned)

Each correlation has a independant threshold for accepting or rejecting the optical flow and the associated picture. This threshold allows us to reject faces that are too tilted along with faces that were not correctly normalized (Fig. 3

shows an example of reject). Those quality measures can also be applied on the line by line warping and multiple band warping methods.

Since this algorithm just requires to improve an existing optical flow and to reconstruct the frontal face, the computation of the frontal face is very fast. This allows us to have a large training set of optical flows. We have build a set of 30 optical flows that reflect all the cases observed earlier.

6   Results and Performances

The multiple band algorithm has been tested in the Surrey database with great results. There were only a few pictures that were not correctly warped. Using constrained warping the result was 100% correct warping.

Constrained warping has also very good result on KRDL database of tilted faces (475 pictures) with more than 97% of correct warping (Fig. 3 and Fig. 4). Within the last 3%, the algorithm whether does not select the best optical flow correctly or doesn't have a registered optical flow adapted to the image. Those problems arise for very tilted faces (35 degree or more).

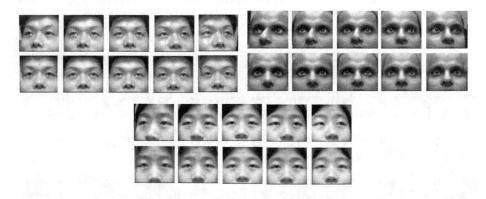

Fig. 4. Results on KRDL database. For each person, the top line is the original image and the bottom line is the result warped image

This algorithm gives also a very fast qualitative and some quantitative information on the left-right pose of the head through the direction of the nose line (Fig. 5). The nose line is obtained by applying half of the reverse optical flow on the symmetry axis of the frontal face. Using the notation of the previous section, that is $NL = OF_{\frac{1}{2}}^{-1}(SA))$, NL being the nose line and SA the symmetry axis of the frontal face. It is very robust and able to detect that a face is very tilted, even if its warped version is not the best one. In a face recognition context it is probably better to reject those faces.

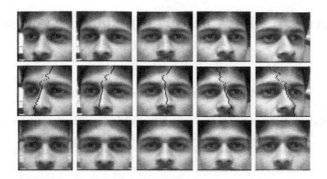

Fig. 5. Results on KRDL database. Top line: Original images; Middle line: Original images with nose axis; Bottom line: Warped images

7   Conclusion and Perspectives

We have proposed an algorithm which creates a frontal face and give an estimation of the horizontal tilt of the face. It uses the natural symmetry of the human face and an average model for the shape of the head. Currently, we are integrating this method within a frontal face recognition system and an automatic pose selection. Current works focus on the measure of the impact of this warping on frontal face recognition algorithms.

References

1. Chellapa, R. and Wilson, C.L and Sirohey, C.: Human and Machine Recognition of Face: A Survey. Proc. of IEEE (1995) **2** 705–740
2. Beymer, D.: Face Recognition Under Varying Pose. International Conference on Computer Vision and Pattern Recognition, Seattle, WA (1994) 756–761
3. Feng, G.C. and Yuen, P.C. and Lai, J.H.: A New Method of View Synthesis under Perspective Projection. The 2nd International Conference on Multi-Modal Interface, Hong-Kong (January 1999) IV-96–IV-101
4. Beymer, D. and Poggio, T.: Face Recognition From One Example View. 5th International Conference on Computer Vision, (1995) **1** 500-507
5. Huang, J. and Shao, X. and Wechsler, H.: Face Pose Discrimination Using Support Vector Machine (SVM). The 14th International Conference on Pattern Recognition, Brisbane, Australia, (August 1998) **2** 154–156
6. Chen, Q. and Wu, H. and Fukumoto, T. and Masahiko, Y.: 3D Head Pose Estimation Without Feature Tracking. International Conference on Automatic Face and Gesture Recognition, Nara, Japan (1998) 88–93

Hierarchical Framework for Facial Expression Recognition

*Ae-Kyung Yang, and Hyung-Il Choi

Vision Laboratory, School of Computing, Soongsil University, Seoul, 156-743, Korea
akyang@media.soongsil.ac.kr
hic@computing.soongsil.ac.kr

Abstract. In recent years, much work on the recognition of facial expression, and the recognition of face and the methods are various. The algorithm for recognizing facial expression includes various preprocessing and core scheme for the detection of facial area, the detection of facial components and the recognition of facial expression. We propose a framework to implement the recognition facial expression through the algorithm which is composed of many steps. The framework allows a substitution and reuse of a step of the algorithm. A step of the algorithm is able to use and update individually. And we also propose an efficient method for each step. First of all, we propose multi-resolution wavelet transform, 2d equilibrium state vector to search facial components at 3rd step. And we propose simple snake method to detect the feature point of a face.

1. Introduction

In recent years, much work on the recognition of facial expression, and the recognition of face and the methods are various. Nowadays, the schemes are applicable for the entrance management system, monitoring system, and web related application such as computer game, electronic trade etc. Especially the facial expression recognition can be grafted with computer graphics. There are many methods for facial expression recognition such as optical flow methods[2] and active model[1]. But existing methods does not have the integrated framework to manage many steps that are parts of the system, and it mean there is no alternative when the side effect occurs because it is substituted a step for another step. So we propose manageable and understandable framework.

We propose a hierarchical framework that can arrange steps used for the facial expression recognition. And the framework can also reduce the time for implementation by coarse-to-fine methods. Our research extracts the features needed for the facial expression recognition. So the hierarchical framework can make that each step is reusable, updated, and the whole system is easy and understandable.

T. Tan, Y. Shi, and W. Gao (Eds.): ICMI 2000, LNCS 1948, pp. 184-190, 2000.

2. Proposed Scheme

2.1 Outline of the Hierarchical Framework

The hierarchical framework is composed of 4 steps, and the structure is tree. From the root node to the leap node, it progresses by coarse-to-fine method. For each step, the extracted features are different each other, and the proposed algorithms are also different, so there is no overlapped technique. At root step, we extract the face area and then we calculate the orientation of face at 1st level step, then we extract the area of facial components using angle of face. At leap node step, we finally extract the feature points in each facial components' area.

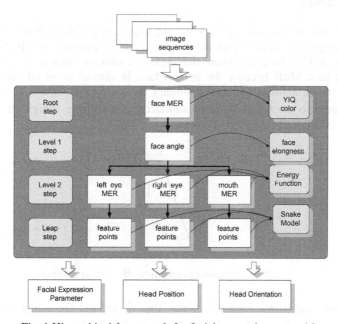

Fig. 1 Hierarchical framework for facial expression recognition

The inner structure of the framework is shown in Fig.1. The results for the implementation of the framework are the feature points, the location, and the orientation of face. We are applicable these results for the facial expression recognition, face recognition, and face tracking etc.

2.2 The Algorithm for Each Step

2.2.1 The Root Node Step

We extract the MER(Minimum Enclosed Rectangle) of face in this step. We use YIQ color space for extraction the MER of face[3]. Initially the skin area is extracted from the input frame, then the MER is extracted by labeling. In this case, the face area is larger than any other label, so the MER area is same largest label area. The extraction of face skin area is implemented by the Marhalanobis distance measure with skin-color model.

2.2.2 1st Level Step

In this step, the orientation of face is calculated from the MER of face. The orientation means the angle of curl and is calculated by elongness and BFD(Boundary Fitting Degree). The elongness means the angle of major axis that is same the angle of ellipse fitted face MER because the human face is similar to an ellipse intuitively. BFD means the similarity of a facial boundary and an ellipse. That is to say, after an ellipse is rotated so much that θ, if the similarity of an ellipse is high, then the angle of an ellipse is selected.

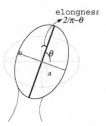

Fig.2 The elongness to calculate θ, face orientation

Fig.2 shows the ellipse rotated so much that θ and put the center on coordinate (x,y). In case of a face, when θ is half of π, the angle of face is zero, so we substitute θ for ($\pi/2 - \theta$).

$$\left(\frac{(x-a)\cos\left(\frac{\pi}{2}-\theta\right)+(y-b)\sin\left(\frac{\pi}{2}-\theta\right)}{a}\right)^2 + \left(\frac{(x-a)\sin\left(\frac{\pi}{2}-\theta\right)-(y-b)\cos\left(\frac{\pi}{2}-\theta\right)}{b}\right)^2 = 1 \tag{1}$$

Using Eq.(1), we can rotate the ellipse so much that θ and put the center on coordinate (x,y). The basis equation is just ellipse equation, however the angle θ is substituted for ($\pi/2 - \theta$) and center coordinate (x,y) is substitutied for (x-a,y-b). The angle ($\pi/2 - \theta$) means the elongness.

The setting of major axis 2b and minor axis 2a to fit with face area is based on fitting ellipse on the face MER at first frame. That is to say, the length major axis 2b is same the height of face MER and the length of minor axis is same the width of face MER. The initial ellipse rotates and compares the similarity by BFD. If the BFD is the maximum, at that time the angle θ is fixed as the orientation of face.

$$BFD = area(ellipse) \cap area(face) \qquad (2)$$

BFD means similarity of the boundary between a face and an ellipse. Using Eq.(2) we can calculate BFD and Eq.(2) means the common area of related ellipse and face area.

2.2.3 2nd Level Step

In this step, each facial component area is extracted by the energy function with texture matching method, and the search area is reduced by the property of 2d equilibrium state vector. Firstly, the texture matching is implemented by multi-resolution gabor wavelet function. The multi-resolution gabor wavelet function has merits that can detect the features that we can not find it because of high or low texture resolution. If the texture resolution is so high, then human being can't find the pattern because of noise, the other way if the texture resolution is so low, then human being recognizes the texture as an object. So we adapt various spatial resolution in the energy function. The equation for energy function is Eq.(3)

$$E_{model} = E_{le} + E_{re} + E_{mo} \qquad (3)$$

$$E_{le} = 1 - \frac{1}{n}\max_i\left(sim(B_i^{le}(\vec{x}))\right)$$

$$E_{re} = 1 - \frac{1}{n}\max_i\left(sim(B_i^{re}(\vec{x}))\right)$$

$$E_{mo} = 1 - \frac{1}{n}\max_i\left(sim(B_i^{mo}(\vec{x}))\right)$$

$$sim(B_i^{le}(\vec{x})) = \text{similarity of } B_i^{le}(\vec{x}) = \max(B_i^{le}(\vec{x}))$$

$$B_i^{le}(\vec{x}) = \frac{1}{N}\left(\max_i(J_j^i(\vec{x}))\right)$$

$$J_j^i(\vec{x}) = I_{res}(\vec{x}) \cdot \Psi_j^i(\vec{x} - \vec{x}') \cdot d^2\vec{x}'$$

$$\Psi_j(\vec{x}) = \frac{k_j^2}{\sigma^2}\exp\left(-\frac{k_j^2 x^2}{2\sigma^2}\right)\left[\exp(i\vec{k}_j\vec{x}) - \exp\left(-\frac{\sigma^2}{2}\right)\right]$$

In Eq.(3) E_{le} is the search space of left eye, E_{re} is for right eye, and E_{mo} is for mouth N means the number of find pixels, res means spatial resolution of image and the values are 2, 4, 16, 256. i means the search block index in each area. j means the feature index of gabor coefficients.

And 2d equilibrium state vector reduces the search space, because 2d equilibrium state vector has following properties. The sum of two vector A, B of horizontal direction is same as the vector C of vertical direction. Fig.3 is shown the 2d equilibrium state vector. The shape of face can be represented the vectors of three direction, if the center of face connects with both eyes and a mouth.

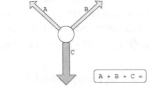

Fig.3 The 2d equilibrium state vector

2.2.4 Leap Node Step

Finally we find the feature points of facial components' area using simplified snake model. The snake model is basically composed of vector of point that constitutes spline, energy function, and optimization algorithm that minimize the energy function as following Eq.4

$$E_{snake}^* = \int_0^1 \left(E_{int}(v(s)) + E_{im}(v(s)) + E_{con}(v(s)) \right) ds \tag{4}$$

In Eq.(4), E_{int} means internal energy generated by bending and discontinuity, E_{im} means image force that is used to find the contour and the edge magnitude is used as image force, and E_{con} means constraint. The energy minimization algorithm guides that can find slow and continuous spline by minimized Eint. E_{int} is defined as Eq.(5)

$$E_{int} = \alpha(s)|v_s(s)|^2 + \beta(s)|v_{ss}(s)|^2 \tag{5}$$

However, our aim is to fine the bending point, so we update the Eq.5 as following

$$E_{snake}^* = 1 - \int_0^1 \left(E_{int}(v(s)) + E_{im}(v(s)) \right) ds \tag{6}$$

3. Experimental Results

We experiment for each step, and we take total 140 frames per behavior. The spatial resolution of a frmae is 320*240, and 32bit per pixel. A woman moves to the left and right slowly with happy and surprise expression. In this paper, the result per 70 frames is shown. We process three frames per second.

Fig.4 is shown as the results of root node and level2 node step. The MER of face and each facial component are shown.

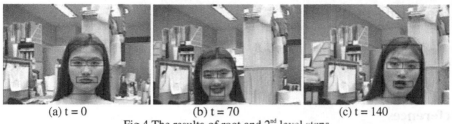

(a) t = 0   (b) t = 70   (c) t = 140
Fig.4 The results of root and 2nd level steps

Fig.5 is shown as the results of level1 node step, the cross mark means major axis and minor axis.

(a) t = 0   (b) t = 70   (c) t = 140
Fig.5 The results of 1st level step

Fig.6 is shown as the results of leap node step, the white dots means active contour and the black dots means the feature points.

(a) t = 0   (b) t = 70   (c) t = 140
Fig.5 The results of leap node step

4. Conclusion

We proposed the hierarchical framework that can integrate each algorithm of step for facial expression recognition. And each algorithm of step can be reused and updated individually. And the steps are understandable for coarse-to-fine design. We also proposed multi-resolution gabor wavelet function, search space reduction using 2d equilibrium state vector, and simplified snake model to find the feature points.

Acknowledgement

This work was partially supported by the KOSEF through the AITrc and BK21 program (E-0075)

References

1. Dae-Sik Jang and Hyung-Il Choi, "Moving Object Tracking with Active Models," Proceedings of the ISCA 7th International Conference on Intelligent Systems, July 1-2, 1998, p212-215
2. Irfan A. Essa and Alex P. Pentland, "Coding, Analysis, Interpretation, and Recognition of Facial Expression," IEEE Transactions on Pattern Analysis and Machine Intelligence, 1995.
3. I.J Ko and Hyung-Il Choi, "A Frame-based model for Hand Gesture recognition," Proc. IEEE Int. Conf. on Image Processing, Lausanne, 1996, Vol.3 pp515-518

Detecting Facial Features on Images with Multiple Faces

Zhenyun Peng[1], Linmi Tao[1], Guangyou Xu[1], and Hongjiang Zhang[2]

[1] Dept. of Computer Sci. & Tech., Tsinghua University, Beijing, 100084, China
{xgy-dcs@mail.tsinghua.edu.cn}
[2] Microsoft Research, China, 5F, Beijing Sigma Center, No.49, Zhichun Road, Haidian District, Beijing, 100080, China

Abstract. This paper presents an approach to detect facial features of multiple faces on complex background and with variable poses, illumination conditions, expressions, ages, image sizes, etc. First, the skin parts of the input image are extracted by color segmentation. Then, the candidate face regions are estimated by grouping the skin parts. Within each candidate face region, an attempt is made to find the eye pair using both Hough Transform and the Principal Component Analysis (PCA) method. If the predicted eye pair is, under the measurement of correlation, close enough to its projection on the eigen eyes space, the corresponding region is confirmed to be a face region. Finally, other facial features, including mouth corners, nose tips and nose holes are detected based on the integral projection algorithm and the average anthropologic measurements of the valid faces.

1 Introduction

Almost all face recognition methods fall into two catalogs[1]: feature-based and image-based. Though the two kinds of approaches are different in many aspects, one of the most essential and critical steps for both of them is facial feature detection. This is evident for the former, while for the latter, features are also needed in navigating the matching process and for the estimation of head poses. In addition, facial feature detection also found applications in gaze determination[2], face compression[3], etc. Besides that, human faces are typical deformable objects. This makes existing feature detection algorithms for rigid objects not suitable for facial feature detection. For these reasons, the problem of facial feature detection has long attracted the attention of many researchers[4]-[13].

Published feature detection methods can be generally summarized as following: (1) Image analysis based on geometry and gray-level distribution. This is the most instinctive approach. It aims to find the points or structures on the images with distinct geometrical shape or gray-level distribution. In [4], bright tokens on the images were detected as the locations of the eyes and mouth corners for registering the head poses. Resifeld et al[5] used the measurement of generalized symmetry for detecting eyes and

T. Tan, Y. Shi, and W. Gao (Eds.): ICMI 2000, LNCS 1948, pp. 191-198, 2000.

mouth corners. Nixon[6] described the pupils and the sclera as circles and ellipses respectively. (2) Deformable template matching. Originally worked out by Yuille et al.[7], this method has been utilized and improved by many researchers[8-9]. (3) Algebraic feature extraction. Besides, wavelet analysis[10] and artificial neural network [11] are also used for facial feature detection.

Among the existing methods and systems, few took into account detecting features of multiple faces on a single image. In this paper, we presents an approach to detecting facial features of multiple faces on complex background and with variable poses, illumination conditions, expressions, ages, image sizes, etc. First, the skin parts of the input image are extracted by color segmentation. Then, the candidate face regions are estimated by grouping the skin parts. Within each candidate face region, an attempt is made to find the eye pair using both Hough Transform and the Principal Component Analysis (PCA) method. If the resulted eye pair is, under the measurement of correlation, close enough to its projection on the eigen eyes space, then the corresponding region is confirmed to be a face region. Finally, other facial features, including mouth corners, nose tips and nose holes are detected based on the integral projection algorithm and the average anthropologic measurements of the valid faces. According to a test on over 1,000 images taken under all kinds of conditions, the presented method is proven to be relatively efficient on both accuracy and speed.

2 Color Segmentation

Color segmentation is performed in CIE XYZ color space as following:

Let $I(x, y)$ be one of the pixels in image I, and $X(x, y)$ be its X color component in CIE XYZ color space. The skin color histogram is calculated as:

$$\forall I(x, y) \in I$$

if $X(x, y) \in [X_j(x, y)*(1-k), X_j(x, y)*(1+k)]$ $k \in [0, 1]$

then $Hist(j) = Hist(j) + 1$

Based on the resulted histogram, the original skin-color map $F(x, y)$ of the image $I(x, y)$ (Fig.1) is extracted by classifying the color in the generalized LHS color space[13]. $F(x, y)$ is defined as:

$$F(x, y) = \begin{cases} 0 & background \\ 1 & skin \end{cases} \quad . \tag{1}$$

The candidate rectangle face regions are predicted in two steps (Fig.2): noise elimination and region merging. The two steps share the same algorithm described as following: (1) divide the image into $K \times K$ regions with size of $M \times N$; and (2) for each region R, modify the color map according to equation (2).

$$\forall (x, y) \in R, F(x, y) = \begin{cases} 1 & if\ S \geq MN/3 \\ 0 & if\ S < MN/3 \end{cases}$$

$$where \quad S = \sum_{(x,y) \in R} F(x, y) \quad . \tag{2}$$

(a)   (b)

Fig.1. Examples of color segmentation (a) Original image (b) Skin-color map

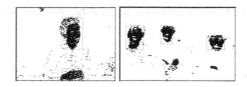

Fig.2. Candidate face regions corresponding to Fig.1. (M=N=8 for noise elimination, M=N=32 for region merging)

3   Representation of Eyes in Eigen-Eyes Space

3.1 Geometrical Calibration of Eyes

In order to utilize KL transform for eyes detection, the areas of interest (AOI) in the input image should have the same size as those in the training images. Therefore, the affine transform is applied to the AOIs to ensure that the irises in input images and training images have the same positions ($E_{L0}(x_{L0}, y_{L0})$, $E_{R0}(x_{R0}, y_{R0})$).

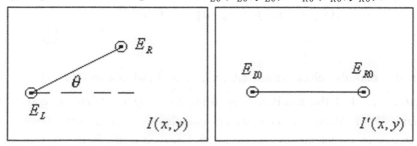

Fig.3. Geometry calibration of eyes

As shown in Fig.3, let $I(x, y)$ and $I'(x, y)$ be the original image and the transformed image respectively. To transform the original iris positions ($E_L(x_L, y_L)$,

$E_R(x_R, y_R)$) to ($E_{L0}(x_{L0}, y_{L0}), E_{R0}(x_{R0}, y_{R0})$), the affine transform is defined as:

$$\begin{pmatrix} x' \\ y' \end{pmatrix} = STR \begin{pmatrix} x \\ y \\ 1 \end{pmatrix} \tag{3}$$

Where, S, T and R are rotation, translation and scaling matrices respectively.

3.2 Eigen-Eyes Space

Let the size of the calibrated AOI be $w \times h = n$ and the training set be $\{i_1, i_2, \ldots, i_m\}$, while $i_i \in R^n, i = 1, 2, \ldots, m$. The eigen-eyes space is calculated as followings.

First, calculate the average eye of the training set as:

$$\mu = \frac{1}{m} \sum_{k=1}^{m} i_i, \qquad \mu \in R^n \tag{4}$$

Then, find the co-variance matrix:

$$R = \frac{1}{m} \sum_{k=1}^{m} (i_k - \mu)(i_k - \mu)^T = AA^T, \qquad R \in R^{n \times n} \tag{5}$$

Where,

$$A = [i_1 - \mu, i_2 - \mu, \ldots, i_m - \mu], \qquad A \in R^{n \times m} \tag{6}$$

Finally, find the eigen values $(\lambda_1, \lambda_2, \ldots, \lambda_r)$ and normalized eigen vectors $(u_1, u_2, \ldots, u_r)$ of the matrix AA^T by using the Singular Vector Decomposition theorem. For reducing the computation complexity, only a part of the vector $(u_1, u_2, \ldots, u_r)$, denoted as $(u_1, u_2, \ldots, u_l)$, is used as the base vector of the eigen-eyes space. While $(u_1, u_2, \ldots, u_l)$ satisfy the following equation:

Where,

$$\sum_{i=1}^{l} |\lambda_i| \geq 0.95 \sum_{i=1}^{r} |\lambda_i| \tag{7}$$

Examples of training eyes and eigen eyes are shown in Fig.4.

(a)

(b)

Fig. 4. (a) Training eyes (b) Eigen eyes

3.3 Representation of Eyes in Eigen-Eyes Space

Let $p \in R^n$ be the normalized input eye pair of size $w \times h$. It can be projected onto the eigen-eyes space as:

$$p = \sum_{i=1}^{l} c_i u_i = U(c_1, c_2, \ldots, c_l)^T \tag{8}$$

Because U is orthogonal, we have

$$(c_1, c_2, \ldots, c_l)^T = U^T p \tag{9}$$

Therefore, p is mapped to the eigen-eyes space as $p' = \sum_{i=1}^{l} c_i u_i$. The representation error is measured by the correlation between p and p':

$$\delta(p, p') = \frac{E(pp') - E(p)E(p')}{\sigma(p)\sigma(p')} \tag{10}$$

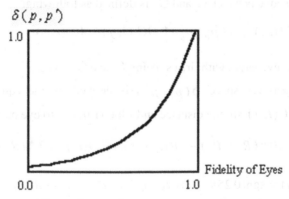

Fig.5. $\delta(p, p')$ vs. the fidelity of eyes

This measurement results in a perfect computation property (Fig.5): if p is really an eye pair, $\delta(p, p')$ is close to 1; otherwise, $\delta(p, p')$ is close to 0. By using this feature (illustrated in Fig.6), non-eye regions can be easily removed out of consideration.

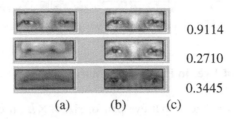

0.9114

0.2710

0.3445

(a)   (b)   (c)

Fig.6. Representation of eyes in eigen-eyes space（a）Original p （b）Projection p' (c)
$$\delta(p, p')$$

4 Facial Feature Detection

4.1 Eyes Detection and Face Region Verification

By using the Hough transform-based eyes detection algorithm we presented in[14] , k candidate irises $C_1, C_2, \cdots, C_k$ can be first found in the AOI under consideration. Let $C_1, C_2, \cdots, C_k$ be the nodes of a complete graph G. A benefit function $BF(i, j)$ for the edge between C_i and C_j is defined as following:

$$BF(i, j) = \delta(p_{ij}, p'_{ij}) * DC(i, j) * AC(i, j) \tag{11}$$

where, p_{ij} is the eye area with irises being C_i and C_j, and p'_{ij} is the projection of p_{ij} on the eigen-eyes space. $\delta(p_{ij}, p'_{ij})$ is defined by the equation（10）; $DC(i, j)$ and $AC(i, j)$ are the distance and tilt constraints to eye pairs:

$$DC(i, j) = sgn(10 * (R_i + R_j) - D(i, j)) * sgn(D(i, j) - 3 * (R_i + R_j)) \tag{12}$$

$$AC(i, j) = sgn(0.25\pi - abs(tg^{-1}((B_i - B_j) / (A_i - A_j))) \tag{13}$$

The iris pair (C_l, C_r) is accepted as the correct eye pair if the following condition holds:

$$BF\ (l,r) = \underset{i,j=1,2,\ldots k}{Max}\ BF\ (i,j) \geq \delta_0 \qquad (14)$$

where, δ_0 is the threshold of the eye fidelity. If $\underset{i,j=1,2,\ldots k}{Max}\ BF\ (i,j) < \delta_0$, the AOI under consideration is considered to be a non-face region.

4.2 Detection of Other Features

Based on the eye positions, other features, including mouth corners, nose tips and nose holds can be detected by using the integral projection algorithm and the average anthropologic measurements of the valid faces. The detailed algorithm is described in[14].

5 Experimental Results and Conclusions

More than 1000 images have been tested using the presented algorithm. The training set includes 80 sample images. By using the Principal Component Analysis method, 55 eigen eye pairs are remained as the base of the eigen-eyes space. All tested images are not included in the training set. The test platform is a PII/350 computer with 128M RAM under Windows NT. The experimental results are concluded as following: (1) correct ratio of face region detection is 89.2%; (2) correct ratio of eyes detection is 92.3%; and (3) correct ratio of other features is 88.3%. The average computation time is 10 sec./image. Fig.7 shows part of the experimental results.

Fig. 7. Experimental results

For eyes detection, the errors fall in two catalogs: (1) the eyes are missed in the AOI; and (2) other features, like brows and nose holds, are detected as the eyes by mistake. 97% of the errors fall in the former, while 100% errors of this kind are due to that one of the irises are not included in the candidate iris set. The latter kind of errors (occupied only 3%) indicate that the brows or nose holds look more similar to eyes. These errors occur when the "eye-style" under consideration is not included in the training set.

References

1. P.Ballard and G.C. Stockman, "Controlling a computer via facial aspect," IEEE Trans. Syst. Man and Cyber., Vol. 25, No. 4, Apr. 1995, pp. 669-677.
2. R. Brunelli and T. Poggio, "Face recognition: Features versus templates," IEEE Trans. Patt. Anal. and Mach. Intell., Vol. 15, No. 10, Oct. 1993, pp1042-1052.
3. D. Reisfeld and Y. Yeshuran, "Robust detection of facial features by generalized symmetry," Proc. 11th Int. Conf. on Patt. Recog., 1992, pp. 117-120.
4. M. Nixon, "Eye spacing measurements for facial recognition," SPIE Proc. 575, Applications of Digital Image Processing VIII, 1985, pp. 279-285.
5. A. L. Yuille, D. S. Cohen and P. W. Halinan, "Feature extraction from faces using deformable templates," Proc. IEEE Computer Soc. Conf. on computer Vision and Patt. Recog., 1989, pp.104-109.
6. X. Xie, R. Sudhakar and H. Zhuang, "On improving eye feature extraction using deformable templates," Pattern Recognition, Vol. 27, No. 6, 1994, pp. 791-799.
7. G. Chow and X. Li, "Towards a system for automatic facial feature detection," Patter Recognition, Vol. 26, No. 12, 1993, pp.1739-1755.
8. C. L. Huang and C. W. Chen, "Human facial feature extraction for face interpretation and recognition," Pattern Recognition, Vol. 25, No. 12, 1992, pp. 1435-1444.
9. M.A. Turk and A.D. Pentland, "Face recognition using eigenfaces," Proc. of Computer Vision and Pattern Recognition, 1991, pp.586-591.
10. B.S. Manjunath, R. Chellappa and C. von der Malsburg, "A feature based approach to face recognition," Proc. IEEE Computer Soc. Conf. on Computer Vision and Pattern Recognition, 1992, pp.373-378.
11. R. Brunelli and T. Poggio, "HyperBF networks for gender classification," in Proc. DARPA Image Understanding Workshop, 1992, pp.311-314.
12. L. Sirovich and M. Kirby, "Low-dimensional procedure for the chracterization of human faces," J. Opt. Soc. Am. A, Vol. 4, No. 3, March 1987, 519-524.
13. Levkowitz H., Color Theory and Modeling for Computer Graphics, Visualization, and Multimedia Applications, Kluwer Academic Publishers, Boston 1997.
14. Z.Y.Peng, S.Y.You and G.Y.Xu, "Locating facial features using threshold images," in the Third Int. Proc. of Signal Processing, Vol.2, 1996, pp.1162-1166.

The Synthesis of Realistic Human Face

Sining Wu, Baocai Yin, Tao Cai

The College of Computer Science,Beijing Polytechnic University, Beijing, 100022 P.R.China

Email: yinbc@bjpu.edu.cn

Abstract: In this paper, we present new techniques for creating a realistic specific human face based on a general 3D face model and several photos of a human. The morphing technique allows interactive alignment of features of the general geometric face model with the features of the multi-direction images of the specific human face, which are pre-provided by the animator. To enhance the realism, we employ a multi-direction texture mapping technique, which gets illumination information from multi-direction images of the specific human face.

Keywords: Face model, Morphing model, Texture-mapping

1. Introduction

At present, the methods for building up the 3D face model of the specific individual are classified into two categories. One is based on direct three-dimensional surface measuring techniques, from digitizing by hand [3] to using digitizers, photogrammetric techniques, or laser scanning techniques [1]. The other is based on the existing 3D face model, which creates a new face model by modifying existing models [2,4,6]. The latter involves two approaches: one starts from an example set of 3D face models and one 2D image or more. A new specific 3D face model is constructed by forming linear combinations of the prototypes to match one 2D image or more [2]. The other, however, based on one general 3D face model with several 2D face images, is morphing the general 3D face model according to the information that gets from the 2D face images [4]. This paper presents the scheme of the latter.

We get feature information from the front, left side, right side and back image of a specific individual, apply "Kriging" technique to morph the general 3D face model, and then get the 3D face model of the specific individual. At the same time we produce a texture image from the foregoing four images, which gets illumination and texture information from multi-direction images of the specific individual. By mapping the texture image on the specific model, we can finish the synthesis of a realistic 3D human face.

T. Tan, Y. Shi, and W. Gao (Eds.): ICMI 2000, LNCS 1948, pp. 199-206, 2000.
© Springer-Verlag Berlin Heidelberg 2000

2. Morphing Algorithm of Human Face Model Based on Kriging Method

Kriging method estimates the unknown points according to the linear combinations of the known points. The result can be represented by following equations.

$$V = \sum_{j=1}^{n} w_j V_j, \quad \sum_{j=1}^{n} w_j = 1 .$$

define error variance as

$$r = V - \overline{V} ,$$

where $\overline{V}$ is the true value. Since these true values are unknown, we can regard $V, V_1, V_2, \cdots, V_n$ as random variables. Then the error variance of random variable r is

$$\delta_r^2 = E(r - E(r))^2 .$$

To minimize δ_r^2, the coefficients $\{w_j\}$ need to satisfy the equation.

$$\min_{w_1, w_2, \cdots, w_n} \quad E(r - E(r))$$

$$\text{Constrain Condition} \sum_{j=1}^{n} w_j = 1 \tag{1}$$

Given that random variable $\overline{V}$ has constant variance, the equation (1) can be simplified

$$\min_{w_1, w_2, \cdots, w_n} \quad E(V - E(V))^2 - 2E((V - E(V))(\overline{V} - E(\overline{V})))$$

$$= \min_{w_1, w_2, \cdots, w_n} \quad \sum_{i=1}^{n} \sum_{j=1}^{n} w_i w_j C_{ij} - 2 \sum_{i=1}^{n} w_i C_{i0}$$

$$\text{Constrain Condition} \sum_{j=1}^{n} w_j = 1$$

where $C_{i0}, C_{ij}, i, j = 1, 2, \cdots, n$ are the respective covariance of the corresponding random variable $V_i, V; V_i, V_j, i, j = 1, 2, \cdots, n$, and $V, V_1, V_2, \cdots, V_n$ are the curved surface interpolation points corresponding to the space mesh points $x, x_1, x_2, \cdots, x_n$. Applying Lagrange's method of multipliers to equation (1), we can get a linear system for $w_i (i = 1, 2, \cdots, n)$.

$$\sum_{j=1}^{n} w_j C_{ij} - C_{i0} + \mu = 0, \ i=1,2,\cdots,n$$

$$\sum_{j=1}^{n} w_j = 1$$

(2)

The equation (2) can be represented in matrix form

$$AW = D \tag{3}$$

where

$$A = (a_{ij})_{(n+1)\times(n+1)}, a_{ij} = C_{ij}, a_{i(n+1)} = a_{(n+1)j} = 1, a_{(n+1)(n+1)} = 0, i,j = 1,2,\cdots n;$$

$$W = (w_1, w_2, \cdots, w_n, \mu)^T, D = (C_{10}, C_{20}, \cdots, C_{n0}, 1)^T.$$

In order to discuss the solution of equation （3）, we write it as:

$$\begin{bmatrix} A & e \\ e^T & 0 \end{bmatrix} W = \begin{bmatrix} D \\ 1 \end{bmatrix} \tag{4}$$

where

$$A = (a_{ij})_{n\times n}, a_{ij} = C_{ij}, i,j = 1,2,\cdots n;$$

$$e = (1,1,\cdots,1)^T, D = (C_{10}, C_{20}, \cdots, C_{n0})^T$$

Many Kriging models are researched for representing the relation between interpolation surface points. The choices here fall into several cases:

Let $C_{ij} = \delta_0^2 - \gamma_{ij}$,

- spherical model

$$\gamma_{ij} = \begin{cases} C_0 + C_1(1.5\dfrac{h}{a} - 0.5(\dfrac{h}{a})^3) & h \le a \\ C_0 + C_1 & h > a \end{cases}$$

- exponential model

$$\gamma_{ij} = \begin{cases} 0 & h = 0 \\ C_0 + C_1(1 - \exp(\dfrac{-3h}{a})) & h > 0 \end{cases}$$

where $h = d_\Delta(x_i, x_j)$, δ_0^2 is the variance of V, and $a > 0, C_0 > 0, C_1 > 0$ are constant.

In our experiment, we set Vi to be the feature point's displacement u_i, using the method represented above to calculate the displacement of other points that are not feature points. We used Inverse distance average model

$$C_{ij} = \overline{C}_{ij} / (\sum_{k=0}^{n} \overline{C}_{ik}^2)^{1/2}$$

where,

$$\overline{C}_{ij} = a(d_\Delta(x_i, x_j))^{-r}, \quad i = 1,2,\cdots,n, j=0,1,2,\cdots,n.$$

The result of Kriging method is shown in Figure 2.

3. Texture Extraction

The Model obtained only by Kriging morphing still needs further refinement because of its lack of detailed features of a specific individual face. Texture mapping will enhance the result in synthesizing realistic human faces [5]. This section presents the synthesis algorithms of the texture image used in our experiments. The steps are as following:

- Set up the mapping relationship between the 3D space of the face model and the 2D space of texture images. Given a point (x,y,z) in the 3D space of the model, its corresponding coordinates in the 2D space of the texture image are (u,θ). There is the following mapping relationship between them:

$$(u,\theta) = (\tan^{-1}(z/x), y).$$

- For each point (u,θ) in the texture space, calculate its color value $I(u,\theta)$.
    a) Calculate the 2D coordinates (u,θ) of point p, according to its 3D coordinates (x,y,z).
    b) Project the point p to the frontal image plane. We can get the 2D coordinates of the projection point (x_z, y_z). The corresponding image color values of this projection point are $I_z(x_z, y_z)$. Similarly, project p to the side image plane, and we can get the 2D coordinates of the projection point (x_s, y_s). The corresponding image color values of this projection point are $I_s(x_s, y_s)$. To the points on the front part of the 3D face model, they should be projected to the front image plane. Likewise, this rule is the same with the other three directions.

c) Calculate the color value $I(u,\theta)$ at (u,θ) in the texture image as following

$$I(u,\theta) = \omega_1 I_z(x_z,y_z) + \omega_2 I_s(x_s,y_s)$$

where $w_1 = \cos^2\theta, w_2 = \sin^2\theta$

The results are as following figures (Figure 11).

4. User Interface

We get the feature point information in 2D human face images by hand. The feature points' coordinates are written exactly in the text file. The morphed face model will be similar with the specific individual in whole figure. However, as for some local features, such as crease in the face, can not be expressed by morphing. We designed an interactive interface in order that the user could adjust this kind of little difference.

5. Results

In order to test our technique, we photographed an associate's face image from four directions, front, back, left and right, with Panasonic Color digital Camera (CP410). Using the technique above, we get a 3D specific face model that resembles our associate (Figure 4,6,8,10).

Because we map the texture onto the triangle meshes, the display of the result goes rapidly. By researching the rules of Chinese pronunciation, we get some parameters to drive the 3D face model to implement lip-motion. With input of a stream of the Chinese Phonetic, we can display 30 frames per minute on the PC with Intel Pentium II. The result satisfies the facial animation purpose.

6. Future Work

Firstly, we are going to get the feature points in 2D human face images automatically. For example, the silhouette of a human face can be extracted using local maximum-curvature tracking (LMCT) algorithm. For the local features in the human face, however, the extraction will be more difficult.

Secondly, we are going to cut down the limiting factors on 2D human face images and make it possible to get information of 3D feature points from arbitrary direction images.

7. Acknowledgment

We thank Jie Yan for providing the general 3D face model, Dehui Kong for advice on face animation. In particular, we thank Jun Miao and Xuecun Chen for their previous work for this project. We thank Chunling Ma, Min Wu, Wanjun Huang and Chunyan Jiang for their help.

References

[1] Cyberware Laboratory, Inc, Monterey, California. 4020/RGB 3D Scanner with Color Digitizer, 1990.

[2] Volker Blanz and Thomas Vetter, A Morphable Model For The Synthesis of 3D Faces, In SIGGRAPH 99 Conference Proceedings.

[3] Frederic I. Parke. Computer Generated Animation of Faces. Proceedings ACM annual conference., August 1972.

[4] F. Pighin, J. Hecker, D. Lischinski, R. Szeliski and D. Salesin. Synthesizing Realistic Facial Expressions from Photographs. In Computer Graphics Proceedings SIGGRAPH'98, pages 75-84, 1998.

[5] Horace H.S.Ip and Lijun Yin. Constructing a 3D individualized head model from two orthogonal views. The Visual Computer 12: 254-266, 1996.

[6] M.J.T. Reinders, B. Sankur and J.C.A. van der Lubbe. Transformation of a general 3D facial model to an actual scene face. IEEE Pages 75-78, 1992.

[7] BaoCai Yin and Wen Gao. Radial Basis Function Interpolation on Space Mesh. ACM SIGGRAPH'97, Virtual Proc. pp150.

[8] Gregory M. Nielson. Scattered Data Modeling. IEEE Computer Graphics and Applications, 13(1):60-70, January 1993.

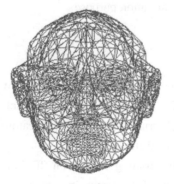

Figure 1: General 3D

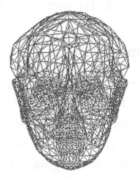

Figure 2:Kriging Morphing Result

Figure 3: Front Face

Figure 4: Kriging Morphing
Result with Texture

Figure 5: Left Side
Face Image

Figure 6: Left side of the
Model with texture

Figure 7: Right
Side Face Image

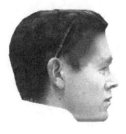

Figure 8: Right side
of Model with

Figure 9: Black
side face image

Figure 10: Synthesis images in several
directions of Model with texture

Figure 11: Texture created from four images: front, left,
right and back

Gravity-Center Template Based Human Face Feature Detection

Jun Miao[1], Wen Gao[1], Yiqiang Chen[1], and Jie Lu [2]

[1]Digital Media Lab, Institute of Computing Technology, Chinese Academy of Sciences,
Beijing 100080, P. R. China
{ jmiao, wgao, yqchen }@ict.ac.cn
[2]Institute of Electronic Information Engineering, Tongji University, Shanghai 200092, P. R.
China

Abstract. This paper presents a simple and fast technique for geometrical feature detection of several human face organs such as eyes and mouth. Human face gravity-center template is firstly used for face location, from which position information of face organs such as eyebrows, eyes, nose and mouth are obtained. Then the original image is processed by extracting edges and the regions around the organs are scanned on the edge image to detect out 4 key points which determine the size of the organs. From these key points, eyes and mouth's shape are characterized by fitting curves. The results look well and the procedure is fast.

Keywords face detection, edge extraction, face feature detection

1 Introduction

Human face and its feature detection is much significant in various applications as human face identification, virtual human face synthesis, and MPEG-4 based human face model coding. As for face image processing , many people have done much work on them. The representative jobs include mosaic technique[1,2], fixed template matching[3], deformable template matching[4], gravity-center template matching [5], "eigenface"[6] or "eigenpicture"[7] scheme, neural network[8,9], and usage of multiple information including color, sound or motion[10,11].

There also have been quite a few techniques reported about face contour and organ geo14metrical feature extraction, such as B-splines[12], adaptive Hough trasmform[13], cost minimiziation[14], deformable template[15,16] , region-based template[17], geometrical model[18]. However, most of them are either complex or time-consuming. Here, we present a simple and fast approach based on the technique of gravity-center template to locate face organs such as eyes and mouth, and detect their key points to characterize their scales and shape.

T. Tan, Y. Shi, and W. Gao (Eds.): ICMI 2000, LNCS 1948, pp. 207–214, 2000.
© Springer-Verlag Berlin Heidelberg 2000

2 Simplified Geometrical Model for Eyes and Mouth

The eyes and mouth have similar geometrical configuration as shown in fig. 1. There are four key points, the highest point P1, the lowest point P2, the left corner point P3 and the right corner point P4, which determine the size (a×b) and the arc-like(R1,R2) shape of the organs. Our goal is to detect out these key points to segment face region and concisely describe the features of eyes and mouth.

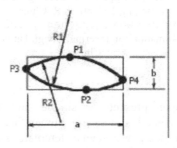

Fig. 1. Simplified model for eyes and mouth

3 Detection System Architecture

We have implemented a facial geometrical feature extraction system for eyes and mouth based on human face gravity-center template technique the author recently suggested [5]. The system consists of there modules : the first one of gravity-center template based human face and its organs location, the second one of four key points extraction, and the last one of geometrical curves fitting.

Inputting an original image, it will be processed by mosaicizing, mosaic edge extracting and feature simplifying for gravity-center template matching to locate human face in an unconstrained background and by adaptive Sobel edge extracting for feature points localization. From this information, geometrical shape of eyelids and lips can be easily fitted.

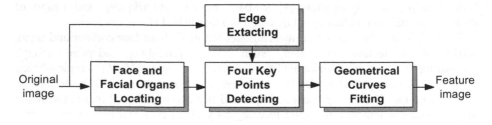

Fig. 2. System architecture

3.1 Face and Facial Organs Localization Based on Face Gravity-Center Template Match

This procedure will produce the location information about face and its organs such as eyebrows, eyes, nose and mouth, which is necessary for further facial feature detection. We use four face gravity-center templates [5] to detect frontal faces. In these templates, there are 4 to 6 small boxes which correspond to the facial organs: eyebrows, eyes, nose and mouth. Once a face in an image is located, simultaneously, its facial organs can be also located. The three steps b.(Image Mosaicizing), c.(Mosaic Horizontal Edge Extracting) and d.(Feature Simplifying and Gravity-Cenetr Template Matching) in Fig.3 illustrate the procedure of face and its organs localization. The detail description on detecting procedures can be referred in reference [5].

a.   b.

c.   d.

Fig. 3. Face and its organs localization

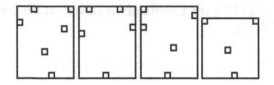

Fig. 4. Face gravity-center templates

3.2 Key Feature Points Extraction

From this step, four key points P1, P2, P3 and P4 illustrated in Fig.1 are expected to be extracted.

The original image is firstly processed with an adaptive Sobel operator introduced in reference [19] to produce a binary edge image. Fig. 5 shows the result processed by this operator.

Fig. 5. Binary edge image using an adaptive Sobel operator

According to the face organs' location information obtained from gravity-center template matching, we can scanned in the edge image in areas around those organs to see if current position is one of four key feature point of P1, P2, P3 and P4 showed in Fig.1. If it is true, its coordinates is to be calculated out using a mechanism of median-pass from the coordinates of the points scanned in rows or columns. The procedure is shown in Fig.6.

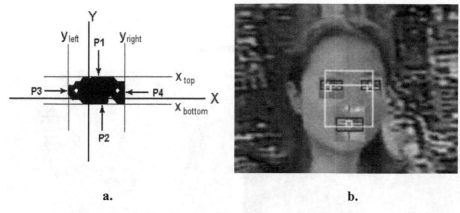

a.   b.

Fig. 6. Key feature points searching (e.g. left eye)

In the upper picture of Fig.6 the black area is cut from the area of the left eye in the Fig. 5, in which the position of the cross point of lines X, Y is obtained from the center of the small box corresponding to left eye. And lines X and Y are the initial positions in horizontal and vertical directions to scan and search the top and bottom sides, the left and right sides of the area. Scan starts from the initial X, Y lines and extend upwards and downwards, leftwards and rightwards, respectively in horizontal and vertical lines. A scan towards the one of the four directions will stop when the number of edge pixels it encounters in the scanning line that is orthogonal to the direction the scan towards is less than a threshold such as 3 edge pixels. In this case, the median value pixel is chosen as one key point which belongs to one of P1, P2, P3 and P4.

3.3 Eyelids and Lips Shapes Description

According to the simplified model for eyes and mouth's, from four key feature points: P1, P2, P3 and P4, a upper and a lower circle arc can be generated to fit the upper and the lower eyelids or lips. The effect is shown in Fig.7.

Fig. 7. Eyelids and lips fitting

4   Some Experimental Examples

Fig.8 to Fig.15 are part of the experiment results.

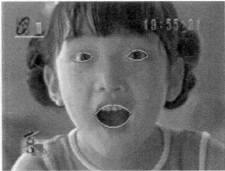

Fig. 8.

Fig. 9.

Fig. 10.

Fig. 11.

Fig. 12.   **Fig. 13.**

 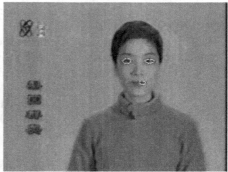

Fig. 14.   **Fig. 15.**

5 Conclusion and Future Work

Compared with some other detecting schemes such as deformable template technique, the approach introduced here is simple and fast (0.28 seconds in average on Pentium-550 PC). Especially for faces with complex background, our system can work as well.

During experiments, we still found some detection results of facial feature for those faces with complex texture and poor light conditions are not as well as those for good quality face images. And for the moment, this method can not characterize lips in more details yet.

Further work will include eyeball, nose and cheek contour detection and the algorithm is expected to be enhanced to improve the system's robustness.

References

1. L.D. Harmon, The recognition of faces, *Scientific American*, **229**(5), 71-82, 1973.
2. G. Z. Yang and T. S. Huang, Human face detection in a complex background, *Pattern Recognition*, **27**(1), 43-63, 1994
3. W. Gao and M. B. Liu, A hierarchical approach to human face detection in a complex background, *Proceedings of the first International Conference on Multimodal Interface'96,* 289-292, 1996
4. A.L. Yuille, Deformable templates for face detection, *J. Cogn. neurosci.* **3**, 59-70, 1991

5. J. Miao, B.C. Yin, K.Q. Wang, et al, A hierarchical multiscale and multiangle system for human face detection in a complex background using gravity-center template, *Pattern Recognition*, **32**(7), 1999

6. M. Turk and A. Pentland, Face recognition using eigenfaces, *Proc. IEEE-CS-CCVPR*, 586-591, 1991

7. M. Kirby and L. Sirovich, Application of the Karhunen-Loeve procedure for the characterization of human faces, *IEEE Trans. PAMI*, **12**(1), 103-108, 1990

8. H.A. Rowley, S. Baluja, and T. Kanade, Neural network-based face detection, *IEEE-PAMI*, **20**(1), 23-38, 1998

9. K.K. Sung and T. Poggio, Example-based learning for view-based human face detection, *IEEE-PAMI*, **20**(1), 39-50, 1998

10. R. Brunelli and D. Falavigna, Person identification using multiple cues, *IEEE Trans. PAMI*, **17**(10), 955-966, 1995

11. C.H. Lee, J.S. Kim and K.H. Park, Automatic human face location in a complex background using motion and color information, *Pattern Recognition*, **29**(11), 1877-1889, 1996

12. C.J. Wu and J.S. Huang, Human faces profiles recognition by computer, *Pattern Recognition*, **23**(3/4), 255-259, 1990

13. X. Li and N. Roeder, Face contour extraction from front-view images, *Pattern recognition*, **28**(8), 1167-1179, 1995

14. K. Lam and H. Yan, Locating and extracting the eye in human face images, *Pattern recognition*, **29**(5), 771-779, 1996

15. A.L. Yuille, D.S. Cohen and P.W. Hallinan, Feature extraction from faces using deformable templates, *Proc. IEEE-CS-CCVPR*, 104-109, 1989

16. C.L. Huang and C.W. Chen, Human face feature extraction for face interpretation and recognition, *Pattern recognition*, **25**(12), 1435-1444, 1996

17. J.Y. Deng and F. Lai, Region-based template deformation and masking for eye-feature extraction and description, *Pattern recognition*, **30**(3), 403-419, 1997

18. S.H. Jeng, H.Y.M. Liao, et al, Facial feature detection using geometrical face model: an efficient approach, *Pattern recognition*, **31**(3), 273-282, 1998

19. S.Y. Lee, Y.K. Ham and R.H. Park, Recignition of human front faces using knowledge-based feature extraction and neuro-fuzzy algorithm, *Pattern recognition*, **29**(11), 1863-1876, 1996

A Improved Facial Expression Recognition Method

Sun Jun   Zhuo Qing   Wang wenyuan

Institute of Information Processing, Automation Department
Tsinghua University 100084
Sunjun@mail.au.tsinghua.edu.cn
Zhuoqing@mail.tsinghua.edu.cn
Wwy-dau@mail.tsinghua.edu.cn

Abstract. We proposed a novel facial expression recognition algorithm based on Independent Component Analysis (ICA) and Linear Discriminant Analysis (LDA). ICA produced a set of independent basis images of expression image, LDA selected features obtained from ICA. Experiments proved the excellent performance of our algorithm.

1. Introduction

The study of human facial expression has a long history. As the progress of computer science, automatic expression analysis has attracted more and more attention. Facial expression has many applications in the interaction between human and machine. Beside, understanding the mechanism of expression recognition is very helpful to the study of psychology and face recognition.

The Facial Action Coding System developed by Ekman and Friesen[3] is regarded as the first rigorous, systematic study of facial movement and its relationship to affective state.Among the most used methods in facial analysis, principal component analysis (PCA) was widely accepted as an efficient way for data presentation and classification. Global PCA (Eigenface) have been applied successfully to recognizing facial identity, classifying gender and recognizing facial expression[4]. Some local versions of PCA were also reported for facial image analysis [5].

While PCA representation is based on the second-order dependencies among the pixels, they are insensitive to the high order dependencies of the image. Independent component analysis (ICA) which is a generalization of PCA captures the high order moments of the data in addition to the second order moments. Since much of the information about facial expression such as the edges are contained in the high order relationships among the image pixels, a face representation based on ICA is more suitable for facial recognition and expression analysis [6].

Both PCA and ICA extract projection coefficients as the feature of the image, but the dimension of feature can be very large (several tens to several hundred), How to select the most effective feature is crucial to recognition. We took a method based on linear discriminant analysis (LDA) to perform feature selection. Experiments show that our methods combined with ICA for facial expression recognition are superior to single PCA and ICA methods.

T. Tan, Y. Shi, and W, Gao (Eds.)· ICMI 2000, LNCS 1948, pp. 215-221, 2000.
© Springer-Verlag Berlin Heidelberg 2000

2. Data Set and Preprocession

The facial expression data was collected from 23 individuals. Each person was asked to perform six basic emotions (happiness, angry, sadness, disgust, surprise and fear) and video stream was recorded. The first frame in the video is neural expression and the last one reaches the apex. We select five emotions from the video stream except for fear since the samples of this expression was not enough and even ourselves can not distinguish the recorded fear expression from surprise.

We take altogether 76 video streams: 14 for anger, 13 for disgust, 19 for happiness, 13 for sadness and 17 for surprise. Each video stream contains four frames. We carried our study on the δ image as in [1] which is the difference between the succedent frame and the first frame. The δ image has many advantages in facial expression analysis since it partly removes the impact of light and facial difference between different individuals.

Accurate image registration is very important in facial expression representation, we manually select three fiducial points (center of two eyes and mouth) in the frame to account for image registration so that the coordinates of these fiducial points were in (32,32), (96,32) and (64, 96) respectively. The size of standard δ image is 128 by 128. All the images were performed histogram expansion and equalization before image registration. Examples of registered δ images are shown in Fig. 1.

Fig. 1. The δ images of five expression.

3. Feature Extraction

Here we use PCA and ICA for feature extraction. PCA is an unsupervised learning method based on the second-order dependencies among the pixels. Here the PCA was obtained by calculating the eigenvectors of the pixelwise covariance matrix of all δ images.

Take X as the δ images matrix, each row of X is a δ image. The eigenvectors was found by decomposing the covariance matrix S into the orthogonal matrix P and diagonal matrix D:

$$S = (X - \overline{X})^T (X - \overline{X}) = PD^T P^T \tag{1}$$

$\overline{X}$ is the mean value of X. The zero-mean δ images are then projected onto the first p eigenvectors in P, producing a feature vector of p coefficients for each δ image.

The independent component representation was obtained by performing "blind separation" on the set of face images [1]. The δ images in the rows of X are assumed to be a linear mixture of an unknown set of statistically independent source images S, where the mixture matrix A is also unknown. The sources are recovered by a learned unmixing matrix W, which approximates A^{-1} and produces statistically independent outputs, U. The relationship between A, U, X and S is as follows:

$$A * S = X \qquad W * X = U \qquad A = W^{-1} \tag{2}$$

Here we took Bell and Sejnowski's algorithm in [2] derived from the principle of optimal information transfer between neurons to get the unmixing matrix W. The algorithm maximizes the mutual information between the input and the output of a nonlinear transfer function g.

Let $u = Wx$, where x is a column of the image matrix X and $y = g(u)$. The update rule for the weight matrix, W, is given by

$$\Delta W = \left(I + y' u^T\right) W \tag{3}$$

$$\text{where } y' = \frac{\partial}{\partial y_i} \frac{\partial y_i}{\partial u_i} = \frac{\partial}{\partial u_i} \ln \frac{\partial y_i}{\partial u_i}$$

We used the logistic transfer function, $g(u) = \dfrac{1}{1+e^{-u}}$, giving $y' = (1 - 2y_i)$.

The unmixing matrix W can be regarded as a linear filter which learns the statistical dependencies between the δ images. The filtered images are the rows of U and are independent with each other. They formed a basis set for expression images. Then each δ image was represented by a set of coefficients a_i which is the rows of A.

Samples of independent components of the δ images are shown in Fig. 2.

Fig. 2. Independent components of δ images.

The number of independent components found by ICA corresponds to the dimensionality of the input. In order to control the number of independent components extracted by the algorithm, we computed the ICA from the first m principle components of the δ images. That is, let P_m denote the matrix containing the first m principal component in its column. The project coefficients of the zero-mean images in X on P_m are defined as $R_m = X * P_m$. A minimum squared error approximation of X is obtained by $X_{rec} = R_m * P_m^T$.

We perform ICA on the P_m to produce a filter matrix W such that:

$$W * P_m^T = U \quad \Rightarrow \quad P_m^T = W^{-1} * U = A * U \tag{4}$$

Therefore

$$X_{rec} = R_m * P_m^T = R_m * A * U \tag{5}$$

Here the rows of $R_m * A$ contained the coefficients for the linear combination of statistically independent sources U that comprised X_{rec}.

4. Linear Discriminant Analysis (LDA) Based Feature Selection

Feature selection and extraction are very important to recognition. Feature selection picks the most discriminant features combination from a large number of features, while feature extraction represents the features in a low dimension space through mapping or transformation.

Local discriminant Analysis is widely used as a feature extraction method. LDA is a linear projection into a subspace that maximizes the between-class scatter S_B while minimizing the within-class scatter S_W of the projected data.

S_B and S_W are defined as follows:

$$S_W = \sum_{i=1}^{C} \sum_{x \in X_i} (x - m_i)(x - m_i)^T \tag{6}$$

$$S_B = \sum_{i=1}^{C} P(\omega_i)(m_i - m)(m_i - m)^T$$

where C is the total number of classes, X_i is the sample set belongs to class i, m_i is the mean of X_i, m is the mean of m_i, $P(\omega_i)$ is the prior probability of class i.

The projection matrix $W_{opt} \left(R^n \mapsto R^{C-1} \right)$ satisfies:

$$W_{opt} = \arg\max_{W} J(W) \overset{\Delta}{=} \arg\max_{W} \frac{\det(W^T S_B W)}{\det(W^T S_W W)} = \{w_1, w_2, \cdots, w_{C-1}\} \tag{7}$$

The $\{w_i\}$ are the solutions to the generalized eigenvalues problem:

$$S_B w_i = \lambda_i S_W w_i \qquad \text{for } i = 1, 2, \cdots, C-1 \qquad (8)$$

Since S_W is always singular, we took a method in [7]: The data is first projected into a lower dimensional space using PCA, thus S_B and S_W become:

$$\tilde{S}_B \overset{\Delta}{=} W_{pca}^T S_B W_{pca} \qquad \text{and} \qquad \tilde{S}_W \overset{\Delta}{=} W_{pca}^T S_W W_{pca} \qquad (9)$$

where W_{pca} is the transformation matrix of PCA.

Performing LDA on $\tilde{S}_B$ and $\tilde{S}_W$ results a linear projection matrix W_{lda}, the final transformation matrix is:

$$W_{opt} = W_{pca} * W_{lda} \qquad (10)$$

Notice that the dimension of the features after LDA is no more than C-1, if C is too small, a lot of discriminant information is discarded during the transformation.

However, since the magnitude of each component of w_i represents its influence on the classification decision, we can use it as a measurement of the validity of the corresponding feature.

Here we performed five LDA for each expression and obtained five discriminant vectors. Then we summed up the magnitude of five vectors and got a new vector -- validity vector.

Since the components of validity vector determine the discriminant power of corresponding feature, we can use this validity vector for feature selection. The p features corresponding to the first p largest components in validity vector are regarded as the optimal combination of p features for classification.

5. Experiments and Results

We performed four experiments to test our algorithm, the first three experiments were compared with our method in experiment 4. Due to the small number of sample data, we used a leave one out cross validation method to generate test set and training set. Five expression feature centers were drawn from the training set. The test set was validated using Euclidean distance as similarity measure and nearest neighbor as classifier.

In experiment 1, 160 PCA coefficients were extracted from δ images, the first p principal components with largest eigenvalues were chosen.

In experiment 2, we perform LDA on the 160 PCA coefficients and p principal components were selected according to the validity vector.

In experiments 3, we first performed ICA on the 160 principal components resulting 160 independent components. Thus for each sample we got 160 ICA coefficients. Since independent components do not have inherent orders, we used method in [1] to compute the class discriminability of each individual independent component. The first p components with largest class discriminability comprised the independent component representation.

In experiment 4, we performed LDA on the ICA coefficients and selected p independent components according to the validity vector.The calculation of ICA out of 160 principal components in experiment 3 and 4 takes 32 minutes on a PII 350 machine.

Recognition results were listed in Table 1. We can see that ICA outperforms PCA since it captured the high order dependencies between image pixels. However, feature selection is also very helpful to improve the recognition ratio.

The eigenvalues of PCA do not have direct relationship with classify discriminability. While select features based on its individually class discriminability as in experiment 3 didn't consider the impact of interaction between features. Our LDA based feature selection not only considers the discriminability of each feature, but also take the relationship of all features into account through within and between scatter matrix S_B and S_W. What's more, since it doesn't require to performing a search process, the speed of feature selection is greatly improved.

Method	Dimension	Recognition ratio
PCA	21	82.4%
PCA + LDA	32	88.4%
ICA	35	91.2%
ICA + LDA	28	95.6%

Table 1. Best performance for each method. Dimension is the p when best recognition ratio was achieved.

6. Discussion

As indicated in the experiment, ICA as a feature extraction method removes the high order dependence among different principal components. However, it also heavily relies on the normalization of the face images. Here the face images are aligned according previously marked eyes and mouse position. Light variance is compensated by histogram equalization and δ images. If the face images are not perfectly aligned, the accuracy will surely drop as in PCA

To implement a full automatic expression recognition system, we should first align the face image, eyes, especially the inner corners of two eyes are the most stable feature among different expression. Locating eyes in face images can be achieved using the gray and gradient information and prior knowledge of face structure. A lot of article has dealt with this topic.

Facial expression recognition is very useful to the interaction between human and machine. First, expression can be regard as a signal input to the machine. Second, expression recognition can help to improve the performance of face recognition which is also very important in human-machine interaction.

We proposed a facial expression recognition algorithm based on ICA and LDA, ICA was used as feature extraction and LDA was used as feature selection. The performance of our method is excellent. We have used this algorithm to analyse the impact of expression to the face recognition.

References

[1]. G. Donato, M. S. Bartlett, J. C. Hager, P. Ekman and T. J. Sejnowski, "Classifying Facial Actions", IEEE trans. PAMI, vol.21, no.10, pp974-989, 1999

[2]. A.J. Bell and T.J. Sejnowski, "An Information-Maximization Approach to Blind Separation and Blind Deconvolution," Neural Computation, vol.7, no.6, pp1129-1159 1995

[3]. P. Ekman and W. V. Friesen, "Pictures for Facial Affect", Human Interaction Laboratory, Univ. of California Medical Center, San Francisco, 1976

[4]. M. Turk and A. Pentland, "Eigenfaces for Recognition," J. Cognitive Neuroscience, Vol.3, no. 1, pp71-86, 1991

[5]. C. Padgett and G. Cottrell, "Representing Face images for Emotion Classification," Advances in Neural Information Processing Systems, vol.9 1997

[6]. M. S. Bartlett, "Face Image Analysis by Unsupervised Learning and Redundancy Reduction," Ph.D thesis. Univ. of UCSD 1998

[7]. P. N. Belhumeur, J. P. Hespanha, and D. J. Kriegman, "Eigenfaces vs. Fisherface: Recognition Using Class Specific Linear Projection," IEEE trans. PAMI, vol.19, no.7, pp711-720, 1997

Face Direction Estimation using Multiple Cameras for Human Computer Interaction

Mamoru Yasumoto[1], Hitoshi Hongo[1], Hiroki Watanabe[1]
and Kazuhiko Yamamoto[2]

[1] Softopia Japan (HOIP) and JST, 4-1-7 Kagano, Ogaki City, Gifu, Japan
{yasu, hongo, watanabe}@softopia.pref.gifu.jp
http://www.softopia.pref.gifu.jp/HOIP
[2] Gifu University, 1-1 Yanagido, Gifu City, Gifu, Japan
yamamoto@info.gifu-u.ac.jp

Abstract. This paper proposes a method for estimating continuous face directions with the discrete face direction feature classes. The feature classes are composed by the linear discriminant analysis of the "four directional features" that are extracted from some people's faces looking in specified directions. Estimation by a single camera gives ambiguous results for non-frontal view of faces. Especially, it is impossible from the back of the face. However, integrating the information from the multiple cameras allows more precise face direction estimation. Experiments show that our proposed method can achieve a mean error of at most $8°$ and a variance of 22.4 for all face directions.

1 Introduction

Nonverbal information such as human face expressions and gestures plays an important role in interactions between human and computer [8, 4]. We are investigating to establish the "Percept-room," which looks at people, interprets human behavior and makes appropriate responses. Since human gaze or face direction is key to detecting what a person is paying attention to, many techniques in this research field have been proposed. However, they have some restrictions such as a small range of direction around the frontal view of a face [1, 6], low precision [9], initialization per user [7] and occlusion tolerance.

We have already proposed a face direction estimation method for unspecified people [10]. Although it determines one of 16 face directions for the coarse classification of hierarchical face recognition, 16 directions are not enough to identify what a user looks at. In this paper, we propose the enhancement of the direction resolution.

2 Face direction estimation

2.1 Coordinates for face direction estimation

Figure 1 shows the coordinate system used in this paper. For the purpose of detecting which object the testee looks at, his/her position $P(r_p, \phi_p)$ and face

T. Tan, Y. Shi, and W. Gao (Eds.): ICMI 2000, LNCS 1948, pp. 222-229, 2000.

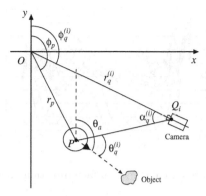

Fig. 1. Coordinates for face direction representation.

direction θ_a should be estimated. On the condition that the testee's position can be given by other methods, we investigate only the face direction estimation here. Equation (1) shows that θ_a is given with the face direction $\theta_q^{(i)}$ observed from the i-th camera fixed at $Q_i(r_q^{(i)}, \phi_q^{(i)})$. That is, the problem is to obtain the face direction estimation $\hat{\theta}_q^{(i)}$.

$$\theta_a = \theta_q^{(i)} + \phi_q^{(i)} - \alpha_q^{(i)}, \tag{1}$$

$$\alpha_q^{(i)} = \tan^{-1} \frac{r_p \sin(\phi_p - \phi_q^{(i)})}{r_q^{(i)} - r_p \cos(\phi_p - \phi_q^{(i)})}. \tag{2}$$

2.2   Feature extraction

First, we detect the face region by means of the skin color in the CIE-L*u*v* color space [3]. If the size of the skin color region is larger than the threshold S_t, it is detected as the face region. Next, the "four directional features"(FDF) [5] are extracted as follows:

1. By applying Prewitt's operator to the detected face region in four directions (vertical, horizontal and two diagonals), four edge images are made.
2. Each edge image is normalized to an 8×8 pixel low-resolution image.
3. 256-dimensional feature vector is made from these four planes of images.

2.3   Estimating continuous face direction

We have investigated the face direction estimation as the coarse classification of hierarchical face recognition[10]. First, the FDF vector $\boldsymbol{x}$ is transformed to $\boldsymbol{y}$ through the coefficient matrix A that is obtained by the linear discriminant

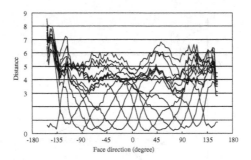

Fig. 2. Square distance to face direction classes.

analysis(LDA)[2]. Second, the distance D_j between $\boldsymbol{y}$ and the mean vector $\bar{\boldsymbol{y}}_j$ of the face direction class $C_j(j = 1..16)$ at 22.5° intervals is calculated, where $D_j = |\boldsymbol{y} - \bar{\boldsymbol{y}}_j|^2$. Finally, the face direction that gives the minimum distance is determined as estimation.

However, estimating one of 16 discrete directions is not sufficient for detecting which object a person looks at. Still, preparing more face direction classes makes the inter-class distance shorter and then estimation performance decreases.

Figure 2 shows the square distance D_j between the feature vector of #1 testee's rotating image to C_j. D_j indicates minimum value at the learned direction d_j. As the face direction varies, D_j increases and either D_{j-1} or D_{j+1}, the square distance for the adjacent class decreases. Although it is not shown here, we also obtained a similar result even for the testee whose data are not used in the LDA process. These results suggest that the continuous face direction could be estimated by integrating the information of D_j. Equation (3) gives the face direction vector $\hat{\boldsymbol{d}}$ with the direction component vector $\boldsymbol{c}_j$ defined by equation (4).

$$\hat{\boldsymbol{d}} = \sum_{j=1}^{16} a_j \boldsymbol{c}_j, \tag{3}$$

$$\boldsymbol{c}_j = \frac{1}{D_j + k} \boldsymbol{u}_j. \tag{4}$$

Where $\boldsymbol{u}_j$ is a unit vector that has the same direction of C_j. k is a positive constant for preventing zero division. We set k to 1 here. When a lack of skin color area disables the face region detection, D_j is set to ∞ and then the length of $\boldsymbol{c}_j$ equals 0. We examine four kinds of settings for a_j. In each setting, the a_j whose index is equal to the indices shown in (s1)–(s4) is set to 1 and the others are set to 0. Here, k is the index for which $|\boldsymbol{c}_j|$ has the maximum value.

(s1) $k - 1, k, k + 1$
(s2) $k - 2, k - 1, k, k + 1, k + 2$
(s3) k, the index of the larger one of $|\boldsymbol{c}_{k-1}|$ or $|\boldsymbol{c}_{k+1}|$
(s4) the indices of the largest three $|\boldsymbol{c}_j|$s

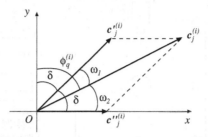

Fig. 3. Decomposition of direction component vector.

3 Coordinated estimation with multiple cameras

In Fig. 2, we assume that the estimation with a single camera becomes worse as the face is looking away from the front of the camera. Apparently, it is impossible to estimate the face direction around the back of the face because of the failure of the face region detection. In the real world, occlusions between the face and the camera might cause faults in estimation: They are caused by one's own arms, furniture, other people passing by, etc... Even in such cases, another camera without occlusion could make the estimation possible.

We propose a multiple-camera coordinated method that gives appropriate estimation for any face direction. For the purpose of coordinating the information from multiple cameras, the j-th direction component vector of the i-th camera $c_j^{(i)}$ is decomposed into two vectors $c'_j^{(i)}$ and $c''_j^{(i)}$ that are nearest and adjacent to $c_j^{(i)}$ as shown in Fig. 3. The length of the decomposed vectors $c'_j^{(i)}$ and $c''_j^{(i)}$ are given as follows:

$$\left|c'_j^{(i)}\right| = \frac{\left|c_j^{(i)}\right| \sin \omega_2}{\sin \omega_1 \cos \omega_2 + \cos \omega_1 \sin \omega_2}, \tag{5}$$

$$\left|c''_j^{(i)}\right| = \frac{\left|c_j^{(i)}\right| \sin \omega_1}{\sin \omega_1 \cos \omega_2 + \cos \omega_1 \sin \omega_2}, \tag{6}$$

$$\omega_1 = \phi_q^{(i)} - \delta \lfloor \phi_q^{(i)}/\delta \rfloor, \tag{7}$$

$$\omega_2 = \delta - \omega_1. \tag{8}$$

Therefore, the length of the j-th direction component vector in the system coordinates is given by equation (9).

$$|c_j| = \sum_{i=1}^{M} \left(\left|c'_j^{(i)}\right| + \left|c''_{j-1}^{(i)}\right| \right). \tag{9}$$

Where M is the number of cameras in coordination.

With the integrated direction component vectors, the multi-camera coordinated face direction estimation is also given by equation (3).

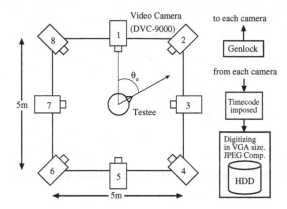

Fig. 4. Multiple-camera system.

4   Experiments

4.1   Data collection

We collected both training data and test data with the multiple-camera system shown by Fig. 4. Since estimating the person's position can be investigated independently of the face direction estimation, as mentioned before, we placed a testee at the center of the studio.

First, as training data, we collected 16 directional face images from ten testees. A testee gazed at the mark placed at 22.5° intervals and 50 images/camera were captured in each face direction by the eight cameras. In the end, we collected 640,000 ($16 \times 50 \times 8 \times 10$) images for training data.

Next, we recorded testee's 360° rotating images as test data. The frame in which the face looks straight at a video camera was detected visually as the reference at an interval of 45°. By dividing every interval equally into 16, we obtained a set of rotating face image data composed of 128 frames. Through the eight cameras, we obtained eight sets of test data for a testee ($8 \times 128 = 1024$ images/testee). Testee #1–10 are the same ten people in making the training data and testee #11 is independent of the training data.

4.2   Estimation results

In order to test the estimation performance, we use the error e of the face direction estimation defined by equation (10) and its variance σ_e^2 of the 128 face directions of the test data.

$$e = \left| \theta_a - \hat{\theta}_a \right|. \tag{10}$$

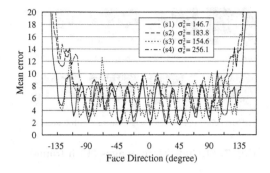

Fig. 5. Mean estimation error of face direction by single camera.

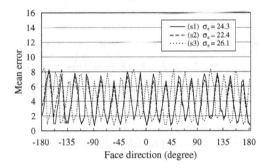

Fig. 6. Mean estimation error of face direction by multiple cameras.

Figure 5 shows the mean estimation error of testee #1–10 at every face direction and σ_e^2 obtained by the single camera estimation. The limits of the single camera cause a significantly larger error outside of $\pm 135°$ in all settings. The setting (s4) is inferior to others since the top three components do not always have matched directions. We judged that this setting is not good for the estimation. Every result has almost $22.5°$ cycles. The cycles of (s1) and (s2) are the same as the period of the face direction class C_j. On the other hand, the cycle of (s3) is half shifted from them. This is because the curve of D_j is similar to each other and the minimum of D_j appears at $22.5°$ intervals as shown in Fig. 2. The processing time for each single camera was 0.3 sec (Pentium III, 550MHz).

Figure 6 shows the results of testee #1–10 by the coordinated estimation of the eight cameras. The error is at most $8°$ and also the estimation around the back of the face is possible. Although not shown by figures, the result for the unknown test data of testee #11 is as good as the above results. It indicates $\sigma_e^2 = 30.5, 24.8$ and 34.2 for each setting. That is, multiple-camera coordination yields a great advantage. Since the phase of (s3) is half shifted to (s1) and (s2), combining them could give a better estimation.

| (a)Camera1 | (b)Camera2 | (c)Camera3 | (d)Camera4 |
| (e)Camera5 | (f)Camera6 | (g)Camera7 | (h)Camera8 |

Fig. 7. Occluded face images (13th frame).

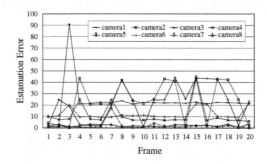

Fig. 8. Estimation error of face direction with occlusion by single camera.

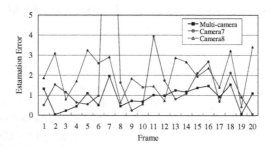

Fig. 9. Estimation of face direction with occlusion by multiple cameras.

Figure 7 shows the 13th frames of the image sequences taken by the eight synchronized cameras. As shown in Fig. 8, the estimation errors through "camera7" and "camera8" without occlusion are small. On the other hand, images through other cameras give gross errors. Figure 9 shows that coordinating the eight cameras can give better results than that given by any single camera. The error of the estimation by multiple-camera is within 2° throughout the sequence.

5   Conclusions

We proposed face direction estimation by coordinating multiple cameras. Experiments showed the advantages of this method:

1. The mean estimation error was at most 8° and its variance is low at any face direction angle.
2. Under a self-occlusion by hand, face direction estimation was successful.

This investigation has given us a good approach to determine where a person is looking.

References

1. Chen, Q., Wu, H., Fukumoto, T. and Yachida, M.: 3d head pose estimation without feature tracking. 3rd International Conference on Automatic Face and Gesture Recognition *FG98* (April 1998) 88–93
2. Fukunaga, K.: Introduction to Statistical Pattern Recognition. Academic Press (1990)
3. Hongo, H. et al.: Focus of attention for face and hand gesture recognition using multiple cameras. 4th International Conference on Automatic Face and Gesture Recognition *FG2000* (March 2000) 156–161
4. Kaneko, M. and Hasegawa,O.: Processing of face images and its application. IEICE Trans. Information and Systems E82-D(3) (March 1999) 589–600
5. Kuriyama, S., Yamamoto, K. et al.: Face recognition by using hyper feature fields. Technical Report of IEICE Pattern Recognition and Media Understanding **99** (Nov. 1999) 105–110
6. Matsumoto, Y. and Zelinsky, A.: An algorithm for real-time stereo vision implementation of head pose and gaze direction measurement. 4th International Conference on Automatic Face and Gesture Recognition *FG2000* (March 2000) 499–504
7. Pappu, R. and Beardsley, P.: A qualitative approach to classifying gaze direction. 3rd International Conference on Automatic Face and Gesture Recognition *FG98* (April 1998) 160–165
8. Pentland, A.: Looking at people: Sensing for ubiquitous and wearable computing. IEEE Trans. Pattern Analysis and Machine Intelligence **22**(1) (January 2000) 107–118
9. Wu, Y. and Toyama, K.: Wide-range, person-and illumination-insensitive head orientation estimation. 4th International Conference on Automatic Face and Gesture Recognition *FG2000* (March 2000) 183–188
10. Yasumoto, M., Hongo, H. et al.: A consideration of face direction estimation and face recognition. Proc. of Meeting on Image Recognition and Understanding *MIRU2000* **I** (July 2000) 469–474

Face Recognition Based on Local Fisher Features

Dao-Qing Dai[1], Guo-Can Feng[1], Jian-Huang Lai[1] and P. C. Yuen[2]

[1]Center for Computer Vision, Faculty of Mathematics and Computing
Zhongshan University, Guangzhou 510275 China
{stsddq,mcsfgc,stsljh}@zsu.edu.cn

[2]Department of Computer Science, Hong Kong Baptist University, Hong Kong
e-mail: pcyuen@comp.hkbu.edu.hk

Abstract. To efficiently solve human face image recognition problem with an image database, many techniques have been proposed. A key step in these techniques is the extraction of features for indexing in the database and afterwards for fulfilling recognition tasks. Linear Discriminate Analysis(LDA) is a statistic method for classification. LDA filter is global in space and local in frequency. It squeezes all discriminant information into few basis vectors so that the interpretation of the extracted features becomes difficult. In this paper, we propose a new idea to enhance the performance of the LDA method for image recognition. We extract localized information of the human face images by virtue of wavelet transform. The simulation results suggest good classification ability of our proposed system.

1. Introduction

Human face displays a range of information relevant to the identification of himself. The extraction of such information plays an important role in such tasks as human-machine interaction applications. This paper is concerned with feature extraction from human face images. The Linear Discriminate Analysis (LDA) of R. A. Fisher[4] is a well known statistic technique for pattern classification tasks and has been recently used for human face recognition. Among the global, face based methods, see e.g. [1] [3] [5] [7] [8] [9], LDA has been proven to be a promising approach for human face recognition[2]. In this paper, we propose a new idea to enhance the performance of the LDA based method. This approach is insensitive to variations in facial expressions and illumination changes. We extract localized information of human face images in both the space domain and frequency domain by virtue of the Wavelet Transform(WT)[6]. From the space-frequency localization property of wavelets, working with the wavelet transformed images help us on one hand to reduce the influence of noise, since the low resolution image is the results of an average processes, and on the other hand to overcome the singularity problem, which was solved[2] by applying a dimension reduction step, the Principal Components Analysis.

The organization of this paper is as follows. In section 2 we review the background of LDA. In section 3 our proposed method, the localized LDA(LLDA) is presented. The last section contains results and conclusions.

T. Tan, Y. Shi, and W. Gao (Eds.): ICMI 2000, LNCS 1948, pp. 230-236, 2000.
© Springer-Verlag Berlin Heidelberg 2000

2.   Reviews of LDA

In this section we give some background on the theory of LDA and discuss the use for face recognition.

LDA is a linear statistic classification method, which tries to find a linear transform T so that after its application the scatter of sample vectors is minimized within each class and the scatter of mean vectors around the total mean vector is maximized simultaneously. Some important details of LDA are highlighted as follows.

Let X be an ensemble of random vectors in R^d

$$X = \{x_1, x_2, ..., x_N\}$$

with c classes $X = \bigcup_{i=1}^{c} X_i$. Let the mean of class i be m_i, i=1,2, ..., c, and let the total mean vector be

$$m = \sum_{i=1}^{c} p_i x_i$$

where p_i is the prior probability of class i.
Let the *within-class* covariance matrix be

$$C_w = \sum_{i=1}^{c} p_i C_i$$

where C_i is the covariance matrix of class i. C_w measures the scatter of samples within each class.

Let the *between-class* covariance matrix be

$$C_b = \sum_{i=1}^{c} p_i (m_i - m)(m_i - m)^t$$

which measures the scatter of mean vectors around the total mean.

After the application of a linear transform A, in the transformed domain $Y=AX$, C_w and C_b become respectively

$$A^t C_w A \quad \text{and} \quad A^t C_b A$$

The Fisher class separability index is defined by

$$I(A) = tr[(A^t C_b A)^{-1}(A^t C_w A)],$$

which measures how much these classes are separated in the feature space.
The Fisher transform T is defined by

$$I(A) = \max_{A} I(A).$$

This requires solving the generalized eigenvalue problem

$$C_b T = C_w T D$$

where D is a diagonal matrix containing the eigenvalues. From linear algebra the number of non-zero eigenvalues is at most $c\text{-}1$.

Once the linear transform T is obtained, the feature vector $T'x_i$ is computed for each i and finally it is assigned to the class which has the mean vector closest to the feature space in the Euclidean distance in this coordinate system. This is equivalent to bisecting the feature space by hyperplanes.

It is known that LDA is optimal if all classes of signal obey multivariate normal distributions with different mean vectors and an equal covariance matrix. But since LDA relies on solving an eigensystem and squeezes all discriminate information into few basis vectors so that the interpretation of the extracted features becomes difficult, it is sensitive to noise, and it requires $O(d^3)$ calculations, which is beyond most of the computer capacity.

In image applications, the random vector x is formed by concantennating pixel intensity values, its dimension equals to the product of image dimensions in the two directions. Feature vectors are obtained by performing inner products with the eigenvectors and probe images.

An implementation difficulty arises in face recognition problem, the within-class scatter matrix $C_w \in R^{d \times d}$ is always singular. This means that there exists linear transform A such that the denominator of the index $I(A)$ is identically zero.

Our LLDA system will try to enhance the performance of LDA. The simulation results suggest good classification ability of our proposed system LLDA.

3. Proposed Method(LLDA)

Traditionally, to represent human faces, LDA is performed on the whole facial image[2]. However, this approach suffers from the limitations mentioned earlier. To resolve these limitations, we proposed a new idea(LLDA) to use LDA - applying LDA on the wavelet subband which corresponds to certain range of space-frequency resolutions.

Wavelet Transform (WT) has been a very popular tool for image analysis in the past ten years[6]. The advantages of WT, such as good time and frequency localizations, have been discussed in many research work. In this paper, Daubechies wavelet D6 is adopted for image decomposition. A two dimensional wavelet transform is derived from two one-dimensional wavelet transform by taking tensor products. The implementation of WT is carried out by applying an one-dimensional transform to the rows of the original image data and the columns of the row-transformed data respectively. An image is decomposed into four subbands LL,LH, HL and HH. The band LL is a coarser approximation to the original image. The bands LH and HL record the changes of the image along horizontal and vertical directions. While the HH band shows the high frequency component of the image. This is the first level decomposition. Further decomposition can be conducted on the LL subband.

In the LLDA system, WT is chosen in image frequency analysis and image decomposition because:(1) By decomposing an image using WT, the resolutions of the subband images are reduced. In turn, the singularity problem of the index $I(A)$ can be overcome by working on a lower resolution image. In earlier works a Principal Components Analysis procedure was used[2], which is an expensive computational step. (2) Wavelet decomposition provides local information in both space domain and frequency domain. The band-pass filter from WT extracts edge information, which helps to interpret the features. (3) The low-pass filter for the band LL of WT plays a role of noise reduction.

A face recognition system is developed to demonstrate the application of the LLDA method. The system follows the face-based approach and it consists of two stages, namely, training and recognition stages. Training stage identifies the representational bases for images in the domain of interest (that is reference images) and converts them into feature vectors. Recognition stage translates the probe image into probe features, and then, matches it with those reference images stored in the library to identify the face image.

4.   Results and Conclusion

Face image database from Yale University is used for these experiments. The images have variations in both facial expression and lighting. There are 165 frontal face images covering 15 individuals takes under eleven different conditions. Figure 1 shows eleven images of one individual. We do not use the last two images for their strong illumination changes.

Fig. 1. The example images of one subject from the Yale database.

For each image, WT is applied to get a mid-range frequency image of size 8x8.

Our first experiment tests the role of eigenvectors, 135 images of 15 persons are used. In Figure 2, the 14 Fisher images, that is eigenvectors corresponding to the 15-1=14 non-zero eigenvalues after proper scaling, are displayed as images. Figure 3 shows the eigenvalue distribution of the 14 non-zero eigenvalues. Table 1 is the recognition rate when the specified eigenvector is used for recognition. The first largest eigenvalue has better discriminate power.

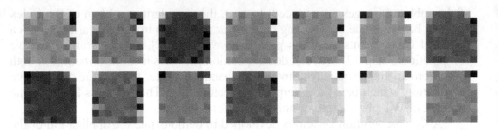

Fig.2. Fisher images: eigenvectors displayed as images.

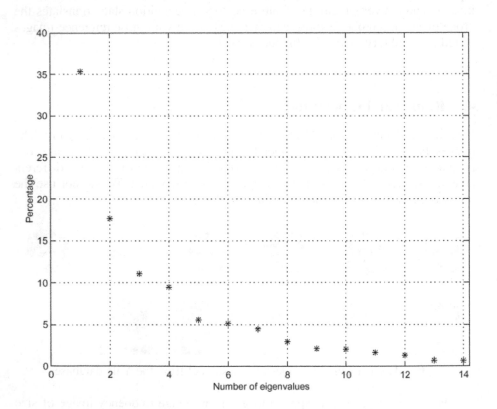

Fig.3. Distribution of the 14 non-zero eigenvalues

eigenvector	1	2	3	4	5	6	7
recognition rate	0.7	0.3667	0.5667	0.5333	0.5333	0.333	0.3667
eigenvector	8	9	10	11	12	13	14
recognition rate	0.4667	0.2667	0.3	0.3	0.2	0.3667	0.3667

Table 1. Recognition rate when one eigenvector is used.

Figure 4 displays the recognition rates when different number of eigenvectors is used. It shows that it suffices to use only seven eigenvectors.

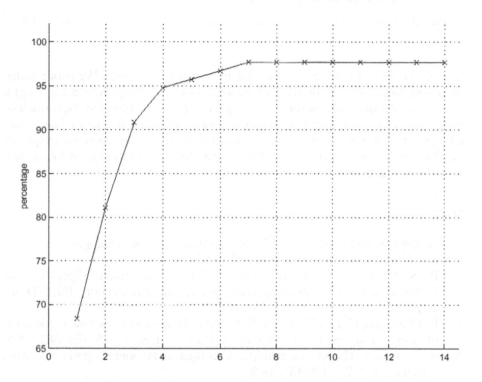

Fig. 4. The recognition rates vs the number of eigenvectors.

The second experiment test the recognition rate. In this case, 105 images of 15 persons are used for training and 135 images of 15 persons are used for recognition. The results are shown in Table 2.

recognition rate	false acceptance rate	false rejection rate
97.8%	0.0%	2.2%

Table 2. Performance of the proposed LLDA system.

In Table 3, we list the recognition rates by the Eigenface method, Principal components analysis(PCA) combined with LDA method and the proposed method. An advantage of the proposed method is that a threshold could be set.

In this paper we proposed LLDA system based on wavelet transform for human face recognition. The experiments show that this system has good classifica-

Methods	Recognition rate	Adaptive threshold
Eigenface	88.6%	
PCA+LDA	94.85	
Proposed method	97.8%	√

Table. 3. Comparison between the eigenface, PCA+LDA and the proposed method.

tion power. The LLDA features are edge information of images. We concentrated on a mid-range subband by the two reasons:(1) it gives edge information, (2) it overcomes the difficulty in solving a singular eigenvalue problem. Nevertheless, there may have other frequency bands, or combination of frequency bands, that can give better performance. Further studies in finding the optimal frequency band(s) for face recognition and adopt the LLDA method to other database will be our future direction.

References

1. D. Beymer and T. Poggio, Face recognition from one example view, Proc. *Int'l Conf. Computer Vision*, 500-507, 1995.
2. P. N. Belhumeur, J. P. Hespanha and D. J. Kriegman, Eigenfaces vs. Fisherfaces: Recognition using class specfic linear projection, *IEEE Trans. on PAMI*, Vol. 19, No. 7, 711-720, 1997.
3. R. Chellappa, C. L. Wilson and S. Sirohey, Human and machine recognition of faces: a survey, *Proceedings of the IEEE*, Vol. 83, No. 5, 705-740, 1995.
4. R. A. Fisher, The use of multiple measures in taxonomic problems, Ann. Eugenics, Vol. 7,   179-188, 1936.
5. X. Jia and M. S. Nixon, Extending the feature vector for automatic face recognition, *IEEE Trans. on PAMI*, Vol. 17, No. 12, 1167-1176, 1995
6. M. V. Wickerhauser, Adapted Wavelet Analysis from Theory to Software, AK Peters, Ltd., Wellesley, Massachusetts, 1994.
7. B. Moghaddam, W. Wahid and A. Pentland, Beyond eigenfaces: Probabilistic matching for face recognition, *Proceeding of face and gesture recognition*, 30-35, 1998.
8. C. Nastar, B. Moghaddam and A. Pentland, Flexible images: matching and recognition using learned deformations, *Computer Vision and Image Understanding*, Vol. 65, No. 2, 179-191, 1997.
9. W. Zhao, R. Chellappa and N. Nandhakumar, Empirical performance analysis of linear discriminant classifiers, *Proceedings of the 1998 Conference on Computer Vision and Pattern*, 164-169,1998

Combining Skin Color Model and Neural Network for Rotation Invariant Face Detection

Hongming Zhang[1], Debin Zhao[1], Wen Gao[1,2], Xilin Chen[1]

[1]Department of Computer Science and Engineering, Harbin Institute of Technology,
Harbin, 150001,China
hmzhang@ict.ac.cn, dbzhao@vilab.hit.edu.cn,xlchen@cti.com.cn
[2] Institute of Computing Technology, Chinese Academy of Sciences, Beijing, 100080,China
wgao@ict.ac.cn

Abstract. Face detection is a key problem in human-computer interaction. In this paper, we present an algorithm for rotation invariant face detection in color images of cluttered scenes. It is a hierarchical approach, which combines a skin color model, a neural network, and an upright face detector. Firstly, the skin color model is used to process the color image to segment the face-like regions from the background. Secondly, the neural network computing and an operation for locating irises are performed to acquire rotation angle of each input window in the face-like regions. Finally, we provide an upright face detector to determine whether or not the rotated window is a face. Those techniques are integrated into a face detection system. The experiments show that the algorithm is robust to different face sizes and various lighting conditions.

1 Introduction

Automatic localization of human faces is significant in many applications, for example, intelligent human-computer interaction, human face animating, face recognition, expression recognition and object-oriented image coding. Face detection is still a challenge because of several difficulties, such as variable face orientations, different face sizes, partial occlusions of faces in an image, and changeable lighting conditions.

Generally speaking, two kinds of methods have been developed to model face pattern, namely, the model-based technique and the feature-based technique. The first one assumes that a face can be represented as a whole unit. Several mechanisms are explored to characterize faces, such as neural network [1], probabilistic distribution [2], support vector machines [3], principal components analysis [4], and local feature analysis [5]. The second method treats a face as a collection of components. Important facial features (eyes, nose and mouth) are first extracted, and by using their locations and relationships, the faces are detected [6][7].

Color is a powerful fundamental cue that can be used as the first step in the face detection process. Many researchers utilize skin color models to locate potential face areas, then examine the locations of faces by analyzing each face candidate's shape and local geometric information [8][9].

T. Tan, Y. Shi, and W. Gao (Eds.): ICMI 2000, LNCS 1948, pp. 237-244, 2000.

Most previous similar systems that detect faces focused on locating frontal faces. So far there are two ways for rotated-face detection. The simplest but computationally expensive scheme would be to employ one of the exiting frontal, upright face detection systems. That is, the entire image is repeatedly rotated at some small increments and the detector is applied to each rotated image [1][7]. A faster and alternative procedure is described in [10], which makes use of a separate neural network to calculate the rotation angle of input window, and rotate the input window for a frontal face detector.

This paper presents a scheme to rotation invariant face detection in color images of cluttered scenes, which combines a skin color model, a neural network, and a frontal face detector. The paper is organized as follows. A brief system overview is given in section 2. In section 3,we analyze the chromatic distribution of human skin and put forward a skin color model to generate face candidates. Section 4 describes the neural network that detects orientation of a face. In section 5,a template-based approach to frontal face detection is provided. At last, the experiments and the conclusion are given.

2   System Overview

As illustrated in Fig.1, the system contains three parts: color-based segmentation, rotation angle examination, and frontal face localization. Initially, the skin model processes the input color image to provide potential face areas. In the gray image relevant to the color image, each 19x19-pixel window at different scales in face-like regions goes through the following procedures. Firstly, the neural network-based rotation detector receives the window as input, and outputs the rotation degree of the window. Secondly, the window is rotated to be upright. Finally, the rotated window is passed to a face detector, and the face detector decides whether or not the window contains an upright face.

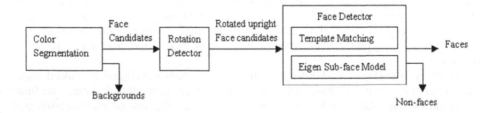

Fig.1. System Overview

3   Color Segmentation

Color information is often effective in image segmentation. For any color, its hue keeps almost constant under different lighting conditions. We consider the YUV color space as interesting color space because it is compatible to human color perception. We first convert the RGB color information in images to YUV color information:

$$Y = 0.299R + 0.587G + 0.114B$$
$$U = -0.147R - 0.289G + 0.436B$$
$$V = 0.615R - 0.515G - 0.100B$$

(1)

As shown in Fig.2, hue is defined as the angle of vector in YUV color space. Proper hue thresholds are obtained according to the observation that the hues of most Asian people and European-American people's skin vary in the range from 105^0 to 150^0 in most cases, as shown in Fig.3.

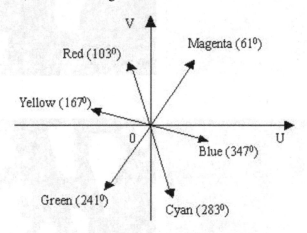

Fig.2. YUV color space

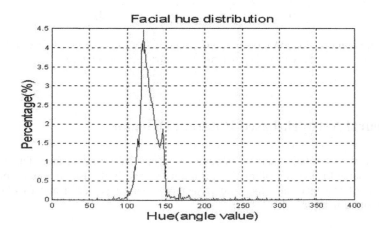

Fig.3. Facial hue distribution

Based on the skin color model, a segmentation procedure is performed on the color image. Firstly, a threshold operation is carried out on the corresponding hue image, and a roughly segmented image is produced, which contains the face-like regions and the background. Then, a morphological operator (e.g. dilation) is applied to fill holes in large connected regions and remove small regions; therefore, the face-like regions are separated. Finally, a labeling algorithm processes the segmented image. As a result, the number of potential face areas and their centers can be found. Examples of such segmentation are given in Fig.4.

Fig.4. Color segmentation(a)A color image in labortary (b)Color segmented image (c) A color image in VCD program   (d)Color segmented image

4   Neural Network-based Rotation Detector

In order to detect rotation angles of faces in images for rotated-face detection, we present a coarse to fine algorithm. A neural network is firstly used to detect the rough orientation of the face, and then the irises localization is performed to obtain the accurate rotation angle.

4.1   Neural Network

Unlike frontal faces, rotated faces in image plane are more difficult to model. To simplify the problem, our approach focuses on the upward faces that rotate less than 90^0 from upright (both clockwise and counterclockwise). It is practical to use neural network for clustering faces of any rotation angles.

The training database is a set of 19x19-pixel gray face images, which are preprocessed. First, we use a 19x19 binary mask in Fig.5 (a) to eliminate boundary pixels in background and hair for each sample. After that, each face pattern is represented as a 283-dimension vector. Then histogram equalization is applied to normalize the changes in lighting. We generate 1235 training examples, and divide them into five groups according to the rotation degrees. Each group represents one rotational orientation, as shown in Fig.5 (b).

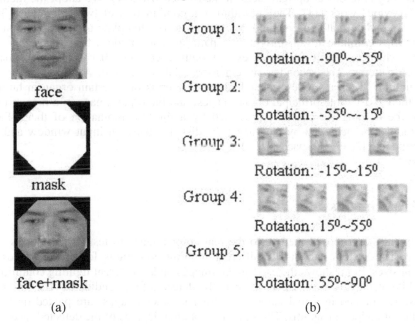

(a)   (b)

Fig. 5. (a)Mask (b) Examples of training data

The input to the neural network is a normalized 283-dimension vector; the output is the group label to which the vector belongs. Note that the network does not require an upward face as input. If a non-face or downward face is encountered, the output will be a meaningless group label.

The architecture of the neural network consists of three layers, an input layer of 283 units, a hidden layer of 5 units and an output layer of 5 units. Every layer is fully connected to the next. Each unit is activated by the sigmoid function, and the network is trained with the back propagation algorithm. We stress that a similar result can be acquired for downward faces by using the same procedure.

4.2   Irises Localization

The classified output from the neural network is insufficient for rotating an input image window to make the face upright, but it initializes the locations of eyes. Therefore the irises are extracted fast to calculate the accurate rotational degree. According to the output group label, we look for the irises in the interest eyes area. If such two points are found, the accurate rotation angle can be obtained.

5   Face Detection Method

The last step is to detect upright faces in the rotated windows. We adopt the method of our previous work [11]. The algorithm is a combination of two techniques. The first is a template matching method. We design a face template according to spatial relationships common to all upright face patterns, and divide a face image into 14 regions, which represent eyebrows, eyes, mouth, cheek, etc. It can eliminate many non-face patterns. The second is an eigenspace-based scheme. A human face is a configuration of organs: eyes, cheek and jaw. In terms of a certain organ in human faces, there exist some principal organ images, and the organ image in a human face among the training set can be represented by a linear combination of them. Face detection can be achieved by measuring the distance between input window and its projection into the eigenespaces.

6   Experiments

We applied the proposed method to the real color images to test its robustness and validity. The system can handle different lighting conditions for upward faces at different sizes. Fig.6 shows the experiment images under different lighting conditions. The white rectangles in the figure indicate the detected face candidates. The first two images are grabbed in a real labortary room.The other images are picked up from web sites and video programs. The sizes and rotation degrees of the detected faces are varied.

We tested our algorithm on 96 still color images, which contain 155 human faces including 46 faces rotated at any degrees. As shown in table.1, the proposed method sometimes fails to detect real faces or detects false faces. This is mainly caused by two reasons. Some faces are largely occluded, so they are often missing detected. Some false detected faces are very similar to face patterns in eigenspaces.

Table 1. Testing results

Number of faces	Correctly detected faces	Miss detected faces	False detected faces
155	125	30	9

Fig. 6. Experimental results for upward faces

We also trained a rotation detector for downward faces. Some experimental results are shown in Fig.7.

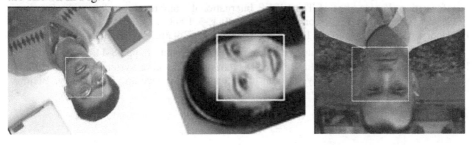

Fig. 7. Experimental results for downward faces

7   Conclusion

In this paper we have presented an approach for the rotation invariant face detection in color images based on a frontal face detection system. We have developed a coarse to fine rotation detector that is neural network-based in combination with shape information. The utilization of color image segmentation improves the speed. We have integrated the presented methods into a face detection system. The system is successful in detecting upward faces at any degree of in-plane rotation in color images. It can be extended into the case of downward face detection.

References

1. Henry A. Rowley, Shumeet Baluja, Takeo Kanade. Neural Network-Based Face Detection. IEEE Translation On Pattern Analysis And Machine Intelligence, Vol.20, No.1, pp29-38, January 1998.
2. Kah-Kay Sung, Tomaso Poggio. Example-Based Learning for View-Based Human Face Detection. IEEE Translation On Pattern Analysis And Machine Intelligence, Vol.20, No.1, pp39-50, January 1998.
3. Edgar Osuna,Rebert Freund, Federico Girosi. Training Support Vector Machines: an Application to Face Detection. Proceedings of CVPR'97, pp130-136, 1997.
4. Menser B. Muller F. Face detection in color images using principal component analysis. Proceedings of 7th International Congress on Image Processing and its Applications. 13-15 July 1999.
5. Penio S Penev, Joseph J Atick. Local Feature Analysis: A general statistical theory for object representation, http://venezia.Rockefeller.edu
6. Kin Choong Yow,Roberto Cipolla. Feature-Based Human Face Detection, CUED/F-INFENG/TR 249, August 1996.
7. Jun Miao, Baocai Yin, Kongqiao Wang, Lansun Shen, Xuecun Chen. A hierarchical multiscale and multiangle system for human face detection in a complex background using gravity-center template. Pattern Recognition, vol.32, no.7, pp.1237-48, July 1999.
8. Sobottka K, Pitas I. A novel method for automatic face segmentation, facial feature extraction and tracking. Signal Processing: Image Communication, vol.12, no.3, and pp.263-81, June 1998.
9. Sun QB, Huang WM, Wu JK. Face detection based on color and local symmetry information. Proceedings of Third IEEE International Conference on Automatic Face and Gesture Recognition. IEEE Comput. Soc. pp.130-135, 1998.
10. Henry A. Rowly, Shumeet Baluja, Takeo Kanade. Rotation Invariant Neural Network-Based Face Detection. Computer Vision and Pattern Recognition, pp.38-44, 1998.
11. Liu Mingbao, Gao Wen. A Human Face Detection and Tracking System for Unconstrained Backgrounds. International Symposium on Information Science and Technology, Beijing, 1996.

Multimodal Integration Using Complex Feature Set[*]

Xiao Bin, Pu Jiantao, Dong Shihai

Graphics Laboratory, Department of Computer Science and Technology, Peking University

Beijing 100871, P.R.China

pjt@graphics.pku.edu.cn

Abstract. Multimodal integration is an important problem of multimodal user interface research. After briefly surveying research on multimodal integration, this paper introduced a natural language processing method to resolve the problem of integration. It introduced the unification operation on complex feature set and extended this operation in order to represent the information from input modalities and implement multimodal integration.

1. Introduction

Multimodal integration is the keystone and nodus of multimodal user interface and its solution is the precondition of multimodal interaction [1]. Currently, the main solutions for this problem are: the integration method based on time designed by Coutaz and Nigay[2]; EDWARD project's method for anaphora resolving [3]; the ATOM model which described HCI sequence and all kinds of grammar constraining information using CFG (Context Free Grammar)[4]; hierarchical integration algorithm[5]. Among these methods, it is difficult to guide the multimodal integration using the semantic constrains between modalities because there is not a consistent describing format for every modality's semantic. This paper introduces the representing method of complex feature set which is used in natural language understanding into the representation of multimodal input information in order to provide a kind of consistent representation form for every modality. Then it extends this unification operation to fulfills multimodal integration so as to make it to take full advantage of the semantic, grammar and time constraints between modalities, finally realize the integration of multimodalities.

The rest of this paper is organized as following: section 2 introduced complex feature set, unification operation, extended unification operation. The 3rd section gives the example that uses this method. Finally, the 4th section summarizes this paper.

---

[*]This work is supported by the National Natural Science Foundation of China, Project Number 69773024 and the Special Research Foundation for Doctorate Discipline in Colleges and Universities of China ,Project Number No. 98000133

T. Tan, Y. Shi, and W. Gao (Eds.): ICMI 2000, LNCS 1948, pp. 245-252, 2000.

2. Extended Unification-Based Multimodal Integration Algorithm

As far as the problem of multimodal integration concerned, we do not think all the modalities are equally important for resolving this problem. We consider speech modality the predominant modality in multimodal HCI because normally it can provide more information than other modalities for expressing user's interaction intent. Some efficiency evaluation experi-ments of multimodal systems have also proved that users prefer speech modality in the multimodal interactive systems which have more input devices, such as speech, pen, keyboard, mouse, and so on[6].

In order to perform the task correctly corresponding to user's intent, the computer should not only transfer the speech signal into text strings, but also understand the meaning of the strings. This requires the supporting of computer natural language understanding. So, in order to solve the problem of multimodal integration, we can start from the text strings that get from speech recognition, and introduce some grammar and semantic representation methods which are used in natural language understanding area to unify the information representation of multimodalities. On the basis of the unifying representation we can realize the understanding of user interaction semantic which would lead to the multimodal integration.

According to this idea, we reviewed some natural language understanding method and choose the complex feature set (CFS) to represent the information from multimoda-lities and used the unification operation on CFS to realize multimodal integration.   Just following the introduction of some related concepts, we will present the corresponding algorithm.

2.1 Basic Concepts

Definition 2.1 Complex feature set (CFS): A is a complex feature set, only if A can be expressed as the following format:

$$\begin{bmatrix} f_1 = v_1 \\ f_2 = v_2 \\ \dots\dots \\ f_n = v_n \end{bmatrix} \quad n \geqslant 1$$

in which f_i represents feature name, v_i represents feature value, and the following conditions are satisfied:

(1) Feature name f_i is atom, feature value v_i is either atom or another complex feature set.

(2) $A(f_i) = v_i$ (i=1,...,n) shows that in complex feature set A, feature f_i's value is v_i.

It is obvious that CFS is a multi-value function in fact. We can use *dom(A)* to represent A's definition domain which is the set of all feature name, and if definition domain is null, we can only get a null CFS which can be represented with Φ.

In addition, we call **Atom Feature** whose value is an atom and **Complex Feature** whose value is a CFS.

Definition 2.2 The unification operation on CFS:

If we use symbol $\overline{U}$ to represent unification operation, then:

1.   If a and b are all atoms, then $a \overline{U} b = a$, only if a = b; otherwise $a \overline{U} b = \Phi$ 。

2.   If a is atom, B is complex feature set , then $a \overline{U} B = \Phi$; and if A is complex feature set, b is atom, then $A \overline{U} b = \Phi$.

3.   If A and B are both CFS, then:

   (1) If $f \in$ dom (A) $\cap$ dom (B), and A(f) = v,   B(f) = v', then

   if v $\overline{U}$ v'= Φ, then $A \overline{U} B = \Phi$;

   if $v \overline{U}$ v' $\neq \Phi$, then $f = (v \overline{U} v') \in A \overline{U} B$;

   (2) If $f \in$ dom(A) – dom (B) and A (f) = v, then $f = v \in A \overline{U} B$.

   (3) If $f \in$ dom(B) – dom (A) and B (f) = v, then $f = v \in A \overline{U} B$.

   From the above two definition we can obtain:

   1.   By introducing appropriate features in CFS, we can use different features to describe words, phrases and syntax and semantic of sentences, and so on. And the syntax and semantic constraints can be expressed by the relations between the features among CFSs. Unification operation can use these constraints to exclude ambiguity and finish the analysis of sentence during the process of unifying small component into larger component. While in multimodal interaction, the interactive means between user and computer not only include natural language but also mouse, keyboard, handwriting and so on. The input primitive of these interactive devices has similar functions when compared with natural language words. Therefore, we can generalize the CFS to represent not only the info of speech modality's natural language sentences, but also the info of the interactive vocabularies from other input modalities such as mouse, keyboard etc. This will establish the basis for every modality's input information.

   2.   Unification operation is suitable for multimodal integration. During the process of multimodal interaction, every modality's input info usually represents partially the overall interaction intent. Here the relations between modalities are complementary. Sometimes redundancy may be occurred [7]. The process of

integration is just the process of getting user's intention from this complementary or redundant information. And the CFS representation of the complementary or redundant partial info coming from different modalities will include corresponding complementary or redundant features, the unification operation on which just realizes representing process of unifying these partial info into an integrated info. The 3rd item's (2)(3) in Definition 2.2 describe the integration of such complementary info, while (1) describes the integration constraint of redundant info.

Therefore, CFS can provide united representing format for every modality's partial info, and unification operation on it   provides a kind of capability for the integration of these partial info. The combination of the above two provides a power means for multimodal integration. But when considering the character of multimodal interaction, simple unification operation can't satisfy the need for integration completely. The (1) in definition 2.2 will succeed only when the values of two atom feature are equal, otherwise the unification will fail. This condition is too strict for the info integration of multimodality, because some features might not be equal completely in multimodal interaction, but their unifications are expected to succeed. For example, when some info is input from two modalities, the time features of the two partial info are not necessarily equal, but if they locate in near time interval (the interval can be obtained from experiment), they are still expected to succeed in unification. Similarly, there are unequal but compatible cases between some semantic genus. Therefore, it is necessary for us to extend the simple unification operation to resolve this problem. Our idea is to weaken the equal judge to compatible judge. However the compatible judge might be different for different atom feature , so we extend unification operation on the basis of introducing a compatible judge function for each atom feature.

2.2   Extend Unification Operation and Multimodal Integration

Firstly, we introduce the corresponding Compatible Judge Function (CJF) $Comp_i()$ for every atom feature f_i. If $A(f_i)=a$, $B(f_i)=b$, then CJF $Comp_i(a,b)$ should judge whether atom a is compatible with atom b according to the character of fi. If compatible, then return a non-null value of a and b's fusion result. If not compatible, return NULL. The realization of function $Comp_i()$, including   the representation of the value after a and b fuse, should be designed by algorithm designer according to the practical condition. Thus, extend unification operation can be defined as following:

Definition 2.3 Extend Unification Operation on CFS:

If we use symbol   to represent unifica-tion operation and A and B are all CFSs, then:

1.If $A(f) = a$, $B(f) = b$，a and b are both atom，and the corresponding CJF of atom feature f is Comp(); then: $a \overline{U} b = Comp\ (a,b)$.

2.If a is atom, B is CFS，then $a \overline{U} B = \Phi$; if A is CFS, b is atom, then $A \overline{U} b = \Phi$.

3.(1)If $f \in \text{dom}(A) \cap \text{dom}(B)$，and $A(f) = v$, $B(f) = v'$, then：

　　if $v \overline{U} v' = \Phi$,then $A \overline{U} B = \Phi$;

　　if $v \overline{U} v' \neq \Phi$, then $f = (v \overline{U} v') \in A \overline{U} B$;

(2)If $f \in \text{dom}(A) - \text{dom}(B)$， and $A(f) = v$, then $f = v \in A \overline{U} B$.

(3)If $f \in \text{dom}(B) - \text{dom}(A)$，and $B(f) = v$, then $f = v \in A \overline{U} B$.

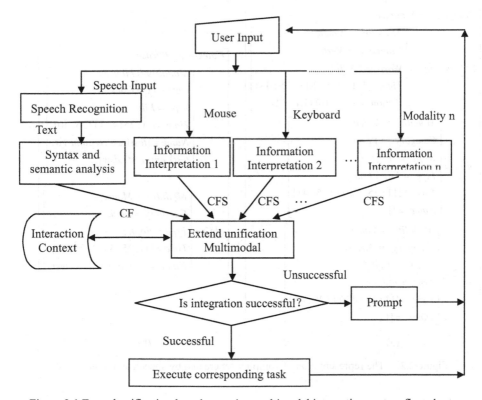

Figure 2.1 Extend unification-based operation multimodal integration system flow chart

According to this definition, the operation can deal with more complex feature, then can be applied in multimodal integration. On the basis of this kind of unification operation, the flow chart of multimodal integration is showed in figure 2.1. Here, we use speech modality as the primary modality of integration. User's input speech(Chinese language)can be transferred to corresponding Chinese character strings after speech recognition. Then syntax, semantic analysis module decompose the strings into words and analyzes the grammar, semantic according to the word knowledge library and got a CFS representation. At the same time, other modalities which receive user's input can also form CFSs corresponding to input info by each interpretation mechanism. Multimodal integration uses extend unification operation to integrate this CFS information. If succeed, then executes the corresponding task. If the complete information of related task can't be obtained from the existing inputs, it tries to find corresponding information from the interaction context, if can not find, reports error or prompts user for the remedy input.

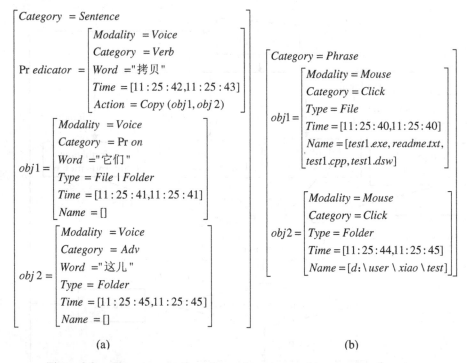

(a)   (b)

Figure 2.2.   The representation of complex set created by speech and mouse

For instance, in the example of the multimodal file manager we have developed, user can say, "Please copy them to this (folder)" (in fact in Chinese "请把它们拷贝到这儿"), at the same time using mouse to select the files and target folder. The CFS can be created as in figure 2.2(a)(b) after system receive such input.

In this figure, (a) is the representation of complex feature set of speech modality information; (b) is the representation of mouse modality information. Feature category represents syntax component, Predicator represents the predicate

$$
\begin{bmatrix}
Category & = Sentence \\[4pt]
Pr\,edicator & =
\begin{bmatrix}
Modality & = Voice \\
Category & = Verb \\
Word & ="拷贝" \\
Time & = [11:25:42, 11:25:43] \\
Action & = Copy\,(obj\,1, obj\,2)
\end{bmatrix} \\[4pt]
obj\,1 =
\begin{bmatrix}
Modality & = Voice \ \& \ Mouse \\
Category & = \Pr on \ \& \ Click \\
Word & ="它们" \\
Type & = File \\
Time & = [11:25:40, 11:25:41] \\
Name & = [test\,1.exe\,, readme\,.txt\,, \\
& test\,1.cpp\,, test\,1.dsw\,]
\end{bmatrix} \\[4pt]
obj\,2 =
\begin{bmatrix}
Modality & = Voice \ \& \ Mouse \\
Category & = Adv \ \& \ Click \\
Word & ="这儿" \\
Type & = Folder \\
Time & = [11:25:44, 11:25:45] \\
Name & = [d:\backslash user \backslash xiao \backslash test\,]
\end{bmatrix}
\end{bmatrix}
$$

Figure 2.3 the result of Extended-Unification

verb of sentence, obj1, obj2 represent the necessary parameters corresponding to predicate verb. It is not difficult to get the meaning of other features from the respective English words, so we don't interpret them here. Although some atom feature values of the two complex feature sets (a)(b) are not equal completely, we can use extend unification   to unify the two complex feature sets by defining their CJFs for the atom features. The result is shown as figure 2.3. At this time the parameters needed for the action "Copy" are all obtained, so system can execute user's action.

We can also deal with some cases when some modality has some error. For example (still based on above example), if the speech recognition result is "Please copy them to" ("请把它们拷贝到"), it is obvious that "this"("这儿") is emitted. But the mouse pointing are correct, then the system can put the feature obj2 in figure 2.2(a) to null, while the representation of mouse modality doesn't change. At such case, extend unification can succeed too, the result (obj2's feature Modality=Mouse，Category=Click，no feature   Word）is similar to the figure2.3. The system can still execute user's action.

3. Instance

Based on the above idea, we develop the application Multimodal File Manager, providing a multimodal user interface to carry out file operations by speech (speaking Chinese) and mouse. When user want do some file operation, he can avoid using the multi-level menus in traditional WIMP interface , instead, directly describes his intent by speech cooperating with other modalities. The system's usability is enhanced and easy to learn.

4. Conclusion

This paper introduces the unification operation acting on complex feature set into the research of multimodality and extends the unification operation according to the actual requirement of multimodal integration. Such a kind of info representation and the corresponding multimodal integration mechanism are provided. The demo instance proved that this method is feasible. Currently, we are doing more research on the way for describing more complex task and dealing method.

REFERENCE

1. Dong ShiHai, Wang Jian, Dai Guozhong: Human-Computer Interaction and Multimodal User Interface, Science Publishing House, 1999

2. Nigay L., Coutaz J.: A Generic Platform for Addressing Multimodal Challenge. In Proceedings of ACM CHI'95 Conference on Human Factors in Computing Systems, 98-105

3. Bos, E.: A multimodal syntax-directed graph-editor. In L. Neal & G. Szwillus, Eds. Structure-based editors and environments. New York: Academic Press

4. Chen Min, Luo Jun, Dong Shihai: ATOM—Task Oriented Multimodal Interface Structure Modal, J. of Computer Aided Design and Graphics, 1996, 9: 61-67

5. Zhang Gao: Multimodal User Interface Develop Method and Environment, Doctoral Dissertation of Software Institute of SAC, 1999

6. Sharon Oviatt, Antonella DeAngeli, Karen Kuhh: Integration and Synchronization of Input Modes during Multimodal Human-Computer Interaction. Proceedings of Conference on Human Factors in Computing System CHI'97, Atlanta, GA, ACM Press, New York, 1997, 410-422

7. Sharon Oviatt: Ten myths of multimodal interaction, Communications of the ACM, Vol. 42, No 11, November, 1999, 74-81

Bilingual Dictionary Based Sentence Alignment for Chinese English Bitext[*]

Zhao Tiejun, Yang Muyun, Qian Liping, Fang Gaolin

Dept. Of Computer Science and Engineering

Harbin Institute of Technology, P. R. China, 15001

tjzhao@mtlab.hit.edu.cn

Abstract. Bitext is a rather "hot" issue among current natural language processing and automatic sentence alignment is the first step towards bitext processing. Following the shift form previous knowledge-poor approach (pure length-based) to current some-what knowledge-rich approach, this paper suggests a bilingual dictionary based sentence alignment method for Chinese English bitext and realizes it through dynamic programming. Experiment on HIT bitext shows that this method has achieved an accuracy as high as 95%, and therefore is worthy of further exploring.

Key words: alignment, bitext, bilingual dictionary

1. Introduction

A bitext is a corpus of text duplicated in two languages, e.g. Hansard bitext in both English and French and HKUST Hansard bitext in both English and Chinese. Bitext has arisen intensive interests among NLP researchers, especially those related to machine translation (MT) [1,2], word sense disambiguation[3,4] and bilingual lexicography[5,6].

Since human translators usually do not include sentence alignment information when creating bitext, it is usually agreed that automatic algorithm for sentence alignment is the first step towards bitext processing (though some may argue that paragraph alignment is worthy of research interest[7]).

The first successful algorithms applied to large bitext are those of Brown et al [8] and Gale and Church [9]. Most of them used models that just compared the lengths of units of text in the parallel corpus. While it seems strange to ignore the richer information available in the text, these methods can still be quite effective. After that, researchers experimented to align sentences with lexical information[10,11,12]. Still these methods avoided using any bilingual dictionary, and turned to all sorts of formula to automatically estimate lexicon correspondences among bitext. All of these algorithms are commonly termed as "knowledge-poor approach" because they do not assume the existence of any linguistic knowledge.

Recently researchers have been employing bilingual dictionaries to improve previous length based alignment algorithm[13,14]. Though all of them restricted the

[*] Supported by the High Technology Research and Development Program of China (306-zd13-04-4)and the National Natural Science Foundation of China(69775017).

T. Tan, Y. Shi, and W. Gao (Eds.): ICMI 2000, LNCS 1948, pp. 253–259, 2000.

involved dictionary entries in small number so as not to burden the dynamic programming too much, these method represents the state-of-the-art for current methods: a move from the knowledge-poor approach that has characterized much statistical NLP work to a knowledge rich approach. And for practical purposes, since knowledge sources like online dictionaries are widely available, it seems unreasonable to avoid their use simply for pure statistics sake.

Such a shift from knowledge-poor approach to knowledge-rich approach can also be observed in the research of Chinese English sentence alignment: from that of Liu et al(1995)[15], that of Liu et al(1998)[16] to those of Wang Bin(1999)[7] and Sun Le et al(1999)[17]. Following this trend, this paper carries out a more "radical" research on sentence alignment purely based on bilingual dictionaries. Viewed together with words correspondence task after sentence alignment, the proposed method will cut the total computation by half if it holds.

2. Design and Implementation of Dictionary-Based Sentence Alignment

2.1 A Formal Description of Sentence Alignment Problem

Generally, the problem of bitext sentence alignment can be formally defined like this[13]:

Suppose S be a text of n sentences of source language, and T be a text of m sentences of target language (translation):

$$S = s_1, s_2, \ldots\ldots s_n$$
$$T = t_1, t_2, \ldots\ldots t_m$$

Let p be a pair of minimal corresponding segments in texts S and T. Suppose p consist of x sentences in S $(S_{a-x+1}, \ldots\ldots S_a)$ and y sentences in T $(T_{b-y+1}, \ldots\ldots, T_b)$, then p can be described as:

$$p = <a, \ x; \ b, \ y;> \ (x, y \ may \ be \ 0)$$

Such p is usually called a sentence *bead*, as described in Brown et al.(1991). Then alignment is to arrange sentences in bitext of S and T into a sequence P of sentence beads:

$$P = p_1, p_2, \ldots\ldots, p_k$$

It is also assumed that each sentence belongs to only one sentence bead and order constraints must be preserved in sentence alignment. Let $p_i = (a_i, x_i; b_i, y_i)$, and:

$$a_0 = 0, \qquad\qquad a_i = a_{i-1} + x_i$$
$$b_0 = 0, \qquad\qquad b_i = b_{i-1} + y_i \quad (1 \le i \le k)$$

Suppose that a scoring function h can be defined for estimating the validity of each sentence bead p_i. Then sentence alignment can be viewed as an optimization problem that finds a sequence P of sentence beads with the maximum total score H:

$$H (P) = H_h (h(p_1), \cdots\cdots, h(p_k))$$

Consequently, various approaches to sentence alignment can be regarded as different decisions on h.

2.2 Calculation of *h* with Bilingual Dictionary

Gale and Church's calculation of h depends simply on the length of source and translation sentences measured in characters. The hypothesis is that longer sentences in one language should correspond to longer sentences in the other language. This seems uncontroversial, at least with similar languages and literal translations.

Using the same length based method also turned good results for Chinese English sentence alignment[15]. However, since English and Chinese belong to 2 totally different families, sentence alignment between them is faced more pressure to overcome the shortcomings—"lack of robustness and error spreading"[11]. In view of previous efforts of lexical information based improvements, we start to consider the practicality of align with bilingual dictionary.

Since translation is more than a dictionary-based transfer process, there should be plenty of misses if we search words' translations of a source sentence in its target sentence. Some previous researches also illustrated that only 1/3 English verbs' dictionary translation could be found in a real corpus[18]. So we conduct a pre-search of dictionary coverage in our bitext as follows:

• English morphological analysis and get the corresponding English-Chinese dictionary translations for each English word;

• Sort the translations of each English word in length ;

• Until all English sentences are searched:

a) For i_{th} English sentence, get (i-1, i, i+1)th Chinese sentences as search target;

b) For each word in English sentence, delete its first found translation in search target;

With a E-C dictionary of 57269 entries (about more than 70,000 translations), we got the basic qualities of HIT Chinese and English bitext(see Table 1). In fact, HIT Chinese and English bitext is formed by texts of College Core English (Book 1-4) and the translations contained in the Teachers' Book(1-4), which are published by Higher Education Press. It contains 1570 paragraphs, 134 texts and 46 units in total.

Total English Sentences	Total Chinese Sentences	Total English Words	Total Translation- found English Words	Dictionary Translation Coverage (%)
5128	5057	84886	43109	50.8%

Table 1. Basics of HIT Chinese English Bitext

Though the combination of several different English-Chinese dictionaries will improve the dictionary translation coverage, we believe 50% coverage is a good point to see if dictionary based alignment is practical. And our algorithm will start from that ground.

To estimate h with translation, the following formula will not escape from researchers' eyes:

$$h\ (alignment\ rate) = \frac{2 \times (\text{Number of aligned words})}{\text{Number of words to be aligned}(English + Chinese)}$$

Application of this formula, however, assumes a module of Chinese segmentation and risks the errors of Chinese segmentation. For the purpose of sentence alignment, the following formula is adopted instead:

2.3 Balancing 2:2 in Dynamic Programming

$$h\,(p) = \frac{\textit{Aligned English words' length } + \textit{Aligned Chinese translations' length}}{\textit{English sentence length} + \textit{Chinese sentence length}}$$

Let $P_i(a_i, x_i; b_i, y_i)$ be the sequence of sentence beads from the beginning of the bitext up to the bead p_i:

$$P_i = p_1, p_2, \cdots, p_i;$$

Then one can recursively define and calculate $H(P_i)$ by using the initial condition $h(a_0, b_0) = h(0,0) = 0$:

$$H(Pi) = \max_{xi,yi} \{h(a_i - x_i, b_i - y_i) + h(a_i, x_i; b_i, y_i)\}$$

where $(x_i, y_i) \in \{(1,0), (0,1), (1,1), (1,2), (2,1), (2,2)\}$ for the moment.

This task can be readily solved with dynamic programming. But such strategy tends to generate a (2,2) rather than (1,1). To balance this tendency, the calculation of $h(p)$ are attached a occurrence probability of (x,y):

$$h\,(p) = \frac{\textit{Aligned English words' length } + \textit{Aligned Chinese translations' length}}{\textit{English sentence length} + \textit{Chinese sentence length}} \times \Pr ob(x_i, y_i)$$

Table 2 lists the value of each *Prob (x,y)*.

Type of (x_i, y_i)	*Prob* (x_i, y_i)
(1, 0) or (0, 1)	0.01
(1, 1)	0.89
(1, 2) or (2, 1)	0.07
(2, 2)	0.03

Table 2. Types of Alignment and Their Occurrence Probability

3. Experimental Results and Discussion

Based on the above algorithm, a Chinese English sentence aligner has been realized to process HIT Chinese and English bitext. In the system, paragraph is set as a hard delimiter that cannot be violated and any paragraph with only one sentence is regarded as a bead. But sentence boundaries are pre-identified and errors are corrected manually before sentence alignment.

From total auto-generated 4725 beads of HIT bitext, 1193 beads belonging to Book 1 are sampled for manual check. The sample corpus contains 34 units, 369 paragraphs, 1,300 English sentences and 1,224 Chinese sentences. It is decided that 1139 of them are correct alignments and the manual correction turns out 1,168 beads as final standard. During the inspection, any wrong bead is counted as an error (note that a distortion of one correct bead will generate 2 errors at least). Table 3 lists the details.

Align Type (E-C)	(1,0)	(0,1)	(1,1)	(1,2)	(2,1)	(2,2)	Total
Aligner Result (Bead Number)	20	1	1017	47	105	3	1193
Error Number	19	1	20	1	13	0	54
Accuracy(%)	5	0	98.0	97.9	87.6	100	95.5
Error Correction Result(Bead Number)	0	0	4	4	8	2	29*
Correct Result (Bead Number)	1	0	1001	50	100	5	1168*

Table 3. Experimental Results and Correct Beads
(*Manual correction find the following exceptions: six (3,1), one (3,2), three
(4,1) and one (5,1), which are not listed in the table.)

From table 3, it can be calculated that 97.5% (recall rate) of the correct alignment can be turned out by our algorithm. What interests us, however, is the error causes. First, we would not blame the improper value of *Prob (x,y)*, since they are independently estimated from other resources. But the experiment indicates that *Prob (x,y)* is dependent on corpus and deserves further modification. Another problem concerns about the 10 exceptions of (x_i, y_i) which are beyond the capability of the algorithm. A more robust mechanism should be integrated into the proposed algorithm in condition that it will not degrade the performance.

Thought the experiment is rather encouraging, there are still several problems that need resolving to facilitate a practical Chinese English sentence aligner. A key problem among them is to what dictionary translation coverage the algorithm can be successful, which decides the adaptability the dictionary based approach. The 50% coverage is rather enough for the experiment purpose. However, large-scale bilingual corpus cannot usually be expected with such high coverage. A further experiment needs to be carried out for detecting the least dictionary coverage requirement.

To solve such drawback, it is possible, on one hand, to improve the lexical corresponding information with 3 strategies: dictionary combination, introducing synonyms and directly estimate uncovered words' correspondences. On the other hand, length-based method could be considered as a compensation, i.e. length serving as a supporting information. All of these researches will be carried on in the near future.

4. Concluding Remarks

With more and more available bilingual dictionaries, bilingual dictionary-based approach seems to be a preferable alternative for Chinese English sentence alignment. Compared with previous length-based or lexicon information-based alignment methods, dictionary-based method possesses a few advantages: e.g.

sentence bead identification from noisy bitext, direct word correspondences calculation and suitable for cross-language information retrieval et al. However, real world problems are rather complex and the reported method still needs perfecting to achieve satisfactory results.

References

1. P. F. Brown, J. Cocke and S. A. Pietra et al, A Statistical Approach to Machine Translation, *Computational Linguistics,* 1990, 16(2):79-85
2. Dekai Wu, Grammarless Extraction of Phrasal Translation Examples from Parallel Texts, *Proc. of the 6ᵗʰ International Conference on Theoretical and Methodological Issues in Machine Translaiton(TMI'95),* 1995: 354-372
3. W. A. Gale, K. W. Church and David Yarowsky, Using Billingual Materials to Develop Word Sense Disambiguation Methods, *Proc. of TMI'92,* 1992:101-112
4. Liu Xiaohu, Research on Methods of Word Sense Disambiguation in English-Chinese Machine Translation, *Ph.D. Thesis, Harbin Institute of Technology,* 1998
5. W. A. Gale, K. W. Church, Identifying Word Correspondences in Parallel Texts, *Proc. of the 4ᵗʰ DARPA Workshop on Speech and Natural Language,* 1991:152-157
6. H. Kaji, T. Aizono, Extracting Word Correspondences from Bilingual Corpora Based on Word Co-occurrence Information, *Proc. of 16ᵗʰ COLING,* 1996:23-28
7. Wang Bin, Automatic Alignment on Chinese-English Bilingual Corpora, *Ph. D. Thesis, Computer Technology Institute of Chinese Academy Science,* 1999
8. P. F. Brown, J. C. Lai and R. L. Mercer, Aligning Sentences in Parallel Corpora, *Proc. of the 29ᵗʰ Annual Meeting of the ACL,* 1991:169-176
9. W. A. Gale, K. W. Church., A Program for Aligning Sentences in Bilingual Corpora, *Proc. of the 29ᵗʰ Annual Meeting of the ACL,* 1991: 177-184
10. M. Simard, G. F. Forest and P. Isabelle, *Using Cognates to Align Sentences in Bilingual Corpora, Proc. Of TMI'92,* 1992: 67-81
11. M. Kay, M. Rocheisen, Text-Translation Alignment, *Computational Linguistics,* 1993, 19(1): 121-142
12. S. F. Chen, Aligning Sentences in Bilingual Corpora Using Lexical Information, *Proc. of the 31ˢᵗ Annual Meeting of the ACL,* 1993: 9-16
13. T. Utsuro, et al, Bilingual Text Matching Using Bilingual Dictionary and Statistics, *Proc. of 15ᵗʰ COLING,* 1994:1076-1083
14. M. Haruno and T. Yamazaki, High-performance Bilingual Text Alignment Using Statistical and Dictionary Information, *Proc. of 34ᵗʰ Annual Meeting of ACL,* 1996:131-138
15. Liu Xin, Zhou Ming, Huang Changning, An Experiment to Align Chinese-English Parallel Text Using Length-Based Algorithm, *Advances and Applications on Computational Linguistics,* ed. by Chen Liwei and Yuan Qi, Tshinghua University Press, 1995: 62-67
16. Liu Xin, Zhou Ming et al, Sentence Alignment Based on Automatic Lexical Information Extraction, *Computer Journal,* 1998, 21(8):151-158

17. Sun Le, Dulin Sun Yufang, Jin Youbin, Sentence Alignment of English-Chinese Complex Bilingual Corpora, *Proc. Of the Workshop of 5th NLPRS*, 1999:135-139

18. Zhao Tiejun, Li Sheng et al, Some Statistic Results Given By A Bilingual Corpus, *Advances and Applications on Computational Linguistics*, ed. by Chen Liwei and Yuan Qi, Tshinghua University Press, 1995: 112-118

A Task Oriented Natural Language Understanding Model[*]

Zeng Yuming   Wang Yongcheng   Wu Fangfang[**]

Shanghai Jiaotong University   Electronic Information School   Shanghai 200030

yumingzeng@263.net

Abstract. A theoretical task oriented natural language understanding(NLU) model is provided in this paper, together with a practical application based on it. The model is based on sentence framework (regular language mode in specific areas), chunks, etc. And it consists of following modules: post-process, NLU, target searching, connection, action. Every module is discussed in detail taking the application as example. The NLU module is of great importance, and we place emphasis on it. The success of our application shows that the model provided in this paper is feasible in specific task areas. The sentence framework, together with chunks, is very important in expressing sentence meaning. Verbs are of great important in sentence analysis. And we can use this model and methods provided in this paper to build many applications in various areas. Further more, the model and methods are worth reference in processing other domains' problems.

Keywords. Human-machine interface, NLU, sentence framework chunk

1. Introduction

Human-Machine interfaces have been an important technology since the appearing of computers. It has made great progress. While what's regretful is it's passive: humans has to learn machines. That is to say, skilled people develop new software, and end users must take great efforts to learn how to use the software. This situation must be changed. We should create positive Human-Machine interfaces: machines learn human. As a tool, a computer should be developed to

---

*The research product is sponsored by national 863 plan (Contract number : 863-306-ZD03-04-1).

**Zeng Yuming, PH.D candidate, interesting in NLP. Wang Yongcheng, Prof. Tutor of PH.D, Wu Fangfang, Master candidate.

T. Tan, Y. Shi, and W. Gao (Eds.): ICMI 2000, LNCS 1948, pp. 260–266, 2000.

be able to learn human, so that people can save much time and energy, and spend them on their work.. We should substitute human's learning machine procedure with machine's learning human procedure.

The procedure that man uses a computer is the procedure that man dispatches his commands to his computer and the computer execute them. Till now natural language is the most popular and convenient way for man to dispatch their commands. So, as a tool, it is more natural for computers to learn natural language, to act on natural language directive from humans. Unfortunately it is impossible for a computer to understand most natural language at present time, but we can make them understand task oriented, or region limited natural language.

In this paper, we developed a task oriented natural language understanding model using mainly sentence framework (regular language mode in specific areas) and chunks. A practical application based on this model is also provided and tested in this paper. Results show that the sentence framework and chunks are very important in expressing sentence meaning, so they are the key problems in sentence analysis. Results also show that verbs usually play the most important role in a sentence, so we can use them to find out sentence framework to some extent during sentence analysis. Result also proves that our model is feasible in specific region.

2. Theoretical Model

Figure 1 shows the schematic diagram of the model.

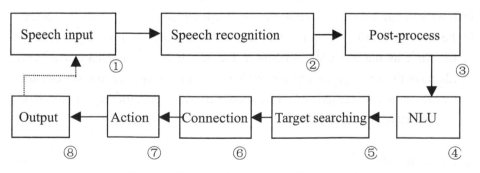

Figure 1 Schematic diagram of the model

The whole procedure can be described as follow: (1) Human speech is received through an input device, such as a microphone. (2) Then we use a speech recognition software (Currently, we use IBM ViaVoice) to recognize the speech, change it to machine-readable text string. (3) Next, we process the text, mainly correct mis-recognition. (4) After that, we try to understand the text, then execute the command. This includes natural language understanding (NLU), target searching, connecting, executing, and output via voice, printing, etc. Sometimes, the output is fed back to the end user to indicate some action. Among all above processing steps, the NLU step is of great importance. During NLU process, we try to find out all available information in the command string. After passing the result through communication channel, the computer startups the proper action module to execute the command.

Now, we will discuss all the above steps in detail. For NLU step, we will put more emphasis.

3. Working Flow

(1), Post-process module

Generally, the speech recognition software is designed for various areas, not limited in one or several specific regions. However no speech recognizer can perfectly recognize all words till now. But with specific information, we can try to recognize more perfectly in some special domain. We can use this information to build a database and use it to improve the performance of the speech recognition software.

For example, if $\{W_1, W_2, ...W_m\}$ is a keyword set for a specific domain, then we can build the database on this set. Now suppose that a word W_n appears in the command string. It isn't in this set, but its pronunciation is very similar to, or the same as the word W_j which is in the set (It is very populare that many words have the same pronunciation in Chinese). In our application, we consider that word W_n is mis-recognized and it should be W_j after retrieving the database, so we substitute W_n with W_j.

(2), NLU module

As mentioned above, NLU in all areas is very difficult, so we only consider a task oriented NLU problem.

Although human commands may vary in forms, they have some regular language modes in most cases. For example, if our application is to create abstracts for documents, we can find those modes such as "... open ... create ... abstract... ", "... create ... abstract for ...", etc. These modes help analyzing the command.

Generally speaking, there are three steps:

1), Extract manually normal language modes used in the specific task as much as possible, and to categorize them according to their functions. And construct a mode-base.

2), Automatically analyze human command, and travel the mode-base to find a proper one which matches the command string.

3), Try to find out all available information in the command string based on the language mode matched out in former step.

We will take creating abstracts as example to explain them.

As to 1), we know language modes are task oriented. When creating abstract for documents, we can categorize human command modes into two groups: (a). Creating abstracts for documents; (b). Changing the abstract length.

As to 2), we use the verb in the command string as the key to find a language mode. If the verb is "open", we will find out a language mode belonging to group (a). Or if the verb is "add", we will find out a mode of group (b).

As to 3), other information related to the specific command. We use the language mode found in the second step to extract these information. Suppose the command mode is to open a document and create an abstract, we have to get the following information: the document (may be described by OS file name, article subject, keyword, author, etc.), abstract length, and abstract bias. If the mode is "...open XXX ... create abstract", we know that "XXX" stands for a document; or if the mode is "... add XXX words ...", we know that "XXX" stands for a number. Note that in oral language, we used to omit some parts, for example, we often omit the document in the "open" command, and sometimes we omit the abstract length, etc. In this case, we can use a default value. For instance, if the document is omitted, people usually refer to the document already opened, or the document that we have created an abstract before. And if the abstract length is omitted, we know that the end-user is not sure or don't care about the abstract length, so we can assign the abstract length with a value according to the document length.

(3), Target searching module

This step is mainly used in our abstract application. After having extracted the document information from the command string, we have to locate the file. As described above, we know a user often refer to a document by it's file name, document subject, keyword, or author name and written time (For instance, we may care about President Jiang Zemin's recent speech). So we must locate the file using this property, not just search the file name.

Now we can locate files in the whole computer's storage (mainly all hard disk drives), and we are about to locate files on the Internet.

As we know, many operating systems' file system is organized as a tree, including Microsoft Windows. So the procedure for locating a file is the procedure for traversing the file system. There are usually two methods for traversing a tree: wide first and depth first. In our practical application, we use depth first searching method.

(4), Connection module

This step is to pass the former result to the action module. When a command is correctly parsed, the command, together with additional information can be achieved. We use the connect module to pass the command and the information to the action module to execute the command. Currently, our application runs under Windows 95 platform, and we use Shared Memory as the connect channel.

In connection step, dispatching commands must be smart. In a complex system there are many action modules that can response to various commands respectively. The connection module must dispatch the commands (including commands related information) to the right action modules. So it must "know" which action module is connected with a specific command.

(5), Action module

Action module is the exact one that executes commands. It accepts commands and commands related information, and response. In our practical application, the action module can create abstracts for documents according to both user specified abstract length and abstract bias. It can also change the abstract length and bias for the abstract.

4. The Practical Application and System Test

As mentioned before, we have built a practical application based on the theoretical model. ABSTRACTOR is one of our products that can automatically create abstracts for documents, and has been taking the lead in the world in quality. Our system can listen to end user's commands, and executing them, using ABSTRACTOR as its action module. The application was exhibited together in Shenzhen and Shanghai and achieved great success.

5. Future Work

In our NLU model, much of our work is based on sentence framework and chunks. In fact, there is much other information we can use to help analyzing commands, such as context, environment, etc. In our future work, we need to integrate this information into the NLU model to make a better one. And it is necessary to expand the single-region model to multi-regions model step by step. Further more, the more intelligent the model is, the wider application it will has.
As to our practical application, on one hand, it should be able to locate files on the internet. On the other hand, other action modules are to be added. For instance, we can add other products of our lab's as action modules.

6. Conclusion

The success of our practical application shows that the natural language understanding model provided in this paper is feasible in task oriented areas. The sentence framework, together with chunks, is very important in expressing sentence meaning. Verbs are of great important in sentence analysis. And we can use this model and methods provided in this paper to build many applications in various areas. Referring to our work, we'd like to emphasis it's application on smart electronic equipment. While there are still much can be done to make better.

References

[1] Lu Chuan "Semotactic Network in Chinese Grammar"   Beijing Shangwu publishing house 1999.

[2] Wang Yongcheng   "Chinese Information Processing" Shanghai Jiaotong Univ. publishing house   1991.12.

[3] Deng Liangdi   "Information Management And Office Automation"   Beijing   enginery industry publishing house 1990.6.

[4] Yu Shiwen "Details of Modern Chinese Grammar information dictionary" Tsinghua Univ. Publishing house1998.4.

Variable Length Language Model for Chinese Character Recognition

Sheng Zhang   Xianli Wu

Engineering Center of Character Recognition
Institute of Automation, Chinese Academy of Sciences, Beijing 100080
{zs,wxl}@hw.ia.ac.cn

Abstract. We present a new type of language model — variable length language model (VLLM) whose length is non-deterministic on the base of 5-gram combined model brought forward previously. Its main advantage lies in that it captures the function of 5-gram combined model and reflects the structural feature of every line in test text as well. Compared to previous language model, the VLLM makes use of current result to determine which kind of language model should be used next and realizes the automatic choice of language model that is always constant before. VLLM also resolves the problem when punctuation marks appear. Based on those improvements, we make experiments and get encouraging result.

1   Introduction

Traditionally, natural language can be viewed as a random process. Each word occurs with certain probability. Formally, let $W_s = <W_1 W_2 ... W_n>$ denote one of possible word strings. Then, the probability of this string is:

$$p(W_s) = p(W_1) \prod_{i=2}^{n} p(W_i \mid W_1 \cdots W_{i-1}) \tag{1}$$

But Eq.(1) is unpractical because of its intricate calculation and occupying very large memory. Since the N-gram [1] language model was brought forward, many language models have been proposed, including trigram model, 3g-gram model ([2], [3]), cache-based language model [4] and so on. They are all Markov models for natural language. Both trigram and 3g-gram models, whose parameters are calculated from a large training corpus, produce a reasonable nonzero probability for every word in the vocabulary. The cache-based model, however, tracks short-term fluctuations in word frequency. Because of inadequacy of the maximum likelihood estimator and the necessity to estimate the probabilities of N-gram word that doesn't occur in the training corpus, the method we use is based on the theory presented by KATZ [5]. Most of them are based on past history, that is: they ignore the information behind the word. So we presented 5-gram combined model [6] which includes both the past and future information. To make a difference, in this paper we call it 5-gram constant model (5-gram CM). This model contains not only all the advantages of language models above, but some new features also. Results in [6] proved its good performance.

T. Tan, Y. Shi, and W. Gao (Eds.): ICMI 2000, LNCS 1948, pp. 267–271, 2000.

However, all above language models, including 5-gram CM, take no consideration of the relationship between current result at time i and next form of the model at time $i+1$. According to current calculated result, we are able to determine the model that should be used next. Theoretically, we assume that the performance of the model should be identical to that of 5-gram CM. After simply introducing 5-gram CM and related issues in Section 2, Section 3 addresses the variable length language model. Section 4 presents experimental results, followed by conclusions.

2   5-gram CM and Related Issues

In fact, in order to have a good performance, the 5-gram CM contains three parts that are trigram model, 3g-gram model and a cache-based model. First, the trigram model is just a simplified version of N-gram when N equates to 3. It is based on the mapping of a history $< W_1,...,W_{i-1} >$ onto the state formed by the two most recent words $< W_{i-1}, W_{i-2} >$. So,

$$p(w_i \mid w_1 \cdots w_{i-1}) \cong p(w_i \mid w_{i-2} w_{i-1})$$

(2)

And $P(W_i|W_{i-2}W_{i-1})$ could be obtained from training corpus by linear interpolation. So,

$$p(w_i \mid w_{i-2} w_{i-1}) \cong q_3 f(w_i \mid w_{i-2} w_{i-1}) + q_2 f(w_i \mid w_{i-1}) + q_1 f(w_i)$$

(3)

where $q3+q2+q1=1$.

Then, 3g-gram model is similar to trigram model but it classifies words into parts of speech (POS) and then chooses every word from classes. That is:

$$p(w_i \mid g_{i-2} g_{i-1}) \cong \sum_{g_i} p(w_i \mid g_i) p(g_i \mid g_{i-2} g_{i-1})$$

(4)

In [6], we present a new method called FAC, and with this method Eq. (3) can be turned into:

$$p(w_i \mid g_{i-2} g_{i-1}) \cong f(w_i \mid g_i)[A_1 f(g_i \mid g_{i-2} g_{i-1}) + A_2 f(g_i \mid g_{i-1}) + e]$$

(5)

where $e=0.0001, A1+A2=1-e=0.9999$. How to calculate $f( )$ and $p( )$ can be found in [6].

Finally, in order to reflect short-term patterns in word use, cache-based [4] model just designs a buffer that contains one hundred words around the word W_i. Now we get Eq. (6).

$$p(w_i \mid g_i) \cong k_m f(w_i \mid g_i) + k_c cache(w_i)$$

(6)

where $Km+Kc=1$.

However, when word W_i couldn't be found in training text, methods presented above are invalid, so we use the method presented by KATZ [5]. Its probability is:

$$D = \frac{n_1}{N}$$

(7)

Here, n_i is the number of the word that only appeared once in training corpus, and N is the total number of training corpus. Now, we can combine all of these and take consideration of the future information, so we can get 5-gram CM.

$$p(w_i \mid w_{i-2} w_{i-1} w_{i+1} w_{i+2}) \cong a_0 p(w_i \mid w_{i-2} w_{i-1}) +$$
$$a_1 p(w_i \mid w_{i-1} w_{i+1}) + a_2 p(w_i \mid w_{i+1} w_{i+2}) \quad (8)$$

where $a0+a1+a2=1$.

We take the first part of Eq. (8) as an example to illustrate the structure of 5-gram combined model. Thus, if word W_i is found in training corpus, then

$$p(w_i \mid w_{i-2} w_{i-1}) = (1-D)\{af(w_i \mid w_{i-2} w_{i-1}) + bf(w_i \mid w_{i-1}) + cf(w_i)$$
$$+ d\{[k_m f(w_i \mid g_i) + k_c cache(w_i)][A_1 f(g_i \mid g_{i-2} g_{i-1}) + A_2 f(g_i \mid g_{i-1}) + e]\} \quad (9.1)$$

else:

$$p(w_i \mid w_{i-2} w_{i-1}) = D \quad (9.2)$$

where $a+b+c+d=1$, $e=0.0001$, $A1+A2=1-e=0.9999$, $Km+Kc=1$

3   Variable Length Language Model

To be more specific, let $W_s = <...W_{i-2} W_{i-1} W_i W_{i+1} W_{i+2} W_{i+3}...>$ denote one of these possible word strings. Now, let's investigate Eq. (8) which only considers word W_i at time i. Based on this Eq.(8), we can know directly that the information of word W_i is related to those of the other four. If we investigate not only W_i but W_{i+1} as well, however, we could find that each calculation of them is not isolated, but related to each other. That is to say: When we calculate the probability of W_i with Eq.(8), if we find $C(W_{i-1}W_i)$ equates to zero, then the equation for next word W_{i+1} at time $i+1$ must be:

$$p(w_{i+1} \mid w_{i-1} w_i w_{i+2} w_{i+3}) = p(w_{i+1} \mid w_i w_{i+2} w_{i+3}) \quad (10)$$

If it does not equate to zero, it should remain as $P(W_{i+1} \mid W_{i-1} W_i W_{i+2} W_{i+3})$. Where $C(W_{i-1}W_i)$ is the frequency of word tuple $< W_{i-1}W_i >$ in training corpus. In fact, if $C(W_{i-1}W_i)$ equates to zero, which means word tuple $<W_{i-1}W_i>$ isn't a bigram, $C(W_{i-1}W_iW_{i+1})$ equates to zero. Otherwise, if $C(W_{i-1}W_iW_{i+1})$ doesn't equate to zero, which means word tuple $< W_{i-1}W_iW_{i+1}>$ is a trigram, $C(W_{i-1}W_i)$ can't equate to zero because word tuple $<W_{i-1}W_i>$ and $<W_iW_{i+1}>$ are both bigrams. Furthermore, we modify the coefficient in the second part of 5-gram CM. From the information point, word W_{i-1} can't provide any information for word W_{i+1}. So we can get variable length language model. Meanwhile, it has another advantage that it can deduce its next formula till the end of the sentence.

On the other hand, when punctuation marks appear, we can take it as a special case of the method above. Because the function of most punctuation is to separate the meaning of one sentence or shows some kind of feeling of the speaker or writer, it seems to us that the meaning in the immediate vicinity of the punctuation is discontinuous in most cases. For instance, the sentence "实际上，我们并不以此为满足。" (In fact, we aren't content with this.) can need some kinds of variable length language models such as: $P(W_i|W_{i-2}W_{i-1})$, $P(W_i|W_{i+1}W_{i+2})$ etc.. In fact, according to different positions where punctuation appears, the VLLM can evolve into many forms such as: $P(W_i|W_{i-2}W_{i-1}W_{i+1})$, $P(W_i|W_{i-1}W_{i+1}W_{i+2})$, $P(W_i|W_{i-2}W_{i-1})$, $P(W_i|W_{I+2}W_{I+1})$, $P(W_i|W_{i+1})$ and $P(W_i|W_{i-1})$. Even though primitive 5-gram CM can do these by searching and calculating and get many kinds of approximate formulas, we think there's a different concept between them.

4 Experiments & Results

In order to inspect the effect of the VLLM, we made several experiments on it in comparison with 5-gram CM. We chose three different recognition rate for the experiment to show different effects. In Table below, "Initial rate" means the rate of these test texts after recognizing, and "Current rate" means the rate of them after processing by 5-gram CM or VLLM. "Improvement" is the difference of "Initial rate" and "Current rate". Table shows that the function of the VLLM is identical to that of 5-gram CM at respective recognition rate. Since the speed is related to both recognition rate and the number of punctuation marks, the elevation in speed is not uniform.

TABLE

GROUP		A	B	C
Total number		8444	11000	26830
Initial rate		95.97%	89.78%	83.89%
Current rate	5-gram CM	98.48%	95.96%	91.88%
	VLLM	98.48%	95.96%	91.88%
Improvement	5-gram CM	2.51%	5.78%	7.99%
	VLLM	2.51%	5.78%	7.99%
Speed (word/sec)	5-gram CM	77	72	65
	VLLM	82	74	68

5 Conclusions

The results listed in the previous chapter seem to confirm our hypothesis. In this paper, we have presented a new language model which contains the advantages of 5-gram combined model and can adapt to different situations automatically. The variable length language model (VLLM) makes use of the mechanism of 5-gram CM without decreasing the improvement of recognition rate and increases the processing speed that is important in practice. Furthermore, the VLLM can also be applied to the post-processing of speech recognition after some minor modifications because the basic ideas of the two problems are same.

References

1. L.R.Bahl, F.Jelinek and R.L.Mercer, "A maximum likelihood approach to continuous speech recognition," *IEEE Trans. Pattern Anal. Machine Intell.*, vol. PAMI-5, pp.179-190, Mar. 1983.
2. A.M.Derouault and B.Merialdo, " Nature language modeling for phoneme-to-text transcription," *IEEE Trans. Pattern Anal. Machine Intell.* vol. PAMI-8, pp.742-749, NOV.1986.
3. ——, "Language modeling at the syntactic level," in *Proc, 7th Inter. Conf. Pattern Recognition*, vol.2, Montreal, Aug, 1984, pp.1373-1375.
4. Roland Kuhn and Renato De Mori, "A cache-based natural language model for speech recognition," *IEEE Trans, Pattern Anal Machine Intell.*, vol. 12, No 6, June, 1990.
5. Slava M. Katz, "Estimation of Probability from Sparse Data for the Language Model Component of a Speech Recognizer" *IEEE Trans, Acoustics, Speech and Signal Processing,* vol. ASSP-35, No 3, Marci, 1987.
6. Sheng Zhang and Xianli Wu, "Language Model of Chinese Character Recognition and Its Application" to appear in *16th IFIP World Computer Congress WCC'2000*, August, 2000.

Automatic Lexical Errors Detecting of Chinese Texts Based on the Orderly-Neighborship[1]

Yangsen Zhang

Dept. of Computer Science, Shanxi University, 030006
Taiyuan,Shanxi Province,China
Zhangys@sxu.edu.cn

Abstract. According to the statistic analysis of common errors found in those texts that are typed-in, OCR-recognized or phonetics-recognized but not be proofread and the characteristics of such texts, we propose a error-detecting principle and error-detecting algorithm based on the orderly-neighborship. Furthermore, Factors that affect performance index of error-detecting system such as recall ratio and accurate ratio are discussed.

1 Introduction

Nowadays, An urgent task to be settled to research *Computer Aided Chinese Text Proofread System* is posed, because manual proofreading is more and more becoming a bottleneck problem of publishing automation with the emerging of electronic documents and electronic publication. [4]

Chinese Text Automatic Proofread, which is an error-text-oriented technology that synthetically apply the analytic and statistic technique of morphology, syntax and semantics in computational linguistic to eliminate the disturb of noisy and to reappear the original content of correct text with high degree of accuracy, is a new research topic in the field of Chinese Information Processing.

It is our research target to obtain interactive computer aided proofreading system at a certain level rather than a complete automatic one because of the limitation of Chinese theoretical research, the particularity of Chinese language and the variety of errors in Chinese texts. [6]

2 Error-Detecting Principle

2.1 Statistic Analysis of Common Errors in Chinese Texts

Manual analysis of 243 real error examples indicates the following ideas:

1. Generally, the errors in Chinese texts are local errors. It is only the surface structure of Chinese character or word that is destroyed in errors of Chinese text (e.g. Miswritten Chinese character can't form a word with neighbor Chinese character or word and have no effects on others). Thus, we can detected the errors occurred in typed-in, OCR-recognized or phonetics-recognized but not be proofread Chinese text

---

[1] Shanxi Province Natural Science Fund Item (981031)

using neighborship and local analysis technique between Chinese characters and phrases without doing overall syntax analysis. [1][2]

2. In a sentence with errors, the point error occurs is usually single Chinese character or single Chinese character strings after segmentation. We named such single Chinese character or single Chinese character as "Sparse string". Statistic shows the circumstance that the point error occurs is "sparse string" reaches 90.3 percent. That is, we'll detect most errors in Chinese text if we can do with "Sparse string" successfully.

3. In general, there is no relationship between single Chinese character in a "sparse string", it means that there is no word-building relationship and neighborship (the relationship between neighbour single Chinese character in order[7]) between the single Chinese character in sparse string.

Thus, we can find most errors in typed-in, OCR-recognized or phonetics-recognized text if we focus on sparse strings and apply neighborship to check neighbour Chinese character after segmentation.

2.2 Error-Detecting Principle of Chinese Texts Based on the Neighborship

According to the discussion above, we construct a lexical error-detecting system base on the neighborship, its construction frame is shown as Figure2.1.

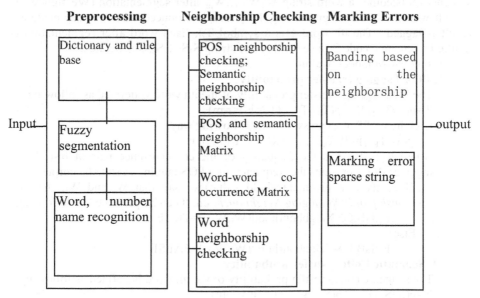

Fig. 2.1: Construction Frame of Lexical Errors Detecting system

In line with the principle of "Long phrase have priority and sparse string is focal point", the processing steps of our lexical detecting system is as following:

(1)Preprocessing

To do segmentation combined FMM (Forward Maximum Match Method) and BMM (Backward Maximum Match Method); to check number, English word and punctuation in Chinese text; to do fuzzy match on long phrase more than three single Chinese character; to mark the "no hit" phrase and the fuzzy matched phrases will be correction candidates.

(2)Neighborship checking

After preprocessing, to check the neighborship of single Chinese characters and words based on the matrix of bi-word-cooccurrence,the matrix of POS tagging neighborship and semantic neighborship between Chinese characters or words.

(3)Banding based on the neighborship

According to the neighborship above,To eliminate spaces between those Chinese characters or words whose neighborship value is under the range of a given threshold value.

(4)Error Marking

To merge the neighbor strings which is sparse string after banding and mark them as an error point.

3   Error Detecting Algorithm Based on POS and Semantic Code Neighborship

We assume that a sentence S is $C_1C_2...C_n$ (Let C_i be the ith Chinese character in the sentence);S becomes a word string $W_1W_2....W_m$ after segmentation (We sign W_i as the ith word); A POS tagging string $T_1T_2...T_m$ is produced when the segmentation result is tagged. (The ith word gets a symbol T_i); Each word after segmentation is marked a semantic code S_i, a semantic code string $S_1S_2...S_m$ is generated.

Using HMM model, we find:

I. POS Tagging transfer probability

The POS tagging transfer probability of the ith word is defined as following:

$P_i=p(T_i/T_{i+1})*...*p(T_{j-1}/T_j)$ $(1<=i<j<=m)$

Each item of P_i is expressed in the following way:

$p(T_k/T_{K+1})= R(T_k/T_{K+1}) /(R(T_k) *R(T_{K+1}))$;

where $R(T_k/T_{K+1})$ is the frequency of the co-occurrence pair of bi-orderly-tagging; $R(T_k)$ presents the frequency of a POS tagging in statistic corpus.

If $P_i>\tau(\tau$ is a given threshold value),we say that W_i and W_{i+1} has the *relationship of POS tagging Neighbourhood.* It can be presented as following:

F_1:IsBiCXNeighbourhood(W_i,W_{i+1})=TRUE

Else

F_1:IsBiCXNeighbourhood(W_i,W_{i+1})=FALSE

II. Semantic Code transfer probability

The semantic code transfer probability of the ith word is defined as following:

$P_i=p(S_i/S_{i+1})*...*p(S_{j-1}/S_j)$ $(1<=i<j<=m)$

Each item of S_i is expressed in the following way:

$p(S_k/S_{K+1})= R(S_k/S_{K+1}) /(R(S_k) *R(S_{K+1}))$;

where $R(S_k/S_{K+1})$ is the frequency of the co-occurrence pair of bi-orderly-semantic code; $R(S_k)$ presents the frequency of a semantic code in statistic corpus.

If $P_i>\sigma(\sigma$ is a given threshold value),we say that W_i and W_{i+1} has the *relationship of semantic code Neighbourhood.* It can be presented as following:

F_n:IsBiSEMNeighbourhood(W_i,W_{i+1},No)=TRUE

Else

F_n:IsBiSEMNeighbourhood(W_i,W_{i+1},No)=FALSE

Where No is the number of a semantic code;n=2 when No=1 and n=3 when No=2;

Let F_1 be bi-POS-neighborship; F_2 be bi-one-semantic-code-neighborship; And F_3 be bi-two-semantic-code-neighborship. We multiple these 3 ways ,use voting mechanism to check the relationship of neighbourhood between Chinese characters and phrases.Let Member=3, the procedure of determining the relationship of neighbourhood between Chinese characters and phrases can be described as the following:

Procedure 1: Neighborship checking procedure using voting mechanism

```
BOOL ISNeighbourhood(W₁,W₂)
BEGIN
    FOR (i=1;i<=Member;i++)
      Ticketᵢ=0;
    FOR (i=1;i<=Member;i++)
      BEGIN
        IF Fᵢ=TRUE Ticketᵢ=1;
        ELSE Ticketᵢ =-1;
      END
    IF ΣTicketᵢ>0 return TRUE;
    ELSE return FALSE;
END
```

W1 and w2 can be banded only if the expression IsNeighbourhood(W_1,W_2) equals TRUE. The banding result is presented by W, banding procedure is following:

Procedure 2: Banding

```
Char* Banding(W₁,W₂)
BEGIN
    W=W₁;
 IF
IsNeighbourhood(W₁,W₂)=TRUE
    W=W₁+W₂;
    ELSE
    W=W₁+space+W₂;
END
```

The procedure of error marking is to pick out the sparse string which can't be banded.So called sparse string is such a string of $W_jW_{j+1} \ldots W_k$ in which $|W_p|=1(j<=p<=k)$,$|W_p|$ is the length of pth phrase. Let $S'=W'_1W'_2 \ldots W'_m$ be a banding string(There is a space between W'_i and W'_{i+1};The length of ith phrase is $|W'_i|$);Let S'' be the error marking result,the Error marking algorithm can be formally described as Procedure3.

Procedure3: Error-marking

```
Char* Marking(W'₁W'₂ …W'ₘ)
BEGIN
    i=0;
    S'='\0';
    WHILE (i<=m)
      BEGIN
        IF (|W'ᵢ|==1)
          BEGIN
            S'=S'+'#';
            WHILE (|W'ᵢ |==1)
              BEGIN
                S'=S'+ W'ᵢ;
                i=i+1;
              END
          END
        ELSE
            S'=S'+W'ᵢ;
      END
END
```

Let $W_1W_2 \ldots W_m$ be the Chinese string ,W' be the banding result and W'' be the error marking result.The following is the detecting program.

Error-Detecting program

```
Char* Detecting(W₁W₂…Wₘ)
BEGIN
    /*Preprocessing(Transformation    of    text    form,
segmentation, number and English word recognition, fuzzy
match of long phrase)*/
    BEGIN
      i=0;
      WHILE (i<=m)
        BEGIN
          IF (IsNeigbourhood(Wᵢ,W ᵢ₊₁)==TRUE)
            W'=Banding(Wᵢ,Wᵢ₊₁);    /*Banding*/
          i=i+1;
        END
        W''=Marking(W');    /*Error-Marking*/
    END
```

4 Experiments on Lexical Errors Detecting System

Based on "General dictionary used by information processing" and "Dictionary of synonym ",we generate bi-one-semantic-code-neighborship,bi-two-semantic-code-neighborship,bi-POS-cooccurrence,Tri-POS-cooccurrence matrix or model combined manual disambiguity and machine statistic on a corpus of 520,000 Chinese character.

Using statistic information we got, we build a experimental system and a simple evaluating system in Visual C++ on the platform of Windows95. Our test set is composed of 133 error point, the test result is shown as following:

Table 1 The comparison of various methods of checking neighborhood

The method for checking neighborhood	Recall ratio(%)	Accurate ratio(%)
Regard all sparse string after segmentation as errors	76.99	14.16
Bi-POS-cooccurrence	61.41	30.34
Tri-POS- cooccurrence	63.72	31.16
Bi-one-semantic-code cooccurrence	55.20	22.81
Bi-two-semantic-code cooccurrence	44.96	25.56
Multiple checking method using voting mechanism mentioned above	72.50	35.20

5 Discuss and the Work ahead

5.1 Discussion of Experiments

1.   Problems of performance index in experiments
Test data above states that in our experiment the recall ratio is relative high but accurate ratio is relative low. In principle, the recall ratio is the prime performance index in Chinese text automatic proofreading system and accurate ratio should be improved in the situation of keeping a steady recall ratio. However, the recall ratio and the accurate ratio are two sides of a coin, that is, in order to detect errors as more as possible, the accurate ratio won't be very high. It is our target to detect real errors as more as we can and to overlook non-errors as fewer as possible.
2.   Factors affect performances index of the system
I. Segmentation error which occur in processing of ambiguity and unknown word leads to accumulated errors in parts of speech tagging and semantic tagging. Banding error happens because of incorrect neighborship being checked.

For example, in sentence "他有一个好的学习惯方法", "学习惯" is an ambiguous field, because of the segmentation error, "学" is likely to be picked out as a error.

There is another example, in sentence "表达了米洛舍维奇的意愿", if "米洛舍维奇" can't be recognized as a name, and algorithm 1 can't band "米洛舍维奇", then sparse string "米洛舍维奇" is marked as a error.
II. Some real neighborship can't be found in the neighborship matrix because of the scale limitation of statistic corpus and the non-completeness of neighborship matrix. Moreover, the neighborship is various at different threshold value which affect the size of errors detected.

For example, in sentence "带领群众撤出危险地带","撤" and "出" should can be banded, but corresponding probability value in neighborship matrix is 0, the system pick "撤出" out as a error.

Another example, in sentence "工人阵级发挥了模范带头作用", the error "阵级" can't be detected when threshold value is very great.

5.2 The Work ahead

1.   Issues on Dictionary[3]
Except for general dictionary, specialized dictionary and temporary dictionary are expected to be added. Moreover, we should emphasize the capability of unknown word recognition(e.g. name, places, terminology and organization name).
2.   Issues on Neighborship model[5][8]
  . To enlarge the statistic scale
  . To get better parameter estimation
  . To do neighborship statistic between words
3.   Issues on boundary define of errors
To strengthen the research on error's boundary definition and make it easy for the next step: Correction.

In general, it is very difficult to research Chinese text automatic error-detecting technology that process texts including errors because theories on corrected large-scaled run text is still not sufficient. On the other hand, focus on special law in texts including errors such as neighborship checking method . To research effective local analysis technology has great potentiality and it becomes a highlight topic and a new direction in Chinese information processing.

Reference

1. I. Dagan and S.Marcus: Contextual Word Similarity and Estimation from Sparse Data. Computer Speech and Language, 9, (1995).123-152
2. K.W.Church and R.L.Mercer:Introduction to the Special Issue on Computational Linguistic Using Large Corpora. Computational Linguistics, 19(1), (1993)1-24
3. P.F.Seitz and U.N. Gupta: A Dictionary for A Very Large Vocabulary Word Recognition System, Computer Speech and Language, 4, (1990)193-202
4. Yangshen Zhang and Bingqing Ding: Present Condition and Prospect of Chinese Text Automatic Proofread Technology. Journal of Chinese Information, 3,(1998)23-32
5. Yong Mu, Cai Sun and Zhensheng Luo, Research on Automatic Checking and Confirmative Correction of Chinese Text. TsingHua Press, (1995) 100-105
6. Cai Sun and Zhensheng Luo, Research on the Lexical Errors in Chinese Text. Journal of the 4th Computational Linguistics Conference: Language Engineering, TsingHua Press (1997), 319-324.
7. Chaojie Qiu, Rou Song and Longgen Ouyang, Statistical Results and Their Analysis of the Neighboring Pairs of Words on Very Large Corpora. Journal of the 4th Computational Linguistic Conference: Language Engineering, TsingHua Press, (1997),88-94
8. Rou Song, Chaojie Qiu etc.: Bi-Orderly-Neighborship and its application to Chinese Word Segmentation and Proof
9. Yangshen Zhang and Bingqing Ding: A Method of Automatic Checking and Correction on English Words Spelling – the Method of Skeleton Key, Computer Development and Application, 2, (1999),9-11

Statistical Analysis of Chinese Language and Language Modeling Based on Huge Text Corpora[1]

Hong Zhang, Bo Xu, Taiyi Huang

National Laboratory of Pattern Recognition
Institute of Automation, Chinese Academy of Sciences
P.O Box 2728, Beijing 100080, P.R China
{ hongzh, xubo, huang }@nlpr.ia.ac.cn

Abstract. This paper presents the statistical characteristics of Chinese language based on huge text corpora. From our investigation, we find that in writing Chinese it is more likely to use long words, while in other language styles the words are shorter. In large text corpora, the number of bigram and trigram can be estimated by the size of the corpus. In the recognition experiments, we find the correlation is weak between the perplexity to either the size of the training set or the recognition character error rate. However, in order to attain good performance, the large training set above tens of million words is necessary.

1 Introduction

Statistical Language model is effective in large vocabulary continuous speech recognition. However, construction of language model needs supporting of large text corpora, which in practice is related to large amount of consumption of human and monetary resource. If there is an empirical research on huge text corpora as the reference of design or selection of the training set for language modeling, the possible waste of resource will be reduced to the minimum.

There are already many text corpora for public purpose of English and other western languages. However the case for Chinese language is not so favorable. Since no previous report of very large Chinese text corpora, this is the first extensive investigation of statistical characteristics of Chinese language on very large text corpora.

The Chinese language is very different from the western languages. The basic unit is Chinese character. There is no separation between words in the Chinese text. Theoretically, any combination of Chinese characters can be defined as a word. In our investigation, the words are defined in a 40k lexicon, in which all Chinese characters are included as the single-character words. Therefore, the out-of-vocabulary problem is circumvented. The words from text are produced by an automatic word

---

[1] The work described in the paper is funded by the National Key Fundamental Research Program (the 973 Project) under No.G1998030504, and the National 863 High-tech Project under No. 863-306-ZD03-01-1.

T. Tan, Y. Shi, and W. Gao (Eds.): ICMI 2000, LNCS 1948, pp. 279-286, 2000.

segmentation algorithm provided by the Chinese Language Model Toolkit, which is developed by the National Lab of Pattern Recognition.

In this paper, the investigation on text corpora of Chinese language includes two aspects: one is the statistics on words and N-grams, the other is the performance of language models on different scale of corpus.

The rest of the paper is organized as the following: Section 2 introduces the language modeling theory briefly. Section 3 presents the statistics of Chinese language. In section 4 we report the recognition experiments of language models. The correlation between the perplexity and the performance of recognizers equipped with the language model is analyzed in section 5. Section 6 is the conclusion of this paper.

2   Language Modeling Theory

The most commonly used language model in large vocabulary speech recognition system is the trigram model [4], which is to determine the probability of a word given the previous two words: $p(w_3|w_1,w_2)$. The simplest way to approximate this probability is by the maximum likelihood (ML) estimate

$$P(w_3 \mid w_1, w_2) = f(w_3 \mid w_1, w_2) = \frac{C(w_1, w_2, w_3)}{C(w_1, w_2)} \qquad (1)$$

where f(|) denotes the relative frequency function, and C( ) denotes the count function.

2.1   Smoothing

Since even in very large corpora of training text, many possible trigram pairs are not encountered. To avoid the language model assign $P(W) = 0$ to strings W containing these gram pairs, it is necessary to smooth the gram pair frequencies, which is the task of smoothing technique.

In the NLPR Toolkit, a smooth method devised by Katz based on Good–Turing estimation [5], was implemented. The basic idea is from the formula (in the case of trigram language model):

$$\hat{P}(w_3 \mid w_1, w_2) = \begin{cases} f(w_3 \mid w_1, w_2) & if \quad C(w_1, w_2, w_3) \geq K \\ \alpha Q_T(w_3 \mid w_1, w_2) & if \quad 1 \leq C(w_1, w_2, w_3) < K \\ \beta(w_1, w_2)\hat{P}(w_3 \mid w_2) & otherwise \end{cases} \qquad (2)$$

Where $Q_T(w_3 \mid w_1, w_2)$ is a Good-Turing type function and $\hat{P}(w_3 \mid w_2)$ is a bigram probability estimate having the same form as $\hat{P}(w_3 \mid w_1, w_2)$

$$\hat{P}(w_3 \mid w_2) = \begin{cases} f(w_3 \mid w_2) & if \quad C(w_2, w_3) \geq L \\ \alpha Q_T(w_3 \mid w_2) & if \quad 1 \leq C(w_2, w_3) < L \\ \beta(1, w_2)f(w_3) & otherwise \end{cases} \qquad (3)$$

2.2 Evaluation Measure

According to the information theory [4], perplexity (PP) is defined to measure the text source complexity as following,

$$PP = 2^{LP} \tag{4}$$

Where *LP* is the logprob, which is defined by

$$LP = \lim_{n \to \infty} -\frac{1}{n} \sum_{i=1}^{n} \log Q(w_i \mid w_1, \ldots, w_{i-1}) \tag{5}$$

Where $Q(w_i/w_1, \ldots w_{i-1})$ denotes the recognizer's estimate of the text production probabilities that is embedded in the language model.

3 Statistics on Large Chinese Text Corpora

3.1 Processing of the Chinese Text Material

As described in section 1, there are no separated signs between the words in Chinese language. In practice, the Chinese words are segmented from the text stream according to a lexicon, in which words are defined as combination of Chinese characters. Besides, a step of preprocessing is also necessary to refine the raw text before the word segmentation, as shown in figure 1.

All the works concerned by this paper of the processing of the text material are under the auxiliary of NLPR Chinese Language Model Toolkit (v 1.0) [6].

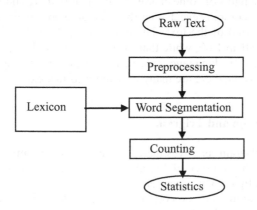

Fig. 1. Processing of Chinese Text Material.

3.2   Average Word Length

The average word length refers to the average length in numbers of Characters of words, which is defined as

$$L_w = \frac{N_c}{N_w} \tag{6}$$

Where L_w is the average word length, N_c is the number of Chinese characters in the text after preprocessing, N_w is the number of Chinese word after word segmentation.

In our investigation, all the Chinese words are defined according to a 40K lexicon, in which the average word length is 2.2266 Nc/W.

Table 1. Statistics of Average Word Length

Classes	Intent	Average Word Length
Newspapers and magazines	People's Daily (81,94,95,96)	1.6527
	Reference News(92)	1.6558
	Daily of Science and Technology (91-96)	1.6527
	The Computer(98)	1.5414
	The Chinese Computer (98)	1.6164
	Chinese Reader's Digests (98, Bound Edition of 200 volumes)	1.5297
Books	Translated Foreign literature	1.4994
	Ancient Chinese literature	1.3519
	Modern Chinese literature	1.4581
	Culture	1.5291
	Techniques	1.5152

The average word lengths of some classes of text material are shown in Table 1.

The statistics of the corpora reveals that the average word length shows prominent feature related to the kind of the text. The average word length of most texts from newspapers are higher than 1.60, while that of most novels and scientific publications are less than 1.5. The main language style of newspapers is writing Chinese, and in books, especially in novels, there are many oral and informal expressions of Chinese.

3.3   Numbers of Bigram and Trigram

The number of N-gram is investigated with increment amount of text. The correlations between the number of bigram and trigram to the size of corpus are shown in figure 2 and figure 3, respectively.

The statistics of N-gram pairs on corpora of totally 2960 million words shows that numbers of bigram pairs and trigram pairs increase with the enlarging of the size of training sets. The increasing speed of pair numbers is a little slow when the training set is larger than 1000 million words. In the log-log scale, there exists an approximate linear relation between number of bigram and trigram pairs and the size of the training set in million words. According to the data we can conclude two coarse estimation equations:

$$\log_{10}(NB) = 0.6127\log_{10}(SC) + 5.7301 \tag{7}$$

$$\log_{10}(NT) = 0.7928\log_{10}(SC) + 6.1053 \tag{8}$$

Where NB is the number of bigram in millions, NT is the number of trigram in millions, and SC is the size of corpus in million words.

In large text corpora, the number of bigram and trigram can be estimated by the size of the corpus. From these relationships, the amount of raw material can be predicted according to the size of the language model, which is often limited by the computing capability of the hardware of the recognizer.

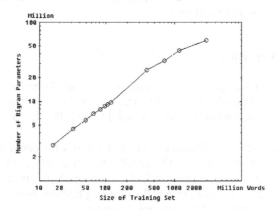

Fig. 2. The number of bigram with different size of corpus.

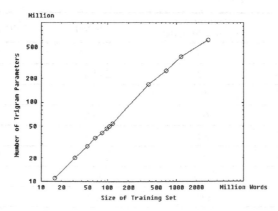

Fig. 3. The number of trigram with different size of corpus.

4   Performance of Language Models

4.1   Corpora for Language Modeling

There are 11 corpora constructed for testing of performance. All the corpora are in an extending style, as shown as the following:

$$C_{i+1} = C_i \bigcup C_{extend}, i = 1,2,...n-1 \qquad (9)$$

where C denotes the corpus, n is the total number of the corpora.

The first 8 corpora are confined within economical affairs from the Economic Daily. The 9th is consisted of pure newspapers. The last two contain newspapers, magazines, books and scientific digests.

4.2   Recognition Experiments

The test corpus is reading speech of economical news from 10 female speakers, which is a sub set of Corpus99 developed by the National Lab of Pattern Recognition.

The recognizer is an embedded system of FlyingTalk, the recognition engine developed by NLPR.

The recognition results in Character Error Rate (CER) concerned with every language model are shown in table 2. The results show that, before the size of training set attains to 1000 million words, the CER decreases with the increasing of training set. When the training set is very large and extending broadly with content, the CER increases and the performance of the recognizer drops.

Table 2. Performance of language models

LM	training set (Million Words)	CER(%)	Perplexity
eco_83_85	16	25.67	389
eco_83_88	32	24.90	287
eco_83_91	49	23.35	244
eco_83_93	65	21.96	211
eco_83_95	82	21.77	195
eco_83_96	96	21.57	185
eco_83_97	106	20.93	178
eco_83_98	118	20.38	166
newspaper	707	18.29	189
balance	1180	22.19	151
general	2960	22.58	169

4.3   Language Evaluations by Perplexity

The language models are evaluated with the economics part of the text corpus from which the transcription of Corpus99 are based, rather than the much larger language model test text. This circumvents the problems caused by any potential mismatch between the language model test text and the recognition task itself.

The values of perplexity of each language model are shown in Table 2. We can find that the correlation between the perplexity and the CER is not very strong.

5   Discussion

5.1   Correlation between CER and Perplexity

In our investigation, the correlation coefficient r is computed according to [1]

$$r = \frac{Cov(X,Y)}{\sqrt{Var(X)Var(Y)}} \qquad (10)$$

The correlation coefficient between CER and the perplexity of language models of the data of section 4 is 0.7626, which shows a loose relation of the two variables.

5.2   Analysis of the Perplexity

Perplexity is a widely used measure of language model quality. However, same as the results in section 4, many recent work by other researchers (for example, in [3]) has demonstrated that the correlation between a language model's perplexity and its effect on a speech recognition system is not as strong as was once thought.

According to equation (5) we know that perplexity is based solely on the probabilities of the words which actually occur in the test text. It does not consider the alternative words which may be competing with the correct word in the decoder of a speech recognizer.

Table 3. Comparison of different recognition results on perplexity computed by Language Model "balance".

Recognition results	With LM balance	With LM newspaper	With LM eco_83_98	With LM eco_83_97
Perplexity computed by LM balance	498	626	572	603

From Table 2 we know that language model "balance" has the lowest perplexity, however, the performance of recognizer with this language model is not as good as it should be according to the perplexity. Table 3 compares the perplexity values of model "balance" on the recognition result texts of recognizer with model "balance", "newspaper", "eco_93_98" and "eco_83_97", the latter 3 models have higher perplexity in Table 2 but lower CER than model "balance". Table 3 implies that, in the space of words array in the decoder of the recognizer, model "balance" has picked the most reasonable result. Or in other words, in the view of model "balance", the more correct results picked by model "newspaper", "eco_93_98" and "eco_83_97", are less probable.

Since the purpose of language model in a recognizer is to pick the correct words from competing words with acoustic similarity in the decoder, the feasible evaluation measure of language model should have the ability to stand for this point. Some researchers already started the exploration of new evaluation measures ([2],[3]). Besides innovation on structure of language model, this is another significant direction in the developing of theory of language modeling.

6   Conclusions

This paper presents the statistical characteristics of Chinese language and its statistical language models based on huge text corpora. From our investigation on the large Chinese text corpora, we find that, the statistics of the corpora reveals that the average word length shows prominent feature related to the class of the text, and the number of bigram and trigram can be estimated by the size of the corpus. In the recognition experiments, we find the correlation is weak between the perplexity and either the size of the training set or the recognition character error rate. The drawback of perplexity exists intrinsically, so new evaluation measures should be explored.

References

[1] Berenson, M., Levine, D. and Mercer, R.L., "Applied Statistics, A First Course." Prentice-Hall International, 1988.
[2] Chen,S., Beeferman, D., and Rosenfeld, R. (1998). "Evaluation Metrics for Language Models." In Proceedings of the DARPA Broadcast News Transcription and Understanding Workshop, 1998.
[3] Clarkson, P. and Robinson, T. "Towards Improved Language Model Evaluation Measures", In proceedings of EUAROSPEECH'99, Sep. 5-9, 1999 Budapest, Hungary.
[4] Jelinek, F. (1998)."Statistical Methods for Speech recognition", The MIT Press, 1998.
[5] Katz, S.M, "Estimation of Probabilities from Sparse Data for the Language Model Component of a Speech Recognizer." IEEE Transactions on Acoustics, Speech and Signal Processing, 35(3): 400 – 401, 1987.
[6] Zhang, H., Huang, T., and Xu, B. (2000), The NLPR Chinese Language Model Toolkit (V1.0) for Large Text corpus, 2000 International Conference on Multilingual Information Processing (2000 ICMIP), Urumqi, China, July 20-25, 2000.

On Building a Simulating Translation Environment for Multilingual Conversation

Jianshe ZHAI, Georges FAFIOTTE

(CLIPS-IMAG / GETA, Université Joseph Fourier - Grenoble 1, France)

{JianShe.Zhai, Georges.Fafiotte}@imag.fr

Abstract. We present a distributed environment (Sim*) for collecting multilingual spoken dialogues through a Wizard of Oz scheme. These speech corpora are then used to build an automatic Speech Translation system. In the future Sim* should develop to offer full support for any combination of resources between human interpreters and automatic Speech Translation systems, in a multimodal and distributed context.

Keywords: Multilingual Dialogue, Spoken Language Translation, Interpreter Simulation, Wizard or Oz, Client/Server.

1   Overview

With the rapid development of Internet technology, people can more and more easily contact with others. However, the current existing net-based oral automatic translation systems are not powerful enough to satisfy the need for communication in different languages. So we carry out stepwise research in simulating speech translation, in order to simulate the real situation of an automatic oral translation, especially in multilingual context. Parallel to participating in the international CSTAR II project (Consortium for Speech Translation Advanced Research, http://www.c-star.org/), we developed the Sim* project. The main idea here is to provide an universal network-based simulation tool for multilingual conversation. We sketched a framework to utilize distributed net-based resources, as well as to create reusable components to ease the building up of a Machine Translation environment. Sim* was developed with the following goals in mind:

- platform independence: to satisfy user with his preferred computer platform;
- multimodality: to provide a set of utilities to aid multilingual conversation, including speech, text, whiteboard, Web-extracted texts or pictures, etc;
- integration facilities: to easily integrate lingware components, e.g. recognition, translation, and synthesis in the platform;
- cooperative work: to ensure all participants to work coordinately and interactively.

T. Tan, Y. Shi, and W. Gao (Eds.): ICMI 2000, LNCS 1948, pp. 287-292, 2000.

2   System Architecture

We constructed Sim* using a classical Client/Server scheme. Being focused on multilingual speech translation, we can provide with interpreters in different languages to serve different speakers. Simply, if there are N clients speaking in N languages, so (N-1)*(N-2) or more bilingual interpreters can be needed. Note that the semantic process of speech translation is in such a pattern as
Speaker A --> Interpreter --> Speaker B.
Let's give a glimpse of Sim* in the following figure: **0** – client C1 wants to talk to client C3,   **1** – to ask the permission to the Server,   **2** – to send C1 speech to the corresponding interpreter I1,   **3** – to send I1's translation to C3.

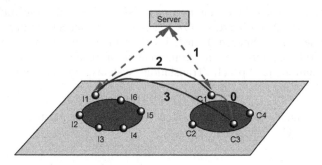

Fig. 1. General architecture

As a grounding for a net-based simulating translation system, the first step is to understand "who are the users ?", "what kinds of content, activities and services do they want ?". So a robust, evolving database of member profiles is a most important element. If handled with integrity, the collection of profiles will provide substantial support to both system and users. We constituted a static hub-to-spoke network topology between the simulated translation server and all other actors in the process (to acquire and record profiles of all connected users or components, to serve for user's requirement and monitor the execution of the system), and dynamically maintained a peer-to-peer topology between two or more communicating users (to lighten the network traffic and improve the performance of the system).

3   Wizard of Oz

Unlike usual videoconference with no translation function, the Sim* interpreter plays an important role. Activating integrated lingware components (recognition, translation, and synthesis) automatic speech translation could be run, whereas human translation is enabled through Wizard of Oz interpreting. Wizard of Oz as interpreters can be used with classical synthesizers in different situations. In our experiment, result of translation was sent to both the sender (for confirmation) and the receivers (for communication). We found that, considering the current technology of automated translation, Wizard of Oz technology is necessary to aid in solving some Machine Translation problems, for example that of disambiguation or proper noun spelling. With Wizard of Oz translation we can simulate real-time spontaneous speech translation and build a basic skeleton, as well as primary reusable components, for a full-fledged net-based translation system.

In the context of multilingual conversing, Sim* could configurate with an appropriate translation component or human interpreter, with respect to specific user demand according to the profile maintained by the system. If several components could offer similar service, Sim* could be used to dispatch user's requirement to them, then receive some results for evaluation or comparison. Therefore, we could enrich the basic knowledge of each component performance, and provide the user with information about component competence in the user's domain of interest.

4   System Features

In an environment of multilingual conversation, broadcasting speech is nearly meaningless. So in Sim*, a user could choose freely one or more correspondents to talk with as needed, with the help of one or more interpreters. This feature, however, can arouse a potential conflict. As well known, every communication group has its share of internal squabbling, and if well handled conflict keeps the group lively and interesting. It can, however, run out of control. With respect to the semantic process of speech translation, we adopted PTTE (Push To Talk Exclusive) Communication protocol to prevent speech turn internal conflicts. In addition, each step associated with speech translation can be carefully tracked. So, an user can know what others are doing from the dynamic display on his interface. Therefore, he (or she) can decide what to do sooner. Generally, there are always sequential and parallel operations in any communication group. In Sim*, the semantic process of speech translation is strictly arranged in order, whereas some similar components or interpreters can do the task concurrently.

Exchanging text message is also important, especially for disambiguition or confirmation. Similar to the speech, text messages are limited in a same language. So an usual monolingual chat should be changed to fit multilingual requirement. Sim* text function can also be used to feedback written translations or to provide some training sentences for a user to practice.

In addition non-text information (such as body language) can be quite important in any situation of conversation. Even though we have text (sentence) function in Sim*, we are still intersted in the role played by non-text information in the translation process. This is because we could sometimes explicitly give a negative answer through body language while making a positive oral answer. Therefore we introduced in Sim* a videoconferencing facility, first using MBone video communication facility in experiment. This combination lets speaker and listener see each other during a conversing session. We found that users felt more respected in this environment, and intended much more to concentrate on the conversation.

The scenarios for the C-STAR II project were domain-oriented in the field of tourism. Web information was often used. We found it is very helpful to provide some public place for all participants to share some information. Therefore, we provided Sim* with a local whiteboard function: in addition to usual whiteboard functions such as writing, drawing, marking, moving an object, etc, Sim* can extract some interesting pictures from a Web site and put them on the whiteboard for sharing.
In the next C-STAR III context this can be very useful when you would like to show one room of a hotel or demonstrate someone how to travel in a city. For instance, in order to explain to a foreign visitor how to get to our laboratory on the Grenoble campus from the railway station, we could use the whiteboard to show the Web campus map, introduce in which building we are located, which bus or tramway line should be taken.

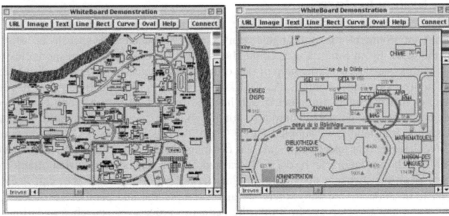

Fig. 2. A part of the campus map as shown on the Sim* whiteboard

Although the multicast communication technology is often used on the Internet to transmit conversation, our chosen strategy is to transmit the original sound files. It allows user or interpreter to replay. It also lightens the needs of materials or reduces the interference of the noise of network. More important is that original speech resources are widely used in research and application of automatic speech translation. We tested the transmitting speed of sound files first on our local network (10Mps). Some results are shown in the following table.

Size of Sound File	PC (P400) to Mac (8500)
1.3 M	4.6 s
930 K	3.9 s
515 K	1.6 s

Table 1.

5   System Applications

Since we focus on multilingual automatic translation, Sim* presently has two main applications. One is to support building speech segment corpora and lexical bases. A first experiment of oral speech collection was done in January 2000 for a Chinese-French cooperative project. Another is to provide a set of utilities to assist automatic analysis and processing of lexical and semantic structures. Web information of France-Meteo is seen as a case study to build an experimental terminology acquiring and consulting system. The system can be designed to find a new word or term from the everyday weather bulletin, pick it up and sort it into the terminology base automatically. In addition, each sentence in the bulletin is tagged with a keyword (term) associated with a special weather event. It could provides fundamental basis on which to build a corpus and translation system.

6   Prospective Development

As a generic simulating tool, it is evident that more information resources should be accessed in future versions of the Sim* platform, for example net-based electronic dictionaries. We may try a pilot real-time connection with some resources being developed in our research group within the UNL project (Universal Network Language, http://www.unl.ias.unu.edu/). We also join some interactive and multimodal features of Sim* into the European NESPOLE project (NEgotiating through SPOken Language in E-commerce, http://nespole.itc.it/). For Sim* itself, we have also considered some asynchronous applications, such as helping in Computer-Aided Language Self-Learning systems (which allows a user to talk with an automatic translation system or to a real interpreter), or off-line translation service (which allows users to be able to leave the talking session for a while).

Conclusion

From the Sim* current prototyping, we got primarily experience about how to constitute a distributed architecture to take use of distributed resources for net-based translation, especially in multilingual situations. We also obtained knowledge about what are the needed primary functions (seen as reusable components) to be satisfied with human's real use in spontaneous spoken language translation. We strongly believe that simulating translation environments will have prosperous research applications.

References

[1] Fafiotte G. & Zhai JS, *A Network-based Simulator for Speech Translation.* Proc. NPLR'99, Beijing, 5-9 Nov. 1999.

[2] Boitet.Ch., GETA's *MT methodology and its current development towards personal networking communication and speech translation in the context of the UNL and C-STAR projects.* Proc. PACLING-97, Ohme, 2-5 Sept. 1997.

[3] Coutaz J., *NEIMO, a Multiworkstation Usability Lab for Observing and Analyzing Multimodal Interaction.* Proc. of CHI'96 Companion, 1996.

[4] Amy Jo Kim, *9 Timeless Principles For Building Community*, Webtechniques, Jan. 1998.

[5] Sycara K.& Pannu A., *Distributed intelligent agents.* IEEE Expert 36-46, 1996.

Approach to Recognition and Understanding of the Time Constituents in the Spoken Chinese Language Translation*

Chengqing ZONG, Taiyi HUANG and Bo XU

National Laboratory of Pattern Recognition, Institute of Automation,
Chinese Academy of Sciences

{cqzong, huang, xubo}@nlpr.ia.ac.cn

Abstract. In the spoken Chinese language, the time constituents occur frequently, especially in the domain of appointment schedule, ticket booking and hotel reservation etc. However, in the current Chinese-to-English Machine Translation (MT) systems, it is still a problem to deal with the time constituents. According to our test results of some commercial Chinese-to-English MT systems, about 57.1% Chinese time constituents are wrongly translated, and 58.3% of the errors are caused by false recognition and misunderstanding of the time constituents. In the paper, we present a new approach to recognition and understanding of the time constituents in the Chinese language, which is integrated by a shallow level analyzer and a deep level analyzer. The shallow level analysis is realized by a Finite State Transition Network (*FSTN*). The time constituents are first recognized by the *FSTN*, and the results are divided into three types. Two types of the results are determinate and one is uncertain. Aimed at the uncertain results, the deep level analyzer checks the semantic context of time constituents, performs necessary phrase structure analysis, and finally decides the type of the time constituents. The approach has been employed in our Chinese-to-English spoken language translation system, which is limited in the domain of hotel reservation. However, the approach is domain-independent. The preliminary experiment has proven that the approach is effective and practical.

1. Introduction

In the Chinese spoken language, the time constituents occur frequently. We ever collected 94 dialogs (3013 utterances) in the domain of hotel reservation[1]. In the total 16470 Chinese words and 18 parts-of-speech of the dialog corpus, the proportion of time words takes about 3.1%, and the proportion of numeral words takes about

---

* The research work described in this paper is supported by the National 863 Hi-Tech Program of China under Grant 863-306-ZT03-02-2, the National Key Fundamental Research Program (the 973 Program) of China under Grant G1998030504 and also the National Natural Science Foundation of China under Grant 69835030.

T. Tan, Y. Shi, and W. Gao (Eds.): ICMI 2000, LNCS 1948, pp. 293-299, 2000.
© Springer-Verlag Berlin Heidelberg 2000

15.6%. According to our preliminary statistical results, about 20% numeral words are related to the time constituents. So, analysis and interpretation of the time constituents are obviously of importance to understand the spoken Chinese language.

In the Chinese language processing, although there are very few papers regarding the processing of time constituents of the Chinese language, it is indeed not a neglectful problem in the Chinese-to-English MT systems. The authors ever tested some commercial Chinese-to-English MT systems (Please forgive me not to mention the concrete names of the systems here with an eye to avoid unwanted trouble.). Unfortunately, the error rate of translation results of time constituents reaches about 57.1% in average. That is only about 42.9% time constituents of the Chinese language are correctly or constrainedly translated into English. In the wrong translated results of time constituents, about 58.3% errors are caused by the false recognition and misunderstanding. The following are two examples that the time constituents are wrongly translated by a commercial Chinese-to-English MT system.

    1). *Input*: 这台计算机已经运行了三分十二秒。(The computer has run for three minutes and twelve seconds.)
    Output: This computer has run 3 <u>to divide into</u> 12 seconds.

    2). *Input*: 五月十八号我将去上海。(I will go to Shanghai on May 18th.)
    Output: <u>May No.18</u> I will go to Shanghai.

Obviously, in the two examples the time constituents are all wrongly translated because they are wrongly parsed. Addressed the problem of recognition and understanding of time constituents, a new approach is proposed in this paper. In our approach, the analysis is performed from shallow level to deep level based on the characteristics of the time constituents in the Chinese language. Remainder of the paper is organized in following way: *Section* 2 gives the basic ideas and describes the related methods in detail. And *Section* 3 shows the experimental results.

2. Approach to Recognition of the Time Constituents

2.1. The Characteristics of Time Constituents in the Chinese Language

In the Chinese language, the time constituent covers a big range. In our system, the time constituents are mainly divided into following six basic types, an extension type and a compounding type as well.

- The Basic Type

(1) Seasons
Such as, 春天(spring), 冬天(winter), 夏天(summer), 秋天(autumn), 初春(the beginning of spring) etc.

(2) Months

一月(份)(January)，二月(份) (February)，…，十二月(份) (December)

(3) Days of Week

Such as, 星期一(Monday), 周二(Tuesday), 星期天(Sunday) etc.

(4) Approximate Time

Such as, 未来(future), 现在(now), 去年(the last year) etc.

(5) Time of Day

Such as, 凌晨(wee hours), 午夜(midnight), 早晨(morning) etc.

- The Extension Type

In Chinese sentences, the words on festivals often act as the role of time constituents. For instance, 元旦(New Year's Day), 春节(Spring Festival) etc. So, in our system, all names of festivals are treated as a special type of time constituents, which is called extension type of time constituents.

- The compounding Type

The compounding type includes the following two type of phrases:
(1) The phrases that are assembled by all time words of the two types mentioned above;
(2) The phrases that consist of numerals and classifiers with time feature.

For example, 星期三上午八点(8 o'clock in Wednesday morning), 二零零零年五月十二日(May 12, 2000), etc.

From the explanation given above we can see that the time constituents in the Chinese language are of the following characteristics: a) The time constituents may consist of only basic time words without any numeral. In this case the time constituents express a determinate time, a period or a moment. There is not any ambiguity. b) The time constituents may also consist of numerals and classifiers. In the case it sometimes causes semantic ambiguity. For example, when a speaker says in Chinese '三号', he or she probably means May 3rd or No. 3. Similarly, '五日' in Chinese may refer to the fifth day of a month, such as May 5th, but it also means 5 days. It is quite different. So, it is a main task of the time constituent analyzer to resolve the ambiguity of compounding type of time constituents.

2.2. Recognition of the Time Constituents

2.2.1 The Shallow Level Analyzer

The shallow level analyzer (SLA) may be described as a Finite State Transition Network (*FSTN*) as follows:

$$FSTN = (Q, \Sigma, \delta) \tag{1}$$

Where, $Q = \{q_0, q_1, \ldots, q_n\} \cup Q'$ ($n \in N$) is a finite and no-empty set of states. q_0 is the initial state. Q' is the set of terminal states. Σ is a finite and no-empty set of time words, numerals and also classifiers with time feature. δ is the transition function. If a state accepts a given word, the state will be transited to another one. In the parsing procedure of a Chinese sentence, if a word W_i ($i \in N$) is recognized as the time word, the word W_i and its position will be inputted to the *FSTN*, and all words after W_i will be analyzed by the *FSTN* until a $W_j \notin \Sigma$ ($j \in N$ and $j > i$) is found. If the string $W_i \ldots W_{j-1}$ is accepted by the *FSTN*, and the end state is T_k ($k=1..3$), T_k is the parsed result of SLA.

Suppose a = a numeral b = 时(o'clock) | 刻(quarter)
  s = 秒(second) n = 年(year)
  c = 周(week)|天(day)|小时(hour) |分钟(minute)
  f = 分(second or cent) d = 点(o'clock or point)
  e = 日(the number of days or one day of month)
  h = 号(one day of month or serial No.)
  $g \in$ {Seasons}∪{Approximate time}∪{Time words of extension type}
  m = 一月(份)(January)|二月(份)(February)| ... |十二月(份)(December)
  w = 星期一(Monday)|周一(Monday)| ... | 星期天/星期日(Sunday)|...
  $t \in$ {Time of day} y = 月(month)

The *FSTN* is described as the following Figure 1:

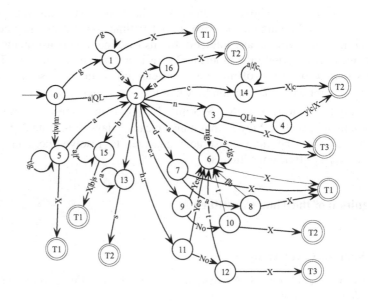

Fig. 1. FSTN for Time Constituent Recognition

In Figure-1, there are 16 states. *State* 0 is the initial state, and the circles with symbols T1, T2 and T3 are terminal states. The terminal state T1 means the recognition result of the time constituent is a determinate time. T2 means the time constituent expresses a period. T3 means the recognition result is unsure and the result will be rechecked and finally decided by the following deep level analyzer. In the *FSTN*, there are four types of letters on directional arcs, which are (1) time related words *a, b, c, d, e, f, g, h, m, n, t, w* and *y*; (2) quantifier *QL*; (3) test condition *e:r* and *h:r*; and (4) *X*. Where, *y* is only accepted when its previous input is *QL*. *QL* mainly refers to the Chinese measure word '个' or a number phrase with the measure word '个'. *X* expresses a word that doesn't belongs to the time words and is not one of *a, b, c, d, e, f, g, h, m, n, t, w* and *y*. *e:r* and *h:r* mean separately the input are *h* and *e*, and the input will be tested by the constraint rule *r*. The rule *r* is designed as a function:

$$Find(\text{L, '月'}) \tag{2}$$

If the analyzer finds the Chinese character '月(month)' within the range from the first word of the time constituent to the current word, the function returns 1 (True). Otherwise, the function returns 0 (False). If the function returns 1, the state will be transited along the arc with sign 'Yes'. Otherwise, the state is transited along the arc with sign 'No'. For example, if *State* 2 is the current state and the input is *e:r*, *State* 2 will be transited to *State* 6 when the test result of constraint *r* is True. Otherwise, *State* 2 will be transited to *State* 10. *State* 9 is just a temporary state.

The time constituents that are not consisted of time words are not recognized by the *FSTN*. Such as, '二战期间(during the World War Two)', '毕业以后(after graduation)' etc. '秒(second)' is the minimum time unit.

2.2.2 The Deep Level Analyzer

The deep level analyzer (DLA) is designed to resolve the ambiguity of T3 from SLA. In the *FSTN*, T3 may be generated by three Chinese characters '年(year)', '号 (serial number)' and '秒(second)'. For example, '十二秒' means 'twelve seconds', which is a segment of time, but '3点5分12秒' means '3 o'clock 5 minutes and 12 seconds', which is a time of day. That is '秒' expresses different meanings in different context. So, DLA will resolve the ambiguity by test of context. Similar with function *Find*, the context test function *C-Test* is designed as follows:

$$C\text{-}Test(Direction:Type, [Range], Search\text{-}target) \tag{3}$$

Where, *Search-target* is the test condition that is probably a Chinese word, a part-of-speech or a semantic feature. *Direction* gives the direction to test the condition. *Direction* = *L* means the system will test the condition on the left side of current position, and *Direction* = *R* means the test operation will be performed on the right side of current position. *Type* is the type of *Search-Target*. *Type* = *W* means the *Search-Target* is a Chinese word. *Type* = *P* means the *Search-Target* is a symbol of part-of-speech or a sign of phrase structure. And *Type* = *S* means the *Search-Target* is a semantic feature. *Range* gives the scope where the condition will be tested, and the *Range* is expressed as $n_1:n_2$. If *Range* is left out, the range is from the left first word of the current position to the first one of the utterance or from the right first word to the last one of the utterance.

To resolve the ambiguity of time constituents, a set of constraint rules are developed, which are mainly expressed by the function *C-Test*. For instance, the following three rules are designed for disambiguation of the Chinese character '秒 (second)':

> IF *C-Test(L:W*, 3:6, 分钟) THEN *T3 => T2*;
> IF *C-Test(L:P*, 3:3, !T) THEN *T3 => T2*;
> IF *C-Test(L:W*, 3:10, 点|时) THEN *T3 => T1*;

Here, we assume the current pointer points to the word behind the Chinese character '秒'.

3. Experimental Results

To evaluate the performance of our time constituent analyzer, 46 types of time expression are tested. In the test sentences with time constituents, 33 time constituents express the time of a day, and 13 time constituents express segments of time. The test results are shown in Table-1.

TYPES OF TIME CONSTITUENTS	T1	T2	T3
Number of time constituents	33	13	-
Results of SLA	27	12	7
Results of DLA	32	14	-
Correct rate (%)	97.8		

Table 1. Experimental Results

From the experimental results we can see our approach to recognition and understanding of the time constituents is effective. The approach has been employed in our Chinese-to-English spoken language translation system[7], but the approach is domain-independent.

However, although each time constituent in an utterance is correctly recognized, sometimes the understanding result may be still wrong. The reason is that in the spoken Chinese language, a time constituent is possibly spoken out separately in the same utterance, or the previous time constituent may be denied or rectified by its posterior ones. So, the system has to recognize the separate related time constituents and merge them into one complete constituent. For example, in the Chinese utterance '大概七月初吧,可能是二号.' the probable time is July 2nd, but the time is expressed by separate two parts '七月初(the beginning of July)' and '二号(the second day of a month)'. Anyway, it is a very hard work to merge the separate time constituents. The problem is being addressed now, and we will report the research results on time.

4. References

[1] Chengqing ZONG, Hua WU, Taiyi HUANG and Bo XU. Analysis on Characteristics of Chinese Spoken Language. In *Proceedings of the 5th Natural Language Processing Pacific Rim Symposium (NLPRS'99)*. Beijing, China. Oct. 1999. Pages 358-362.

[2] Zhiwei FENG. The Computer Processing of Natural Language (in Chinese). Shanghai Foreign Language Education Press. 1996.

[3] Bernd Hildebrandt, Gernot A. Fink et al. Understanding of Time Constituents in Spoken Language Dialogues. *Proc.* ICSLP'94. Pages 939-942.

[4] B. Hildebrandt, H. Rautenstrauch, G. Sagerer. Evaluation of Spoken Language Understanding and Dialogue System. *Proc.* ICSLP'96. Pages 685-688.

[5] James Allen. Natural Language Understanding (2nd ed.). The Benjamin/ Cummings Publishing Company, Inc. 1994.

[6] Zhu Dexi. The Chinese Grammar (in Chinese). The Commerce Press. 1997.

[7] Chengqing ZONG, Bo XU and Taiyi HUNAG. An Improved Template-Based Approach to Spoken Language Translation. To appear in ICSLP'2000, Beijing, China. Oct. 2000.

KD2000 Chinese Text-To-Speech System

Ren-Hua Wang Qinfeng Liu Yu Hu Bo Yin Xiaoru Wu

University of Science & Technology of China

P.O.Box 4, Hefei, 230027 P.R.CHINA

Email: rhw@ustc.edu.cn

Abstract. This paper presents the new progress we made in recent two years on Chinese text-to-speech towards higher naturalness. The results can be summarized as follows: 1). Aim at the different characteristics between philology and phonetics, a kind of hierarchy process model for Chinese TTS system has been proposed. Five layers are defined to label the sentence in the text analysis. 2). A prosodic generating model is built for selecting appropriate unit with higher accuracy. 3). Extracting prosodic parameter from the unit base and then normalizing them to form a prosodic parameter base. 4). Proposed an effective method for processing the special text, which integrates external descriptor rules and model matching. With these progresses a new Chinese text-to-speech system named KD2000 has been developed. The improved performance has been confirmed by evaluation test.

1   Introduction

In the TTS system based on waveform concatenating technique, a sequence of units are selected from the unit base and then are concatenated to generate new speech after appropriate prosodic modification. Most of Chinese TTS system use syllables as basic synthesis units[1]. Since the signal processing technique is not so advanced, the quality of synthesis speech will descend dramatically if the range of prosodic modification on the syllables is too great. For bypassing this problem, several tokens for one syllable are recorded, cut out from real utterances, and then all of them are stored as basic units in a unit base in advance. During period of speech synthesis, some appropriate units are selected from the unit base and concatenated after fine modifying, even no modifying. The quality of output speech can be improved dramatically comparing with completely depending on modifying unit. How to design and get these units from real speech have been introduced in a former article [2] in details.

In this way, selecting appropriate unit for a syllable contributes much more than modifying it to form high quality speech. But another problem arises, when a few of units are stored in the unit base for each syllable, that is how to pick out most appropriate unit from the unit base. For resolving the problem, both statistical method and some specific rules are required to analyze input text for obtaining enough information, which is essential to conclude the closely linked degree between syllables in a context and then generate similar prosodic parameter to real speech. By the prosodic parameter we can pick out correct unit for each syllable with more accuracy.

This paper presents some progresses we made in solving the above problems. With the progress a new Chinese text-to-speech system, named KD2000, has been

T. Tan, Y. Shi, and W. Gao (Eds.): ICMI 2000, LNCS 1948, pp. 300–307, 2000.

developed. KD2000 manifests very good performance, especially, higher naturalness comparing with the existed Chinese TTS system. The remained part of this paper is organized as follows: section two introduces the method of hierarchical text analysis, section three introduces the prosody generation and unit selection, section four introduces special text processing, final the performance evaluation is discussed, and the test results is given in the last part.

2   Text Processing in KD2000 Chinese TTS System

2.1   The Purpose of Text Analysis

In TTS system how to make the output speech with high naturalness is always the most crucial problem. By text analysis, enough and correct information on the interior syntax structure and the prosody layer structure of the text in a given context will be extracted, which directly determines whether the output speech has high quality.

Along with the progress of speech synthesis technique, abundant and essential prosodic information are going to play more and more important role in the improvement of the quality of synthesis speech. So the comprehensive syntax structure and the text's meaning should be analyzed and extracted from the input text to promote the quality of output speech of TTS system.

One of our efforts is to put forward a kind of accurate and feasible text-analysis strategy, which can correctly analyze and describe each language layer's information and mapped them into prosodic level, establish an effective and general model, and then to provide more valuable information to speech synthesizer.

2.2   The Method of Hierarchical Text Analysis

"Chinese is a pile of concepts". In fact, not only in Chinese, but also in other natural language, it is always true that the small language units are assembled into bigger units to express specific meaning. However this constitutional relationship based on layers is more obvious in Chinese. So we define some distinctive layers and adopted them as different standards to process text. With this idea, all layer's information can be organized in a form of parsing tree, in which higher level's information is obtained from ones of lower levels. Then the closely linked degree between the syllables in a context can be obtained from the sentence's parsing tree. Compared with other methods, our method can describe natural language in more details. Additionally, it is easier to process and expand with the method.

To realize hierarchy processing, firstly we need a labeling system to distinguish each layer of text prosodic structure. But the existent labeling systems are all built for either philological purpose or prosodic one. Due to this reason, we put forward our own hierarchy labeling system with reference to TOBI labeling system and some others' Chinese prosody labeling method, which give attention to both of philological and prosodic application.

We define the following layers:

L0: syllable layer, In Chinese, it represents a Chinese character.

L1: speech foot layer, it is well known as the layer of rhythm.

L2: sub—phrase layer, syntax words combine with mono-syllable word, and appear as one pronouncing unit.

L3: master—phase layer, it is relatively independent syntax unit.

L4: Breath—group layer, it happens to take a breath in a long sentence.

L5: Sentence layer, it is the top layer of all the concerned layers.

The establishment of this labeling system makes hierarchy processing to be feasible. And it is of great significant in labeling text, sharing resources and transmitting text with prosodic information.

Segmenting sentence into small words, which may be just a little bigger than a Chinese character, is the base of further advanced analysis. The accuracy of segmenting sentence directly effects the result of whole text analysis. In this step, the most important problem is to avoid generating false meaning owing to improper segmentation. Because of Chinese text lacking for obvious separate symbols, meaning unclear is hard to be avoided. Such as in the phase of "人\民\生\活\水\平", "民生" "活水" all are words. Then how to syncopate a sentence is the most important key in our text analysis model.

Because of being lack of ability to understand the meaning of the input text, the base of sentence segmentation strategy at present is still depend on the syntax lexicon with attributes of the part of word. It is possible to select the best segmentation path from all possible segmentation paths at the help of appropriate evaluating functions and Full-Expended-Words-Segmentation-Net we set up.

The following is a example:

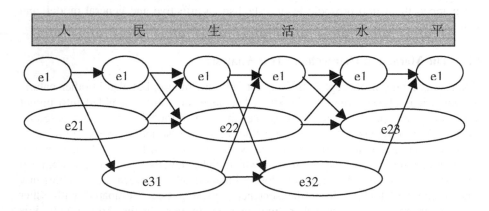

Fig. 1.   Full-Expended-Words-Segmentation-Net

Considering both performance and cost, we adopt A* algorithm.

On the assumption that segmentation is $W=w_1,w_2,...,w_n$,

$P_{cut}(W_i)$ is word's Segmented Word-Frequency(SWF);

$P_{form}(W_i)$ is word's Form Word-Frequency(FWF).

Then the path's evaluating function is defined as:

$$F(W) = \sum_{i=1}^{n} \ln \frac{P_{cut}(W_i)}{Total_{cut}} + \ln \frac{P_{form}(W_i)}{Total_{form}}. \tag{1}$$

WF is gained from abundant language materials (forty million words) with the statistical method. SWF is the total appearing times of each lexicon word in all of the statistical material, which had been cut into words in advance. FWF is the total appearing times of each lexicon word in the rude statistical materials.

Along with segmentation, we label the attributes of the part of words and process unregistered words and polyphones, which can't be gotten from the lexicon, with some rules only.

Now that the information of hierarchy is especially useful to the synthesis speech, how to generate correct layers and the descriptive information of each layer becomes one of our main research subjects. Because no idea can be mentioned before about adopting information of several levels to improving output speech, we are fishing for the better way gradually.

As far as present speech synthesizer is concerned, experiment show that the layers of L2 and L3 are determinants to generate correct prosodic parameter. So to define the two layers and the other essential information in the levels is our main aim too.

As the junction of philological level and speech level, Sub—phrase layer carries a lot of syntax and prosodic information. In the view of prosodist, L2 is the reflection of prosody word, and the information of it has great significance to the naturalness of speech. In the opinion of philologist, L2 is mainly determined by philological knowledge. So we adopt describing rules and corresponding costs to generate the optimizing path for L2. The rules are originated from both of mature syntax rules and statistical information. Because of the different contribution of each rule a cost coefficient are set up to embody the factor.

The decision of L3 depends on the language understanding, which is also a difficult subject. We can avoid the problem at certain degree by adopting the statistical method and the information of L2. We introduce C45 Decision Tree to generate L3 and it's lingual and prosodic information at the help of the L2 information. The C45 Decision Tree is built up by C45 arithmetic and the training text material which have been segmented and labeled by some lingual experts in advance. In the limit of maximum number of the words, at each division among L2 layers we calculate the probability in which it belongs to the boundary of L3.

The division with the greatest probability is selected as the boundary of L3 layer.

3   Prosody Generation and Concatenation with Unit Selection

Correct prosody generation will be especially helpful to unit selection and to prosodic modification. At the help of concept of "perception-quantizing" [2] and the existed unit base, using the template of quantified prosodic unit and base intonation contour we can generate the rude prosody contour and then select the best unit with the guidance of the prosody contour to generate the high naturalness speech.

3.1   Prosody Generation Using the Template of Quantified Prosodic Unit and Base Intonation Contour

Based on two features of prosody in Chinese mandarin: the prosody chunks (prosody words), which have the same prosodic structure (including tone-class, rhythm), manifest relative stable prosodic mode in the continuous speech; the prosodic mode shows different hierarchical character depending on its context, we can simplify the prosody generation into two steps: building up the stable template of prosody in the prosody word and the generation of base intonation contour, which presents the hierarchical feature.

3.1.1 Normalize Prosody Parameters

To build the stable template of prosody, we must first get accurate prosodic parameter of "perception-quantizing" units from the corpus, including duration, pitch contour, energy and so on, to build the prosodic parameter base; then concatenate the prosodic parameters of "perception-quantizing" units together under prosodic rules. When recording the "perception-quantizing" units, the speaker can't always read the text as required, for example she possibly read too fast or too strong sometimes, we can't extract the parameter from the units directly. To eliminate these kind of disaccords aroused from unstable speaking style of the speaker, some normalization must be carried out before storing the prosodic parameter into the parameter base.

Due to different prosodic position, different PinYin, different Tone-Class of the syllable in Chinese, it has different duration feature. The syllables with the same Tone-Class but different PinYin should have stable duration ratio between them when they are in different prosodic positions, because the prosodic position makes the same influence to each syllable with different PinYin. So we can normalize the duration using the following formula:

$$\frac{\alpha_i S_i}{\alpha_1 S_1 + \alpha_2 S_2 + ... + \alpha_i S_i + ... + \alpha_n S_n} = \frac{A_i}{A_1 + A_2 + ... + A_i + ... + A_n} = \beta_i. \quad (2)$$

where S_i is the duration of a syllable S (with assured PinYin and Tone-Class) in a special prosodic position i $(0<i<n)$, A_i is the average duration of syllables (which have same Tone-Class and prosodic position with S), α_i is the normalize parameter that we want to get.

In the same time, the syllables with the same Tone-Class and prosodic position should have similar pitch contour in the value level. So just simply moving the pitch contour up or down can resolve the problem of pitch contour disaccord well.

3.1.2 Build Prosody Contour with Normalized Prosodic Parameter

In KD2000 system, the full way of prosody generation can be summarized as following:

A. The stable template of prosody is corresponding to the prosody of the L2 layer in text prosodic structure. So to each L2 layer of the text prosodic structure generated by text analysis, we use appropriate template of prosody to build the L2 layer prosody chunk. In this session, the prosodic parameter will be picked out from parameter base and be filled in the prosodic template. The rude prosodic contour of each L2 layer can be formed.

B. After we get each L2 layer prosody chunk's prosody contour, we will process the whole sentence prosody chunk from the bottom to the top of the text prosodic structure. The main step of this process is concatenating two L2 layer prosody chunk to one union. First we change the prosodic template units of each L2 layer prosody chunk's prosody contour around the concatenate position to appropriate template unit, because in detached L2 layer prosody chunk's prosody contour, they are in the head or the tail position, but now they are in middle position of the new connective prosody contour. Then we modify connective template units' duration feature by connective rules. At last, when the later connective template unit has voiced consonant or no consonant in beginning, we should smooth the pitch contour between the two template units, to get the natural pitch contour. If the union prosody chunk is not at the top of the text prosodic structure, it will be processed with the other prosody chunk around it to get a bigger prosody chunk, until to the top of the text prosodic structure.

C. The next step is to use the base intonation contour to refine the prosody contour generated above. The base intonation contour is used to define the hierarchical feature of the template of quantified prosodic unit according to its position in the continuous speech streams. It comes of simulating the parameters obtained from analysis of the whole feature (including pitch, loudness, syllable duration) of the prosodic chunks at the different prosodic position (including position in speech and accent position in each level) in natural speech streams. In this system, the base intonation is based on the sentence type of statement, presenting most usual mandarin phonetics phenomena, include declining, downstep and accent. The effect produced by the base intonation contour make the prosody contour well reflect the hierarchical feature of Chinese mandarin to gain higher naturalness.

3.2 Selecting the Unit from the Unit Base

With the guidance of the high naturalness prosody contour generated, the most appropriate unit should be selected from the unit database, and then be concatenated to form the final speech. During the period of selecting the unit, we adopt the cost of object and cost of connection as the basis of selecting appropriate unit. The cost of object is decided by the distances of F0, duration between the prosodic parameter generated and the prosodic parameter from parameter base. The cost of connection is decided by the co-articulation of the two selected units. The cost of object is used to select the best compatible unit in unit base to fit the generated prosodic parameter, and then improve the naturalness of the synthesis speech. Cost of connection is used to eliminate the co-articulation influence partially at the junction of two syllables, when the last one of the two syllables has voiced consonant. With the two costs being adopted, we can select the unit with high accuracy. In this way, we can wipe off the influence of the disaccord in the recording database, make more efficient use of the corpus, and achieve the best speech quality.

4   Special Text Processing

The purpose of special text processing, which is also named text normalization, is to get enough information for pronouncing special text correctly in Chinese TTS. The way of reading a special text sometimes is extremely different from that of reading Chinese characters, in which we only need to read every Chinese character orderly according to the pronunciation in a lexicon. In this paper we put forth a special text processing method based on external descriptor rule and model matching.

4.1   Top-down Analysis and Hierarchical Text Processing

The method of top-down analysis and hierarchical text processing is adopted. Different type of text will be processed on different level, such as punctuation mark, number and so on. According to the rules, which are defined in advance, particular marks will be decided whether or not they should be processed at the current level. If they should be processed by current rule, the marks will be labeled as a child level of current level.

4.2   Integration of External Descriptor Rule and Model Matching

According to different priority of the rules, the input mark will be matched with the every model which are described in a external form. If matching succeeds, the marks will be pronounced following the way given in this model. A few of distinctive and describable rules are summarized based on the intrinsic properties of some kind of marks such as decimal fraction, telephone number and so on, and then the rules are saved in a external form. Based on the rules we can decide which class the input marks belong to and how they should be pronounced.

4.3   The Enumerating Method is Adopted to Bypass Some Problem

There are some marks whose pronunciation can't be defined by a perfect model, for example, the pronunciation of number sometimes must be defined only by fully semantic understanding. An enumerative method is used to bypass the difficulties at a certain extend. At same time a external interface is provided for user to add some specific item which is helpful to get pronunciation of some ambiguous special text.

Additionally for avoiding mistakes due to mixing SBC case and DBC case or capital letter and lowercase, we'll normalize all input special marks into DBC case and capital letter after filtering out all kind of control marks and random code.

4.4   Preliminary Experiment

To provide a set of quantitative evaluations to this method, we have performed a series of experiments. Around 40 sentences, all of which contain special text, are extracted from different kinds of articles by three persons, they don't know why the sentences are needed. When the special texts in the sentences are processed, only one mistake happen. The experiments indicate that special text can be processed accurately in the method we proposed.

5   System Evaluation

The developed KD2000 Chinese text-to-speech system runs on the platform of common PCII266 with 80MB disk capacity for storage of the unit base. The system can convert any Chinese text to speech with very high naturalness. With client/server mode the text in clients can be sent to a speech synthesis server, and be synthesized into high quality speech there in real time. The maximum number of the clients permitted is 100 and the text of every client can be different each other. In the end of November 1999, an assessment of the performance of KD863 and KD2000 text-to-speech systems was presided over by 863 experts group in HeFei. The test was based on the subject valuation. The test comment is as follows: "partially due to a integrated method about analyzing text and marking prosodic parameters being put forth, The whole performance of KD2000 has been improved much more. The naturalness of KD2000 has made much more progress over that of KD863 and the quality of the synthesis speech of KD2000 has been close to real speech. KD2000 had kept ahead among international Chinese Text-to-Speech research". And also in the end of November, 1999, a informal comparing assessment of the performance of KD863 and KD2000 was held in HeFei. The testing materials include 100 typical sentence. To each sentence, the output speech of KD863 and KD2000 were played random. Five trained listeners hear the output speech and then evaluate the rhythm accuracy, sentence intelligibility and naturalness of synthesis speech of KD863 and KD2000. The test result is listed on table 1.

Table 1. TTS System Evaluation for KD863 and KD2000

	Rhythm Accuracy	Sentence's intelligibility	Naturalness
KD863	82%	91%	3.1
KD2000	93%	97%	3.8

The evaluating standard is as same as what have been presented in the former article [3]. It should be mentioned that KD863 get the first grade in a national assessment about the naturalness of synthesis speech in the end of March 1998.

References

1. R.H.Wang ,"Overview of Chinese Text-To-Speech Systems", *Keynote of ISCSLP98, Singapore, 1998.*

2. R.H.Wang, Q.Liu, D.Tang "A New Chinese Text-To-Speech System with High Naturalness", *Proc. of ICSLP96, p1441-1444, USA,1996.*

3. R.H.Wang, Q.Liu, Y.Teng, Deyu Xia. "Towards a Chinese Text-to-Speech system with higher naturalness", *Proc. ICSLP98, p2047-2050, Sydney, 1998.*

Multimodal Speaker Detection Using Input/Output Dynamic Bayesian Networks

Vladimir Pavlović[1], Ashutosh Garg[2], and James M. Rehg[1]

[1] Compaq Cambridge Research Lab
Cambridge, MA 02142
{vladimir,rehg}@crl.dec.com
[2] ECE Department and Beckman Institute
University of Illinois
Urbana, IL 61801
ashutosh@ifp.uiuc.edu

Abstract. Inferring users' actions and intentions forms an integral part of design and development of any human-computer interface. The presence of noisy and at times ambiguous sensory data makes this problem challenging. We formulate a framework for temporal fusion of multiple sensors using input–output dynamic Bayesian networks (IODBNs). We find that contextual information about the state of the computer interface, used as an input to the DBN, and sensor distributions learned from data are crucial for good detection performance. Nevertheless, classical DBN learning methods can cause such models to fail when the data exhibits complex behavior. To further improve the detection rate we formulate an *error-feedback* learning strategy for DBNs. We apply this framework to the problem of audio/visual speaker detection in an interactive kiosk application using "off-the-shelf" visual and audio sensors (face, skin, texture, mouth motion, and silence detectors). Detection results obtained in this setup demonstrate numerous benefits of our learning-based framework.

1 Introduction

Human-centered user-interfaces based on vision and speech present challenging sensing problems. Multiple sources of information, including high-level application-specific information, must be combined to infer the user's actions and intentions. Statistical modeling techniques play a critical role in the design and analysis of such systems. Dynamic Bayesian network (DBN) models are an attractive choice, as they combine an intuitive graphical representation with efficient algorithms for inference and learning. Previous work has demonstrated the power of these models in fusing video and audio cues with contextual information and expert knowledge both for speaker detection and other similar applications [3, 5, 4, 2].

Speaker detection is a particularly interesting example of a multi-modal sensing task with application in video conferencing, video indexing and human-computer interaction. Both video and audio sensing provide important information in a multi-person and noisy scenarios. Contextual information or the state of the application is another important component because it often governs the type of interaction. This naturally

T. Tan, Y. Shi, and W. Gao (Eds.): ICMI 2000, LNCS 1948, pp. 308-316, 2000.

leads one to consider a special DBN architecture known as input/output DBN (Figure 2(b)). The state of the application forms an input to the system which is, along with the sensory outputs, used to determine the state of the user. We are interested in network models that combine "off-the-shelf" vision and speech sensing with contextual cues.

Estimation of the DBN model parameters is a key step in the design of the detection system. Strengths of the DBN arcs in Figure 1, from context to task variables to sensors, can be automatically learned from data using standard maximum-likelihood (ML) learning schemes, similar to [3]. However, it is often the case that the chosen model structure only approximately represents the data. To circumvent this drawback we introduce a learning algorithm for DBNs that uses *error-feedback* to improve recognition accuracy of the model. In error-feedback DBNs (EFDBNs) strengths of DBN arcs are iteratively adjusted by focusing on data instances incorrectly detected by the previous models.

This paper demonstrates that modeling the contextual states of the application as an input to a DBN together with the learning of continuous sensor distributions can enhance performance of DBNs which fuse temporal data from weak multimodal sensors. We also show how EFDBN learning strategy yields significant improvements in detection accuracy. We present these results in the context of a network architecture of Figure 1 which infers the state of the speaker who actively interacts with the Genie Casino game. Our evaluation of the learned DBN model indicates its superiority over previous static [7] and dynamic [3, 5] detection models.

2 Speaker Detection

An estimate of the persons state (whether s/he is or isn't a speaker) is important for the reliable functioning of any speech-based interface. We argue that for a person to be an active speaker, s/he must be expected to speak, face the computer system and actually speak. Visual cues can be useful in deciding whether the person is facing the system and whether he is moving his lips. However, they are not capable on their own to distinguish an active user from an active listener (listener may be smiling or nodding). Audio cues, on the other hand, can detect the presence of relevant audio in the application. Unfortunately, simple audio cues are not sufficient to discriminate a user in front of the system speaking to the system from the same user speaking to another individual. Finally, contextual information describing the "state of the world" also has bearing on when a user is actively speaking. For instance, in certain contexts the user may not be expected to speak at all. Hence, audio and visual cues as well as the context need to be used jointly to infer the active speaker.

We have analyzed the problem of speaker detection in a specific scenario of the Genie Casino Kiosk. The Smart Kiosk [6] developed at Compaq's Cambridge Research Lab (CRL) provides an interface which allows the user to interact with the system using spoken commands. This version of kiosk simulates a multiplayer blackjack game (see Figure 4 for a screen capture.) The user uses a set of spoken commands to interact with the dealer (kiosk) and play the game. The kiosk has a camera mounted on the top that provides visual feedback. A microphone is used to acquire speech input from the user.

We use a set of five "off-the-shelf" visual and audio sensors: the CMU face detector [8], a Gaussian skin color detector [10], a face texture detector, a mouth motion detector, and an audio silence detector. A detailed description of these detectors can be found in [7]. Contextual input provides the state of the application (the blackjack game) which may help in inferring the state of the user.

2.1   Bayesian Networks for Speaker Detection with Continuous Sensors and Contextual Input

We adopt a modular approach towards the design of the Bayesian network for speaker detection. We have designed modules for vision and audio tasks separately which are then integrated along with the higher level information.

The graph in Figure 1 shows the vision network for this task. This network takes the output of the vision sensors and outputs the query variables corresponding to visibility and the frontal information of the user. Face detector gives a binary output whereas the output of the texture detector is modeled as a conditional Gaussian distribution whose parameters are *learned* from the training data. We contrast this to the cases studied in [3, 5] where all sensors had binary outputs.

The audio network combines the output of the silence detector and the mouth motion detector. These detectors provide continuous valued output as the measure of the silence and mouth motion respectively. The audio network selected for this task is shown in Figure 1. The output of the audio network corresponds to the probability that the audio in the application corresponds to the user present.

Once constructed, the audio and visual networks are fused to obtain the integrated audio–visual network. The contextual information acts as an input to the model with the sensory observations as the output. The state of the user (e.g. speaker vs. nonspeaker) forms the state of the model and needs to be inferred given the observations and the inputs. The final network obtained is shown in Figure 1.

The final step in designing the topology of the speaker detection network involves its temporal aspect. Measurement information from several consecutive time steps can be fused to make a better informed decision. This expert knowledge becomes a part of the speaker detection network once the temporal dependency shown in Figure 2(a) is imposed. Incorporating all of the above elements into a single structure lead to the input/output DBN shown in Figure 2(b). The input/output DBN structure is a generalization of the input/output HMM [1]—here the probabilistic dependencies between the variables are governed by the BN shown in Figure 1. The speaker node is the final speaker detection query node.

The use of continuous valued sensor outputs allows the network to automatically learn optimal sensor models and, in turn, optimal decision thresholds. In the previous work [3, 5] sensor outputs were first discretized using decision thresholds set by expert users. Here all continuous sensory outputs are modeled as conditional Gaussian distributions, as shown in Figure 3. The learned distributions allow soft sensory decisions which can be superior to discrete sensory outputs in noisy environments. Indeed, results outlined later in this paper show that improved performance is obtained using this model.

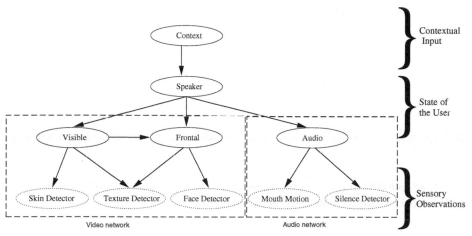

Fig. 1. Integrated audio-visual network.

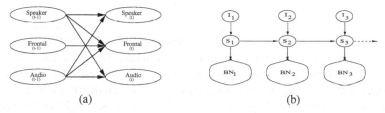

(a)   (b)

Fig. 2. (a) Temporal dependencies between the speaker, audio, and frontal nodes at two consecutive time instances. (b) The input output DBN, with contextual information as input and the sensory observations as outputs.

3 Learning Dynamic Bayesian Networks

Dynamic Bayesian networks are a class of Bayesian networks specifically tailored to model temporal consistency present in data. In addition to describing dependencies among different static variables DBNs describe probabilistic dependencies among variables at different time instances. A set of random variables at each time instance t is represented as a static BN. Out of all the variables in this set temporal dependency is imposed on some. Thanks to its constrained topology efficient inference and learning algorithms, such as forward-backward propagation and Baum-Welch, can be employed in DBNs (see [5] for more details.)

However, classical DBN learning algorithms assume that the selected generative model accurately represents the data. This is often not the case as the selected model is only an approximation of the true process. Recently, Schapire et al. [9] have proposed a method called *boosting* aimed at improving the performance of any simple (classification) model. In particular, their *Adaboost* algorithm "boosts" the classification on a set of data points by linearly combining a number of weak models, each of which is trained to correct "mistakes" of the previous one. In a similar spirit we formulated the

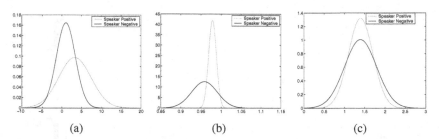

Fig. 3. Learned continuous sensor distributions: (a) silence, (b) skin texture, and (c) mouth motion.

Error Feedback DBN framework [5]. Here we extend this framework to handle the case of continuous sensory outputs and contextual input.

3.1 Error Feedback DBNs

Consider the training data $D = \{(s_1, y_1, i_1), ..., (s_T, y_T, i_T)\}$ of duration T, where s denotes DBN states, i are the inputs, and y are the measurements, and the DBN shown in Figure 2(b). The goal of DBN learning is to, given data D, obtain the DBN model $\Theta = (A, B, \pi)$, (where A is the transition probability matrix dependent on input i, B is the observation matrix which maps s_t to y_t and π is the initial distribution of s_0) which minimizes the probability of classification error in s on dataset D. EFDBN algorithm for this setting can be formulated as follows.

Given: $D\{(s_1, y_1, i_1), ..., (s_T, y_T, i_T)\}$;

Assume all states are detected equally well, $P_D^{(1)}(t) = 1/T$;
For $k = 1, ..., K$
- Train static BN with s_t as the root node to obtain B_k. Use $P_D^{(k)}$ as the weight over the training samples.
- Use the DBN learning algorithm to obtain A for fixed B_k.
- Use the learned DBN, $\Theta = (A, B_t, \pi)$ to decode $(\hat{s}_1, ... \hat{s}_T)$ from $(y_1, ..., y_T)$ and $(i_1, ..., i_T)$.
- Update:
  if $\hat{s}_k = s_k$ then
  $$P_D^{(k+1)}(t) \propto P_D^{(k)}(t) \exp(-\alpha_k)$$
  else
  $$P_D^{(k+1)}(t) \propto P_D^{(k)}(t) \exp(\alpha_k)$$
The final DBN model is $\lambda = (A, B, \pi)$

where $B = \dfrac{\sum_{k=1}^{K} \alpha_k B_k}{\sum_{k=1}^{K} \alpha_k}$

The algorithm maintains a weight distribution defined over the data. It starts by assigning equal weight to all the samples. As the algorithm proceeds, the weight of correctly classified samples is decreased whereas that of misclassified ones is increased. Details of the original algorithm can be found in [5].

4   Experiments and Results

We conducted three experiments using a common data set. The data set comprised of five sequences of a user playing the blackjack game in the Genie Casino Kiosk setup. The experimental setup is depicted in Figure 4. The same figure shows some of the

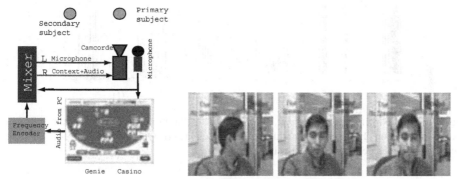

Fig. 4. Data collection setup for Genie Casino kiosk and three frames from a test video sequence.

recorded frames from the video sequence. Each sequence included audio and video tracks recorded through a camcorder along with frequency encoded contextual information (see Figure 4.) The visual and audio sensors were then applied to audio and video streams. Examples of individual sensor observations (e.g., frontal v.s. non frontal, silence v.s. non silence, etc.) are shown in Figure 5. Abundance of noise and ambiguity in these sensory outputs clearly justifies the need for intelligent yet data-driven sensor fusion.

4.1   Static Bayesian Network

The first experiment was done using the static BN of Figure 1 to form the baseline for comparison with the dynamic model. In this experiment all samples of each sequence was considered to be independent of any other sample. Part of the whole data set was considered as the training data and rest was retained for testing. During the training phase, output of the sensors along with the hand labeled values for the hidden nodes (speaker, frontal and audio) were presented to the network.

During testing only the sensor outputs were presented and inference was done to obtain the values for the hidden nodes. Mismatch in any of the three (speaker, frontal, audio) is considered to be an error. An relatively low average accuracy of 77% is obtained (see Figure 6 for results on individual sequences.) The sensor data (as shown in Figure 5) is noisy and it is hard to infer the speaker without making substantial errors. Figure 7(a) shows the ground truth sequence for the state of the speaker and (b) shows the decoded sequence using static BN. However, this detection accuracy is superior to 68% obtained in [3, 5] when discrete sensors and non-input context were used.

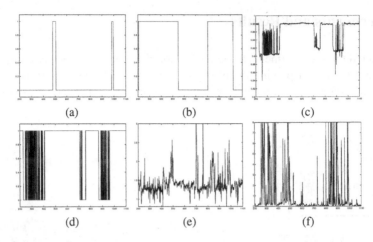

(a)   (b)   (c)

(d)   (e)   (f)

Fig. 5. (a) Ground truth for the speaker state: 1 indicates the presence of the speaker. (b) Contextual information: 1 indicates user's turn to play. (c),(d),(e),(f) Outputs of texture, face, mouth motion and silence detectors, respectively.

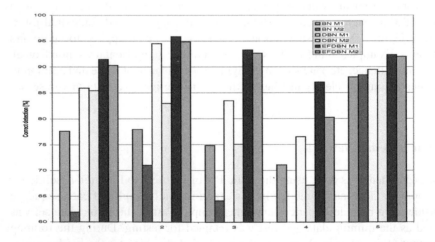

Fig. 6. A comparison between the results obtained using static BN, DBN, and EFDBN. M1 and M2 denote the models with continuous and discrete sensory inputs, respectively.

4.2 Input/Output DBN

Second experiment was conducted using the input/output DBN model. The standard ML learning algorithm described in Section 3 was employed to learn the dynamic transitional probabilities among frontal, speaker, and audio states. During testing phase a temporal sequence of sensor values was presented to the model and Viterbi decoding

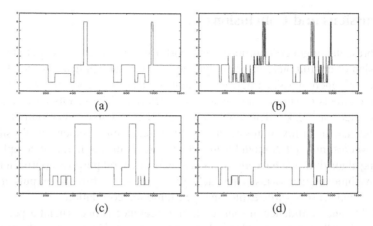

Fig. 7. (a) True state sequence. (b),(c),(d) Decoded state sequences by static BN, DBN, EFDBN, respectively. (state 1 - no speaker, no frontal, no audio; state 2 - no speaker, no frontal, audio; state 3 - no speaker, frontal, no audio; state 8 - speaker, frontal, audio)

was used to find the most likely sequence of the speaker states. Overall, we obtained the accuracy of the speaker detection of 85%, an improvement of $\approx 10\%$ over the static BN model. An indicative of this can be seen in actual decoded sequences in Figure 7. The improved performance by the use of DBN stems from the inherent temporal correlation present between the features. Again, the use of continuous sensors and input context produced significant improvement compared to 80% rate of [3, 5].

4.3 Error-Feedback DBN

Our final experiment employed the newly designed EFDBN framework for continuous sensors and contextual application input. The learning algorithm described in Section 3.1 was used. For a training sequence, we used EFDBN to estimate the parameters which minimized the classification error. A leave-one-out crossvalidation resulted in the overall accuracy of 92%. Figure 6 summarizes classification results on individual sequences. We see that for all the sequences, an improvement of $5 - 10\%$ over the best DBN result is obtained. While the improvement is less dramatic over the 90% detection rate of the EFDBN with discrete sensors and contextual measurement [5], it still remains significant.

The DBN model learned using the EFDBN framework was also applied to the prediction of hidden states. An overall accuracy of 88% was obtained. This indicates, together with the previously noted results, that EFDBN significantly improves the performance of simple DBN classifiers. In comparison of the results with the ones reported in [3, 5], we observe that significant improvement in performance is obtained. This can be attributed to the use of continuous sensory observations and input/output DBN structure.

5   Discussions and Conclusions

We have presented a general purpose framework for learning input/output DBN models for fusion of continuous sensory output and contextual, application-specific input. The framework encompasses a new error-feedback learning procedure which can circumvent the effects of simple models and complex data. The results obtained for the difficult problem of speaker detection where a number of noisy sensor outputs need to be fused indicate the utility of this algorithm. Significant improvements in classification accuracy over a simple DBN model were achieved without sacrificing of complexity of the learning algorithm. We have also demonstrated a general purpose approach to solving man-machine interaction tasks in which DBNs are used to fuse the outputs of simple audio and visual sensors while exploiting their temporal correlation.

Reliability and confidence of sensors during inference is one crucial aspect of sensor fusion tasks which was not addressed in this framework. For instance, the number of skin colored pixels in the whole image can be used as a measure of the reliability of the skin sensor and hence weigh its contribution relative to other sensors. Future research will focus on incorporating sensor reliabilities into our current framework. Another interesting opportunity in this DBN framework arises as a consequence of modeling application-specific context as input. Namely, one can study methods of designing contextual input which will force the user from its present state (e.g., non-speaking) to a new desired state in a number of steps. These opportunities become even more significant when the number of system states becomes large, a case often encountered in dialog systems.

References

[1] Y. Bengio and P. Frasconi, "An input-output HMM architecture," in *Advances in Neural Information Processing Systems 7*, pp. 427–434, Cambridge, MA: MIT Press, 1995.

[2] M. Brand, N. Oliver, and A. Pentland, "Coupled hidden markov models for complex action recognition," in *Proc. IEEE Conf. on Computer Vision and Pattern Recognition*, (San Juan, PR), pp. 994–999, 1997.

[3] A. Garg, V. Pavlovic, J. Rehg, and T. S. Huang, "Audio–visual speaker detection using dynamic Bayesian networks," in *Proc. of 4rd Intl Conf. Automatic Face and Gesture Rec.*, (Grenbole, France), pp. 374–471, 2000.

[4] S. Intille and A. Bobick, "Representation and visual recognition of complex, multi-agent actions using belief networks," Tech. Rep. 454, MIT Media Lab, Cambridge, MA, 1998.

[5] V. Pavlovic, A. Garg, J. Rehg, and T. S. Huang, "Multimodal speaker detection using error feedback dynamic Bayesian networks." To appear in Computer Vision and Pattern Recognition 2000.

[6] J. M. Rehg, M. Loughlin, and K. Waters, "Vision for a smart kiosk," in *Proc. IEEE Conf. on Computer Vision and Pattern Recognition*, (Puerto Rico), pp. 690–696, 1997.

[7] J. M. Rehg, K. P. Murphy, and P. W. Fieguth, "Vision-based speaker detection using bayesian networks," in *Proc. IEEE Conf. on Computer Vision and Pattern Recognition*, (Ft. Collins, CO), pp. 110–116, 1999.

[8] H. Rowley, S. Baluja, and T. Kanade, "Neural network-based face detection," in *Proc. IEEE Conf. on Computer Vision and Pattern Recognition*, (San Francisco, CA), pp. 203–208, 1996.

[9] R. E. Schapire and Y. Singer, "Improved boosting algorithms using cofidence rated predictions." To appear in Machine Learning.

[10] J. Yang and A. Waibel, "A real-time face tracker," in *Proc. of 3rd Workshop on Appl. of Comp. Vision*, (Sarasota, FL), pp. 142–147, 1996.

Research on Speech Recognition Based on Phase Space Reconstruction Theory

Chen Liang[1], Chen Yanxin[2], Zhang Xiongwei[1]

[1] Department of Electronic Information Engineering ICE, PLAUST, Nanjing 210016
xwzhang@public1.ptt.js.cn
[2] Center of network management, Tibet military area, 850000
ch_l@21cn.net

Abstract. Based on Takens theory, time delay method is used to reconstruct the phase space of speech signal in this paper. In hyper dimensional phase space, similar sequence repeatability (RPT) and the entropy information of speech are calculated, and they are applied to speech recognition. The result proves that speech signal shows some geometric property between the stochastic Gauss noise and deterministic Lorenz attractor. For speech phonemes, the RPT parameters of same kind show some similarity, and of different kind show some different characteristics as well. The method proposed in this paper provides a new way for the non-linear analysis of speech recognition.

1   Introduction

Now, research results show that speech signal is a kind of chaotic signal. According to non-linear theory, chaotic signal has stochastic feature although it is produced by deterministic system, and it shows the characteristics of time-domain irregularity and broad frequency as well as stochastic signal. Conventional Fourier transform can't distinguish the two kind of signal effectively. Therefore, how to distinguish speech from stochastic signal and abstract characteristics of speech signal to recognize speech phonemes, have become important more and more in the field of speech recognition.

Recently, with the development of non-linear theory such as chaos and fractal, they are more and more applied to analysis of time series and abstraction of characteristic parameters. In non-linear processing, phase space reconstruction is an effective means to discover intrinsic features. Based on Takens phase space reconstruction theory, in this paper, a new non-linear method named similar sequence repeatability (RPT) is provided to analyze time-domain characteristics of speech, and study the non-linear features in the hyper dimensional phase space. Then, the entropy information is calculated to reflect time-domain characteristics and recognize speech signal.

RPT can reflect the similar degree of trajectory points in phase space, and discover the geometric structure of system trajectory in hyper dimensional phase space. When the distribution of trajectory points are regular, the RPT has large value, otherwise small RPT value indicates that the distribution are disorderly, and trajectory

T. Tan, Y. Shi, and W. Gao (Eds.): ICMI 2000, LNCS 1948, pp. 317-324, 2000.

points are independent each other. Therefore PRT can be used to check the deterministic or stochastic feature, and the value distribution can be used as recognition parameter of speech phonemes.

This paper is arranged as following: In section 2, the phase space reconstruction theory is studied. In section 3, RPT applies to the study of non-linear feature of speech signal. Section 4 presents the entropy characteristic of RPT curves. Finally conclusions are given in Section 5.

2   Phase Space Reconstruction

2.1   Phase Space Reconstruction Theory

By the analysis of attractor (phase space), the dynamic features of phonation system can be found out. But the differential eguation of system is unknown, we impossibly get the trajectory by numerical analysis. What we can utilize is only time series. Takens theory is the basis of phase space reconstruction using time series. As to speech series $\{X_i\}_{i=1}^{N}$, a d-dimension vector group is produced as following:

$$Y_1 = \{X_1, X_{1+\tau}, X_{1+2\tau}, \ldots\ldots, X_{1+(d-1)\tau}\}^T$$
$$Y_2 = \{X_{1+p}, X_{1+p+\tau}, X_{1+p+2\tau}, \ldots\ldots, X_{1+p+(d-1)\tau}\}^T \cdots\cdots\cdots\cdots$$
$$Y_M = \{X_{1+Mp}, X_{1+Mp+\tau}, \ldots\ldots, X_{1+Mp+(d-1)\tau}\}^T$$

Of which, p is the delay time between first component of adjacent vectors, τ is the delay time between adjacent components of the same vector; $M = [N - (d-1)\tau - 1]/p$, is the total number of vectors, d is the dimension of embedded space, which meets *Takens* inequality:

$$d \geq 2p + 1$$

p is the number of differential eguations, which are used to describe dynamic system corresponding to speech signal. In principle, the dynamics information can be restored in the meaning of differential homeomorphism by embedding origin system trajectory in d-dimension phase space. Supposing a discretional vector $X(t_i)$ in origin space W, $X(t_i) \in W$, mapping $X(t_i)$ into embedded space V, $Y_i = \{X_i, X_{i+\tau}, X_{i+2\tau}, \ldots\ldots, X_{i+(d-1)\tau}\}^T$, $Y_i \in V$. If set the last $(d-m)$ coordinates constant, i.e. project Y_i to $(X_i, X_{i+\tau}, \ldots\ldots, X_{i+m\tau})$ m-dimension space, we obtain vector group $Y_i = [X_i, X_{i+\tau}, \ldots\ldots X_{i+m\tau}]^T, i = 1, 2, \ldots\ldots M$. This is the basic idea of m-dimension reconstruction.

Phase space wants system variables independent, but actual speech signal isn't strict chaotic signal. Therefore delay time τ is the key factor which influences space reconstruction. In consideration of time correlation of samples, there are two methods of self-correlation and Average Mutual Information (AMI). The first zero-cross point of self-correlation only shows the linear independence, while AMI emphasizes the generalized independence in stochastic meaning. That is, AMI can reflect non-linear independence of two variables. In this paper, we adopt AMI to determine τ.

2.2   Abstraction of Delay τ

Let $\qquad S = \{s_1, s_2, \ldots\ldots, s_n\}$,   where $s_i = x_i$;

Let $\qquad Q = \{q_1, q_2, \ldots\ldots, q_n\}$, $\qquad$ where $q_i = x_{i+\tau}$.

Entropy $\qquad H(Q) = -\sum_i p(q_i) \log p(q_i)$;

$$H(S) = -\sum_j p(s_j) \log p(s_j).$$

$$H(Q,S) = -\sum_{i,j} p(q_i, s_j) \log p(q_i, s_j).$$

For associated system (S, Q), define AMI as:

$$I(S,Q) = H(S) + H(Q) - H(S,Q)$$

For discrete time system, we adopt section-overlapping methods to calculate associated distribution. Define N_{ij} as the number of points, which fall into (i, j) section. The probability is:

$$P_{sq}(s_i, q_j) = N_{ij} / N$$

$$p_s(s_i) = \sum_j p_{sq}(s_i, q_j)$$

$$p_Q(q_i) = \sum_i p_{sq}(s_i, q_i)$$

$I(S,Q)$ is the function of τ, curve $I(\tau) \sim \tau$ is AMI curve. The optic delay τ is the point that is corresponding to the first local minimal value.

Fig.1 calculates the self-correlation (a) and AMI (b) of Y component of Lorenz attractor (8000 samples). For self-correlation curve, $\tau = 60$, while $\tau = 12$ for AMI curve. (c) (d) are the phase space maps corresponding to $\tau = 60$ and $\tau = 12$. From Fig.1, it can be found that Lorenz attractor is outspread enough, and reflects dynamic features better when adopting the method of AMI.

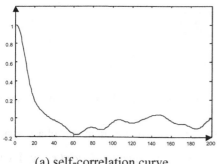

(a) self-correlation curve

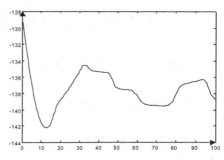

(b) average mutual information curve

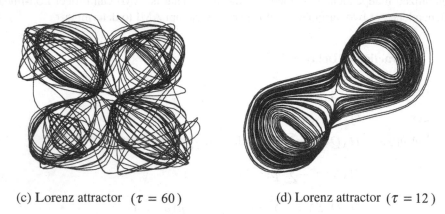

(c) Lorenz attractor $(\tau = 60)$ (d) Lorenz attractor $(\tau = 12)$

Fig. 1. 2-dimension phase space reconstruction using self-correlation (a)(c) and AMI (b)(d)

3   Similar Sequence Repeatability

According to the features of chaotic signal, we abstract the RPT from signal series to reflect time-domain characteristics. Experiment shows that RPT can distinguish different signals well.

3.1   Idea of RPT and Its Calculation

In phase space, more regular the points, more repetitive the points of trajectory in space, so we can detect the deterministic or stochastic feature of signal, and the RPT value can be used as recognition parameter by phase space reconstruction.

First, we reconstruct phase space using Takens theory, then determine τ and embedded dimension d using the methods AMI and False Nearest Neighbor (FNN). Get $\mathbf{y}_i = \{ x_i, x_{i+\tau}, ..., x_{i+(d-1)\tau} \}$, $i=1,2,..., M$. Convenient for comparison, let d=5,10,20,30. Defining referenced point $\mathbf{r}_j = (r_{j1}, r_{j2}, ..., r_{jd})$, the similarity can be measured by calculating unitary correlative coefficient $cor(\mathbf{r}_j, \mathbf{y}_i)$. Where:

$$cor\,(\mathbf{r}_j, \mathbf{y}_i) = \frac{\mathrm{cov}(\mathbf{r}_j, \mathbf{y}_i)}{\sqrt{\mathrm{cov}(\mathbf{r}_j, \mathbf{r}_j)}\sqrt{\mathrm{cov}(\mathbf{y}_i, \mathbf{y}_i)}}$$

$$\mathrm{cov}(\mathbf{r}_j, \mathbf{y}_i) = \frac{1}{d-1}\sum_{k=1}^{d}(r_{jk} - m_{\mathbf{r}_j})(x_{i,(k-1)\tau} - m_{\mathbf{y}_i})$$

Where: $m_r = \dfrac{1}{d}\sum\limits_{k=1}^{d} r_k$. The more similar the vectors $\mathbf{y}_i$ and $\mathbf{r}_j$, the closer to

1. When $\mathbf{y}_i$ and $\mathbf{r}_j$ match completely, $cor(\mathbf{r}_j, \mathbf{y}_i)$=1. In d-dimension phase space $Y = \{\mathbf{y}_i\}$, which includes M vectors, every vector $\mathbf{y}_i$ calculates $cor(\mathbf{r}_j, \mathbf{y}_i)$ with

referent vector r_j. If $cor(r_j, y_i) > TH$ ($0 < TH < 1$), scaler CNT (j) (j represents the jth referenced vector) pluses 1. Therefore RPT (j) is given by $RPT(j) = \dfrac{CNT(j)}{M}$. We set every vector as referenced vector, and vectors of trajectory are compared with it in sequence. RPT curve is drawn by sorting RPT (j) from small to large. The curve reflects the time-domain feature itself well.

3.2   RPT Characteristic of Speech Signal

Speech can be regarded as produced by deterministic system, and the vectors in phase space are similar to referenced vector. Contrarily, vectors produced by stochastic system have small similarity. Fig.2(a~d) compare the images, respectively representing voiced, unvoiced, Y component of Lorenz attractor and stochastic Gauss noise. Where, X-coordinate represents sample point of vector, Y-coordinate represents referenced vector point. Fig.3 (a~d) also show the curves of them when d=5,10.From Fig.2 and Fig.3, it is easy to find that Y component of Lorenz has deterministic feature obviously, while Gauss noise has stochastic feature. For speech signal, the similar degree is larger than stochastic system and less than deterministic system. Of which, voiced has deterministic feature more obviously than unvoiced.

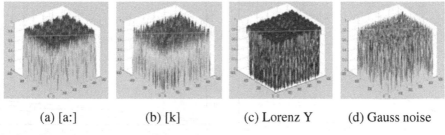

| (a) [a:] | (b) [k] | (c) Lorenz Y | (d) Gauss noise |

Fig. 2.   RPT of vectors with different referenced vectors in phase space

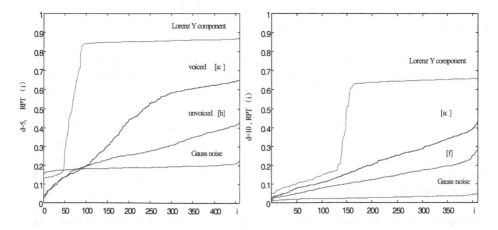

Fig. 3. RPT characteristic curve of different signals when d=5,10

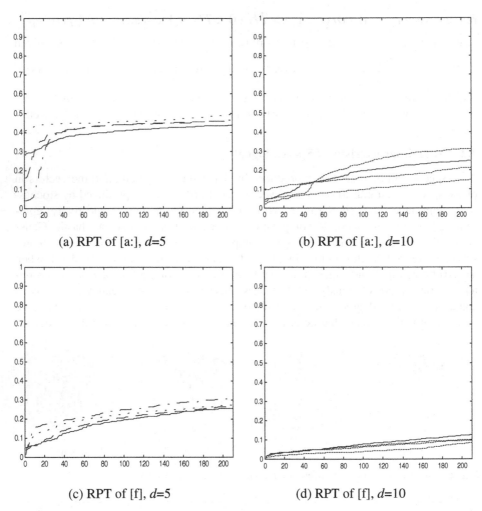

(a) RPT of [a:], d=5   (b) RPT of [a:], d=10

(c) RPT of [f], d=5   (d) RPT of [f], d=10

Fig. 4. RPT curves of [a:] and [f] in different embedded dimensions

When embedded dimension d=5,10, Fig.4 (a~d) give 4 pronunciations of [a:] and [f], by the same male in different environment. Threshold $TH = 0.9$. RPT curve reflects similar degree of vectors and referenced vector. From Fig.2~4, we can find:

(1)   For stochastic system, because of relatively small correlation, the RPT image that is generated by comparing vectors with every referenced vector has no regularity, and RPT value is small. On the contrary, for deterministic system, RPT value is obviously great, and the image has distinct regularity. General speaking, the RPT curve of speech signal is higher than stochastic system and lower than deterministic system.

(2)   RPT curve of voiced is higher than unvoiced. That is to say, voiced data have better correlation than unvoiced. The aspect is determined by phonation mechanism.

(3)   Embedded dimension d is one of important parameters, which would determine the RPT value. Experiment results show that higher embedded dimension corresponds to less RPT value.

(4)   Threshold TH influences RPT value, larger TH responds to less RPT value.

(5)   RPT curves of speech signal have gradient knowing from that of Lorenz attractor or gauss noise. The difference shows that vectors have different similar degree comparing with different referenced vector.

(6)   RPT curves have favorable comparability when pronouncing the same phoneme even in different environment.

(7)   RPT value has relatively great difference, with the change of phonemes.

4   Entropy of Phonic RPT

Entropy is used to measure the indeterminacy of system's state. Assume that stochastic variable A is made up of numerable state sets $\sum_i A_i$ $(i = 1,2,...,n)$, then the entropy is given by

$$E(A) = -\sum_{i=1}^{n} p_i \ln p_i$$

When $p_1 = p_2 = ... = p_n$, $E(A)$ achieves maximum value, the stochastic variable is in maximum indeterminacy. Because RPT is a kind of frequency statistics, $\sum_{i=1}^{M} RPT\ (i) \neq 1$. First, we unit $RPT\ (i)$ by:

$$rpt\ (i) = \frac{RPT\ (i)}{\sum_{i=1}^{M} RPT\ (i)}$$

Therefore entropy is defined as $E(d) = -\sum_{i=1}^{M} rpt(i) \lg rpt(i)$

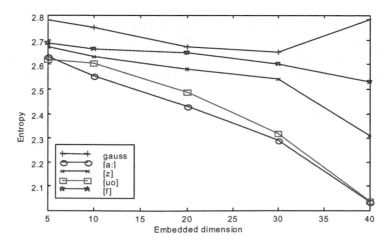

Fig. 5. entropy of different signals

Fig.5 gives entropy value of some speech phonemes, comparison with the entropy of Gauss noise. Sampling frequency is 8000Hz, precision is 16bits. Embedded dimension d=5,10,20,30,40.

Seen from Fig.5, the entropy of Gauss noise is maximal. The aspect indicates that the correlation of Gauss noise is minimum, and the system's state is the most uncertain. While the entropy of speech seems relatively small, this show speech signal is different from stochastic system. Of which, because voiced signal has better similarity and deterministic feature than unvoiced, the entropy of unvoiced is greater than that of voiced. Therefore, it is obvious that the entropy information reflects the essential difference between speech signal and stochastic noise.

5   Conclusion

Characteristic model is an important premise in the establishment of classified recognition system. According to the self-similar feature of speech signal, we calculate the RPT of speech, plot the RPT characteristic curves as well as images, and abstract the entropy information. The results prove that speech signal shows some geometric property between the stochastic Gauss noise and deterministic Lorenz attractor. The experiment results also confirm that RPT and its entropy information can represent the time-domain features of speech signal effectively. In the mean time, making use of the different RPT value, we can distinguish voiced and unvoiced from speech signal, even can recognize speech phonemes. The method is simple but effective, and provides a new analysis measure in the respect of speech characteristic abstraction.

Reference

1. Passamante A,Bromely D, Farrel M E. Time series characterizaion using the repeatability of similar sequences. Physica D,1996;96:100-109
2. Chen Xiangdong, Song Aiguo, Gao Xiang, Lu Jiren. The Study of ship radiated noise based on phase space reconstruction theory. ACAT ACUSTICA, Vol.24, No.1 :12-18

Estimating the Pose of Phicons for Human Computer Interaction

Daniela Hall* and James L. Crowley

Project PRIMA, Lab. GRAVIR - IMAG
INRIA Rhône-Alpes
655, ave. de l'Europe
38330 Montbonnot St. Martin, France

Abstract. Physical icons (phicons) are ordinary objects that can serve as user interface in an intelligent environment. This article addresses the problem of recognizing the position and orientation of such objects. Such recognition enables free manipulation of phicons in 3D space.

Local appearance techniques have recently been demonstrated for recognition and tracking of objects. Such techniques are robust to occlusions, scale and orientation changes. This paper describes results using a local appearance based approach to recognize the identity and pose of ordinary desk top objects. Among the original contributions is the use of coloured receptive fields to describe local object appearance. The view sphere of each object is sampled and used for training. An observed image is matched to one or several images of the same object of the view sphere. Among the difficult challenges are the fact that many of the neighborhoods have similar appearances over a range of view-points.

The local neighborhoods whose appearance is unique to a viewpoint can be determined from the similarity of adjacent images. Such points can be identified from similarity maps. Similarity maps provide a means to decide which points must be tested to confirm a hypothesis for correspondence matching. These maps enable the implementation of an efficient prediction–verification algorithm.

The impact of the similarity maps is demonstrated by comparing the results of the prediction–verification algorithm to the results of a voting algorithm. The ability of the algorithm to recognize the identity and pose of ordinary desk-top objects is experimentally evaluated.

Keywords: Object Recognition, Appearance-Based Vision, Phicons

1 Introduction

Phicons provide an important new mode for man machine interaction in an intelligent environment [5]. Phicons are physical objects whose manipulation can serve as a numerical interaction device. For example, an environment may be configured so that grasping an object determines a functional context for interaction, and the manner in which the object is manipulated determines input parameters. Another example would be to use a phicon as a 3D space mouse in which the position and orientation of the object determines the position and orientation of the cursor. The use of the phicon needs

---

* Daniela.Hall@inrialpes.fr

T. Tan, Y. Shi, and W. Gao (Eds.): ICMI 2000, LNCS 1948, pp. 325-331, 2000.
© Springer-Verlag Berlin Heidelberg 2000

no explication and appears for this reason much more natural to the user. When coupled with steerable cameras, the use of computer vision to sense the manipulation of phicons can permit such input at any location within the intelligent environment. This should allow phicons to become a graspable, easy to use, human computer interaction mode.

This article describes an algorithm for pose estimation of manipulated phicons based on local appearance. Local appearance techniques are robust to occlusions[2] and to scale changes[1] and they have recently been demonstrated for the recognition of objects. During the matching process of an observed image to a set of view sphere images, many ambiguous points are present. We propose a method to determine characteristic points for a particular view, which avoid ambiguous points and enables an efficient prediction–verification algorithm for view point determination.

2   Pose Recognition

The problem of pose recognition can be transformed to the problem of determining the relative position between the camera and the object. Although the illumination of an object changes in an uncontrolled environment while it is turned in space, the recognition of the view point of an object serves as the basis for the recognition of phicon manipulation.

The experiments in this article are based on an object recognition algorithm using coloured receptive fields [4]. The basis for the receptive field are Gaussian derivatives in order to profit from scalability and the possible orientation to any arbitrary direction. These two properties allow recognition independent from scale and orientation [6, 3].

Coloured receptive fields can be adapted to the recognition problem by the selection of different derivatives in the luminance and chrominance channels. An example of a coloured receptive field used for this application is shown in figure 1. The receptive field is oriented vertically with standard deviation $\sigma = 2.3$. In the coloured receptive field technique the luminance channel is maintained and complemented with two channels based on chrominance. The RGB coordinate system is transformed to the Luminance/Chrominance space according to equation (1). The chrominance channels are described using colour-opponent receptive fields. Luminance is known to describe object geometric structure while chrominance is primarily used for discrimination.

$$\begin{pmatrix} l \\ c_1 \\ c_2 \end{pmatrix} = \begin{pmatrix} \frac{1}{3} & \frac{1}{3} & \frac{1}{3} \\ -\frac{1}{2} & -\frac{1}{2} & 1 \\ \frac{1}{2} & -\frac{1}{2} & 0 \end{pmatrix} \begin{pmatrix} r \\ g \\ b \end{pmatrix} \tag{1}$$

During the training phase the local appearance of each point neighborhood within a training image is measured by the receptive field normalized to the local scale and oriented to the dominant direction. The receptive field response is stored in a hash table for fast access together with identification of the image and coordinates of the center of the point neighborhood. This supplemental information enables recognition of objects and their pose.

The object recognition algorithm is extended to view point recognition as follows. Images of an object from different view points serve as training base. The object is

Fig. 1. Vertically oriented coloured receptive field with $\sigma = 2.3$ (1st order Gaussian derivative in the luminance channel, 0th and 1st Gaussian derivatives in the two chrominance channels, and 2nd order Gaussian derivatives in the luminance channel).

Fig. 2. Images of latitude 58° from the sampled view sphere.

segmented from the background, in order to consider only object information. Pose is determined by recognizing which image of the view sphere most closely resembles the observed image. For this purpose the View Sphere Database[1] has been created, which contains images of a dense sampling of a geodesic view sphere for 8 objects.

3   Prediction–Verification with Similarity Maps

An unexpected difficulty occurs when using a relatively dense sampling of the view sphere: Adjacent view-points demonstrate similar appearance. As a results, the recognition process must take into account the geometric constraint between several observed points and their correspondences in the training images. The question for the design of a good prediction–verification algorithm is: Which points lead to reliable recognition? How can these points be found?

Reliable recognition can be performed when one or several discriminant or characteristic points for a particular view are found. A characteristic point for a particular view $V_A(\varphi, \lambda)$ defined by latitude φ and longitude λ of object A is a point that is different in $V_A(\varphi, \lambda)$ than in all other views and allows to distinguishes $V_A(\varphi, \lambda)$ from other views. The most ambiguities and confusions during a matching process occur between neighboring views. A useful approximation for the determination of characteristic points for $V_A(\varphi, \lambda)$ is to restrict the search of characteristic points of $V_A(\varphi, \lambda)$ to neighboring view images $V_A(\varphi_k, \lambda_k)$.

Receptive fields provide a description of point neighborhoods. The distance of the receptive field responses in the descriptor space is a measure for the similarity of point neighborhoods. This similarity property is used to determine a similarity map for all training images which indicates if a particular point is characteristic for this training image or not (see figure 3).

---

[1] The View Sphere Database: Information and data is available at http://www-prima.inrialpes.fr/Prima/hall/view_sphere.html.

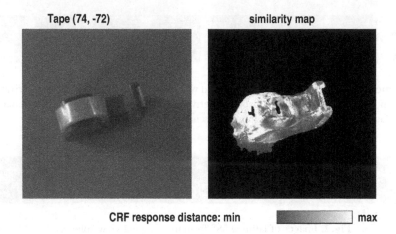

Fig. 3. Image and its similarity map. The map is computed by measuring the similarity of corresponding points between the image and its neighboring images. The images are pre-segmented and the background is not considered.

To determine the similarity map of an image $V_A(\varphi, \lambda)$, a structure is trained with all neighbor images of $V_A(\varphi, \lambda)$. Then point correspondences are searched between the points in $V_A(\varphi, \lambda)$ and neighboring views $V_A(\varphi_k, \lambda_k)$. This is done by simply searching the closest receptive field response to the observed response within the trained structure. For each point, the distance of the observed receptive field response and the trained receptive field response is measured. These values form the similarity map. Small values signify high similarity of a particular point within the neighboring images. Such a point is ambiguous and can not serve to identify the pose. Whereas higher values signify that no similar correspondence has been found in the neighboring images. Such points are characteristic and can serve to distinguish the view $V_A(\varphi, \lambda)$ from its neighboring views.

Let $V_A(\varphi_0, \lambda_0)$ be the image of object A from view point with latitude φ_0 and longitude λ_0. Let $V_A(\varphi_k, \lambda_k), k = 1 \ldots u$ be the u direct neighbor images of $V_A(\varphi_0, \lambda_0)$ of object A. Let

$$m_{V_A(\varphi_l, \lambda_l)}(i, j) = \mathrm{CRF}(V_A(\varphi_l, \lambda_l), i, j) \tag{2}$$

be the coloured receptive field (CRF) response of image $V_A(\varphi_l, \lambda_l)$ at position (i, j). The characteristic $c_{V_A(\varphi_0, \lambda_0)}$ of a point (i, j) in image $V_A(\varphi_0, \lambda_0)$ is

$$c_{V_A(\varphi_0, \lambda_0)}(i, j) = \min_{x, y, k} \left(\| m_{V_A(\varphi_0, \lambda_0)}(i, j) - m_{V_A(\varphi_k, \lambda_k)}(x, y) \| \right) \tag{3}$$

$\|.\|$ is the Mahalanobis distance in CRF space taking into account the distribution of the CRF responses. The point characteristic $c_{V_A(\varphi_0, \lambda_0)}$ forms the similarity map $S_{V_A(\varphi_0, \lambda_0)}$.

$$S_{V_A(\varphi_0, \lambda_0)} = ((c_{ij}))_{i=0 \ldots n, j=0 \ldots m} \text{, with } c_{ij} = c_{V_A(\varphi_0, \lambda_0)}(i, j) \tag{4}$$

where $((.))$ forms an image.

The computation of the similarity maps is a method to determine characteristic points for images over the view sphere. If a characteristic point in an observed image can be found at a predicted position, the object and its pose can be reliably determined. The following prediction–verification algorithm uses the precomputed similarity maps to determine characteristic points and search them in the observed image to verify a hypothesis.

In a first step a hypothesis list for an image is obtained by applying the receptive field to a point neighborhood and searching similar receptive field responses in the structure. For each hypothesis of the hypothesis list characteristic points are searched in the hypothesis image $V_{Obj}(\varphi, \lambda)$ using the similarity map $S_{V_{Obj}(\varphi,\lambda)}$. For each selected characteristic point a region of interest in the observed image is computed, that fulfills the spatial geometric constraints required by the hypothesis image. Within this region of interest the corresponding point is searched by minimizing similarity and spatial distances. If the similarity is below a threshold, the confirmation counter of the hypothesis is incremented. The hypothesis with the highest confirmation counter is returned.

This algorithm uses several points to verify the hypothesis. Only those points are considered for testing for which the geometric constraint is fulfilled. The algorithm is fast, because the point within the region of interest is returned, whose receptive field response minimizes the distance in the descriptor space. Note that the characteristic points of the training images can not be found by any other interest point extraction technique, because the characteristic of a point depends only on the variance of features in the training images.

4   Experimental Results

Since the goal of this project is phicon recognition in an intelligent environment, the database used in the experiments contains 8 different objects from an office environment such as scissors, a tape, and a pen representing natural phicons. The view sphere is sampled according to a geodesic sphere with a spherical distance of $\frac{\pi}{10}$ between images. This corresponds to a sampling of 90 images per half sphere. A test set of a 100 images is constructed such that a test image has exactly 3 nearest neighbors among the training images. Each image $V_{Obj}(\varphi, \lambda)$ is identified by its latitude φ and longitude λ with respect to the view sphere.

In order to illustrate the utility of the similarity maps, the results of two different pose recognition algorithms are compared. A voting algorithm serves as benchmark. The best hypothesis is determined by considering the votes of 4 direct spatial neighbor points of each hypothesis of the hypothesis list. The hypothesis with the maximum number of votes is returned. The results of the vote algorithm are then compared to the results of the prediction–verification algorithm described above. The difference between the two algorithms consists in the choice of the verification or voting points. In the voting algorithm the voting points are located north, south, east and west of the test point. Whereas in the prediction–verification algorithm, the verification points are selected according to their characteristics $c_{V_A(\varphi,\lambda)}$ obtained from the similarity map $S_{V_A(\varphi,\lambda)}$ of the hypothesis image $V_A(\varphi, \lambda)$.

The view point of the test images $V_{Obj}(\alpha, \beta)$ is recognized by matching the observed image to one or several of the training base. The result of the recognition process is evaluated by computing the spherical distance of the hypothesis view point $V_{Obj}(\varphi, \lambda)$ and the view point $V_{Obj}(\alpha, \beta)$ of the observed image. In table 1 the percentage of images whose pose is detected correctly is displayed. For each algorithm the percentage of correct pose determination with two different precisions are given. The higher precision corresponds to the percentage of images for which one of the closest neighbors is found. The resulting precision is a spherical distance of $< \frac{\pi}{16}$. The other percentage is measured with a precision of $< \frac{\pi}{9}$, which is sufficient for many applications. Lower precision results in higher reliability of the results.

Object	recognition by vote		recognition by pred-verif	
	$d < \frac{\pi}{16}$	$d < \frac{\pi}{9}$	$d < \frac{\pi}{16}$	$d < \frac{\pi}{9}$
Eraser	0.52	0.68	0.66	0.81
Pen	0.52	0.67	0.75	0.86
Scissors	0.35	0.48	0.47	0.65
Sharpener	0.46	0.53	0.67	0.80
Stapler	0.29	0.40	0.48	0.65
Tape	0.33	0.75	0.80	0.90
Protractor	0.54	0.68	0.57	0.70
Vosgienne	0.58	0.75	0.59	0.76

Table 1. Comparison of the pose recognition results of a voting algorithm and prediction–verification using similarity maps. Displayed is the percentage of images for which the correct pose has been detected with a precision of either $< \frac{\pi}{16}$ or $< \frac{\pi}{9}$.

The results confirm the superiority of the prediction–verification algorithm. The recognition rates shown in table 1 are remarkable, because the change in view point between test and training images is up to 9 degree in latitude and up to 36 degree in longitude. The object is illuminated with 2 diffuse spot lights. In more than half of the images specularities are present. Specularities change according to the Lambertian rule and alter the appearance of the object. Both algorithms can deal with this difficulty due to the locality of the receptive fields.

5   Conclusions and Outlook

This article proposes a prediction–verification algorithm for a reliable recognition of camera view points on an object. This serves as the basis for pose recognition of ph-icons during manipulation in an intelligent environment. The problem of ambiguities between images of the training set is solved by computing similarity maps allowing a determination of characteristic points for a particular view and leading to a reliable recognition.

The next step will be to test the performance of this algorithm for the recognition of the manipulation of objects. The problem of partial occlusions and rapidly changing orientation in 3D space will be adressed. The use of phicons that can be freely manipulated will have an significant impact on man machine communication, since any physical object can than serve as input device.

References

1. O. Chomat, V. Colin de Verdière, D. Hall, and J.L. Crowley. Local scale selection for gaussian based description techniques. In *European Conference on Computer Vision (ECCV 2000)*, Dublin, Ireland, June 2000.
2. V. Colin de Verdière. *Représentation et Reconnaissance d'Objets par Champs Réceptifs*. PhD thesis, Institut National Polytechnique de Grenoble, France, 1999.
3. W.T. Freeman and E.H. Adelson. The design and use of steerable filters. *Transactions on Pattern Analysis and Machine Intelligence*, 13(9):891–906, September 1991.
4. D. Hall, V. Colin de Verdière, and J.L. Crowley. Object recognition using coloured receptive fields. In *European Conference on Computer Vision (ECCV 2000)*, Dublin, Ireland, June 2000.
5. H. Ishii and B. Ullmer. Tangible bits: Towards seamless interfaces between people, bits and atoms. In *Computer Human Interfaces (CHI '97)*, Atlanta, USA, March 1997.
6. T. Lindeberg. Feature detection with automatic scale selection. *International Journal of Computer Vision*, 30(2):79–116, 1998.

An Approach to Robust and Fast Locating Lip Motion

Rui Wang[1], Wen Gao[1,2],   Jiyong Ma[2]

[1] Department of Computer Science, Harbin Institute of Technology, Harbin,150001,China
[2] Institute of Computing Technology, Chinese Academy of Sciences, Beijing , 100080, China
{rwang,jyma,wgao}@ict.ac.cn

Abstract. In this paper,we present a novel approach to robust and fast locating lip motion.Firstly, the fisher transform with constraints is presented to enhance the lip region in a face image. Secondly, two distribution characteristics of the lip in human face space are proposed to increase the accuracy and and real-time implementation performance of lip locating. Experiments with 2000 images show that this approach can satisfy requirements not only in real-time performance but also in reliability and accuracy.

1   Introduction

The problem of fast and robust locating lips becomes more and more important in multimode interface such as lip reading, facial expression recognition, lip motion synthesis and Synthetic/Natural Hybrid Coding (SNHC) in MPEG4. However, locating and tracking lip motion accurate are very difficult and challenging because the lip contours of different people vary greatly and are affected by lip variations, head movement and illuminant conditions while a speaker is uttering. The typically traditional system utilizes manual approach to accurately locate the contours of the lips. Lipsticks are often used to accurately locate the contours of the lips under a specific illuminant condition. Some researchers used the LED put on lips to track the lip motion. However, these approaches cannot satisfy the need of practical application. Therefore, it is necessary to develop approaches to locating and tracking lip under natural light conditions.

To solve this problem, many approaches were proposed to describe the lip shape such as deformable template [2], snake [3] and active shape models [4], etc. Among them, deformable template is widely accepted due to its advantages of simplicity, easy use for locating lips, etc. The grayscale image was used in early research work. Because the difference between grayscale information of lip color and that of skin color is small, the accuracy of detecting and locating lips is not high, and the requirement of illuminant conditions is rigorous. With the increase of computer speed, colors of human face skin and lips[1] were paid more and more attention. Colors provide richer information than grayscale，So it is possible to improve the accuracy of lip locating with the help of color information. One key technique is the transform from the RGB color images to other color spaces such as YUV space, HSV space, YIQ space and rgb space, etc. After the transform the intensity information is omitted. One or two components, which have more discriminant power to separate the skin

T. Tan, Y. Shi, and W. Gao (Eds.): ICMI 2000, LNCS 1948, pp. 332–339, 2000.

color and lip color, are selected and used to detect and locate lips. The approach using linear-scale range of lip color to locate lip during is not accurate. The approach based on the probability distributions of lip color and skin color was proposed to increase the accuracy of detecting and locating lips. Because the lip region must be in a face image, the decision on if a region is a lip region becomes a two-class discriminant problem, namely, face or lip. The Bayessian decision is usually used. However, the threshold setting becomes a key problem, a bigger threshold makes more false reject rates and a smaller one makes more false acceptance rates. Although color information indeed reduces the illuminant effects, it cannot completely exclude the illuminant effects and the individual variations. Therefore, the locating approach has no adaptivity.

Fisher transform is a kind of linear discriminant transform. It uses the principle of maximizing the ratio of inter-class distance to within class distance and transforms the multi-dimensional space to one-dimensional space. The approach can not only enhance the discrimination between lip color and skin color but also enhance the lip contours. Therefore, it increases the accuracy of locating lips. And the approach uses the lip color and skin color discriminant information, which provides a certain degree of invariance to changes in illumination. Although the relative distributions between lip color and skin color are considered in the approach, the locating approach has no adaptivity either. Therefore, other approaches need to be developed to realize locating lips adaptively.

In summary, the Fisher's discriminant analysis itself only considers the distributions of lip color and skin color. But it doesn't consider the geometrical distribution of lip contours in a face. It will be advantageous for lip locating using the geometrical distribution of lip contours in a face. The conventional approaches for lip locating don't solve the problem of locating inner lips. Because the inner lip is subject to the effects of teeth, black hole and tongue, it is difficult to locate the inner lips accurately. Although, some systems utilize the penalty function to increase the accuracy of locating inner lips, this problem of accurately locating inner lips is not completely solved. Resolving the accurately locating inner lips becomes a great challenge.

Fisher transform with constraints is proposed to enhance the lip region of a face image. Two techniques are proposed to facilitate locating lips. One is the relative area of the lip to that of the face region is almost invariant for a specific person, this characteristic is used to adaptively set threshold to distinguish the lip color and skin color. The other is the relative area of upper lip or lower lip to that of the square whose length equals to the distance of two lip corners is almost invariant, based on whicha linear correlation between parameters of inner lip and those of outer lip can be obtainedand used to predict parameters of the inner lip by parameters of the outer lip. The local minimums of cost function can be overcome by using this technique. And the locating accuracy is increased. The deformable template is used to describe the lips and locate the lips. As shown in Fig.1, the outer lips are described as two quartics and the inner lips are described as two parabolic curves. The cost function consists of three terms, i.e. potential fields, integrals and penalty terms [5].

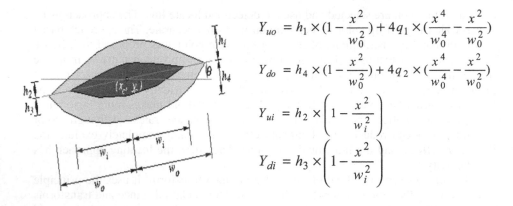

$$Y_{uo} = h_1 \times (1 - \frac{x^2}{w_0^2}) + 4q_1 \times (\frac{x^4}{w_0^4} - \frac{x^2}{w_0^2})$$

$$Y_{do} = h_4 \times (1 - \frac{x^2}{w_0^2}) + 4q_2 \times (\frac{x^4}{w_0^4} - \frac{x^2}{w_0^2})$$

$$Y_{ui} = h_2 \times \left(1 - \frac{x^2}{w_i^2}\right)$$

$$Y_{di} = h_3 \times \left(1 - \frac{x^2}{w_i^2}\right)$$

Figure. 1. The Deformable Template and Its Associated Curves

2   Fisher Transform with Constraints

Fisher's discriminant analysis is a kind of approach to distinguishing between lips and skin. Its basic idea is to transform the original data to maximize the separability of the two classes. The objective is to maximize the criterion function $J(w) = w^t S_b w / w^t S_w w$ so that an optimal one-dimensional projection direction W is obtained. The RGB space is transformed into YIQ space. Two components, Q and phase angle FI, are set to a vector x which is used in Fisher transform to distinguish the skin color and lip color. The Fisher discriminant analysis is as the following:

1. Compute the mean color of each class, k=1,2

$$m_k = \frac{1}{n_k} \sum_{x \in \chi_k} x \tag{1}$$

2. Compute the within class scatter matrices k=1,2

$$S_k = \sum_{x \in \omega_k} (x - m_k)(x - m_k)^T \tag{2}$$

3. Compute the optimal Fisher discriminant vector:

$$W = S_w^{-1}(m_1 - m_2) \tag{3}$$

Where k=1 corresponds to the skin color class; k=2 corresponds to the lip color class;

$$S_w = S_1 + S_2 \tag{4}$$

Because there are other color classes in a face besides the skin color and the lip color, for example, the eye color, eyebrow color, nostril black color, etc, and the colors of these parts are very close to the lip color after fisher transform. This affects the accuracy of lip detecting. Therefore, some constraints must be developed. The colors of pixels not occupied in the skin area and the lip area are replaced with the average skin color. This is achieved by choosing the pixels, whose color is far from the lip color and skin color. The Fisher transform with the constraints is as the following:

$$y = \begin{cases} x^T W & \text{if } x \in \mathcal{E}_{lip} \text{ or } x \in \mathcal{E}_{face} \\ m_1^T W & \text{else} \end{cases} \quad (5)$$

Where y is the image after Fisher transform, $\mathcal{E}_{lip}$ denotes the set of lip color, $\mathcal{E}_{face}$ denotes the set of face color. The Fisher transform with constraints reduces the effects of other components in a human face on lip locating. The images transformed by the Fisher transform are shown in Fig.2. The lip color is enhanced. The skin and lip colors are well separated.

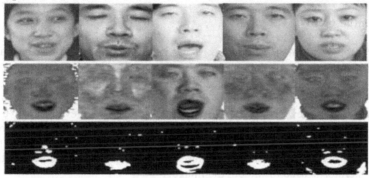

Figure .2. The Detection results of different people under different illuminant conditions

3   The Adaptive Threshold Setting

One fact is the relative area of the lip in color to that of a face region is almost invariant   for a specific person, this characteristic is used to adaptively set threshold to distinguish the lip color and skin color. Note that the lip area has been enhanced by using Fisher transform, i.e. the lip area is the brightest area in a face. The threshold to separate the lip area and skin area can be determined according to the above observation and statistics. The threshold is used to binarize the enhanced face image. Suppose that the color distributions of skin colors and lip colors are Gaussians. The mean and standard deviation of lip colors are μ and σ respectively. The ratio of lip area to skin area is within 4%-7% according to our statistic. Although the ratio is different for different person, but its change range is small. The ratio value is set to 5% in our system, and the corresponding threshold is $\mu + 1.65\sigma$ according to statistics.

As shown in Fig.2, the lip area can be detected for three peoples under different illuminant conditions and lip states. The detecting accuracy of regions of interest (ROI) is 100%. This approach to setting threshold has very good adaptivity for different environments. This approach not only can increase the accuracy of locating lips but also has a very fast speed. It can achieve real-time implementation. The outer lips are easily found by using this approach, including the key points of outer lips such as the top point of upper outer lip and the lowest point of the outer lower lip.

4 Lip Location by Deformable Template

The initial template is computed by the above approach. Adapting the template to match the lip contour involves changing the parameters to minimize cost function. The cost function in our system includes information of lip edges, space constraints, etc.

4.1 Potential Fields

The lip contours are main information for template locating. Since the edges are mostly horizontal, the vertical gradient of the image is used as the potential field, i.e. 3×3 Prewitt operator is used to enhance the edges of the image after Fisher transform. The advantages of the Prewitt operator are not only it is sensitive to horizontal edges, but also it differentiates between "positive" and "negative" edges. A positive edge is one where the image intensity of the pixels above the edge is higher than that of pixels below it, and vice versa. A template that uses this difference will not confuse the upper and the lower lip. In order to ensure the continuity of the cost function, the Gaussian operator is used to smooth the image.

4.2 Cost Function

The cost function includes four curve integrals, one for each of the four lip edges. Taking into account positive and negative edges, it is as the following form:

$$E_1 = -\frac{1}{|\Gamma_1|} \int_{\Gamma_1} \Phi_e(\vec{x}) ds \qquad E_2 = \frac{1}{|\Gamma_2|} \int_{\Gamma_2} \Phi_e(\vec{x}) ds \qquad (6)$$

$$E_3 = -\frac{1}{|\Gamma_3|} \int_{\Gamma_3} \Phi_e(\vec{x}) ds \qquad E_4 = \frac{1}{|\Gamma_4|} \int_{\Gamma_4} \Phi_e(\vec{x}) ds \qquad (7)$$

Where $|\Gamma_i|$ is the lengths of the curves, $\Phi_e(\vec{x})$ is the potential field. Each normalization term is the length of the curve. This facilitates the choice of the weighting coefficient.

The cost function also includes a number of penalty terms. The constraints make sure that the parameters of the template stay within reasonable limits. For example, the constraint $h_1 > h_2$ makes sure that the inner lip is not higher than that of the outer lip, the constraint $a_1 < (h_1 - h_2) < a_2$ makes sure that the thickness of lips is within reasonable limits. If the constraint is broken, the cost function will increase quickly. In addition, there are other penalty terms which make the template be within reasonable limits.

In summary, the cost function is as the following

$$E = \sum_{i=1}^{4} C_i E_i + \sum_j K_j E_{penity_j} \qquad (8)$$

where, C_i and, K_j are weighting factors.

The outer lip contours becomes clear after Fisher transform and the initial position of the template obtained by the adaptive threshold setting approach discussed in section 3. Therefore, the outer lip contours are easily matched by using the deformable template approach discussed above. As shown in table 1, the accuracy of locating the outer lips can reach as high as 96%without any match error. But the inner lip is subject to the effects of teeth, black hole and tongue, this enables locating inner lips inaccurately. To solve this problem, the initial parameters of the inner lips need to be set. The linear correlation between outer lips and inner lips is used to predicate the parameters of the inner lip by those of the outer lip.

5   Predicting the Positions of Inner Lips

It is assumed that the height(h2) of the inner lip has a linear relation with the parameters(W_0, h1, q1) of the outer lip. Similarly ,it is assumed that the height(h3) of the inner lip has a linear relation with the parameters(W_0, q2, h4) of the outer lip. Therefore, we have the following equations

$$h_2 = A_1 h_1 + B_1 q_1 + C_1 W_0 \qquad (9)$$

$$h_3 = A_2 h_4 + B_2 q_2 + C_2 W_0 \qquad (10)$$

The multivariate regression is used to estimate the parameters in the above equations. To verify the linear assumption, the test results of predicted inner lip parameters h2, h3 with 200 lips are shown in the Fig.3. The bold line is the true value and the thin line is the predicted value. From the figure, it is seen that the predicted value is very near the true value. Therefore, the assumption is hold.

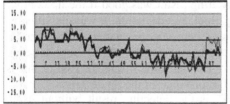 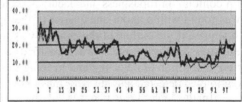

Figure.3. The comparison between the actual values and predicted values of the inner lip parameters. The left one is for the parameter h2, and the right one is for h3

By using the above-obtained linear relationship between h2 and h3, the initial inner lip parameters can be calculated for the inner lip deformable template. This increases the accuracy of inner lip locating and avoids the issue of local minimum. As shown in Fig.4, the original images around the lips are shown at the top. The lip locating results without using linear prediction is shown at the middle, in which it is obvious that the inner lip locating is not accurate. The lip locating results using linear prediction is shown at the bottom, in which the locating result is very accurate. The locating speed is also increased using the predicted parameters of inner lips.

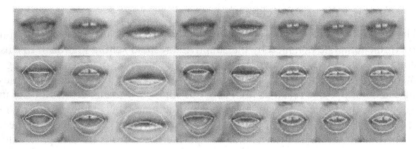

Figure .4. The results obtained by the linear prediction

6   Experiments

Experiments were conducted to compare the accuracy of detecting regions of interest (ROI) and lips with and without linear prediction. The experiment environment is: CPE-3000 image grab card, JVC TK-1070 color camera and MIMTRON MTV-33-1 CB color camera, the frame rate is 25 frames per second, the CPU of computer is Pentium-II 300. To show the advantages of the approaches proposed in this paper, experiments were conducted using the active shape model and the Fisher transform. The comparison between them were conducted by using the image enhancement approach with Q, FI components. The image database consists of images from 4 males and 5 females, the total images are 2000, which were collected under different illuminant conditions and different utterances. The subjective evaluation approach to locating lips [4] was used. The results of this approach are Good, Adequate and Miss. The approach is an important model for evaluating different approaches for locating lip contours in robustness and accuracy. A locating result was good if the lip contour was found within about one quarter of the lip thickness deviation. It was classified as Adequate if the outline of the contour was found between one quarter and half the lip thickness deviation and it was classified as a Miss otherwise. The average locating time is the sum of detecting and locating time. The detection means the ROI of lips is completed detected and the error range is within one quarter the lip deviation.

In the table 1,the first row of the result is the detecting result without using predicting, instead the second row is shown the detecting result using predicting. The ROI detection speed is 18-20 frames per second. Average locating lip contours time is 6-8fram/s. And the outer lip locating accuracy is 96%.

Table 1. Comparison among different approaches

Inner lip locating			Average
Good	Adequate	Miss	locating Time
70%	5%	25%	6-8fram/s
84%	13%	3%	6-8fram/s

7   Conclusion

In the paper, the Fisher transform with constraints is proposed to enhance the lip color region from the face region.   And the presented adaptive threshold setting approach is very effective to separate the lip color and skin color, which enables fast detecting of lips possible. The multivariate regression approach, which uses the parameters of the outer lips to predict the parameters of the lips, greatly increases the locating accuracy of the inner lips and locating speed. Experiments have shown that the approach is efficient in robustness, accuracy, speed and real-time performance.

References

1.   Marcus E,Hennecke, K,Venaktesh,Prasad, David G.Stork, "Visionary Speech: Looking Ahead to Practical Speechreading Systems", In Speechreading by Humans and Machines,Volume150 of NATO ASI Series F: Computer and Systems Sciences,Berlin,1995, Springer.
2.   A. L. Yuille, P. Hallinan, D.S.Cohen, Feature extraction from faces using deformable templates. Int. Journal of Computer Vision, Vol.8 (2), pp.99-112, August 1992.
3.   M.Kass,A.Witkin,and D.Terzopoulos. Snakes: Active contour models. Int. Journal of Computer Vision, pp.321-331, 1988
4.   J.Luettin,Neil A.Thacker,S.W.Beet, Locating and Tracking Facial Speech Features. International Conference on Pattern Recognition. Vienna, Austral, 1996.
5.   M.E.Hennecke,K.V.Prasad,D.G.Stork, Using Deformable Templates to Infer Visual Speech Dynamics. In 28th Annual Asilomar Conference on signals, Systems and Computers. IEEE, November 1994
6.   Michael Vogt. Fast Matching of a Dynamic Lip Model to Color Video Sequences under Regular illumination Conditions. Nato ASI Series F, Vol.150, D.G.Stork, M.E.Hennecke, pp.399-407, 1996
7.   Tarcisio Coianiz, Lorenzo Torresani, Bruno Caprile. 2D Deformable Models for Visual Speech Analysis. In Speechreading by Humans and Machines, Springer-Verlag 1995, Proceedings of the NATO Advanced Study Institute: Speechreading by Man and Machine, 1995.
8.   Ying-li Tian,Takeo Kanade,J.F.Cohn. Robust Lip Tracking by Combining Shape, Color and Motion. Proceedings of the 4th Asian Conference on Computer Vision (ACC'00), Janary, 2000
9.   Robert Kaucic, Andrew Blake, Accurate, real-time, Unadorned Lip Tracking. Pro 6th Int. Conf. Computer Vision,1998

Multi-level Human Tracking[+]

Hongzan Sun, Hao Yang, and Tieniu Tan

National Laboratory of Pattern Recognition, Institute of Automation,
Chinese Academy of Sciences, P.O. Box 2728, Beijing 100080, China
{sunhz, yanghao, tnt}@nlpr.ia.ac.cn

Abstract. In this paper, we describe a system on automatically tracking people in indoor environment. Parametric Estimation methods are adopted in training images to obtain means and covariance of the background pixels. They are processed in pixel classification step to extract moving people and eliminate their shadows. Our tracking algorithm consists of two levels: region level and blob level. The region level tracks a whole human by utilizing the position information of the detected persons. The EM algorithm is adopted at the blob level to segment human image into different parts. When region level tracking fails, the system switches to blob-level tracking. Average Bhattacharyya distances between corresponding blobs of consecutive frames are calculated to get the correct match between different regions. Potential applications of the proposed algorithm include HCI and visual surveillance. Experimental results are given to demonstrate the robustness and efficiency of our algorithm.

1 Introduction

Human tracking plays an important role in many applications of computer vision, such as human-machine interface and visual surveillance. It is prerequisite for action recognition and intelligent behavior analysis. As the human body is a non-rigid form, human tracking is more difficult than tracking of a rigid body.

[+] This work is funded by research grants from the NSFC(Grant No. 59825105) and the Chinese Academy of Sciences.

T. Tan, Y. Shi, and W. Gao (Eds.): ICMI 2000, LNCS 1948, pp. 340-348, 2000.

Many vision-based systems for detecting human motion have already been proposed. The system described in [1][2] extracts human regions from input images by background subtraction, and obtains the correspondence between frames based on position, velocity and intensity information. Haritaoglu et.al. [3]utilize shape information of foreground regions to track different parts of a person. Pfinder [4] tracks single person under constrainted environments with color.

Here we propose a new strategy to deal with this problem. People are tracked at two levels: region level and blob level. Simple shape information of foreground regions acquired based on background model is used in region level to deal with common situation, where no occlusion or sudden motion occurs. When the region level module fails, we match the people in consecutive frames at the blob level. Firstly we match different blobs by utilizing color information. Then average Bhattacharyya distances between corresponding blobs are calculated to get the correct match between adjacent frames.

The paper is organized as follows: Section 2 describes the human detection step that includes the background model and the shadow elimination algorithm. Section 3 details the algorithm of region level tracking and describes the switch condition between two levels. The blob tracking algorithm is described in Section.4. Section 5 presents the experimental results of our system and discusses the deficiency of our algorithm. Finally, in Section 6, we draw a conclusion and discuss possible future work.

2 Human Detection

Two stages are performed in human detection step: 1) Estimation of the background model from training image sequences. 2) Classification of pixels to eliminate shadows and extract the regions of people. Detection of human motion regions can reduce the search space, and make the system robust and efficient. When the boundaries of a person have no intersect with those of the image, we can justify that a person enters our view and our system begins to track it.

We estimate the background image with a five-second long image sequence captured with a static camera. Under common situation, our method [5] can acquire mean and covariance of each pixel in the image accurately even motion occurs in the sequence. Like the method proposed by Horprasert et.al. [6], we separate the difference between current image and background image into brightness distortion and chromaticity distortion. Both are used in pixel classification step to extract foreground regions and eliminate their shadow. For our system is mainly for indoor environments, only simple rule was used to classify moving regions into human and non-human. Here we adopt the ratio between the height and width of the region. Fig.1. gives an example.

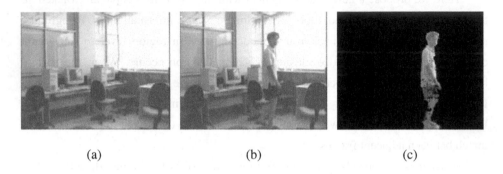

(a)   (b)   (c)

Fig. 1. . An example of human detection. (a) Background image. (b) Frame 50. (c) Human detection result of Frame 50.

3 Region Level Tracking

Currently our system contains only one static camera to track humans. Once a new object enters its view and is identified as people, region level tracking module begins to work. As the human body is deformable, and light variations exist in the environment, it is hard to choose features to ensure the robustness of tracking. Here we choose the median coordinate of the region instead of its centroid. It is more robust to changes in shape of the silhouette of a person, as explained by Haritaoglu et.al. [3]. We explore Bayes' rule to find the correct match between adjacent frames. Let X_t denote a feature vector of a subject at time t, and θ denote the feature parameters corresponding to the tracked object at time t-1. The *pdf of* the geometric features are modeled as

$$P(X_t \mid \theta) = \frac{1}{\sqrt{2\pi}\delta} \exp[-\frac{(X_t - \overline{X}_t)^2}{2\delta^2}] \tag{1}$$

where $\overline{X}_t$ is the position of median point in the subject of interest in the previous

frame, and δ is the maximum value of $\left| X - \overline{X} \right|$ for all candidates in the current

frame. We find the maximum value of Equation (1) to get the correct match. If we assume that human motion is constant, we can predict the position of human from the following equation

$$\overline{X}_{(t-1)} - r_{t-1}\overline{X}_{(t-2)} = \overline{X}_t / r_t - \overline{X}_{(t-1)} \qquad (2)$$

where $\overline{X}_{(t-1)}$ and $\overline{X}_{(t-2)}$ are the location of the median point in the previous two

frames and r_t denotes the area ratio of the tracked human body between time t and

$t-1$.

In our system, region-level tracking is simple and fast, and it can get correct results at many situations. Observed a common situation, we can find occlusions between moving peoples and between moving people and static objects often occur. Our region level tracking fails to deal with those situations, especially it may induce two or more subjects corresponding the same subject. Then the system switches to blob-level tracking module, which can find the correct match when in the situation that motion model assumption is violated or occlusions occur.

4 Blob Level Tracking

The blob level tracking module of our system includes three parts: region segmentation, blob matching, and region matching. Firstly each moving region was separated into different parts based on color. Then we can get the correspondence between parts of different regions according to position information. Finally, we calculate average Bhatacharyya distances between moving regions. Correct match between regions can be obtained according to the minima of the distance.

4.1 EM Segmentation

Once a person is detected, we segment the person into regions of similar color. We use the EM algorithm [7] to fit a Gaussian mixture model to the color distribution of the person, which is given by:

$$M(x) = \sum_{i=1}^{k} \frac{\omega_i}{(2\pi)^{d/2}|\Sigma_i|^{1/2}} e^{-\frac{(x-\mu_i)^T \Sigma^{-1}(x-\mu_i)}{2}} \tag{3}$$

where ω_i is the weight assigned to the i th Gaussian and d is the dimension of our space (three in this case). Each Gaussian fitted by the EM algorithm now represents one of the classes of the full person. Finally, a maximum likelihood computation is done to assign each pixel to its correct class.

Given a color image and a mask representing the foreground region, the mixture of Gaussians is computed by finding the correct parameter $\omega_i, \mu_i, \Sigma_i$ for each of the k Gaussians in the mixture model using EM algorithm. Details are as follows.

1. Initialise the k Gaussians so that each of Σ_i is the identity matrix, each of the ω_i is 1/k, and the k centres are taken to be some small random step away from a randomly selected point of the data set X. Here we choose k=3.

2. For each point $x \in X$, and for each i between 1 and k, compute the weighted likelihood $\omega_i g_{[\mu_i, \Sigma_i]}(x)$ using the last estimate of the parameters $\omega_i, \mu_i, \Sigma_i$. For each point x, let $L_i(x)$ be the likelihood assigned to it by the i th member of the mixture:

$$L_i(x) = \omega_i g[\mu_i, \Sigma_i](x) \tag{4}$$

where

$$g[\mu_i, \Sigma_i](x) = \frac{\omega_i}{(2\pi)^{d/2}|\Sigma_i|^{1/2}} e^{-\frac{(x-\mu_i)^T \Sigma^{-1}(x-\mu_i)}{2}} \tag{5}$$

Let S_x be the sum

$$S_x = \sum_{i=1}^{k} L_i(x)$$

(6)

and

$$P_i(x) = \frac{L_i(x)}{S_x}$$

(7)

We think of the point x as ready to be split up between the k Gaussians according to the k fractions $P_i(x)$. In consequence, we have that

$$\sum_{x \in X} \sum_{i=1}^{k} P_i(x) = |X|, \text{ the cardinality of } X.$$

3. Now re-estimate the ω_i^+ as

$$\omega_i^+ = \frac{\sum_{x \in X} P_i(x)}{|X|}$$

(8)

and the centres μ_i^+ as the weighted centroids

$$\mu_i^+ = \frac{\sum_{x \in X} P_i(x)x}{\sum_{x \in X} P_i(x)}$$

(9)

and finally, the i th covariance matrix, Σ_i^+ is re-estimated as

$$\Sigma_i^+ = \frac{1}{\sum_{x \in X} P_i(x) - 1} \sum_{x \in X} P_i(x)[(x - \mu)(x - \mu)^T]$$

(10)

4. Run through steps 2 and 3 until there is no significant change in the parameters.

After the EM algorithm converges, we can obtain the different parts of the human body. Figure 2 shows an example of initial segmentation.

4.2 Blob Matching

To measure the color variation caused by the changes of lighting or the pose of the human body, we must first get the correspondences between different blobs. Here we obtain the correspondences according to the minimum Bhattacharyya distance (described in the next section) between blobs of different frames. It works well enough here for there are only four regions in our experiments. More complex algorithms need be investigated in the future.

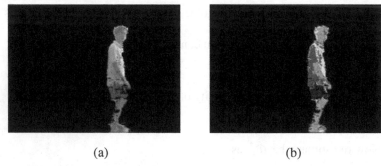

(a)   (b)

Fig2. Initial segmentation results: (a) The frame after human detection
(b) The frame after EM segmentation.

4.3 Average Bhattacharyya Distances

We choose the Bhatacharyya distance [8] to measure the change of color distribution between corresponding blobs. It is given by:

$$J_b = -\ln \int [p(x \mid \omega_1) p(x \mid \omega_2)]^{1/2} dx \qquad (11)$$

When $p(x \mid \omega_1)$ and $p(x \mid \omega_2)$ satisfy Gaussian distribution, it becomes

$$J_b = \frac{1}{8}(\mu_i - \mu_j)^T [\frac{\Sigma_i + \Sigma_j}{2}]^{-1}(\mu_i - \mu_j) + \frac{1}{2} \ln \frac{\left|\frac{1}{2}(\Sigma_i + \Sigma_j)\right|}{[\left|\Sigma_i\right| \left|\Sigma_j\right|]^{1/2}} \qquad (12)$$

and the mean is

$$\overline{J_b} = \sum_{i=1}^{k} J_{bi} \tag{13}$$

where J_{bi} represents the Bhatacharyya distance between the corresponding i th blobs. Then we can classify and track different people based on their color information by finding the minima of $\overline{J_b}$. It is assumed that different people have different color, which is often satisfied under common situation. The algorithm may be used to track people after occlusions occur.

5   Experimental Results and Discussion

We built a prototype system with a PC which has a single 450MHz processor. The region level module runs very fast and can work for real-time applications. The blob level module has a speed of five frames per second for much more computation is required.

The average Bhattacharyya distances were calculated between the same people at different frames and different people at the same frame. Table 1. gives an example. It can be seen that the distances between color distributions can be used to classify different people in spite of changes in pose and lighting.

6   Conclusion and Future Work

In this paper, we proposed a new algorithm for human tracking. Bhattacharyya distance was adopted to measure the appearance change of human image caused by variations in pose and lighting. Experimental results show that color information can be utilized to classify different people as people often have different color and no sudden light variations occur. Future work includes extending the system into multi-camera tracking.

Table 1. Comparison of inter-frame and intra-frame Bhattacharyya distances.

Frame	1	2	3	4	5	6	7	8	9
a)	0.0058	0.0127	0.0043	0.0079	0.0238	0.0155	0.0302	0.0052	0.0273
b)	1.5832	2.3735	1.4587	1.3567	1.7533	1.4566	2.3444	1.8412	2.6841

(a) The distance between current frame and its next frame of corresponding human region.

(b) The distance between different human regions of current frame.

Reference

[1] Q. Cai and J.K. Aggarwal, "Tracking Human Motion Using Multiple Cameras," *Proc. Int'l Conf. Pattern and Recognition*, pp. 68-72, Vienna, Austria, Aug. 1996.

[2] Q. Cai and J.K. Aggarwal, "Tracking Human Motion in Structured Environments Using a Distributed-Camera System", *IEEE Tran. PAMI*, Vol 21, No. 12, November 1999.

[3] I. Haritaoglu, D. Harwook, and H. Davis, "W4: Who, What, When, Where: A Real-Time System for Detecting and Tracking People," 1998.

[4] C.R. Wren, A. Azarbayejani, T. Darrell and A. Pentland, "Pfinder: Real-time Tracking of Human Body," *IEEE Transactions on Pattern Analysis and Machine Intelligence*, July 1997, vol 19,no 7, pp. 780-185.

[5] H.Z. Sun, T. Feng and T.N. Tan, "Robust extraction of moving objects from image sequences," In *Proceedings of ACCV2000*.

[6] T. Horprasert, D. Harwood, L. S. Davis, "A Robust Background Substraction and Shadow Detection," In *Proceedings of ACCV2000*.

[7] M.D. Alder, "An Introduction to Pattern Recognition: Statistical, Neural Net and Syntactic Methods of Getting Robots to See", http://ciips.ee.uwa.edu.au/~mike/PatRec/node95.html, September 1997.

[8]A Introduction to Pattern Recognition (In Chinese), Zhaoqi Bian and Xuegong Zhang., Tsinghua University Press, 2000.

Region-Based Tracking in Video Sequences Using Planar Perspective Models

Yuwen He, Li Zhao, Shiqiang Yang, Yuzhuo Zhong

Department of Computer Science and Technology, Tsinghua University, Beijing 100084, China
heyw@media.cs.tsinghua.edu.cn, li.zhao@263.net, yangshq@tsinghua.edu.cn

Abstract. Object tracking can be used in many applications using motion information. This paper proposes a method of region-based tracking using planar perspective motion models. Planar perspective models can represent motion information of the plane rigid motion in the sequence properly. And in many cases the real object's motion can be represented using planar perspective motion models approximately. The method estimates model parameters on three pyramid levels, and it is based on the reliable estimation of planar perspective models in the region to be tracked. The calculation on three pyramid levels can accelerate the speed of estimation. Gauss-Newton and Levenberg-Marquadet are combined to estimate the models' parameters. During the tracking process there is some noise, so robust estimation is used in the parameters' estimation of models. Some experimental results are shown at the end of paper. From the experiments' result the tracking method is effective in those complicated situations.

Keywords: motion estimation, tracking, planar, parameter models

1 Introduction

Region-based tracking is a general problem in motion analysis. It can be applied in many areas relative to HCI, such as tracking of human face. In the past time Meyer [7] studied the tracking of moving region with affine models. The affine models can not describe the exact motion field because of its approximation. The tracking process relies on two steps. The first step is motion region detection, and the second step is tracking sequentially. The work about global motion estimation that we have finished is related to motion region detection [13]. This paper only studies the tracking of a predefined moving region. We presume the region to track was initially extracted manually.

The method used in the paper is based on the analysis of motion field in the video sequence. Because the approach is dependent on motion models, to select the proper motion models is very important. If the models can describe the motion filed of motion region properly, then tracking will probably succeed. Planar perspective models are applied to the tracking because it can describe very complex motion filed.

T. Tan, Y. Shi, and W. Gao (Eds.): ICMI 2000, LNCS 1948, pp. 349-355, 2000.

It is superior compared to the affine models. The affine models are used in region-based tracking by Meyer [7]. The method in [7] is only available when the camera is static, and the motion filed in [7] is relatively simple. On the other hand the estimation of motion modals is based on dense estimation. During the tracking much noise occurs, so robust estimation is very necessary for the tracking in the long sequence. Konrad [1] uses Levenberg-Marquadet method to estimate the parameters of motion models. But the result is sensitive to the noise because they do not use robust statistics method in the estimation.

This paper is organized as follows. In the next section the principle of region-based tracking is explained, and the method used in this paper is shown also. Experimental evaluation about the tracking algorithm is given in section 4. A short discussion and conclusion are presented in section 4.

2   Principle and Method

This section is divided into two parts. The planar perspective models used in the tracking will be introduced in the first part, which is stated in theory. Region based tracking algorithm is stated in the second part.

2.1   Planar Perspective Models

Region-based tracking is based on the motion models of the velocity field of motion. We first setup the planar perspective models as follows. To a rigid body if it undergoes a small rotation, the following equation is tenable.

$$\begin{bmatrix} \dot{X}_1 \\ \dot{X}_2 \\ \dot{X}_3 \end{bmatrix} = \begin{bmatrix} 0 & -\Omega_3 & \Omega_2 \\ \Omega_3 & 0 & -\Omega_1 \\ -\Omega_2 & \Omega_1 & 0 \end{bmatrix} \begin{bmatrix} X_1 \\ X_2 \\ X_3 \end{bmatrix} + \begin{bmatrix} V_1 \\ V_2 \\ V_3 \end{bmatrix}. \tag{1}$$

or $\dot{X} = \Omega \times X + V$, here X is the vector representing the position. $\Omega = [\Omega_1 \; \Omega_2 \; \Omega_3]^T$, Ω is the angular velocity vector. $V = [V_1 \; V_2 \; V_3]^T$, V is translation velocity vector. Here the subscript 1 represents horizontal axis. The subscript 2 represents vertical axis, and the subscript 3 represents optical axis. The position of pixel in the image plane can be got according to the perspective equation 2.

$$x_1 = f \frac{X_1}{X_3} \; , \; x_2 = f \frac{X_2}{X_3}. \tag{2}$$

Then the velocity of pixel in the image plane can be got by differential of equation 2.

$$v_1 = \dot{x}_1 = f(X_3\dot{X}_1 - X_1\dot{X}_3)/X_3^2 \; , \; v_2 = \dot{x}_2 = f(X_3\dot{X}_2 - X_2\dot{X}_3)/X_3^2 . \tag{3}$$

Substitute $\dot{X}$ in equation 1 for the velocity in equation 3, then get the following equation:

$$v_1 = f(\frac{V_1}{X_3} + \Omega_2) - \frac{V_3}{X_3}x_1 - \Omega_3 x_2 - \frac{\Omega_1}{f}x_1 x_2 + \frac{\Omega_2}{f}x_1^2 \ , \ v_2 = f(\frac{V_2}{X_3} - \Omega_1) - \frac{V_3}{X_3}x_2 + \Omega_3 x_1 + \frac{\Omega_2}{f}x_1 x_2 - \frac{\Omega_1}{f}x_2^2 \ . \tag{4}$$

When the projection is normalized, set f=1 then the equation 4 is transformed as:

$$v_1 = (\frac{V_1}{X_3} - \frac{V_3}{X_3}x_1) - \Omega_1 x_1 x_2 + \Omega_2(1 + x_1^2) - \Omega_3 x_2 \ , \ v_2 = (\frac{V_2}{X_3} - \frac{V_3}{X_3}x_2) - \Omega_1(1 + x_2^2) + \Omega_2 x_1 x_2 + \Omega_3 x_1 \ . \tag{5}$$

If there is a plane constraint for the moving region, then a plane equation exists:

$$k_1 X_1 + k_2 X_2 + k_3 X_3 = 1 \ , \ 1/X_3 = k_1 X_1/X_3 + k_2 X_2/X_3 + k_3 = k_1 x_1 + k_2 x_2 + k_3 \ . \tag{6}$$

In the equation 6 the projection has been normalized. Then X_3 in the equation 5 can be substituted according to the equation 6.

$$v_1 = a_1 + a_2 x_1 + a_3 x_2 + a_7 x_1^2 + a_8 x_1 x_2 \ , \ v_2 = a_4 + a_5 x_1 + a_6 x_2 + a_7 x_1 x_2 + a_8 x_2^2 \ . \tag{7}$$

$$a_1 = k_3 V_1 + \Omega_2 \ , \ a_2 = k_1 V_1 - k_3 V_3 \ , \ a_3 = k_2 V_1 - \Omega_3 \ , \ a_4 = k_3 V_2 - \Omega_1 \ ,$$

$$a_5 = k_1 V_2 + \Omega_3 \ , \ a_6 = k_2 V_2 - k_3 V_3 \ , \ a_7 = -k_1 V_3 + \Omega_2 \ , \ a_8 = -k_2 V_3 - \Omega_1 \ . \tag{8}$$

The equation 7 is the planar perspective models that it is accurate models for the planar rigid motion field. The planar perspective models are also available for the local plane if the smooth curved surface is expanded in first-order terms of Taylor series. If the depth of the region to be tracked varies slightly compared with the distance between itself and camera, the planar perspective models are also available.

2.2 Region-Based Tracking Algorithm

We assume that the planar perspective models can describe the velocity filed of the region to be tracked. Then our tracking method is based on this planar perspective models defined by equation 7. We will define some denotations as follows. I_t represents the image at time t. $\vec{\theta}$ represents the parameter vector of the planar perspective models described by equation 7. Let $\vec{\theta} = [a_1 \ a_2 \ a_3 \ a_4 \ a_5 \ a_6 \ a_7 \ a_8]^T$. The problem in region-based tracking is how to estimate the value of $\vec{\theta}$ between the image at time t-1 and next image at time t using the dots in the region to be tracked. The parameters' estimation is the dense estimation using the least square method. We use $I_t'(\vec{\theta})$ to represent the predicted image at time t using the parameter $\vec{\theta}$.

Expand the image I_{t-1} in Taylor series at $\vec{\theta}_i$ and drop the second items, we can get the following equation:

$$I_{t-1} = I_t'(\vec{\theta}_i) + \frac{\partial I_t}{\partial \vec{\theta}}\Big|_{\vec{\theta}_i}(\vec{\theta} - \vec{\theta}_i) \tag{9}$$

We can calculation the increment of parameter vector according equation 9. If parameter vector $\vec{\theta}$ has a initial value $\vec{\theta}_0$, then $\vec{\theta}$ can be approached step by step using the equation 9 iteratively. Gauss-Newton and Levenberg-Marquadet iterative methods are all available to calculate this kind of question. But all this nonlinear methods are based on the least square method. The least square method is sensitive to the noise. During the tracking there are many outliers that do not meet the planar perspective models, so the estimation method must consider robust method to reduce the influence of the noise. We introduce the Geman-McLure function as error function ,

$$\rho_{GM} = \frac{r^2}{\sigma^2 + r^2} \text{ , where } r \text{ is the residue, } \sigma \text{ is a scale factor that can be}$$

computed as follows: $\sigma = 1.4826 \times \underset{i}{median} \left| r_i \right|$.

The energy function of the least square method is $E = \sum_{i \in \mathrm{Re}\,gion} \rho_{GM}(r_i)$. Because the residue of outliers is large, Geman-McLure function can reduce the influence compared with error function r^2. Figure 1 shows the curves of two functions: r^2, ρ_{GM}. In the figure 1 when r is large, ρ_{GM} approaches 1 but r^2 has not the upper limit.

In the estimation parameters of planar perspective models, three-level pyramid is used in order to accelerate the estimation speed. Gauss-Newton method is applied to the top level of pyramid, and Levenberg-Marquadet method with robust statistics is used on the other two levels. In order to reduce the influence of the outliers, some pixels will be excluded by the means [1] as follows. Before the iterative calculation on every level of pyramid, a histogram of the absolute value of error term $\delta I / \delta t$ is calculated. The top 10% of the errors in the histogram will be excluded. The framework of tracking algorithm is shown in figure 2.

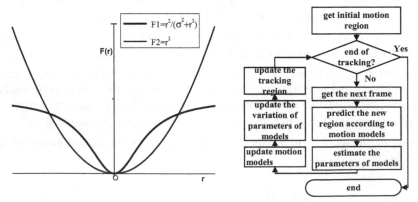

Fig. 1. two error functions   **Fig. 2.** framework of tracking algrithm

In figure 2 because we need trace the change of parameters of motion models in order to describe dynamic attribution of the motion field with time, we update the variation of parameters of models at every times of tracking.

3   Experimental Evaluation

Some testing sequences of MPEG-4 standard sequences are selected to do the experiments. These sequences can be divided into two classes. The one is that camera is static and some regions are moving, such as container sequence and mother-daughter sequence. The other is that camera is moving, such as foreman sequence and coastguard sequence.

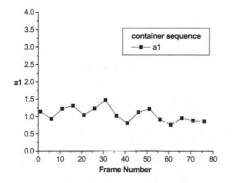
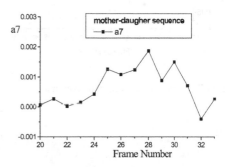

Fig. 3. a1 of container sequence   **Fig. 4.** a7 of mother-daughter sequence

In container sequence the tracking region is a ship, which undergoes translation in the horizontal direction. We know that a1 is correlative to horizontal translation from equation E-8. Figure 3 shows a1 values estimated during the tracking process of container sequence. It indicates that the ship is translating in the horizontal direction. In mother-daughter sequence the tracking region is the face of mother, which is rotating mainly. From frame 20 to frame 32 the mother's head rotates round a vertical axis continuously. We can know that a7 is correlative to the rotation round vertical axis from equation E-8. The values of a7 estimated are shown in figure 4 during the tracking process of container sequence. It indicates that the region is rotating round a vertical axis continuously. The frame 20 and frame 31 of mother-daughter sequence are shown in figure 6 and figure 7. The tracking region is enclosed with a white loop.

Fig. 5. frame 20 of mother-daughter sequence   **Fig. 6.** frame 31

Fig. 7. a3 and a5 of foreman sequence   **Fig. 8.** a1 of coastguard sequence

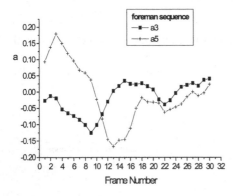

Fig. 9. frame 1   **Fig. 10.** frame 10   **Fig. 11.** frame 101   **Fig. 12.** frame 110

In foreman sequence the tracking region is the face of the man, whose head rotates randomly including rotation round vertical axis, horizontal axis and optical axis. The value of a3 and a5 components of motion modals estimated in foreman sequence is shown in figure 7. The frame 1 and frame 10 of foreman sequence are shown in figure 9 and figure 10. In coastguard sequence the tracking region is the white ship, which translates in horizontal direction. The value of a1 of motion models is shown in figure 8. The frame 101 and frame 110 of coastguard sequence are shown in figure 11 and figure 12. The value of a1 is varied when the origin of image coordinates. In our estimation the origin is the left-top position of image, not the center of image, so the value of a1 is larger than the value expressed in equation 8. In foreman and coastguard sequences the camera is moving also, so the value of motion models estimated is relative to the camera, not relative to the absolute still ground.

4   Conclusion and Discussion

There are two classes in the testing sequences. One is that camera is static and the other is that camera is moving. The motion of moving region in the test sequences includes rotation, translation and the two combined. From the experimental results in section 3 we can draw the conclusion that our new region-based tracking method is effective when there are the large translation motion and noise in the sequence. Because the method based on models is dependent on the dense estimation, if there are not enough dots in the tracking region the method will be disabled. The defect of region-based tracking is that the region edge got from tracking will become vague. So if the accurate margin of moving region got from tracking is required, a contour revision is necessary according to the motion and gradient information. In the following time we will carry out the relevant research about motion margin determination. Our method is based on motion models. So it will be unavailable when the situation is complex and there are not very good motion filed models to describe that motion filed. But there are many deformable motion objects in real situations, thus the tracking of deformable object's motion is another question to be solved.

Reference

1. Janusz Konrad, Frederic Dufaux., Digital Equipment Corporation, Improved global motion estimation for N3, the meeting of ISO/IEC/SC29/WG11,No. MPEG97/M3096 , San Jose, February 1998.
2. G.Wolberg, Digital Image Warping, Los Alamitos, CA:IEEE Comp.Soc.Press, 1990.
3. ISO/IEC WG11 MPEG Video Group, MPEG-4 Video Verification Model Version 12.1, ISO/IEC/JTC1/SC29/WG11,No.MPEG98/N2552,Roma, December 1998.
4. Yaxiang Yuan, Numeric Method of Nonlinear Programming, Book Concern of Science and Technology of Shanghai, 1993.
5. Yujin Zhang, Image Engineering, Publishing house of Tsinghua University, 1999.
6. A.Murat Tekalp, Digital Video Processing, reprinted by Tsinghua University of China, 1998.
7. Frangois G. Meyer, Patrick Bouthemy, Region-based Tracking Using Affine Motion Models in Long Image Sequences, CVGIP:Image Understanding, Vol.60,No.2, Septermber,pp119-140,1994.
8. W.Enkelmann, Investigations of multigrid algorithms for the estimation of optical flow fields in image sequences, Comput. Vision Graphics Image Process. 43,1988, 150-177.
9. F.Leymarie and M.D.Levine, Tracking deformable objects in the plane using an active contour model, IEEE Trans. Pattern Anal. Mach.Intell. June 1993, 617-634.
10. F.Meyer and P.Bouthemy, Region-based tracking in an image sequence, in Proceedings of ECCV-92, Italy(G.Sandini,ed.), pp.476-484, Springer-Verlag, Berlin/New York, 1992.
11. S.Negahdaripour and S.Lee, Motion recovery from image sequences using only first-order optical flow information, Int.J.Comput.Vision 9(3),1992,163-184.
12. R.J.Schalkoff and E.S.McVey, A model and tracking algorithm for a class of video targets, IEEE Trans, PAMI PAMI-4(1), 1982, 2-10.
13 Yuwen He, Shiqiang Yang, Yuzhuo Zhong, "Improved Method for Global Motion Estimation Used in Sprite Coding", Hawaii Meeting of ISO/IEC JTC1/SC29/WG11, MPEG99/M5626, December 1999, Hawaii, America.

A Novel Motion Estimation Algorithm Based on Dynamic Search Window and Spiral Search

Yaming Tu[1], Bo Li[2] and Jianwei Niu[3]

Dept. Of Computer Science, Beijing University of Aeronautics and Astronautics
Beijing 100083, China
tuyaming@263.net[1], boli@maindns.buaa.edu.cn[2]

Abstract. Motion estimation is a key technique in MPEG and H.263 encoder, due to its significant impact on the bit rate and the output quality of the encoded sequence. The full search algorithm (FS), which is considered to be the optimal, is computational intensive, and traditional fast algorithms still need to be improved in both computational complexity and matching accuracy. This paper proposes a novel algorithm, DSWSS, which is based on dynamic search window and spiral search. It performs as accurately as full search algorithm, but needs much less computation.

1. Introduction

Motion estimation (ME) is central to video compression. For easy implementation, ME in MPEG and H. 263 are based on the block matching algorithm (BMA). The full search algorithm (FS)[1] is the most accurate one among all the BMAs, because it matches all possible displaced blocks within the search area. Nevertheless, FS has not been a popular choice because of the high computational complexity involved. As a result, many fast search algorithms were proposed, such as the three-step search (3SS) and the cross search (CS) [2] in the early time, the dynamic search window adjustment and interlaced search (DSWA)[3], the new three-step search (N3SS)[4] and the four-step search (4SS) [5] in the later time. In July, 1999, the diamond search (DS) [6] was accepted by MPEG-4 international standard and incorporated in MPEG-4 Verification Model (VM) [7]. Recently, a Part 7 has been created in MPEG-4 to contain all optimized non-normative modules, and DS was removed from the Reference Software [8].

These fast algorithms can substantially reduce the computational complexity, and each has its own merits and drawbacks. For example, 3SS uses a uniformly allocated search pattern in its first step, which is not very efficient for catching small motions. DSWA adaptively adjusts search step to remedy the problem of 3SS to a certain degree. N3SS, 4SS and DS employ the characteristics of center-biased motion vector distribution, and start with a smaller search grid pattern. On the whole, the number of search points reduced, and the matching accuracy increased. However, both computational complexity and matching accuracy still need to be improved in application.

T. Tan, Y. Shi, and W. Gao (Eds.): ICMI 2000, LNCS 1948, pp. 356-362, 2000.

2. The Framework of Fast BMA

In BMA based ME, the current frame is divided into a number of blocks. For each block, the reference frame is searched in order to find the closest one to it, according to a predefined block distortion measure (BDM). This closest block, also called the optimum, is then used as the predictor for the current one, and the displacement between the two blocks is defined as the motion vector (MV) associated with the current block. Usually, a block is denoted by its first pixel (point), and the location of the current block is chosen as the center of the search window in the reference frame. As shown in Figure 1, the block size is M×M pixels and the maximum displacement of an MV is ± d in both horizontal and vertical directions. The MV (u, v) is obtained by finding a matching block within a search window of (2d+1)×(2d+1) in the reference frame. Usually M is equal to 16.

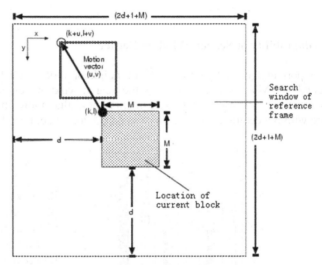

Fig. 1. The MV (u, v) is obtained by finding a matching block within a search window of (2d+1)×(2d+1) in the reference frame.

Almost all fast algorithms are based on an important assumption, that is, the BDM increases monotonically as the checking point moves away from the global minimum BDM point. Actually, this assumption is true only in a local area. In case of multi-extremum distribution of BDM, it is more susceptible to being trapped into local optima.

An important fact is that the distribution of motion vector is center-biased. As shown in Figure 2, x-axis refers to search window size centered with (0,0), and y-axis refers to the percentage of optima located in the window of corresponding size based on FS. It shows that more than 80 percent of optima are located in 7×7 window centered with (0,0).

This paper proposes a novel search algorithm, DSWSS, which adaptively changes both the center and the size of the search window. Therefore, it can significantly reduce search time with little loss in accuracy.

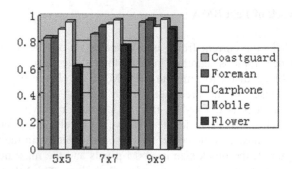

Fig. 2. The distribution of motion vector is center-biased: more than 80 percent of optima are located in 7×7 window centered with (0,0).

2.1 The Adaptability of Search Window Center

Another important fact is that the MVs of neighbor blocks are highly correlated. Using neighbor's MV instead of (0,0) as the search window center, the distribution becomes more center-biased. Our experiments show that more than 90 percent of optima are within a central 5×5 area after the prediction of center (cf. Figure 3).

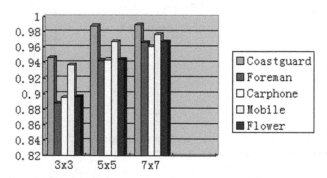

Fig. 3. After the prediction of search window center, the MV distribution becomes more center-biased

But how to make use of neighbor's motion vector? As shown in Figure 4, block O is the first block of a frame, and no prediction is conducted. Block B and D can be predicted only by their left blocks, and block A and E by their top blocks. For such block as C, which makes up the majority, can be predicted by A, B and D. Experiments show that using the mean vector of A and B can predict C better. For easy implementation in practice, the MVs of the collocated block in reference frame are not used in the prediction.

O	B	D								
A	C									
E										
...										

Fig. 4. Blocks in a frame: O stands for the first block of a frame, B and D stand for the top most blocks, A and E stand for left most blocks. A is the left block of C, and B is the top block of C.

2.2 Search Strategy and the Adaptability of Search Window Size

Since the prediction of the center makes the MV distribution becomes more center-biased, it is possible to find optimum or sub-optimum in only a small area. Obviously, in this small area, using spiral search is more likely to get optimum than using gradient search. Furthermore, we can check every other point outside of the first loop to reduce search point number (cf. Figure 5).

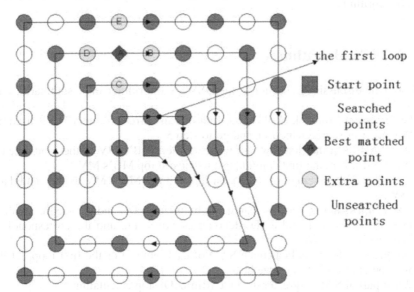

Fig. 5. The search pattern of DSWSS: after spiral search, four or three extra searches are conducted.

The motion varies with the type of video sequences. If fixed search window size is used, it will lead to either excess search or insufficient search. Here we use the following stopping criteria to make window size adaptive:

(1) If no better point is found in current loop than the optimum found so far, stop the spiral search; else continue until the boundary of the search window.

(2) After the spiral search as in (1), check the four (or three) points around the current optimum, such as point B, C, D, E around A in Figure 5.

2. 3 Block Distortion Measure

Due to its simplicity and its similar performance as MSE, MAD (also called MAE) is widely used as BDM. Instead of MAD, we use SAD (Sum of Absolute Difference) here, which is defined as:

$$SAD(i, j)= \sum_{m=0}^{N-1} \sum_{n=0}^{N-1} | I_2(m,n) - I_1(i+m, j+n) |$$

Where $-W \leq i, j \leq +W$, and W is the size of search window, N is the size of block. It is obvious that SAD is the same as MAD in measuring block distortion, and the calculating process can be stopped when partial SAD is greater than the minimum SAD found so far. Therefore, a lot of search time can be saved.

In summary, our new algorithm is adaptive in search window. Its center can be chosen by the prediction instead of a fixed one, and the size is adapted through stopping search at appropriate time. Furthermore, through adjusting the step size in spiral search, it can cover the points in all directions without complex computation. As such, it overcomes the blind search in FS and the local-optimum in many other fast search algorithms.

3. The DSWSS Algorithm

According to the discussion in section 2, the DSWSS algorithm can be summarized as follows:
1) If the current macroblock (MB) is the first one in current frame, (0,0) is used as CSW (center of search window), and go to step 5.
2) If current MB is at the first row, use its closest left MB's MV as the candidate of CSW; else if it is at the first column, use its closest top MB's MV.
3) Otherwise, use the average of its closest left and top MB's MV as the candidate of CSW.
4) Compare blocks at (0,0) and the candidate by their SADs, and choose the smaller SAD as the initial minimal SAD (denoted as MinSAD), and the corresponding point as the initial CSW.
5) Search the next loop, and calculate SAD of each point. For the first loop, let its stepsize be 1; else let its stepsize be 2.
   a) When partial SAD is greater than the MinSAD, stop calculating;
   b) If the total SAD of current point is smaller than MinSAD, then assign it to the MinSAD, and set the Fmin (MinSAD updated flag) to 1
6) When current loop ends, if the Fmin=1, go to step 5; else continue step 7.
7) Refine the optimal point. If it is the same as the initial CSW, continue step 8; else check extra four (or three when it is in the second loop) points around (cf. Figure 5), and then go to step 8.
8) Conduct half pixel search.

4. Results and Comparisons

The DSWSS was compared with FS, 3SS and 4SS under the same condition. We used MPEG-4 encoder VM8.05. Maximum search area is (±15, ±15), and six test video sequences involving different kinds in table 1. We didn't include DS algorithm in our simulation, but we can compare it with DSWSS through the results in document [6] in some ways.

Table 1. Brief description of the test video sequences

Sequence	Size	Kinds of motion
Miss American	QCIF	Low amount of movement
Coast-Guard	QCIF	Medium spatial detail and low amount of movement or vice versa
Foreman	QCIF	object translation and panning
Carphone	QCIF	fast object translation
Flower Garden	CIF	fast panning with high motion activity
Mobile & Calendar	SIF	High spatial detail and medium amount of movement or vice versa

For each sequence, every other frame from the first 80 frames are selected, and only I and P frames are included. The average search points, average MSE and average PSNR are shown in Table 2 to Table 4.

Table 2. Average search points

Sequence	FS	3SS	4SS	DSWSS
Miss American	961	33	17.24	10.06
Coast-Guard	961	33	20.12	13.26
Foreman	961	33	19.65	14.63
Carphone	961	33	18.15	13.63
Flower Garden	961	33	20.04	13.68
Mobile & Calendar	961	33	19.29	13.50

Table 3. Average MSE

Sequence	FS	3SS	4SS	DSWSS
Miss American	8.26	8.65	8.49	8.26
Coast-Guard	81.57	113.56	106.22	83.55
Foreman	47.71	73.35	53.77	50.83
Carphone	41.25	46.10	44.84	42.37
Flower Garden	254.86	548.09	334.76	262.27
Mobile & Calendar	348.59	556.96	402.50	355.30

Table 2 shows that the three fast algorithms need much fewer search points than FS, and DSWSS needs still fewer than 3SS and 4SS. Table 3 and 4 show that in terms of average MSE and PSNR, DSWSS is more accurate than either 3SS or 4SS and is very close to FS. From the figures we can also conclude that DSWSS is more robust when tested by different kinds of video sequences. From the results of document [6], we can also conclude that DSWSS is better than DS.

Table 4. Average PSNR

Sequence	FS	3SS	4SS	DSWSS
Miss American	44.13	44.05	44.07	44.12
Coast-Guard	31.71	31.27	31.39	31.70
Foreman	34.61	33.48	34.04	34.39
Carphone	37.23	36.83	36.89	37.06
Flower Garden	34.56	32.30	33.97	34.14
Mobile & Calendar	25.66	24.11	25.19	25.60

In a word, DSWSS can reach almost the same search accuracy as FS through very little computation, and it is more efficient, more effective and more robust than the other fast search algorithms.

Acknowledgements

This research was partially supported by the NSFC Grant 69935010 and 69974005, the National High Technology Program, the New Star Project in Science and Technology of Beijing and Foundation for University Key Teacher by the Ministry of Education.

Reference

1. MPEG-4 Video (ISO/IEC 14496-2) VM8. 05，1997
2. A. Murat Tekalp. Digital video processing, Prentice Hall, 1995
3. W. Lee, J. F. Wang, J. Y. Lee, and J. D. Shie. "Dynamic search-window adjustment and interlaced search for block-matching algorithm". IEEE Trans. CASVT. Vol. 3，pp. 85-87, Feb, 1993
4. R. Li，B. Zeng and M. Liou, " A new three-step search algorithm for block motion estimation" , IEEE Trans. CASVT，Vol. 4，No. 4，pp. 438-442，Aug. 1994
5. L. M. Po and W. C. Ma，"A novel four-step algorithm for fast block motion estimation", IEEE Trans. CASVT，Vol. 6，No. 3，pp313-317, Jun. 1996.
6. Shan Zhu and K. -K. Ma, "A new diamond search algorithm for fast block matching motion estimation," *Int'l. Conf. on Information, Commun. And Signal Proc.* (ICICS'97), Singapore, pp. 292-296, 9-12 Sept. 1997.
7. ISO/IEC JTC1/SC29/WG11 N2932, "MPEG-4 Video Verification Model version 14.0," pp. 301-303, October 1999.
8. "Text of MPEG-4 OM v1.0 for part 7", ISO/IEC JTC1/SC29/WG11 N3324, Noordwijkerhout, March 2000

Determining Motion Components Using the Point Distribution Model

Ezra Tassone, Geoff West and Svetha Venkatesh

School of Computing
Curtin University of Technology
GPO Box U1987, Perth 6001 Western Australia
Ph: +61 8 9266 7680 Fax: +61 8 9266 2819
{tassonee,geoff,svetha}@computing.edu.au

Abstract. The Point Distribution Model (PDM) has proven effective in modelling variations in shape in sets of images, including those in which motion is involved such as body and hand tracking. This paper proposes an extension to the PDM through a re-parameterisation of the model which uses factors such as the angular velocity and distance travelled for sets of points on a moving shape. This then enables non-linear quantities such as acceleration and the average velocity of the body to be expressed in a linear model by the PDM. Results are shown for objects with known acceleration and deceleration components, these being a simulated pendulum modelled using simple harmonic motion and video sequences of a real pendulum in motion.

Keywords: Point Distribution Model, temporal model, computer vision, shape analysis

1 Introduction

A number of computer vision techniques have been devised and successfully used to model variations in shape in large sets of images. Such models are built from the image data and are capable of characterising the significant features of a correlated set of images whilst still allowing for significant divergences without constructing overly complicated or specific representations. These are known as deformable models.

One specific model is the Point Distribution Model (PDM) [4] which builds a deformable model of shape for a set of objects based upon the coordinates of the object in an image and the use of Principal Component Analysis (PCA) on this data. Importantly, the PDM provides a linear model of the variations in shape. This paper proposes an extension to the PDM in which the model is not constructed solely from image data. We propose a re-parameterisation of the model and determine quantities such as acceleration and velocity based upon an image sequence of a moving object. Often these variables are derived from specialised functions increasing the complexity of the model required. The approach described in this paper differs as it allows these non-linear factors to be described compactly and accurately by the linear PDM. Further, it explicitly

T. Tan, Y. Shi, and W. Gao (Eds.): ICMI 2000, LNCS 1948, pp. 363-370, 2000.
© Springer-Verlag Berlin Heidelberg 2000

describes a temporal sequencing of shapes and not only the range of variation within those shapes.

2   Background

The Point Distribution Model is built from the coordinates of objects in a set of images and then performing PCA on this data. This enables the class of objects described by the model to be linearised into the form of a mean shape plus a number of modes of variation which show the deviations from the mean for all points on the object [4]. The use of PCA enables the dimensionality of the model to be reduced as only the most significant modes of variation are added into the model, those which will allow for a high proportion of the variance in the shapes to be represented. Additionally, new shapes may be derived from the model and also techniques such as the Active Shape Model [3] used to fit shape parameters to previously unseen images.

The PDM has typically been used on static images such as that found in medical imagery However tracking using the PDM has been developed for a walking person [2], using the control points of a B-spline of the boundary of the moving person and a Kalman filter for determining shape parameters. The PDM has also been used to track and classify sequences of hand gestures [1].

Further applications of the model involving the re-parameterisation of the PDM has also been achieved in the domain of motion. One such application is the Cartesian-Polar Hybrid PDM which specifically considers objects which may pivot around an axis [5]. Points which undergo angular motion are mapped into polar coordinates, while other points remain as Cartesian coordinates and this allows for a more accurate representation of the motion. Other research has described flocks of animal movement by adding parameters such as flock velocity and relative positions of other moving objects in the scene to the PDM (in addition to the standard image coordinates) [6]. All of these require heuristics or explicit models of shape to aid the PDM.

3   The Point Distribution Model

3.1   Standard linear PDM

The construction of the PDM is based upon the shapes found in a training set of data [4]. Each shape is modelled as a set of n "landmark" points placed on the object, which generally indicate significant features of the modelled shapes. Hence, each shape is represented as a vector of the form (for the 2D version of the model):

$$\mathbf{x} = (x_0, y_0, x_1, y_1, \ldots, x_{n-1}, y_{n-1})^T$$

To derive proper statistics from the set of training shapes, the shapes are aligned using a weighted least squares method, in which all shapes are translated, rotated and scaled to correspond with each other. From the set of aligned shapes, the mean shape is calculated (N_s being the number of shapes in the training set):

$$\overline{\mathbf{x}} = \frac{1}{N_s} \sum_{i=1}^{N_s} \mathbf{x}_i \tag{1}$$

The difference dx_i of each of the aligned shapes from the mean shape is then taken and the covariance matrix $\mathbf{S}$ derived:

$$\mathbf{S} = \frac{1}{N_s} \sum_{i=1}^{N_s} \mathbf{dx}_i \mathbf{dx}_i^T \tag{2}$$

The modes of variation of the shape are then found from the derivation of the unit eigenvectors, $\mathbf{p}_i$, of $\mathbf{S}$:

$$\mathbf{Sp}_i = \lambda_i \mathbf{p}_i \tag{3}$$

The most significant modes of variation are aligned to the eigenvectors with the largest eigenvalues. The total variation of the training set is derived from the sum of all the eigenvalues, and the smallest set of eigenvectors that will describe a certain proportion (typically 95% or 99%) of the variation is chosen.

Hence, any shape, $\mathbf{x}$, in the training set can be derived from the equation:

$$\mathbf{x} = \overline{\mathbf{x}} + \mathbf{Pb} \tag{4}$$

where $\mathbf{P} = (\mathbf{p}_1 \mathbf{p}_2 \ldots \mathbf{p}_m)$ is a matrix with columns containing the m most significant eigenvectors, and $\mathbf{b} = (\mathbf{b}_1 \mathbf{b}_2 \ldots \mathbf{b}_m)^T$ being the set of linearly independent weights associated with each eigenvector. The weights would generally lie in the limits (for the model representing 99% of the variation):

$$-3\sqrt{\lambda_i} \leq b_i \leq 3\sqrt{\lambda_i} \tag{5}$$

3.2   Modified PDM for Motion Components

While it would be possible to simply use the standard PDM in order to build some model of deformable shape for a sequence of images of a moving body, this paper proposes instead a re-parameterisation of the PDM. In this method, the image coordinates of the body are not directly used as input for the PDM but are processed and used to derive other measures for input into the model. By doing this, factors more appropriate for the analysis of motion can be used hence increasing the utility of the model.

To build the PDM, a number of frames of an object in motion are taken. Each new movement of the body from frame to frame represents a new "shape" for the training set of the model. After extracting the boundary of the object, a subset of n points is selected for use in developing the model. However as the focus of this work is to describe motion components, the temporal sequencing of the shapes and the relative movement of the points on the shapes is what is used to to re-parameterise the PDM.

To achieve this a set of three frames is considered at a time with the movement of a point from the first to second frame being one vector and the movement from the second frame to the third being a second vector. From these vectors, the relevant motion components and thus parameters for the PDM can be calculated. Clearly there are many potential motion components that can be modelled, however this extension will focus on the following three components:

1. **Angular velocity**, $\Delta\theta_0$ — the change in angle between the vectors, with a counter-clockwise movement considered a positive angular velocity and a clockwise movement a negative velocity.
2. **Acceleration**, a_0 — the difference in the Euclidean norm between the vectors with the norm of the first vector being v_{i-1} and that of second v_i ie. $v_i - v_{i-1}$.
3. **Linear velocity**, v_0 — this is the norm of the second vector v_i.

The acceleration and velocity are measured on a frame-by-frame basis, that is the velocity measures the change in displacement per frame and the acceleration component hence measures the change in velocity per triple of frames. This data is calculated for every one of the n points of the object leading to a new vector representation for the PDM:

$$\mathbf{x} = (\Delta\theta_0, a_0, v_0, \ldots, \Delta\theta_{n-1}, a_{n-1}, v_{n-1})^T$$

This process is repeated for all triples of consecutive frames in the sequence. In this way all frames in the sequence are included, however this reduces the number of vectors in the training set to $n - 2$. After this adjustment to the model, the PDM can be built in the standard way. A mean vector will be derived plus a set of eigenvectors representing the deviation from the mean which are all weighted by the limits of the eigenvalues. However, the effect of the re-parameterisation ensures that it is the mean angular velocity, acceleration and linear velocity that are calculated in the mean vector. The modes of variation hence represent the deviations from the mean for a set of motion components, giving a compact and linear model. It is also significant that this characterisation encapsulates the temporal sequencing of the motion with the changes in parameters modelled on a frame to frame basis. This is substantially different from the PDM which incorporates no temporal information in the model. This property of the modified PDM would also allow for the derivation of absolute shape coordinates for a given frame via a transformation derived from the constructed PDM model applied to the absolute coordinates of the previous frame.

4   Experimental Results

The model was tested with both a simulated (modelled under assumptions of simple harmonic motion) and a real-world pendulum. Results of these experiments are presented here.

4.1   Simulated Pendulum

To simulate the motion of the pendulum, the motion is modelled as in Figure 1 with a rotation through an angle θ, a string of length l and an initial displacement of A. Due to the model, all displacement is along the x-axis and y remains constant at a value of 0. Using the following equations for an ideal pendulum :

$$\ddot{\theta} = \frac{-g}{l} \sin\theta \qquad \text{where for small } \theta, \sin\theta \approx \theta \tag{6}$$

$$\frac{x}{l} = \tan\theta \Rightarrow \theta = \arctan(\frac{x}{l}) \tag{7}$$

where $x(0) = A$ and $\theta(0) = \arctan(\frac{A}{l})$. Using second-order differential equations, the displacement x as a function of time can be derived:

$$x(t) = l\tan(\arctan(\frac{A}{l})\cos\left[\sqrt{\frac{g}{l}}t\right]) \tag{8}$$

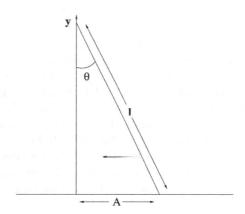

Fig. 1. Motion of Simulated Pendulum

Various simulations were run over a specified time period with a sampling rate of 0.2 units of time. Clearly the smaller the units of time sampled the more accurately the displacement of the pendulum can be captured. An example of simulated pendulum motion with an initial displacement of 10 units and a pendulum length of 10 is now presented. This time period of this motion was 6.6 units of time, with each sample being taken at 0.2 units.

A modified PDM was constructed for this sequence which consisted of the pendulum swinging from an initial displacement to a maximum negative displacement and then almost back to the starting position again. Typically, an overall PDM is constructed for all images however as pendulum movement has definitive acceleration and deceleration components it was more beneficial to construct a set of PDMs, each for a separate segment of the motion. Hence A

was designated as the initial position and the final position, B the central position at $x = 0$ and C the position at the other extreme of motion (although due to sampling the positions are not completely accurate). The PDMs for the segments are presented in Table 1 with the models designed to explain 95% or more of the variance. Note that the mean angular velocity is always zero as no change in angle occurs due to the displacement only taking place along the x-axis.

Table 1. Mean values and number of modes of variation generated for all PDMs

Segment	$\overline{\Delta\theta}$	$\overline{a}$	$\overline{v}$	modes
$A \rightarrow B$	0	0.1792	1.4001	1
$B \rightarrow C$	0	-0.1907	1.1905	1
$C \rightarrow B$	0	0.1680	1.4176	1
$B \rightarrow A$	0	-0.2023	1.1586	1
$A \rightarrow C$	0	-0.0053	1.3128	2
$A \rightarrow C \rightarrow A$	0	0.0048	1.2534	2

As can be seen from the table, the PDM has accurately modelled the motion in the segments showing a positive acceleration for the first segment approaching the centre position and the segment swinging back away from the extreme negative x position. Deceleration is also modelled correctly in the inverse segments. Average velocity for each segment has also been determined, and the components have a similar magnitude indicating correct modelling. However, one possible flaw in the model is that while the PDM correctly averages an acceleration of zero in those models having both accelerated and decelerated motion also increases the number of modes of variation required. The model compensates with more modes to generate an accurate representation.

4.2   Real Pendulum

The model was also validated against the motion of a real-world pendulum, where a number of frames of motion were captured on video and then image processing techniques applied to generate acceptable inputs for the development of the PDM. The grayscale images were thresholded so that only the mass at the end of the string was segmented from the scene, and these processed images then chaincoded to generate the boundary of the mass. To further simplify the analysis, the center of the mass was then calculated and this point used as the coordinates of the object for processing into the motion components. Two trials are presented, again using A as the initial and final position, B as the central position and C the other extreme displacement. However the motion differs from the simulated scenario in that angular velocity is now a changing component and also in that the real pendulum suffers energy loss over time reducing the amplitude of its motion. Again, the PDM correctly illustrates the periods of acceleration, deceleration and zero acceleration as well as average velocity. The average angular velocity has also been computed, with a change in direction also bringing about a sign change in the angular velocity for Pendulum 1 but not Pendulum 2. Finally, the eigenvalues of the constructed PDMs were large indicating high variance in the models. A future issue will to scale the factors to reduce the variance, this has been explored by [6].

Table 2. Pendulum 1 - Mean values and number of modes of variation generated for all PDMs

Segment	$\overline{\Delta\theta}$	$\overline{a}$	$\overline{v}$	modes
$A \to B$	0.0474	2.7536	15.2530	2
$B \to C$	0.0533	-3.0978	15.8466	1
$C \to B$	-0.0532	2.9729	18.3597	1
$B \to A$	-0.5842	-3.2524	13.9831	1
$A \to C$	0.0515	0	16.1704	2
$A \to C \to A$	-0.1180	-0.0621	16.0574	2

Table 3. Pendulum 2 - Mean values and number of modes of variation generated for all PDMs

Segment	$\overline{\Delta\theta}$	$\overline{a}$	$\overline{v}$	modes
$A \to B$	0.0489	5.3087	21.7405	2
$B \to C$	0.0336	-3.1498	17.5748	2
$C \to B$	0.0091	3.8045	23.9307	2
$B \to A$	1.0064	-3.1461	17.8907	2
$A \to C$	0.0436	0.0739	19.9164	2
$A \to C \to A$	0.1238	0.0370	19.6889	2

5 Conclusion

This paper has described an adaptation of the Point Distribution Model for a sequence of images taken of a moving object. Image data is used to derive quantities for inclusion into the model, these being angular velocity, acceleration and linear velocity. The construction of a PDM from these factors thus leads to a linear model of the motion components, which describes mean values for these quantities along with deviations from the mean via the modes of variation of the PDM. The model also incorporates temporal sequencing in the modelling of changes in parameters. Results show that the modified PDM can accurately describe the motion components of both a simulated and real pendulum. Further work will aim to build the model of the motion components and determine the most appropriate parameters for an unseen sequence of motion via a technique such as the Active Shape Model. Research will also be done in building different models for sequences of motion and then using appropriate techniques to choose the model that best represents the motion.

References

[1] T. Ahmad, C. Taylor, A. Lanitis, and T. Cootes. Tracking and recognising hand gestures using statistical shape models. In *Proceedings of the British Machine Vision Conference, Birmingham, UK*, pages 403–412. BMVA Press, 1995.

[2] A. Baumberg and D. Hogg. An efficient method of contour tracking using Active Shape Models. In *1994 IEEE Workshop on Motion of Non-rigid and Articulated Objects*, 1994.

[3] T. Cootes and C. Taylor. Active shape models - 'smart snakes'. In *Proceedings of the British Machine Vision Conference, Leeds, UK*, pages 266–275. Springer-Verlag, 1992.

[4] T. Cootes, C. Taylor, D. Cooper, and J. Graham. Training models of shape from sets of examples. In *Proceedings of the British Machine Vision Conference, Leeds, UK*, pages 9–18. Springer-Verlag, 1992.

[5] T. Heap and D. Hogg. Extending the Point Distribution Model using polar coordinates. *Image and Vision Computing*, 14:589–599, 1996.

[6] N. Sumpter, R. Boyle, and R. Tillett. Modelling collective animal behaviour using extended Point Distribution Models. In *Proceedings of the British Machine Vision Conference, Colchester, UK*. BMVA Press, 1997.

Automatic Recognition of Unconstrained Off-Line Bangla Handwritten Numerals

U. Pal and B. B. Chaudhuri

Computer Vision and Pattern Recognition Unit, Indian Statistical Institute
203, B. T. Road, Calcutta - 700 035, India.
{umapada,bbc}@isical.ac.in

Abstract. This paper deals with an automatic recognition   method for unconstrained off-line Bangla handwritten numerals. To take care of variability involved in the writing style of different individuals a   robust scheme is presented here. The scheme is based on new features obtained from the concept of water overflow from the reservoir as well as topological and statistical features of the numerals. If we pour water from upper part of the character, the region where water will be stored in the character is imagined as a reservoir of the character. The direction of water overflow, height of water level when water overflows from the reservoir,   position of the reservoir with respect to the character bounding box, shape of the reservoir etc. are used in the recognition scheme. The proposed   scheme is tested on   data collected from different individuals of various background and we obtained an overall recognition accuracy of about 91.98%.

1 Introduction

Recognition of handwritten numerals has been a popular topic of research for many years[9][10]. It has many application potentials such as automatic postal sorting, automatic bank cheque processing, Share certificate sorting, recognition of various other special forms etc. The earlier systems that had been developed were confined to recognize only constrained handwritten numerals. But, modern systems aim at recognizing numerals of varied size and shape. Several commercial systems on Latin script based numeral recognition are available in the market [1][9][11].

Most of the previous studies on numeral recognition are reported for non-Indian languages, like English, Arabic, Chinese, Japanese and Korean [1][9]. In these papers various approaches have been proposed for numeral recognition. Some techniques use structural information, such as stroke features, junctions, concavities, convexities, endpoints, contours etc. [2][3][9]. Statistical information (for example moments) are also used by some researchers [6]. In some pieces of work Fourier and Wavelet Descriptors have been applied [7][12]. Among others, Syntactic approach [6], Relaxation matching based approach [8], Neural networks and Fuzzy rules are reported for numeral recognition [4].

To the best of our knowledge, there is only one piece of work on Bangla numeral recognition [5]. In that work, curvature features like curvature maxima,

T. Tan, Y. Shi, and W. Gao (Eds.): ICMI 2000, LNCS 1948, pp. 371-378, 2000.
© Springer-Verlag Berlin Heidelberg 2000

curvature minima and inflectional points etc., detected after thinning the numeral are used for the recognition. A two-stage feed forward neural net based recognition scheme is used for classification. The size of data set was very small. The experiment was done on 10 samples of each numeral and 90% recognition accuracy was obtained.

In this paper, we propose a thinning and normalization free automatic recognition scheme for unconstrained off-line Bangla handwritten numerals. Since different individuals have different writing styles, the topological and other properties of numerals may vary with writing. To take care of variability in writing styles, the scheme is based on new features obtained from the concept of water overflow from the reservoir as well as topological and statistical features of the numerals. The direction of water overflow, height of water level when water overflows from the reservoir, position of the reservoir with respect to the character bounding box, shape of the reservoir etc. are used in the recognition scheme. Using these features a tree classifier is employed for recognition.

Bangla is the second-most popular language and script in Indian sub-continent and fifth-most popular language in the world. About 200 million people of eastern India and Bangladesh use this language. To get an idea about the actual shapes of Bangla numerals, a set of printed Bangla numerals is shown in Fig.1. Note that two numerals zero(০) and four (৪) are identical with zero and eight of Roman script.

One	Two	Three	Four	Five	Six	Seven	Eight	Nine	Zero
১	২	৩	8	৫	৬	৭	৮	৯	০

Fig.1. Bangla printed numerals are shown.

2 Preprocessing

Document digitization for the experiment of the system has been done by a flatbed scanner (manufactured by HP, Model no ScanJet 4C). The digitized images are in gray tone (256 level) and we have used a histogram based thresholding approach to convert them into two-tone images. The digitized image may contain spurious noise pixels and irregularities on the boundary of the characters, leading to undesired effects on the system. Also, to improve recognition performance, broken numerals should be connected. For preprocessing we use the method described by Cheng and Yan [3].

A connected component labeling scheme is used to get some topological features and to define the bounding box (minimum upright rectangle containing the component) of the numeral. Before going to actual recognition scheme we check whether a component is an isolated numeral or touching numeral (when two or more numerals touch one another and a single connected component is generated, we call this component as touching numerals). To check a touching numeral the following points are observed: (i) if the number of hole in the component is 3 or more the component is a touching numeral; (ii) if the number of hole in the component is 2 and if the holes are side-by-side then the component is a touching numeral. It may be noted that in Bangla there is a character (four) which has two holes but the holes are in top-down   position; (iii) if   (i) or (ii)   are not true then we use the   width information of touching numerals.   In general, a touching numeral will have high bounding box width with respect to height, large number of boarder pixels per unit height as well as high accumulated curvature per unit height. A discriminate plane, based on these feature, is used to identify isolated and touching numerals. If the component is identified as isolated then our recognition scheme is applied on it.

3 Feature Selection and Detection

We consider features obtained from the concept of the water overflow as well as topological and statistical features for recognizing Bangla numerals. The features are chosen with the following considerations   (a) Independence of various writing styles of different individuals (b) Robustness, accuracy and simplicity of detection. The features are used to design a tree classifier where the decision at each node of the tree is taken on the basis of presence/absence of a particular feature.

The important features considered in the scheme are based on   (i) existence of holes and its number (ii) position of holes with respect to its bounding box (iii) ratio of hole length to height of the numerals (iv) center of gravity of the holes (v) number of crossings in a particular region of the numeral (vi) convexity of hole, etc.

If in a component there is no hole we use the concept of water overflow from reservoir to get holes and other topological properties. The principle is as follows. If we pour water from the above of the numeral, the position where water   will store is considered as a reservoir. The shape of the reservoir in the numeral is considered as a hole. The direction of water overflow, height of water level when water overflows from the reservoir, position of the reservoir with respect to the character bounding box, shape of the reservoir etc. are noted for the use in the classification tree. For illustration please see Fig.2.

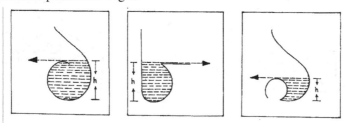

Fig.2. Reservoir (hole) and   water overflow   direction from the reservoir(hole) is shown. Reservoir area in the numeral is shown by small line segments. Here 'h' is the reservoir (hole) height.

For the detection of the features a component labeling algorithm is executed. Most of the above features are detected during the time of component labeling.

Because of page limitation details of feature detection procedure is not provided here. Detection procedure of some features will be discussed in the recognition section.

4 Recognition of Handwritten Numerals

As mentioned earlier, we consider features obtained from the concept of the water overflow as well as topological and statistical features for recognizing Bangla numerals. In our scheme, the features obtained from the concept of water overflow from reservoir play an important role in numeral recognition. A tree classifier, using these features, is employed for the recognition of handwritten numerals. A part of the decision tree is shown in Fig.3.

The design of a tree classifier has three components: (1) a tree skeleton or hierarchical ordering of class labels, (2) the choice of features at each non-terminal node, and (3) the decision rule at each non-terminal node.

Ours is a binary tree where the number of descendants from a non-terminal node is two. Only one feature is tested at each non-terminal node for traversing the tree. The decision rules are mostly binary e.g. presence/absence of the feature. To choose the features at a non-terminal node we have considered the feature which can represent the tree in a optimum way.

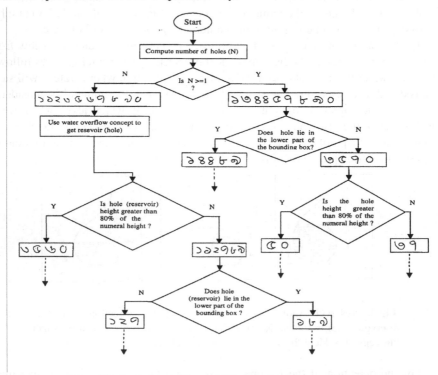

Fig.3. A part of the tree classifier is shown.

In Bangla printed numeral set, there are five numerals which have hole. These numerals are 8(four), ৫(five), ৭(seven), ৮(eight) and ০(zero). But in handwritings we may not see the holes in these numerals. Conversely, in some handwritings we may get holes in some numerals where no hole is there in their printed form. For example, see Fig.4 where typical examples of handwritten Bangla numerals are shown. It may be noted that   some persons write the numeral ১(one) with a hole while others write it without hole.   To make the system robust we try to use some features which are present in almost all styles of handwritings.

From a handwritten Bangla data set of 10000 numerals we noted that most people try to make a hole in the numerals if the actual numeral has a hole. For example, in numeral ৮(eight) there is a hole in the lower half position. From our experiment on the above data set we note that 65% individual give a hole   in the numeral and   in the writing of 35% individuals there are no hole but a hole-like structure is present. Here we use   the   feature based on the   principle of water overflow from reservoir by which we can detect hole-like structure.

Fig.4. Typical examples of Bangla handwritten numerals.

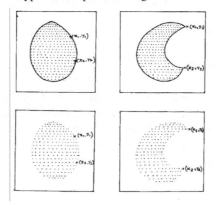

Fig.5. Example of recognition using convex set property. Here dots in the numeral represent the hole. (x_1, y_1) and (x_2, y_2) are the rightmost points of the first and third strip of the hole. For the numeral zero the line segment obtained by joining these two points lies entirely inside the hole whereas for five it goes outside the hole.

To give an idea about the use of this hole-like information in the recognition process consider the numerals shown in Fig.2. Because of their common feature these numerals belong in the same node of the classification tree of Fig.3. Note that the position of reservoir for all these numerals is lower half of the bounding box and because of this feature they have been classified in the same group. For further classification the water overflow direction from reservoir is noted. For the numerals one(১) and nine(৯) the overflow direction is left whereas it is right for the numeral eight(৮). For the classification of the numerals nine and one we use the number of crossing in a portion where discriminate shape of these numerals is observed . We consider left portion of the hole (reservoir) and compute row-wise number of crossing in the left side of the hole(reservoir). For the numeral one (১) the maximum number of crossing in row-wise scans will be only one whereas for nine (৯) the maximum number of crossings will be two.

To recognize some numerals the ratio of hole height to the numeral height is used. For example, consider the subgroup containing numerals ৫(five) and ০(zero). Here the ratio of hole height to numeral height is more than 0.80.   For the classification of these numerals we use convex set property of the points inside the hole. The points inside the hole of the numeral ০(zero) is nearly convex whereas points inside the hole of the numeral ৫(five) are highly non-convex. For example, see Fig.5. To test the near convexity we have chosen two points on these numerals. We divide the hole of the numeral into three horizontal strips and find the rightmost points $(x_1 y_1)$ among the points of the first strip of the hole and $(x_2 y_2)$ from the points of the third strip of the hole. These points are shown by $(x_1 y_1)$ and $(x_2 y_2)$ in Fig.5. The line segment obtained by joining these two points always lie on the hole for the numeral zero whereas it goes outside the hole for the numeral five. For faster recognition we do not consider all points on the line segment obtained by joining the points $(x_1 y_1)$ and $(x_2 y_2)$. We test only the points (Xm_i , Ym_i) obtained by the following equations.

$$Xm_i = [(m_i x_1 + (d-m_i)x_2]/d \qquad\qquad Ym_i = [(m_i y_1 + (d-m_i)y_2]/d$$

Where $mi = 2, 4, \text{---------} d-2$ if d is even
$= 2, 4, \text{---------} d-1$ if d is odd,

and d is the Euclidean distance between the points $(x_1 y_1)$ and $(x_2 y_2)$.

Many other features are used in the recognition scheme. Because of the page limitation their details are not provided here. Note that in our classification scheme we do not use any stroke-like features because in hand-written text, stroke features may change widely from person to person.

5 Results and Discussion

We applied our separation scheme on 10000 different numerals obtained from different individuals of different professionals. We noted that the data sets contain varieties of writing styles. We noted that the accuracy rates of the recognition scheme is 91.98%.   The rejection rate of the system is about 2%. This scheme does not depend on the size and style of the handwritten numerals.   The confusion rates of the numerals obtained after the experiment on 10000 numerals are presented in Table.1.

From the   Table   it   may be noted that the numeral ৮(eight) has maximum recognition rate of 97.06%. This is because of its different water overflow direction as compare to its group numerals one or nine. Also, we noted that the most confusing numeral pair is ৩(three) and ৬(six). From the experiment we noted that about 8.49% cases these numerals confuse. Their similar shapes rank their confusion rate at the top position. Next confusion pair is one and two and their confusing rate is 7.9%. Some typical images where confusion noted during experiment are given in Fig.6.

Table 1. Recognition result obtained from the experiment on 10000 handwritten numerals.

Digit is recognized as $\longrightarrow$

	১	২	৩	8	৫	৬	৯	৮	৭	০
১	91.30	2.51	1.51	0.00	0.00	0.00	0.00	1.01	1.67	2.01
২	5.39	91.02	1.20	0.00	0.00	0.00	0.00	0.00	2.40	0.00
৩	1.61	0.00	89.78	1.08	1.08	3.23	0.00	0.00	2.69	0.54
8	1.13	0.00	0.56	94.92	0.00	0.00	0.00	2.26	0.00	1.13
৫	0.47	0.00	1.87	0.93	89.72	2.80	0.93	0.00	0.00	3.27
৬	1.44	0.00	5.26	0.00	3.35	88.52	0.00	0.00	0.00	1.44
৯	0.48	0.00	2.90	1.45	0.00	0.00	95.17	0.00	0.00	0.00
৮	0.82	0.00	0.00	1.15	0.00	0.00	0.00	97.06	0.96	0.00
৭	4.26	1.42	0.71	0.00	2.84	0.00	0.00	0.00	90.78	0.00
০	2.10	0.00	1.40	0.00	2.80	2.10	0.00	0.00	0.00	91.61

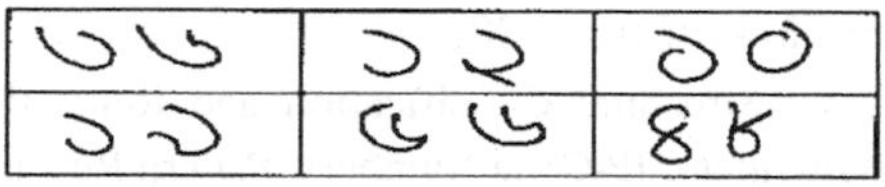

Fig.6. Example of some confusing Bangla handwritten numeral pairs.

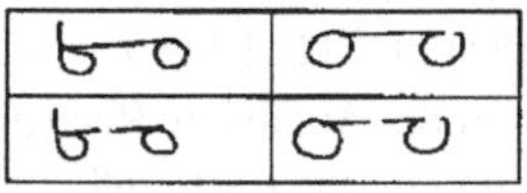

Fig.7. Example of touching Bangla numerals and their segmented form.

Methods based on thinning may face many problems due the protrusions obtained after thinning. Also, the method based on contour following approach will not work properly because of various writing styles. For example, if a touching numeral is segmented in a form shown in Fig.7, then the method based on contour tracing may fail. Our recognition method will properly recognize these segmented numerals without any modification in the system.

We do not use any normalization or thinning mechanism. We use here some simple feature which are easy to compute and hence our system is very fast. The drawback of the proposed method is that it will fail if there is a break on the contour used as the boundary of the reservoir. In that case water cannot be filled up properly to get reservoir (hole) and hence misrecognition will occur. But many other methods (for example those based on contour following) will also fail for this type of situation. We note that such cases are very rare. We can use smearing technique to remove some of these situations where size of the break point area is small.

References

1. Amin A.: Off-Line Arabic Character Recognition: The State of the Art. Pattern Recognition. **31** (1998) 517-530

2. Cao J., Ahmadi M., Shridhar M.: Recognition of Handwritten Numerals with Multiple Feature and Multistage Classifier. Pattern Recognition. **28** (1995) 153-160

3. Cheng D., Yan H.: Recognition of Broken and Noisy Handwritten Characters Using Statistical Methods Based on a Broken Character-Mending Algorithm. Optical Engineering. **36** (1997) 1465-1479

4. Chi Z., Wu, j., Yan H.: Handwritten Numeral Recognition Using Self-Organizing Maps and Fuzzy Rules. Pattern Recognition. **28** (1995) 59-66

5. Dutta A., Chaudhuri S.: Bengali Alpha-Numeric Character Recognition Using Curvature Features. Pattern Recognition. **26** (1993)1757-1770

6. Duerr B., Haettich W., Tropf H., Winkler G.: A Combination of Statistical and Syntactical Pattern Recognition Applied to Classification of Unconstrained Handwritten Numerals. Pattern Recognition. **12** (1980) 189-199

7. Granlund G. H.: Fourier Preprocessing for Handprinted Character Recognition. IEEE Trans on Computers. **21** (1972) 195-201

8. Lam L., Suen C. Y.: Structural Classification and Relaxation Matching of Totally Unconstrained Handwritten ZIP-Code Numbers. Pattern Recognition. **21** (1988) 19-31

9. Mori S. Suen C. Y., Yamamoto K.: Historical Review of OCR Research and Development. Proceedings of the IEEE. **80** (1992) 1029-1058

10. Suen C. Y.: Computer Recognition of Unconstrained Handwritten Numerals. Proceedings of the IEEE. **80** (1992) 1162-1180

11. Suen C. Y., Berthod M., Mori S.: Automatic Recognition of Handprinted Characters. The state of the Art. Proceedings of the IEEE. **68** (1980) 469-483

12. Wunsch P., Laine A. F.: Wavelet Descriptors for Multiresolution Recognition of Hand-printed Characterss. Pattern Recognition. **28** (1995) 1237-1249

A Novel Algorithm for Handwritten Chinese Character Recognition

Feng Qi, Minghua Deng, Minping Qian, and Xueqing Zhu

Department of Mathematics, School of Mathematical Sciences
Peking University, Beijing, 100871, P.R.China
{qf_cn, dengmh}@263.net

Abstract. 2-D HMM methods have recently been applied to handwritten Chinese character recognition(HCCR) with much practical prospect, however, the time consuming is a great obstacle for its widely using. To overcome this weakness a novel block-based ICM algorithm, which will decode image block by block other than pixel by pixel, is proposed. Experiments shows that it can work much better than the traditional one for HCCR with a little higher accuracy and much higher speed of recognition. It offers a great potential for HCCR when using truly-2D HMM.

1 Introduction

There have been a large number of methods for on-line Chinese character recognition(CCR) reported recent years[4, 5]. Some of them almost have made their own practical system for business. However, off-line handwritten CCR still have a lot of work to do. One of the most difficult problems in the field is to construct or find an appropriate model for it. While considering its practicability, that is, its accuracy and speed, it become even more difficult.

A survey of off-line CCR reveals that Chinese have the following characteristics: (1). Chinese character is a kind of structural character, this means that it is composed of smaller units such as strokes in two-dimensional space while western alphabets can make a detour on this problem; (2). Chinese has a large alphabet, this means there is a large number of different categories which make the recognition even more difficult; (3). off-line characters can only present a binary image without any other information, such as stroke segment; (4). there might be large variability among different writers, which makes the feature-extraction difficult.

Due to the success of speech recognition systems using hidden Markov model (HMM)[6], it is natural to extend HMM from 1-D to 2-D for character recognition. However the previous attempts have not yet achieved satisfactory results because of the difficulty of constructing a real 2-D model with appropriate computation complexity[7] and utilizing structural information completely. Since Chinese has strong structural characteristics, any local feature may have certain continuity or uniformity, this means pixels with similar feature (noted as block)

T. Tan, Y. Shi, and W. Gao (Eds.): ICMI 2000, LNCS 1948, pp. 379-385, 2000.

can be processed as a unit. In this paper a novel algorithm which is based on 2-D HMM and block technique is proposed. It has appropriate computational complexity and take full advantage of structural information.

2   Preliminary of Markov Random Field

It is well known that there are two kind of causal 2-D Markov chains in the literature: the Markov mesh random field(MMRF) and the nonsymmetric half plane(NSHP) Markov chain[8]. We will adopt the latter and construct hidden NSHP Markov chain model for character pattern representation.

2.1   Markov Random Field(MRF) and NSHP Markov Chain

Let us consider a random field $X = \{X_{m,n}\}$ defined on a finite $M \times N$ integer lattice $L = \{(m, n) : 1 \leq m \leq M, 1 \leq n \leq N\}$ and a system of neighborhoods $\Omega = \{N_l, l \in L\}$ on L, such that each N_l consists of a certain number $| N_l |$ of neighbors of point l, not including l. Denote $X(N_l) = \{X_k, k \in N_l\}$

Definition 2.1: $X = \{X_l, l \in L\}$ is a MRF if

$$P[X_l = x | X_k, k \in L, k \neq l] = P[X_l = x | X(N_l)] \tag{1}$$

That is, for each X_l, the dependence between the random variable is determined only by random variables in its neighborhood N_l.

Definition 2.2: $X = \{X_{i,j}, (i, j) \in L\}$ is a NSHP Markov chain model if

$$P[X_{i,j} = x | X_{k,l}, (k, l) \in N_{i,j}] = P[X_{i,j} = x | X_{k,l}, (k, l) \in \Psi_{i,j}] \tag{2}$$

where $N_{i,j} = \{(k, l); 0 \leq k < i \ or \ (k = i, l < j)\}$ and $\Psi_{i,j} = \{(i, j - 1), (i - 1, j)\}$.

It is easy to see that NSHP Markov chain model is a kind of MRF.

2.2   Hidden NSHP Markov chain model(HNSHPMCM)

It is assumed that $X = \{X_{i,j} : (i, j) \in L\}$ is a NSHP Markov chain with known transition probability $P[X_{i,j} = x | X_{k,l}, (k, l) \in \Psi_{i,j}]$ where $\Psi_{i,j} = \{(i, j - 1), (i - 1, j)\}$, and initial probability $P[X_{1,1}]$. In a HNSHPMCM, there is an output symbol array of random variables Y, which is a probabilistic function of X. Following the similar notational framework introduced by Rabiner[6] for 1-D HMM, the elements of 2-D HMM can be formally defined as follows:

- T: the number of states
- $X = \{X_{m,n}\}$: the sequence of states, $X_{m,n} \in \{1, 2, ..., T\}, 1 \leq m \leq M, 1 \leq n \leq N$
- $Y = \{Y_{m,n}\}$: the observation sequence of states, $1 \leq m \leq M, 1 \leq n \leq N$

- **A** $= \{a_{i,j}^k\}$: the transition probability distribution of the underlying Markov Random Field. Here, $a_{i,j}^k$ is the probability of $X_{m,n} = k$, given $X_{m,n-1} = i$ and $X_{m-1,n} = j$, that is

$$a_{i,j}^k = P[X_{m,n} = k | X_{m,n-1} = i, X_{m-1,n} = j]$$

where $1 \leq i, j, k \leq T$. At the boundary $a_{i,j}^k$ has different configurations from those at interior sites. So it is specified as follows:

$$a_{i,j}^k = \begin{cases} P[X_{m,n} = k | X_{m,n-1} = i, X_{m-1,n} = j] & m > 1, n > 1 \\ P[X_{1,n} = k | X_{1,n-1} = i] & m = 1, n > 1 \\ P[X_{m,1} = k | X_{m-1,1} = j] & m > 1, n = 1 \\ P[X_{1,1} = k] & m = 1, n = 1 \end{cases} \quad (3)$$

- **B** $= \{b_j(O_k)\}$: the output symbol probability distribution, where $b_j(O_k)$ is the probability of output symbol O_k, given current state j, that is

$$b_j(O_k) = P[Y_{m,n} = O_k | X_{m,n} = j], j \in 1, 2, ..., T$$

- $\varPi = \{\pi_i\}$: the initial state probability vector, i=1,2,...,T; $\pi = P[X_{1,1} = i]$, the initial state probability given that the state number on site (1,1) is i.

Generally HNSHPMCM is represented by a compact notation of parameter set $\lambda = (\varPi, A, B)$. HNSHPMCM can model the set of observation using these probabilistic parameters as a probabilistic function of an underlying Markov Random Field whose state transitions are not directly observable. Given the form of HNSHPMCM above, we can find that three problems have to be addressed when using HNSHPMCM for CCR.

- Evaluation problem
- Parameter re-estimation problem
- Decoding problem.

The first problem is to compute probability of the observation under the pattern λ, that is, $P[Y|\lambda]$, the second problem will arise during training phrase, and the third one is to find the physical meaning of the observations. Practically the decoding problem is the basis of the other two, and the efficiency of decoding affects the speed of recognition system deeply, so the third problem will be discussed extensively in the following sections.

2.3   Using HNSHPMCM for CCR

It is well known that the principle of a recognition system is shown as follows:

$$\lambda^* = \arg\max_\lambda P[Y|\lambda] = \arg\max_\lambda \sum_X P[Y|X, \lambda]P[X|\lambda]$$

which means, Y is to be recognized to pattern λ^* with maximal likelihood $P[Y|\lambda]$.

It is clear that using the formula directly will result in an enormous computational complexity which will be $O(T^{M \times N})$. Generally the forward and backward algorithms are adopted to reduce the computation complexity from exponential to polynomial times, however it is still time consuming in the above 2-D model. Generally people use different mutation of the formula rather than itself directly. We take two steps to approximate the probability function $P[Y|\lambda]$: 1)decoding state: given a certain λ, get a state array X^* which maximize $P[Y|X, \lambda]P[X|\lambda]$; 2)compute the likelihood under this state configuration X^* as the similarity between the sample and the pattern.

The decoding method we used before is the famous ICM (iterative condition mode) algorithm[9], it decide the state of image pixel by pixel given certain model parameters λ and observation, it can be stated as follows:

ICM algorithm for deciding states of image pixel by pixel

1. Give an initial value of state to $X = \{X_{i,j}\}$.
2. From i=1 to M, j=1 to N do

$$X'_{i,j} = \arg \max_{X_{i,j}} \{\log P[Y_{i,j}|X_{i,j}] + \log P[X_{i,j}|X_{i,j-1}, X_{i-1,j}]$$
$$+ \log P[X_{i,j+1}|X_{i,j}, X_{i-1,j+1}] + \log P[X_{i+1,j}|X_{i+1,j-1}, X_{i,j}]\}. \quad (4)$$

and $X' = \{X'_{i,j}\}$
3. Compare X and X', if different then $X = X'$ return step 2 else end.

3   Block-based ICM algorithm

To use HNSHPMCM in CCR, it is necessary to evaluate the speed of recognizing which is determined by the corresponding decoding algorithm. This section discusses the problems that arised when using ICM algorithm on our model, and propose a novel block-based ICM algorithm.

While many results on applying MRF in image processing have been reported recently, some attempts are made to extend it to CCR[2, 3]. However there are still some obstacles in its widely using on CCR. The main reason is that efficient algorithms of parameter estimation especially decoding does not exist. We can see that the speed of recognizing system is much in demand than that of training system. At the same time, classical ICM algorithm as demonstrated above is pixel-based, which means that it decide the state of image pixel by pixel. However, for the writing continuity of characters some pixels with similar feature can be viewed as a unit for decoding. This idea can propose a reasonable assumption that: *if we split the character image into certain little blocks according to some appropriate criteria, such as pixels' continuity or similar feature, the pixels in the same block should be assigned to the same state together when using ICM algorithm*. If the character image is split into γ blocks $\{B_k\}, k = 1, 2, ..., \gamma$. We can get the following theorem.

Theorem 3.1: Under the assumption above,

$$\arg\max_t P[X_{i,j} = t, Y_{i,j}, \ (i,j) \in B_k | X_{L \setminus B_k}, Y_{L \setminus B_k}]$$

$$= \arg\max_t \prod_{(i,j) \in B_k} P[Y_{i,j} | X_{i,j}] P[X_{i,j} = t | X_{i,j-1}, X_{i-1,j}].$$

Proof. The proof is straightforward:

$$P[X_{i,j} = t, Y_{i,j}, \ (i,j) \in B_k | X_{L \setminus B_k}, Y_{L \setminus B_k}]$$

$$= T \cdot \left(\prod_{(i,j) \in B_k} P[X_{i,j} | X_{i,j-1}, X_{i-1,j}] P[Y_{i,j} | X_{i,j}] \right) / \{ T \cdot \left(\sum_{X_{k_1, l_1}} \cdots \sum_{X_{k_{NB_k}, l_{NB_k}}} \right.$$

$$\left. \prod_{m=1}^{NB_k} P[X_{k_m, l_m} | X_{k_m, l_m - 1}, X_{k_m - 1, l_m}] P[Y_{k_m, l_m} | X_{k_m, l_m}] \right) \}$$

$$= \{ \prod_{(i,j) \in B_k} P[X_{i,j} | X_{i,j-1}, X_{i-1,j}] P[Y_{i,j} | X_{i,j}] \} / G$$

where

$$NB_k = \sharp\{(i,j); (i,j) \in B_k\}$$

$$(k_i, l_i) \in B_k, i - 1, 2, \cdots, ND_k$$

$$T = \prod_{(k,l) \in L \setminus B_k} P[X_{k,l} | X_{k,l-1}, X_{k-1,l}] P[Y_{k,l} | X_{k,l}]$$

$$G = \sum_{k_1, l_1} \cdots \sum_{k_{NB_k}, l_{NB_k}} \prod_{m=1}^{NB_k} P[X_{k_m, l_m} | X_{k_m, l_m - 1}, X_{k_m - 1, l_m}] P[Y_{k_m, l_m} | X_{k_m, l_m}]$$

Since G is the same one for every k, it is easy to find that the theorem is true.

Remark 1. This theorem give us a proposal on computing. Classical ICM algorithm compute the probability on pixels one by one, we can see, however, it can be computed on block level. From this theorem, we get a block-based ICM algorithm as follows:

1. split character image into γ blocks $B_i, i = 1, 2, ..., \gamma$ which are made up of pixels with similar feature.
2. Give an initial value of state to $X = \{X_{i,j}\}$
3. From i=1 to γ the new state of B_k, S'_k, can be assigned as follows

$$S'_k = \arg\max_{t=1,2,...T} \{ \sum_{(k,l) \in B_k} \log P[Y_{k,l} | X_{k,l} = t] + \log P[X_{k,l} = t | X_{k,l-1}, X_{k-1,l}]$$

$$+ \log P[X_{k,l+1} | X_{k,l} = t, X_{k-1,l+1}] + \log P[X_{k+1,l} | X_{k+1,l-1}, X_{k,l} = t] \}$$

4. Compare X and X', if different then $X = X'$ return 3) else 5)

5. Compute the probability of $\log P[Y|\lambda]$ as follows:

$$\log P[Y|\lambda] = \sum_{i,j} \log P[X_{i,j}|Y_{i,j}] + \log P[X_{i,j}|X_{i,j-1}, X_{i-1,j}]$$
$$+ \log P[X_{i,j+1}|X_{i,j}, X_{i-1,j+1}] + \log P[X_{i+1,j}|X_{i+1,j-1}, X_{i,j}].$$

Experimental results shows that it boost the recognizing speed obviously.

4 Experimental results

To evaluate the practical performance of the proposed approach for handwritten Chinese character recognition, several experiments were carried out. We implemented two recognition system with the Block-Based ICM(B-B ICM) and Pixel-Based ICM(P-B ICM) respectively, using Visual C++ on a MS-Windows 98 platform of Celeron400 PC. The size of column of the processed character samples is from 50 to 100, and row is from 50 to 140, so the characters are not in the same size.

4.1 Experiments

It is natural that B-B ICM recognizing system is compared with P-B ICM system on their accuracy and speed. The two experiments are done on 10 sets of 1-order frequently used Chinese character Library which is composed of 3755 Chinese characters released by CNSI(Chinese National Standard Institution) with two systems respectively. Table 1 shows the results:

Table 1. Performance comparison using B-B ICM and P-B ICM

	Accuracy	Speed	Iter_times(ave)
P-B ICM	71.46%	3.19s/1	6.4723
B-B ICM	72.36%	1.31s/1	3.0109

5 Analysis on experiments

In fact, the proposed approach of this paper reduce the times of iteration largely. The main reason is that it avoid too much surge which exists in pixel-based ICM algorithm. Our experiments prove this as the Table.1 shows.

At the same time, due to the continuity of pixels in the same block, it is easy to see that during one iteration the times of multiplication of pixel-based ICM is just 3*T*M*N while some multiplication can be omitted in block-based ICM algorithm.

6 Conclusion

The large computation of 2-D models is perhaps the largest difficulty in the application of character recognition. The proposed approach in this paper shows that it is possible to reduce the application greatly. Deep survey of the properties of Chinese characters reveals that many computation can be omitted as long as some pixels can be sure to be decided to the same state. More stiring results are prospective if this approach is connected to excellent stroke-extraction algorithm.

References

1. T.H.Hildebrand, W.Liu, Optical recognition of handwritten Chinese characters: Advances since 1980. Pattern Recognition **26(2)** (1993) 205-225.
2. M.H.Deng, Research and Implementation of Handwritten Chinese Character Recognition Based on Hidden Markov Random Field. Ph.D thesis, Peking University (1997).
3. H.S.Park and S.W.Lee, A truly 2-D hidden markov model for off-line handwritten character recognition. Pattern Recognition **31(12)** (1998) 1849-1864.
4. C.Tappert,C.Y.Suen and T.Wakahana, The state of the art in on-line handwriting recognition. IEEE Trans. Pattern Analysis Machine. Intel. **12(8)** (1990) 787-808.
5. T.Wakahana, H.Murase and K.Odaka, On-line handwriting recognition. Proc.IEEE. **80(7)** (1992) 1181-1194.
6. L.R.Rabiner, A tutorial on hidden Markov models and selected applications in speech recognition. Proc. IEEE. **77(2)** (1989) 257-286.
7. S.-S.Kuo and O.E.Agazzi, Keyword spotting in poorly printed documents using pseudo 2-D hidden Markov models. IEEE Trans. Pattern Analysis Machine Intel. **16(8)** (1994) 842-848.
8. F.-C.Jeng and J.W.Woods, On the Relationship of the Markov Mesh to the NSHP Markov Chain. Pattern Recognition Letter. Vol. 5 (1987) 273-279.
9. J.Besag, On the Statistical Analysis of Dirty Picture. J. Roy. Statistics Soc. Ser.B **48(3)** (1986) 259-302.

An HMM Based Two-Pass Approach for Off-Line Cursive Handwriting Recognition

Wenwei Wang, Anja Brakensiek, and Gerhard Rigoll

Department of Computer Science, Faculty of Electrical Engineering
Gerhard-Mercator-University Duisburg
Bismarckstrasse 90, 47057 Duisburg, Germany
{wwwang, anja, rigoll}@fb9-ti.uni-duisburg.de

Abstract. The cursive handwriting recognition is a challenging task because the recognition system has to handle not only large shape variation of human handwriting, but also character segmentation. Usually the recognition performance depends crucially upon the segmentation process. Hidden Markov Models (HMMs) have the ability to model similarity and variation among samples of a class. In this paper we present an extended sliding window feature extraction method and an HMM based two-pass modeling approach. Whereas our feature extraction method makes the resulting system more robust with word baseline detection, the two-pass recognition approach exploits the segmentation ability of the Viterbi algorithm and creates another HMM set and carries out a second pass recognition. The total performance is enhanced by combination of the two pass results. Experiments of recognizing cursive handwritten words with 30000 words lexicon have been carried out and show that our novel approach can achieve better recognition performance and reduce the relative error rate significantly.

1 Introduction

Optical Character Recognition (OCR) carries out automatic conversion from scanned images of machine printed or handwritten text into a computer processable format. While the recognition of machine printed characters is considered as a mature technique today, the recognition of handwritten characters, especially cursive text, remains a challenging task, because the system has to handle not only large shape variation and variable length of the human handwriting, but also character segmentation and language modeling. Generally strategies for cursive word recognition can be roughly subdivided into two different categories: segmentation-explicit and segmentation-implicit approaches.

As pattern classifier Hidden Markov Models (HMMs) have been successively applied in the areas of speech recognition [6] and on-line handwriting recognition [7]. HMMs are stochastic models which can deal with pattern variation and noise and represent probably the most powerful tool for modeling time varying dynamic patterns, such as the cases in speech recognition and on-line handwriting recognition. During recent years HMMs have been also used to the problem

T. Tan, Y. Shi, and W. Gao (Eds.): ICMI 2000, LNCS 1948, pp. 386–393, 2000.

of off-line cursive script recognition in various ways [1][2][9]. Besides the ability of coping with dynamic property and noise among human handwriting, HMM based approaches can perform recognition and segmentation in a single step, thus avoid segmenting words into characters, what is very difficult and vulnerable to noise and small variation.

The main difficulty of HMMs' application to off-line recognition of cursive words is to produce a consistent sequence of feature vectors from the input word image [2] [3]. With our extended sliding window feature extraction method presented shortly ago in [9], such a consistent sequence of feature vectors can be created. This sequence of feature vectors is tolerant to the baseline detection error to some extent. An enhancement of recognition performance has thus been achieved.

The general approach presented in this paper is to exploit the inherent segmentation ability of an HMM classifier which has been trained for vertical feature vector sequence and to create a second HMM set for horizontal feature vector sequence. Utilizing the first pass recognition not only for transcription but also for segmentation hypothesis, a second pass recognition can be carried out. The total system recognition performance is thus enhanced by combination of the two pass recognition results.

In the following section we first shortly describe the most important modules of an HMM based word recognition system, especially the extended feature extraction. Then we present in Section 3 the new two-pass HMM based approach for cursive handwritten word recognition. In Section 4 we provide experimental results and in Section 5 conclusions.

2   HMM Modeling and Word Image Feature Extraction

A good introduction to HMMs can be found in the paper by Rabiner [6]. In our previous work [1][9] some methods about HMM based handwriting recognition, including feature extraction and modeling techniques are described.

The selection of a suitable feature extraction method is one of the most important factors to achieve high recognition performance in all OCR systems [8]. The main difficulty of HMMs' application to off-line cursive handwriting recognition is to produce a consistent sequence of feature vectors from the input word image. In conventional HMM based methods, a sequence of thin fixed-width vertical frames are extracted as feature vectors from the image[1][3][5]. The extracted feature is sensitive to the error of the preprocessing steps e.g. baseline detection. In the paper [9] we presented an extended sliding window method in order to decrease the influence of the baseline detection error. As shown in Fig. 1 the new feature vector is generated with:

$$f_e = (f_{1-}, f_1, f_{1+}, f_{2-}, f_2, f_{2+}, ..., f_{4-}, f_4, f_{4+})^T \tag{1}$$

$f_j = f(z_j), f_{j-} = f(z_j^-), f_{j+} = f(z_j^+).$ z_j is the jth zone from top to bottom, $f(z_j)$ expresses some coding function, for example black points percentage of the zone. We get the zones z_j^- by the means of assuming an up-shift of the

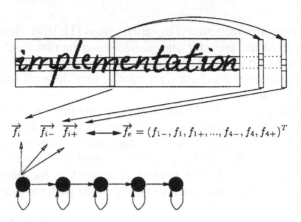

Fig. 1. Feature extraction of sliding window extended zone coding

two baselines locally and applying the same zone dividing method. The zones z_{j+} are generated while assuming a down-shift of the baselines. Obviously the new extended feature vector f_e contains more distinctive information about the character shape and has higher discrimination power than the original feature vector $f_i = (f_1, f_2, f_3, f_4)^T$, thus for describing the character image the new extended feature vector could be tolerant to the error of baseline detecting to some extent. Optionally the discrete cosine transform coefficient (DCT) of the vector f_e with the form in Eq.(2) can be taken as feature vector.

$$c(k) = \begin{cases} \dfrac{1}{\sqrt{N}} & k = 0 \\ \displaystyle\sum_{n=0}^{N-1} \sqrt{\dfrac{2}{N}} \cos\dfrac{\pi(2n+1)k}{2N} f(n) & 1 \le k \le N-1 \end{cases} \tag{2}$$

As shown in Fig. 1, discrete HMMs with linear left-to-right topology are applied. Usually one HMM is created for each character of the alphabet. Providing a sufficient number of examples for that model the parameters of an HMM can be determined by the robust Baum-Welch algorithm. The various states of a character model can model different parts of the character. The state output probabilities model the shape variation of that part. One important feature of the HMM approach is that a word model can be built by concatenating the appropriate letter models. By this means the recognition system requires only a small number of models to represent even a large recognition task dictionary.

After the models have been trained, an unknown pattern can be classified by the maximum likelihood decision:

$$\lambda^* = \arg\max_i \left(P(O \mid \lambda_i) \right) \tag{3}$$

A given observation symbol sequence O will be classified to class λ^*, namely the word class associated with the compound word HMM yielding the highest likelihood. In the HMM scheme the likelihood $P(O|\lambda_j)$ is determined by the well-known Viterbi algorithm. The core of the algorithm is expressed in the following recursion:

$$\phi_j(t) = \max_i\{\phi_j(t-1)a_{ij}\}b_j(o_t) \tag{4}$$

a_{ij} is the state transition probability from state i to state j, and $b_j(o_t)$ is the observation probability in state j, o_t is one symbol of the observation sequence $o = o_1 o_2 \ldots o_T$, T is its length, and N is the number of states in the HMM. By the means of searching the recognition network, the best path representing the most likely state sequence and the approximation of the likelihood can be determined: $P(O|\lambda) \approx \phi_N(T)$. Segmentation information between each character model can then be obtained by tracing the route in the best path.

3   Two-Pass Recognition: Combination of Vertical HMMs and Horizontal HMMs

The extended feature vector described in Section 2 is dependent on the baseline, although it is tolerant to error of baseline detection to some extent, compared with conventional methods. Baseline estimation is actually not trivial. If the baseline fluctuation is too large, this method will also come to it's limit. Hitherto we have not yet utilized the segmentation ability of the Viterbi algorithm. The two-pass strategy exploit this ability to enhance the recognition performance. Its core is using the set of HMMs for vertical features trained in the first pass (V-HMMs) not only for transcription, but also for segmentation. We can then create another set of HMMs for horizontal features (H-HMMs) which will be used in the second recognition pass to verify the result of the first pass.

The horizontal feature vector is extracted as shown in Fig. 2. In the image of a word, this requires that the boundary between characters is already determined. For training of H-HMMs we want to use the same train data set as in training of V-HMMs. Because we have the transcription of the word image, a recognition network can be created and is matched against feature vectors from the word image itself. Utilizing the alignment function of the Viterbi algorithm we can get the accurate position of each character in the training word image. Then the horizontal feature vectors will be generated from rows of the "character image". Accordingly the Baum-Welch algorithm is used again to estimate the parameters for H-HMM for each character, which has the same topology as the V-HMM. The H-HMMs for horizontal features are insensitive to baseline estimation error and vertical shift of word image.

The same feature extraction process in horizontal direction must be applied to the unknown test words. An severe issue exists in the second pass recognition for testing words, because the correctness of the transcription and boundary information between characters is not known yet. If in the first pass the transcription is correct, the segmentation positions are most likely also correct, then the second pass recognition can verify the hypothesis of the first pass. If in the first pass the transcription is not correct, the segmentation positions may be false

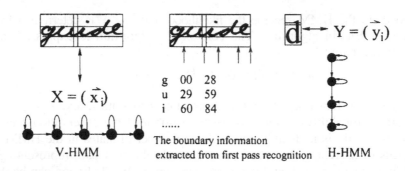

Fig. 2. Feature extraction and HMM modeling for two directions

or correct, then the second pass recognition can not simply serve as verification of the first pass hypothesis.

This problem is solved by the means shown in Fig. 3. The recognition rate with Top N hypotheses (e.g. N=3) is obviously higher than recognition with Top 1. In the first pass, three hypotheses will be made, each of them together with their segmentation position information will be used to generate row feature vectors which will be matched to H-HMMs. In the second pass, the H-HMMs are used to make Top 3 hypotheses too.

In Elms' work[4] a similar two-pass recognition approach is applied for degraded machine print character recognition, combining the results of the two passes under the assumption that the results are independent. We think this assumption can not conform to the reality, at least in the case of cursive handwriting recognition.

To make a Top 1 hypothesis from above *3+9* hypotheses we propose a decision criterion expressed in Eq.(5). The log recognition scores of the three hypotheses in the first pass recognition will be adjusted according to whether the three hypotheses of the second pass can match and verify the hypotheses of the first pass.

$$ScoreV[I] = (1 - (nHTN + 1 - J) * \alpha / (nHTN + 1)) * ScoreV[I] \qquad (5)$$

$ScoreV[I]$ is the log likelihood probability of the I th hypothesis of the first pass, $nHTN$ is the number of the hypotheses, and α is an adjustment control constant, in the later experiment $\alpha = 0.1$ is set. If a hypothesis appears also in the second pass, then J represents its rank, if the hypothesis doesn't appear in the second pass, then set $J = nHTN+1$.

4 Experiments

We applied the HMM based two-pass recognition approach to the recognition of cursive handwritten words. The same database as described in [9] is used. Three writers have each written about 2000 words, four fifth of them taken as training

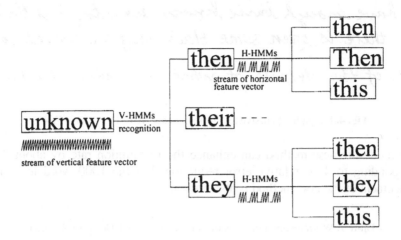

Fig. 3. Two-pass recognition with Top 3 hypotheses

set, the other one fifth as testing set. In Fig. 4 three lines of the sample text, one line for each writer, are shown.

We have carried out two experiments: one-writer mode experiment for each of the three writers (WWW, LLC, ABR), and multi writer mode experiment for the three writers (WLA) together. For each of 52 English letters one HMM is built, whereas for each word in the task dictionary the word model is built by concatenating the appropriate letter models. Experiments have been done with two lexicons of different size. One of them has 1000 words, the other has 30000 words.

Almost the same preprocessing algorithms including text line extraction and word isolation, slant correction and baseline detection as described in [9] are applied. The extended feature extraction method (seeing Eq.(1)) is used, but in this experiment its DCT coefficient, according to Eq.(2), is taken as feature vector, which has achieved higher recognition rate for large scale lexicon (30000 words) than the original feature vector.

The experiment results of our new HMM based recognition system are reported in Table 1 and Table 2. The column "ABR" is the recognition rate for data set of writer ABR. While applying Top 1 recognition with 1000 word lexicon in the first pass, 6 from 387 test words are misclassified, corresponding to a recognition rate of 98.45%. With Top 3 we get a recognition rate of 99.74%. Row 3 shows the results obtained with our new two-pass approach. For the writer ABR the recognition rate is 98.97%. In the multi-writer mode, the training sets of each writer are mixed together to train the HMMs, whereas the testing sets of each are taken together to assess the performance. The column " WLA" presents the results obtained in this mode.

Comparing row 1 and row 3 in Table 2, the new two-pass method outperforms the conventional one-pass method averagely about 1.22% for 30000 word lexicon in one-writer mode, this corresponds to a 48% relative error reduction. In multi-

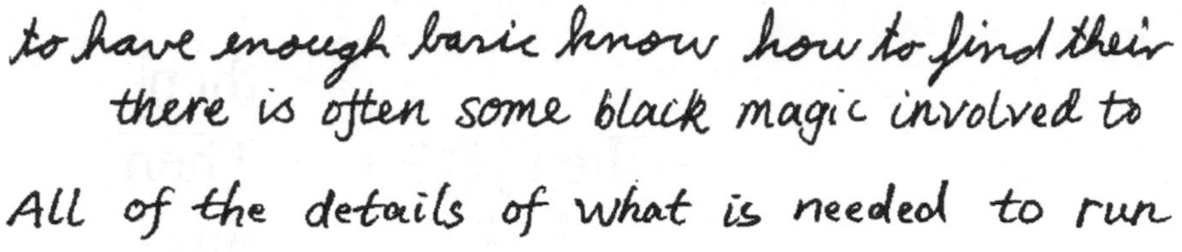

to have enough basic know how to find their

there is often some black magic involved to

All of the details of what is needed to run

Fig. 4. Example of cursive text from data set ABR, LLC and WWW

writer mode the new method can enhance the recognition rate by about 2.06% corresponding to 43% relative error reduction. For the 1000 word lexicon the enhancement of the new method is also obvious.

Table 1. Word recognition rates (accuracy in %) for 1000 words lexicon

Method	WWW	LLC	ABR	Average	WLA
1 pass/top 1	98.20	98.64	98.45	98.41	97.10
1 pass/top 3	99.48	99.66	99.74	99.63	99.35
2 pass/top 1	98.97	99.32	98.97	99.06	97.85

Table 2. Word recognition rates (accuracy in %) for 30000 words lexicon

Method	WWW	LLC	ABR	Average	WLA
1 pass/top 1	97.16	97.62	97.67	97.47	95.23
1 pass/top 3	99.48	99.66	99.74	99.63	98.60
2 pass/top 1	98.45	98.98	98.71	98.69	97.29

5   Conclusion and Outlook

In this paper we presented for cursive handwritten word recognition an extended sliding window feature extraction method and an HMM based two-pass modeling approach. The extended feature extraction makes the resulting system more robust with word baseline detection. Exploiting the segmentation ability of the well known Viterbi algorithm, the two-pass modeling approach utilizes the first pass recognition not only for transcription but also for segmentation hypothesis, thus a second HMM set can be created and a second pass recognition can be performed. The total system recognition performance is enhanced by combination of the two pass recognition results. A recognition rate of 97.29% has been achieved for recognizing cursive handwritten words in a 3-writer mode with a 30000 word lexicon. The relative error rate reduction is over 40% compared to only one pass recognition.

Our future work includes applying the multi-pass approach to a public larger database and improving the system by making preprocessing and feature extraction more robust with the multi-pass approach. We believe that the multi-pass recognition approach is very important to handwriting recognition, conforming to the recognition mechanism of human brain to some extent, and can be generalized.

References

1. A. Brakensiek, A. Kosmala, D. Willett, W. Wang and G. Rigoll, "Performance Evaluation of a New Hybrid Modeling Technique for Handwriting Recognition Using Identical On-Line and Off-Line Data", *Proc. International Conference on Document Analysis and Recognition (ICDAR)*, Bangalore, India, 1999, pp.446–449.
2. H. Bunke, M. Roth and E. G. Schukat-Talamazzini, "Off-line Cursive Handwriting Recognition Using Hidden Markov Models", *Pattern Recognition*, Vol. 28, No. 9, 1995, pp.1399–1413.
3. Wongyu Cho, Seong-Whan Lee and Jin H. Kim, "Modeling and Recognition of Cursive Words with Hidden Markov Models'", *Pattern Recognition*, Vol. 28, No. 12, 1995, pp.1941–1953.
4. A. J. Elms, S. Procter and J. Illingworth, "The advantage of using an HMM-based approach for faxed word recognition", *International Journal on Document Analysis and Recognition*, Vol. 1, No. 1, 1998, pp.18–36.
5. D. Guillevic and Ching Y. Suen, "HMM Word Recognition Engine", *Proc International Conference on Document Analysis and Recognition (ICDAR)*, Ulm, Germany, 1997, pp.544–547.
6. L. R. Rabiner and B. H. Juang, "An Introduction to Hidden Markov Models'", *IEEE ASSP Magazine*, Vol. 3, No. 1, jan. 1986, pp.4–16.
7. Gerhard Rigoll, Andreas Kosmala and Daniel Willett, "A New Hybrid Approach to Large Vocabulary Cursive Handwriting Recognition", *Proc. International Conference on Pattern Recognition (ICPR)*, Brisbane, 1998, pp.446–449.
8. Trier, A. K. Jain and T. Taxt, "Feature Extraction Methods for Character Recognition--a Survey'", *Pattern Recognition*, Vol. 29, No. 4, 1996, pp.641–662.
9. Wenwei Wang, Anja Brakensiek, Andreas Kosmala, and Gerhard Rigoll, "HMM based High Accuracy Off-line Cursive Handwriting Recognition by a Baseline Detection Error Tolerant Feature Extraction Approach". *Proc. 7th International Workshop on Frontiers in Handwriting Recognition (IWFHR)*, Amsterdam, Netherland, September 2000.

On-Line Recognition of Mathematical Expressions Using Automatic Rewriting Method

T. Kanahori[1], K. Tabata[1], W. Cong[2], F.Tamari[2], and M. Suzuki[1]

[1] Graduate School of Mathematics, Kyushu University 36,
Fukuoka, 812–8581 Japan
suzuki@math.kyushu-u.ac.jp
[2] Fukuoka University of Education
729 Akama, Munakata-shi, Fukuoka, 811–41 Japan,
tamari@fukuoka-edu.ac.jp

Abstract. This paper describes our system of on-line recognition of mathematical expressions. Users can input mathematical expressions by handwriting. As soon as a character is written, it is rewritten by neat strokes in an appropriate position and size automatically. This *Automatic Rewriting Method* improves the accuracy of the structure analysis of the written mathematical expressions. The written mathematical expressions can be output into files in the notation of LaTeX and MathML. By this handwriting interface, the system realizes a very easy intuitive methods to input mathematical expressions into computer.

1 Introduction

Recently, the use of computer and network is becoming widely spread. However, it is true that the user interfaces of current computer systems are not convenient to input mathematical expressions (see [1], [2] and [3]). For example, the widely used data format TeX requires some learning to master the notations, and it is not easy to understand the meaning of the written expressions by the TeX source at a glance. To realize easier treatment of mathematical expressions of various formats, we are developing a new handwriting interface to input mathematical expressions into computer. Users can input mathematical expressions by handwriting in the *handwriting area*, and edit the input expressions on the *display area*. The edited mathematical expressions can be output into a file in the notation of LaTeX or MathML.

Presently, this system can recognize all the alphanumeric characters, almost all Greek letters, and other symbols frequently used in mathematical expressions, and can analyze the expressions used in high school mathematics or in the first course of university mathematics, including fractions, square roots, subscripts, superscripts, integrals, limits, summations and the function names 'lim', 'log', 'cos', 'sin', 'tan', etc. The matrices are excluded at the moment. The structure of mathematical expressions may have the nested structures. However, deeply nested structure leads to the increase of small size characters and naturally increases the errors of the recognition.

T. Tan, Y. Shi, and W. Gao (Eds.): ICMI 2000, LNCS 1948, pp. 394-401, 2000.

In this paper, we describe our method of the recognition of handwritten characters and the algorithms of *Automatic Rewriting Method*, and report the performance of our experimental system.

2   Outline of The Experimental System

The main window of the system consists of two areas (see Figure 1). The upper area is the display area and the lower is the handwriting area. In the handwriting area, users write mathematical expressions by using a mouse, a data tablet or a pen display.

As soon as a character is written, it is recognized and rewritten by neat strokes in an appropriate size and position automatically. Pushing the button ⟨OK⟩, the expression is analyzed, and the result is displayed in type setting form in the display area.

In the display area, we can edit mathematical expressions using ordinary editing operations: selection, cut, copy, paste and delete. The users can save them into a file in the notations of LᴬTᴇX or MathML.

Fig. 1. The experimental system

3   Character Recognition

In this section, we describe our method of character recognition in our handwriting interface. We implemented two different methods of character recognition, to get recognition results by *voting*. One of the recognition methods uses the distribution of the 8-direction elements of the strokes on 3×5 meshes of the he character rectangle (Section 3.1). The other method uses the matching of segmented stroke sequence (Section 3.2). Each of the two recognition methods returns three ordered candidates with costs. The *voting cost* of the candidate is taken to be the ratio of its cost to the third candidate's, and the final recognition results is determined by the ascending order of the sum of their two voting costs of the two recognition methods.

The characters and symbols recognized in this system include all the alphanumeric characters, some Greek letters, and other symbols frequently used in mathematical expressions (see [3]).

3.1   Direction Element Feature

To extract this feature, we first take a normalized coordinate system which squarely converts the bounding rectangle of a character, and subdivide the bounding rectangle into 3×5 meshes (3 meshes in the horizontal direction, and 5 meshes in the vertical direction).

Let $d_0, d_1, d_2, \ldots, d_7$ be the directions taken from the x-axial direction to the backing every 45 degree (see Figure 2). Given a segment of length L of direction between d_i, d_{i+1} making the angle θ_1, θ_2 respectively with them, the contribution of the segment to the directions d_i, d_{i+1} are defined by

$$L_i = \frac{L\theta_2}{\theta_1 + \theta_2}, \quad L_{i+1} = \frac{L\theta_1}{\theta_1 + \theta_2},$$

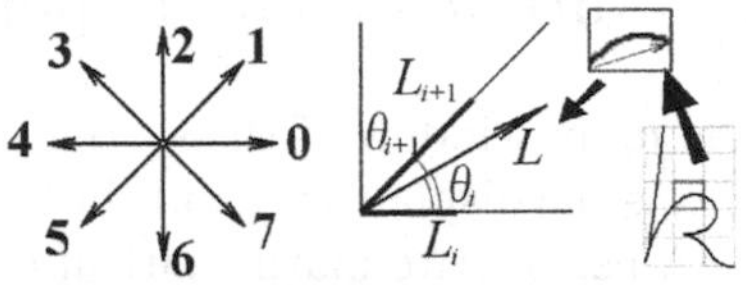

respectively, where $L_8 = L_0$ and $d_8 = d_0$. L_i is called the *direction component* of the segment to d.

Fig. 2. The direction component

Calculating the direction components of each segment which constitutes the strokes of a character, these components are distributed into the 3×5 meshes defined above. Thus we obtain a feature vector of dimension $3 \times 5 \times 8 = 120$, which we call the *direction element* feature([4]). Since we take a coarse mesh (3×5), this feature vector has a robust property for the distortion of the character shape.

3.2 Matching of Segmented Stroke Sequence

If a written stoke has some small loops, they are modified to cusps (see Figure 3). After this modification, the written stroke is segmented at extreme points on the vertical coordinate. The stroke is segmented at cusp points again (see Figure 3). Segment strokes which can be regarded as a straight line are classified into 8 patterns (8 directions in Figure 2), and segment strokes which are winding down strokes are classified into 10 patterns (see Figure 4). Segment strokes which are winding upstrokes are also classified similarly. Hence, there are $8 + 10 \times 2 = 28$ patterns of segment strokes. The character recognition is done by the matching of the sequences of segment stroke patterns thus obtained.

To calculate the recognition cost, the following features are used: segment strokes' aspect ratios, the positions of their bounding rectangles, the directions of their original vectors v_o and terminal vectors v_t (see Figure 5). The cost of a written stroke for a candidate is determined by the sum of the differences of these features between the corresponding segment strokes.

Fig. 3. Modification and Segmentation   **Fig. 4.** Downstrokes   **Fig. 5.** v_o, v_t

4   Automatic Rewriting Method

The distortion of input characters and the turbulence of the positions or the scales of characters usually cause serious difficulties in the structure analysis of mathematical expressions, in which the positions and the scales of the characters have special meanings. An error of a character recognition or of the segmentation of the strokes into character units leads sometimes to a fatal error of the structure analysis of the mathematical expression. The labors for the correction of this kind of errors disturb seriously the smooth input of mathematical expressions.

Our automatic rewriting method is introduced in order to overcome this difficulty. In this method, whenever the pen is up, the strokes are recognized. Each recognized character is rewritten by neat strokes in an appropriate size and position automatically (see Figure 1). By this rewriting, the user can identify each recognition error immediately when it occurs, and can correct it easily.

In this section, we shall explain the algorithms to determine characters and to select appropriate positions and sizes of the determinate characters.

4.1   Determination of Characters

When a stroke is written, it is necessary to decide whether a character is fully written up or not at each time the pen is up in our method. To explain the Determination Algorithm for this operation, we classify the results of the character recognition into two groups. One is the group of characters, named *extendable characters*, which can be extended to other character by adding some strokes. For example, 'F' is extendable to 'E', 'C' to 'G' or 'd', and '=' to '≠'.

The other is the group of characters, named *unextendable characters*, which can not be extended to any other character by adding strokes. For example, 'E' and 'G' are unextendable.

Each extendable character has extendable areas, where a next stroke is expected to be pushed down (see Figure 6). The determination algorithm of character unit proceeds as follows:

1. Let S be empty. (S means the sequence of untreated strokes.)
2. When a pen is up, add the written stroke to S
3. Let R be the recognition result considering S as one character.
4. If R is unextendable, then output R as the determined character of the strokes S and go to the step 1.
5. Wait for the next stroke N to be written. If N started from R's extendable area within 2 seconds, then go to step 2, else output R as the determined character of the strokes S and go to step 1.

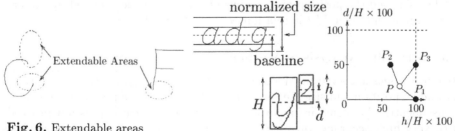

Fig. 6. Extendable areas

Fig. 7. Decision of Position and Sizes

4.2   Determination of Positions and Sizes of Characters

A structure of mathematical expression can be represented by a tree structure. Corresponding to it, the notion of *parent (or child) character* is introduced. For the expression $x^a + y^{b'}$, for example, 'x' and y are the parent characters of 'a' and b with the relation 'superscript', and x is also the parent of '+' with the relation 'same line'. The position and the size of an input character are determined together with its parent character.

Let C be an input character, and D be the candidate of the parent character of C. Then, the *relation point* $P(C, D)$ is defined by

$$P(C, D) := (h/H \times 100, d/H \times 100),$$

where h is the normalized size of C, H is the normalized size of D and d is the distance of the baselines of C and D (see Figure 7). The *cost* between C and D on **Relation** i, $c_i(C, D)$, is defined by

$$c_i(C, D) := d(P_i, P(C, D)) \quad (i = 1, 2, 3),$$

where **Relation 1** means that C and D are on the same line, **Relation 2** means that C is either the superscript or the subscript of D, **Relation 3** means that C is either the numerator or the denominator of the fraction on the same line as D's, and P_i are ideal relation points on **Relation** i. In this system, $P_1 = (100, 0)$, $P_2 = (60, 50)$ and $P_3 = (100, 50)$.

The algorithm to find the parent character M of the new input character C and to obtain their relation R is as follows:

1. Let M and R be NULL. If C is the first input character in the mathematical expression, then quit.
2. Let D be the input character just before C, *min* be a positive number which is large enough, and set $n=0$.
3. If D is NULL then quit.
4. If the cost $c_1(C, D)$ is small enough, then let $M = D$ and $R = $ **Relation 1** and quit.
5. Let $m = \text{argmin}\{k \neq n | c_k(C, D)\}$. If $c_m(C, D)$ is smaller than *min*, then let $min = c_m(C, D)$, $M = D$, and $R = $ **Relation** m. Let further **Relation** n be the relation between D and its parent character, and replace D by the parent character of D. Go to the step 3.

The size and the position of the input character C is determined by the pair of its parent character M the relation R thus obtained.

4.3   Structure Analysis of Mathematical Expressions

By our Automatic Rewriting Method, the recognition result of each input character and its position and size are corrected by the user, as soon as an error occurs. Consequently, the structure analysis of the mathematical expression proceeds with very high accuracy. In practice, no error occurs in the structure analysis expect for the case where the input rule is ignored intentionally.

5   Experimental Results

To evaluate the efficiency of Automatic Rewriting Method and the performance of our recognition engine, the experiments were carried out using thirty writers who had got their hands in writing mathematical expressions, but never used our system in the following order:

Experiment 1 First, we explained how to use our system, following the manual prepared for this experiment (for about 10 minutes for each writer). Then, each writer wrote the following mathematical expressions (1)~(7),

$$(1)\quad ax^2 + bx + c = 0, \quad (2)\quad \frac{s+t}{p+q}, \quad (3)\quad e^{-\frac{x^2}{2}}, \quad (4)\quad x = \frac{-b \pm \sqrt{b^2 - 4ac}}{2a},$$

$$(5)\quad \sum_{i=1}^{n} i = \frac{n(n+1)}{2}, \quad (6)\quad \lim_{x \to 0} \frac{\sin x}{x}, \quad (7)\quad \int_a^b \log x\, dx,$$

and we counted the number of times of error correction actions of the writer. For the correction of character recognition error, the writers are instructed to switch the recognition result with its next candidate by clicking on the character, and to delete last one character by pushing the button ⟨Delete⟩ only if there is no correct result among the three candidates. On the other hand, for the case the size or the position of the rewritten character is not appropriate, or the character segmentation is wrong, the writers are instructed to push first the button ⟨Delete⟩. In this way, the error correction actions are classified by their causes into four groups: **1. order**: the first candidate of the character recognition was not correct, **2. candidate**: no correct result was found among the three candidates, **3. position**: the position or the size of the rewritten character was wrong, **4. segmentation**: one character was recognized as several characters. For each written character, we logged its recognition cost, the distance between baselines of the character and its rewritten character. We did this process 6 times. Finally, we reshuffled the above expressions and did the similar experiment as 7th process.

Experiment 2 After Experiment 1, each writer wrote the following expressions (i)~(iv), which are more complicated than Experiment 1's. Then, we counted and logged in the similar way to Experiment 1. We did this process 3 times.

$$(i)\quad \lim_{n \to \infty} \left(1 + \frac{1}{n}\right)^n = e, \quad (ii)\quad \int_0^\infty \frac{\log x}{1 + x^2}\, dx, \quad (iii)\quad \sum_{n=0}^{\infty} \frac{(-1)^n}{(2n+1)^2},$$

$$(iv)\quad \int_0^\infty e^{-a^2 x^2}\, dx = \frac{\sqrt{\pi}}{2|a|}$$

Experiment 3 After Experiment 2, we told each writer to write as fast as possible each the expressions (i)~(iv). For each expression, we counted the time from the first stroke's beginning until the whole expression exactly written up, including the time to correct.

Experiments 1∼3 are intended to evaluate the effects of Automatic Rewriting Method. After these experiments, we did the following experiment to evaluate the recognition rate of our recognition engine.

Experiment 4 For the training, each writer wrote all alphabets and numerics 3 times. Then, each writer wrote 20 characters (capital:small:numeric = 8:8:4) which we had chosen randomly, and the number of recognition errors is counted. We did this process 5 times, and classified the number of errors into two cases: 1. the correct result was not the first candidate, but in the first three candidates, 2. the correct result was not in the first three candidates. Moreover, we also classified lower-case letters into two classes according as the writers had written in Experiment 1 or 2, or not, and counted the number of errors for each classes.

Figure 8 shows the result of Experiment 1 and 2. In Figure 8, the x-axis means the process times of the experiment, and the y-axis means ratio of correction time to the total number of characters. For example, the ratio of the case **order** is calculated by $(1/N)\sum_{i=1}^{7} e_i$, where $e_i(i = 1, \cdots, 7)$ is the number of errors **order** at the expression (i), and N is the total number of characters in all of the expressions $(1)\sim(7)$.

Figure 8 shows that the numbers of correction except **order**'s decrease as the number of writing increases. The **order** correction increases from 4th time. This means that the concentration's slip or the habituation of the writer causes rough writing. But, the decrease of the other corrections shows that this system can cope with the 'rough writing'. Moreover, the **position** correction (or **segmentation**) which leads a fatal error of the structure analysis of mathematical expressions is about 3 times (resp. 4 times) per 100 characters at 7th writing. Hence, the Automatic Rewriting Method is effective to on-line recognition of mathematical expressions.

We expected another advantage of Automatic Rewriting Method as below: The rewritten characters may serve as a model of the feasible character forms and the size for the recognition, and leads the user to a neater writing style. However, there was no major difference in the log of the costs and the distances between baselines from the 1st writing to the 7th in Experimental 1 and 2. Therefore, we could not see the advantage from this point of view.

Figure 9 shows the result of Experiment 3. It shows that, everyone wrote up all the expressions (i)∼(iv) between 1.5 minutes and 4 minutes. For 5 experts of TEX, who had written at least 5 articles of mathematics using TEX, we counted the times to input the same expressions in the notation of TEX. Then, the fastest time was about 1.5 minutes, the latest was about 3.5 minutes, and the average was 1 minute and 58 seconds. This shows that beginners can input complicated mathematical expressions smoothly with simple training, by using our interface, as well as inputting of the experts in the notation of TEX.

Table 1 shows the result of Experiment 4. In Table 1, the 1st row shows the recognition rate of the first candidate and the 2nd row shows the rate in the first three candidates. The 4th column 'appear' (resp. the 5th column 'not appear') means that the rate is for characters which appeared (resp. did not appear) in Experiment 1 or 2. The 1st rate of 'appeared' characters is worst.

This shows the similar result to Graph 8, 'rough writing' by the concentration' slip or habituation.

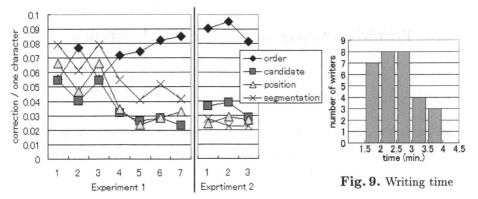

Fig. 8. The ratios of correction

Fig. 9. Writing time

	capital	small	numeric	appear	not appear	total
at the first candidate	95.5%	91.7%	94.5%	90.9%	93.4%	93.8%
in three candidates	97.8%	97.1%	97.3%	97.2%	96.6%	97.4%

Table 1. The rates of current result

6   Conclusion

In this paper, we introduced our system of on-line recognition of mathematical expressions. We proposed Automatic Rewriting Method to improve the accuracy of the structure analysis of handwritten mathematical expressions and to realize an easy and prompt correction method of recognition errors. We emphasized that an easy correction method of the recognition results is extremely important to realize smooth writing of mathematical expressions. From the experimental result, In the experiment, we observed the efficiency of our method to input mathematical expressions by handwriting smoothly with minimum training.

References

1. D. Blostein and A. Grbavec, "Recognition of Mathematical Notation", *Handbook of Character Recognition and Document Image Analysis*, (1997) 557-582.
2. T. Sakurai, Y. Zhao, H. Sugiura and T. Torii, "A Front-end Tool for Mathematical Computation and Education in a Network Environment", *Proc. 3rd. Asian Technology Conference in Mathematics*, Springer (1998) 197-205.
3. H. Okamura, T. Kanahori, W. Cong, R. Fukuda, F. Tamari and M. Suzuki, "Handwriting Interface for Computer Algebra Systems", *Proc. 4th. Asian Technology Conference in Mathematics*, December (1999) 291-300.
4. N. Sun , T. Tabara, H. Aso and M. Kimura, "Printed Character Recognition Using Directional Element Feature ", **IEIEC J74-D-II**, 3, 1991, 330-339 (in Japanese).

On-Line Hand-Drawn Symbol Recognition Based on

Primitives Separation and Fuzzy Inference[1]

GONG Xin, PEI Jihong and XIE Weixin

No. 202 Staff Room, School of Electronics Engineering, Xidian University, Xi'an, Shaanxi
Province.   Postal code: 710071

jhpei@mail.xidian.edu.cn

Abstract. In this paper, a universal recognition method for hand-drawn
symbols is presented which based on primitives separation and fuzzy inference.
This technique first separates input figure symbol into primitives using a
single-band integral algorithm, then a fuzzy inference framework is used to
combine these primitives into complicate symbols. The features used in the
inference process are very simple, so the figure set can be easily extended. The
method was tested under regular input circumstance with an average
recognition rate of 90 percent reported.

Keywords: On-Line, Hand-drawn, Figure recognition, Fuzzy inference

1. Introduction

With the introduction of new peripheral equipments of computer, especially the
electronic tablet, many researches have been done to facilitate the interaction between
human and computer [1], [3]. Since the concept of pen computer was proposed in
1960's, most research team focused their interests in the automatic recognition of
handwriting text, and lots of exciting achievements have been gained.. All these
works greatly improve the efficiency of human-machine interface. However, as an
important part of handwriting symbols, figure symbols have aroused little attention in
the field of handwriting recognition although they are widely used. Unlike the text

---

[1] This work is sponsored by national pre-research project.

T. Tan, Y. Shi, and W. Gao (Eds.): ICMI 2000, LNCS 1948, pp. 402–409, 2000.
© Springer-Verlag Berlin Heidelberg 2000

symbols, figure symbols do not have strokes as the basic drawing unit, nor do they have fixed sequence in drawing. So it is more ambiguous to recognize handwriting figure symbols than the text symbols in many cases.

The object of this paper is to present a method for recognizing figure symbols based on primitives separation and fuzzy inference. Our method can process the handwriting figure symbols irrespective of drawing sequence and can extend the figure set easily. The rest of this paper is organized as follows. Section 2 introduces the preprocessing and feature extraction of the input signal. In Sect. 3, we present the algorithm to separate figure symbols into primitives. The fuzzy inference based on these primitives will be discussed in sect. 4. Section 5 gives the result of our experiment and a conclusion will be made in sect. 6.

2. Preprocessing and Feature Extraction

Prior to any recognition, the acquired data is preprocessed in order to remove spurious noise and smooth input points. A simple linear interpolation algorithm is first used to produce an adjacent sequence of points, each of these points then be averaged with its neighborhoods, finally the entirely sequence is resampled to yield an equal spaced point string. We call this preprocessed point string between a pendown and a penlift a component, so a component can be expressed as $P = \{Pt_1, Pt_2, \cdots\cdots, Pt_n\}$.

After preprocessing, the features for recognition should be extracted, which include the boundary rectangle, arc length, chord length and rotating angles. The boundary rectangle represents the minimum rectangle enclose a component, arc length means the total length along the pen trajectory, $la = \sum_{i=1}^{n-1} |P_{i+1} - P_i|$, and the chord length

equals the distance between the start and end point of a component, $lc = |P_n - P_1|$.

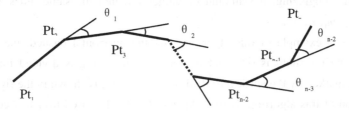

Fig. 1 A simple component and its rotation angle

Rotation Angles describe the direction change in pen moving, which is an important feature. A clockwise rotation results a positive angle and counterclockwise negative, so a component produces an angle string $\theta = \{\theta_1, \theta_2, \cdots\cdots, \theta_{n-3}, \theta_{n-2}\}$ $-\pi < \theta_i < \pi$.

3. Primitive Extraction

As we mentioned before, usually there is no strokes as the basic drawing units of the figure symbols. But many complicated figures can be subdivided into simpler parts like straight-line, curve, circle, ellipse and so on, which we call primitives. The process of primitives separating involves corner detection and primitives recognizing, a single-band integral algorithm is used to combine these two tasks together.

3.1 Single-Band Integral

This algorithm is used to process the rotating angle string, it can smoothes and separates the raw data into different part based on the angle value and magnitude of angle change. The algorithm is as follows:

Step 1: Calculate the sign of each value in the angle string according to the threshold function, $S(\theta) = \begin{cases} -1 & \theta < T \\ 0 & -T < \theta < T \\ 1 & \theta > T \end{cases}$, where T is a predefined threshold.

Step 2: Divide the whole angle string into different blocks so that the values have same sign in succession belong to the same block.

Step 3: Add all the values in each block to produce the resultant integration string $S = \{S_1, S_2, \cdots, S_m\}$, where m is dependent on the property of the angle string. For the convenience of recognizing, the amount of values fall into the same block is also recorded as $N = \{n_1, n_2, \cdots, n_m\}$.

Fig.2 shows an example of this algorithm. As the horizontal dashed line in the figure shows, the threshold is 0.52 radian, so the input string is divided into five blocks that are marked by the vertical dashed line. The result is shown in the figure.

The intention of this algorithm is twofold, the threshold is used to detect corners, and the sum of each block can be used to recognize primitives. Because the input string represents the rotation angle sequence, the peaks in it mark the position of corners in the input component. As for the proportions between corners, a straight-line part means a rotating angle sequence fluctuating near zero, while a curve part leads to a sequence with same sign away from zero. This fact means a larger sum denotes a curve part, a smaller sum imply a straight-line part, so in the Fig. 2(c), the first and second blocks represent the straight-line part, while the fifth block belong to the curve part.

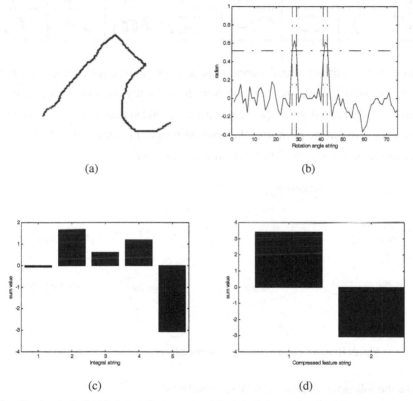

(a)   (b)

(c)   (d)

Fig. 2. A single-band integral example (a)original input handwriting component. (b)the component's rotation angle string. (c)integral value string. (d)second order integral value string.

Let the output integral string be the new input string and apply the algorithm second time, can we get the second order integral string, which we called compressed feature string. This string describes the times of direction change in the drawing of the input component, the drawing direction changes twice in this case as shown in Fig.2 (d).

3.2 Primitive Separation and Recognition

As shown in Table 1, we adopt seven primitives as the basic construction units for a more complicated figure symbol.

Table 1. Primitives

Name	sc	se	Sw	sk	sn	ss	sv
Shape	○	⬭	∼	∽	ᴨᴨᴨ	╱	C

The first five primitives usually appear as a single component, which we call stand-alone primitives, while the last two primitives mostly as a composition of a component so is mentioned as composite primitives. Noticed that a closed component always means a stand-alone primitive, the relation between component and primitive can be expressed as a hierarchical tree structure shown in Fig. 3.

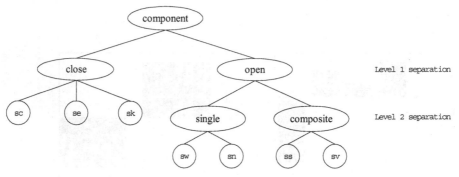

Fig. 3. Hierarchical tree structure of primitives and components

We take the following steps to recognize primitives:

Step 1: Separate closed component. If $\dfrac{lc}{la} < 0.15$ and $lc > 30$, the input component can be treated as closed.

Step 2: Stand-alone primitive recognition. Because the figure's property is ambiguous, the stand-alone primitives are defined by some fuzzy shape grammars. Let primitive sc be out example, its properties are closed, no direction change in drawing and each $\dfrac{S_i}{n_i} > 0.052$ for sure that there is no straight-line part in it. Other primitives can be defined similarly.

Step 3:Composition primitives separation and recognition. Based on the discussion in sect. 3.1, we can define fuzzy rules listed in Table 2 for dividing and recognizing straight-line and curve parts. The fuzzy membership functions for these rules are shown in Fig. 4. After all the primitives be separated and recognized, each primitive's boundary rectangle is recorded, the start and end point of straight-line primitive is also recorded, for curve primitive, its mid point is added.

Table 2. Fuzzy Rules for Primitives Recognition

Rules	S_i	n_i	Type
R1	Small	not-too-short	ss
R2	Large	not-too-short	sv
R3	Large	short	corner point

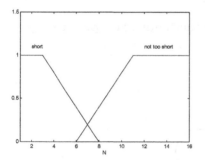 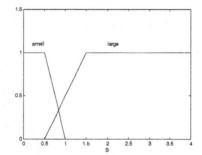

Fig. 4. Fuzzy membership function shapes (a)S_i fuzzy membership; (b) n_i fuzzy membership

4. Fuzzy Inference of Primitives

Because primitives can be combined to build more complicated figure symbols, their relation must be processed. To handle this kind of task, fuzzy rules are widely used [2]. We define six types of relation between these primitives as shown in Fig. 5.

Fig. 5. Relation between primitives (a)conjoint relation (b) conterminous relation (c)intersected relation (d)separated relation (e)overlapped relation (f)containment relation

In Fig. 5, the first three relations are between composite primitives, these three relations describe the end-to-end point, mid-to-end point and mid-to-mid point relation. The last three relations are between components or stand-alone primitives, which describe the relation between the boundary rectangles of these units.

We use fuzzy logic language to describe a complicate figure symbol as shown in Fig. 6.

```
symbol 20 = {ss, ss, sv, sv, sc}

start

(p1, p2, p3, p4) = sort (LtoR, TtoB, ss, ss, sv, sv);

verify(p1 = sv & p2 = ss & p3 = ss & p4 = sv);

verify(IsClosed(p1, p2, p3, p4));

verify(IsUpto(boundrect(sc), boundrect(p1, p2, p3, p4)));

end
```

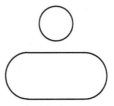

Fig. 6. The No. 20 standard figure symbol

In this definition, reserved token *symbol* lists all the primitives required for a specified symbol. Before any other definition, primitives are first sorted using the *sort* routine in which the primary and secondary sorting order are designated. After sorting is completed, primitives are recorded in an order of from top-left to bottom-right. Next step is to make sure that each primitive is the right type what the symbol required. The routine *verify* judges if the situation is true. If the primitive types are correct, we verify they are closed end to end by *IsClosed* routine that check if the according primitives are conjoined in right order. Finally, the symbol is confirmed if the primitive circle is above the ellipse like symbol combined by four primitives.

Using this kind of definition, new figures can be added easily. Noticed that parameters required, mainly the boundary rectangle, is very simple to be extracted, we also provided a visual edit environment that can build figure library using a drag and drop method.

5. Experimental Results

The method was tested on a figure library comprise 900 symbols, the testing machine using a Pentium II central processing unit. Each tester draws about one hundred figure symbols, the average recognition speed is about 0.02 second, and average recognition rate can be 90 percent under the regular drawing circumstance.

6. Conclusions

This paper has presented a universal framework for modeling and recognition of handwriting figure symbols. In order to make the method robust to large figure set, a primitive set is defined as the basic units. Complicated figure symbols are constructed using fuzzy rules. The standard figure set can be extended easily.

Reference

[1]. Rejean Plamondon and Sargur N. Srihari "On-Line and Off-Line Handwriting Recognition: A Comprehensive Survey" *IEEE Trans. Pattern Analysis and Machine Intelligence,* Vol. 22, No. 1, pp 63-84, January 2000.

[2]. Marc Parizeau and Rejean Plamondon "A Fuzzy-Syntactic Approach to Allograph Modeling for Cursive Script Recognition" IEEE Trans. Pattern Analysis and Machine Intelligence, Vol. 17, No. 7, pp 702-712, July 1995.

[3]. Yannis A. Dimitriadis and Juan Lopez Coronado "Towards an Art Based Mathematical Editor That Uses On-Line Handwritten Symbol Recognition" Pattern Recognition, Vol. 28, No. 6, pp 807-821, 1995.

[4]. Liyuan Li and Weinan Chen "Corner Detection and Interpretation on Planar Curves Using Fuzzy Reasoning" IEEE Trans. Pattern Analysis and Machine Intelligence, Vol. 21, No. 11, pp 1204-1210, Novermber 1999.

Integration MBHMM and Neural Network for Totally Unconstrained Handwritten Numerals Recognition

Dong Lin[1] , *ShanPei Wu[2]* , and *Yuan Bao-Zong[1]*

[1]Institute of Information Science, North Jiaotong University, Beijing 100044, P.R. China
Telephone:86-010-63240626, E-mail:dong_lin@126.com
[2]Beijing University of Posts & Telecommunication, Beijing 100088, P.R. China

Abstract: In this paper we present a method of Multi-branch two dimensional HMM (hidden markov model) for handwritten numeral recognition and another method of Neural network for handwrittern numeral recognition and then integrated the two method into a totally unconstrained handwritten numeral recognition system. The system is composed of a horizontal super state multi-branch two dimensional HMM, a vertical super state multi-branch two dimensional HMM and a Neural network. The integrated recognition system recognition rate is higher than the three subsystem. The data base of handwriting digits used in this paper was collected at Beijing Postal Center, the digits was scanned from letters zip code, altogether 4000 hand writing digits, 2000 are used for training set, 2000 are used for testing set. The training set recognition rate is 99.85%, the testing set recognition rate is 98.05%. If we used more complex decision strategy the system performance will be better.

1 Introduction

In recent years, handwritten character recognition research work was attract many researcher's attention, many new method was proposed. The recent work have been directed toward more sophisticated system [2],[4]. The HMM was successfully applied in speech recognition, In recent year HMM method was used in Handwritten character recognition and get good result. Neural network method for handwritten character recognition was done by many researchers [10],[11]. The above two method have some advantages in many aspect. So in this paper we integrated the two method into a totally unconstrained handwritten numeral recognition system, The system inherent the two methods advantages and perform better then the individual one, The paper was organised as follow, second section was discussed MB 2D HMM method recognition subsystem. The third section introduced the neural network method. The section four is integrate the MB 2D HMM method and neural network into a system, The section five is the concluding remarks. The system is showed as Fig 1.

T. Tan, Y. Shi, and W. Gao (Eds.): ICMI 2000, LNCS 1948, pp. 410–417, 2000.
© Springer-Verlag Berlin Heidelberg 2000

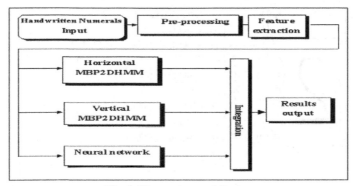

Fig. 1. The system architecture

2 MB2DHMM

In current hand writing recognition study, the main method is doing the alignment and matching work separately. This method have some disadvantages, such as in alignment there are some information been loosing during the alignment, in pattern matching the lost information is unrecoverable. HMM (hidden markov model) method used in speech recognition have got so many progress, it became the major method deal with the speech recognition problem. In recent years HMM was used in character recognition, such as keyword spotting, In HMM method, it joined alignment and matching together. So the result of HMM method is better than conventional 's [11]. In speech recognition, speech signal is 1D signal, and HMM is very suitable for 1D signal. But in character recognition the signal of character is 2D signal, we use conventional HMM is not suitable. 2D HMM is the choice to process 2D signal, but it calculation is NP problem, so we must work out a way to deal with it. 2D Pseudo Hidden Markov Model is the way to solve our problem. It can preserve the main 2D characteristic of the signal, and calculation is much lighter than full 2D HMM. Multi-branch Hidden Markov Model is deduce from HMM, it changes transition probability from one branch to multi-branch. This change is very suitable process such as multi-speecher speech recognition. It gets a good result in Chinese digits recognition [12]. In this paper we use multi-branch 2D HMM for hand writing digits recognition. Because In hand writing digits recognition we must find more robust model to deal the variation of the signal.

2.1 MB 2D PHMM

MB 2D PHMM (multi-branch two dimension pseudo hidden markov model) is composed of a master HMM embedded a slave HMM. The structure of the MB 2D PHMM is showed as Fig 2.

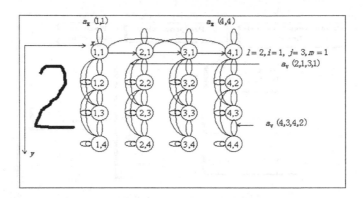

Fig. 2. The structure of the MB 2D PHMM

In master's HMM every state of HMM is a slave HMM. In simplify We use horizontal (x) as master HMM. Vertical is used as slave HMM. The multi-branch is denoted every state's translation have multi rout. In conventional HMM the translation probability a_{ij}. In multi-branch HMM the translation probability is $a_{ij}(k)$. k is denote the branch. For describe clearly. We define the variable as follow.

1) Horizontal state number $N = \{N_x\}$

$$(1)$$

2) Horizontal translation probability $A_x = \{a_h(k,l); 1 \leq k, l \leq N_x\}$

$$a_h(k,l) = 0; k > l, k < l - 1$$

$$(2)$$

3) Horizontal initial probability $\Pi = \{\pi_l; 1 \leq l \leq Nx\}, \pi_l = P(q_1 = S_l)$

$$(3)$$

4) Vertical direction HMM $\Lambda = \{\lambda_j^l; 1 \leq j \leq Ny\}, \lambda_j^l = \{q_{1x} = S_j^l\}$

$$(4)$$

λ_j^l Include parameter as follow:

a) Vertical direction state number N_y^l

b) Vertical translation probability

$$A_y^l = \{a_v(l,i,j,m); 1 \leq i, j \leq N_y^l, 1 \leq m \leq K_{y\max}\}$$

$$a_v(l,i,j,m) = 0; 1 \leq l \leq N_x, i > j, i < j - 1, 1 \leq m \leq K_{y\max}$$

$$(5)$$

c) Vertical translation branch number $K_{y\max}$

d) Vertical direction observe probability

$$B^l = \{b_j(l,m,O_{xy}); 1 \leq l \leq N_x, 1 \leq j \leq N_y^l, 1 \leq m \leq K_{y\max}\}$$

$$b_j(l,m,O_{xy}) = P(O_{xy} | q_{xy} = S_j^l(m))$$
$$(6)$$

e) Vertical direction initial probability

$$\Pi^l = \{\pi_j^l; 1 \le l \le Nx, 1 \le j \le N_y^l\}, \pi_j^l = P(q_{1y} = S_j^l) \quad (7)$$

Multi-branch 2D PHMM's parameter can be described as follow.

$$MBPHMM, \eta = (N, A, \Pi, \Lambda)$$
$$(8)$$

2.2 MB 2D HMM Training

For training the model, we want to reestimation the following parameter. $a_h(k,l)$, $a_v(l,i,j,m)$, $b_j(l,m,O_{xy})$

training algorithm we use multi-branch 2D viterbi algorithm. Algorithm as follow.

Step 1. Divide all hand writing digits into equal segments. Then account all observe vector histogram. Calculate initial observe probability. Put horizontal direction translation probability initial value. Put every slave HMM translation probability initial value.

Step 2. Use multi-branch 2D viterbi algorithm calculate the model's probability, at same time get the best path. Use best path to account histogram of each state, and each state observe vector's histogram.

Step 3. Model parameter reestimation. Reestimate the following parameter. $a_h(k,l)$, $a_v(l,i,j,m)$, $b_j(l,m,O_{xy})$ such as :

$$\hat{a}_h(k,l) = \frac{counter\_a_h \times \delta(q_{x-1} - s_k) \times \delta(q_x - s_l)}{counter\_a_h \times \delta(q_x - s_k)}$$
$$(9)$$

Step 4. Decided if model is convergent. If convergent then finished the training. If model not convergent then goto step 2.

2.3 Experiment and Result

The feature is pixel value of binary data. Data size is 32x32 pixel. We construct hand writing digits 0~9 10 model separately. In recognition stage use 10 model alignment and matching, then select max probability model as the recognition digits. From the experiment we can see MB 2D PHMM could take a new method to solve the character recognition. It use simple feature pixel value get a such result, we may see this method have potential ability. Compared to conventional 2D PHMM the recognition rate of MB 2D PHMM is higher, that is say MB 2D PHMM could describe the property of the model much suitable. The recognition rate of the training set and recognition set is different about 2%, that mean the MB 2D PHMM is robust. As the feature is simple, if we use the more complex feature that can have rotate invariant and move invert the result may be better. Our study of MB 2D PHMM is primitive. Later we would do more work on HMM used for OCR.

3 Neural Network Method

In handwritten character recognition one important work is to extract features, if select suitable feature it can compress the useless information of the pattern and remain the meaningful information. In this part some new feature extraction method used in neural network classifier.

The wavelet transformation is a very useful tool in signal processing, it can decompose the signal at different frequent scale, and remain the time dome characteristic of the signal. The fractal geometry is a recently year new area of research. Use fractal geometry do some signal processing work we can see [1].

Our method used image signal from four direction do projection, the four direction is horizontal, vertical, left diagonal and right diagonal, then do orthogonal wavlet transformation decompose the projection signal at different frequency scale. At the each frequency scale calculate the projection signal fractal dimension as global feature of the image. Used different frequent dome projection signal fractal dimension as feature have some advantages, it can absorb the stroke length different caused influence, and it is rotate invariant, it can get the handwritten numeral global information. In the local feature extraction we used Kirsch edge detector, in [2] we can see it is efficient.

3.1 Feature Extraction

The feature extraction in this paper is dived into two part. The one part is global feature the other is local feature. The global feature is captured the image main characteristic, it play main role in dived the pattern into several part. The local feature is reflected the image local character, it is important to dived the sample to a certain pattern.

3.1.1 Global Feature Extraction

In handwritten numerals image the projection of the image hold the global information of the pattern. So first we do horizontal, vertical, left diagonal and right diagonal projection, and then for each projection curve do orthogonal wavelet transformation decomposed the curve at different frequency scale.

3.1.1.1 Fractal Dimension Calculation

In case of calculation convenience we use boxcounting method to calculate fractal dimension. In n dimension Euclidean space sub set F, the boxcounting fractal dimension D_B define as follow:

$$D_B = \lim_{\delta \to 0} \frac{\log N_\delta(F)}{\log(\frac{1}{\delta})}$$

(10)

where, δ is the cubic edge length, $N_\delta(F)$ is the number of covered cube which edge length is δ.

We obtain totally 60 global D_B feature derive from image projection after wavelet transformation.

3.1.2 Local Feature Extraction

Handwritten numeral is put one-dimensional structure to the two-dimensional space. So line segments reflect main part local information of the pattern. Kirsch edge detector is adequate for local detection of a line segment. Kirsch defined a nonlinear edge enhancement algorithm as [9].

In this paper our data base is 32×32 normalised input image. We compressed it into 6×6 feature map. So we have $4 \times 6 \times 6$ local feature vectors.

3.2 Neural Network Classifier

In this paper we used three-layer multi-layer perceptron (MLP) neural network. The input layer consists of $4 \times 6 \times 6$ local feature maps and 60 global feature vectors. In the hidden layer there are 100 units. The output layer is composed of 10 units. So the neural network has 314 units and 21,400 independent parameters. The neural network is trained with fast back propagation algorithm.

3.3 Experimental Results and Analyse

Data size is 32×32 pixel. The data base which was the same as above section, used multilayer perceptron (MLP) neural network as classifier, used fast BP training algorithm. The recognition rate of training set is 99.55%. The testing set recognition rate is 96.5%.

From the training and testing recognition results we can see that this new features can represent the handwritten numeral global and local character fairly good. Compared to our earlier work [10], that used Two-dimensional PHMM method to recognise handwritten numerals used same data base, the recognition rate is higher then earlier one.

4 Integrated System

Since the above two method have their own advantages. So we try to combine the above two method into one system. The MB 2D HMM method was used in two way. First one we used horizontal super state MB 2D HMM. Second one we used vertical super state MB 2D HMM. The two model have different character, and can complement each other. The third one is the neural network we discussed in section three. We used the same training set as for the development stage of the method 1 and 2. The combination rule is a very simple vote rule. The decision rule can be stated as follows.

$$D(x) = \begin{cases} i, \text{when } D_1(x) = D_2(x) = D_3(x) = i; \\ -1, \text{otherwise} \end{cases}$$

(11)

where $D(x)$ is stand for the decision variable. I is recognition index. Used the decision rule we get the result, the training set recognition rate is 99.85%, the testing set recognition rate is 98.05%.From the result we can see that although the decision rule is very simple but the performance of the system is better than individual one. That mean the integrated system is inherent the advantage of each subsystem good quality.

5 Conclusion

At the present time, we used integrated system combined three subsystems which was used two methods, HMM and neural network method, gets results better than individual subsystem. It is seem if we incorporate more subsystem and more method, the results will be more good. Hidden Markov model method have is advantages that is it have dynamic property which is more suitable to solve such as speech and handwritten character recognition problem. Neural network have very good learning ability, if we use some more sophistic feature as input, the performance will be better. In integrated system the decision rule is very important factor, in our system we just used voting rule as decision rule, if we incorporate human knowledge into the decision rule, our system will perform more good than present one.

References

1. Wang. Y, et al.: Fractal dimensions studying of random sequence. Proc of ICSP'96, Beijing, China, vol.1, (1996) 257-261
2. Seong-Whan Lee.:Off-line Recognition of Totally Unconstrained Handwritten Numerals Using Multilayer Cluster Neural Network. IEEE Trans. PAMI, vol.18, no.6, (1996) 648-652
3. T.Mai and C.Y. Suen.:Generalized Knowledge-Based System for the Recognition of Unconstrained Handwritten Numerals. IEEE Trans. Systems, Man and Cybernetics, vol. 20, no.4, (1990) 835-848

4. C.Y. Suen, C. Nadal, T.A. Mai, R. Legault, and L. Lam.:Recognition of Handwritten Numerals Based on the Concept of Multiple Experts. Proc. of first Int Workshop on Frontiers in Handwriting Recognition, Montreal, Canada, (1990) 131-144
5. A. Krzyzak, W. Dai, and C.Y. Suen.:Unconstrained Handwritten Character Classification Using Modified Backpropagation Model. Proc. of first Int Workshop on Frontiers in Handwriting Recognition, Montreal, Canada,(1990) 155-166
6. Y. Le Cun, et al.:Constrained Neural Network for Unconstrained Handwritten Digit Recognition.   Proc. of first Intl Workshop on Frontiers in Handwriting Recognition, Montreal, Canada, (1990) 145-154
7. S. Knerr, L. Personnaz, and G. Dreyfus.:Handwritten Digit Recognition by Neural Networks with Single-Layer Training. IEEE Trans. Neural Networks, vol.3.no.6 (1992) 962-968
8. S. Mallat.:Multiresolution approximation and wavelet orthonormal bases of $L^2(R)$. Trans. Am. Math. Soc. 315, (1989) 68-87
9. W.K. Pratt.:Digital Image Processing. Wiley, New York (1978)
10. Dong Lin, et al.:Two-dimensional PHMM for Handwriting Digits Recognition.   Proc of ICSP'96, Beijing, China, vol.2, (1996) 1316-1320
11. Sy-shuaw Kou and Oscar E. Agazzi.:Keyword spotting in Poor printed Documents Using Pseudo 2-D Hidden Markov Models. IEEE Trans.PAMI vol.16.no.8 (1994) 842-848
12. Xixian Chen, Yinong Li, Xiaoming Ma and Lie Zhang .:On The Application of Multiple Transition Branch Hidden Markov Models to Chinese Digit Recognition. ICSLP94. (1994) 251-254

Aspect Ratio Adaptive Normalization for Handwritten Character Recognition

Cheng-Lin Liu, Masashi Koga, Hiroshi Sako, Hiromichi Fujisawa

Multimedia Systems Research Dept., Central Research Laboratory, Hitachi Ltd.
1-280 Higashi-koigakubo, Kokubunji-shi, Tokyo 185-8601, Japan
{liucl, koga, sakou, fujisawa}@crl.hitachi.co.jp

Abstract. The normalization strategy is popularly used in character recognition to reduce the shape variation. This procedure, however, also gives rise to excessive shape distortion and eliminates some useful information. This paper proposes an aspect ratio adaptive normalization (ARAN) method to overcome the above problems and so as to improve the recognition performance. Experimental results of multilingual character recognition and numeral recognition demonstrate the advantage of ARAN over conventional normalization method.

1 Introduction

The diversity of writing styles is a major source of difficulty in handwritten character recognition. To alleviate this problem, various normalization techniques have been proposed to reduce the shape variation of character images. Size normalization is most popularly used and was reported to significantly improve the recognition accuracy [1, 2]. Other normalization techniques, such as moment normalization [3], shear transformation, perspective transformation [4], nonlinear normalization [5, 6, 7], etc, were proposed to solve the variations remained by size normalization. They can also be combined to solve multiple sources of variation.

Despite that normalization reduces the shape variation so as to improve the recognition accuracy, it also eliminates some useful information which is important to discriminate characters of similar shape. The geometric features useful for discrimination include the relative size and position of character image in text line, aspect ratio and slant, etc. It was suggested that these features are stored before normalization and then used in post-processing to resolve ambiguities [8]. We feel that integrating these features into segmentation-recognition process other than in post-processing should give better performance. In addition to information loss, the conventional normalization method may give rise to excessive shape distortion, particularly when long-shaped characters are scaled to square shape.

This paper proposes an aspect ratio adaptive normalization (ARAN) method to overcome the above problems of conventional normalization. The aspect ratio of character image is a useful feature inherent in some characters. Rather than considering it in decision stage, we incorporate it into normalization procedure to control the aspect ratio of normalized images. We will show that even though this strategy sounds quite simple, the improvement of recognition accuracy is meaningful,

T. Tan, Y. Shi, and W. Gao (Eds.): ICMI 2000, LNCS 1948, pp. 418-425, 2000.
© Springer-Verlag Berlin Heidelberg 2000

particularly for numeric characters, as alone or in multilingual character recognition.

In ARAN, the aspect ratio of normalized image is a continuous function of the aspect ratio of original image. This strategy is opposed to the aspect ratio-preserving normalization by thresholding the aspect ratio, where the mapping function is not continuous. Aspect ratio thresholding does not perform well because the border between long-shaped characters and block characters is ambiguous. ARAN alleviates the shape distortion of conventional normalization and also, the original aspect ratio is reflected in the normalized image. ARAN can be combined with any other normalization technique, where ARAN is to control the aspect ratio of normalized image while other normalization techniques are to reduce shape variations.

2 Aspect Ratio Adaptive Normalization

2.1 Linear/Nonlinear Normalization

In our experiments, ARAN is embedded in linear normalization and nonlinear normalization, i.e., the images are transformed linearly or nonlinearly with aspect ratio controlled by ARAN. For ease of implementation, the normalization procedure is accomplished by inverse mapping, i.e., a pixel in normalized image inherits the value of a sampled pixel in original image. ARAN can also be implemented by sophisticated interpolation techniques, which were shown to be able to improve the recognition performance [9], while a detailed mathematical treatment of normalization can be found in [10].

Denote the width and height of normalized image by W_2 and H_2 respectively, the coordinates of pixels in original image are mapped to the coordinates in normalized image by

$$T_X(x) = W_2 P_X(x) \qquad T_Y(y) = H_2 P_Y(y)$$

where $P_X(x)$ and $P_Y(y)$ are scaling functions. The scaling functions of linear normalization are

$$P_X(x) = \frac{x}{W_1} \qquad P_Y(y) = \frac{y}{H_1}$$

where W_1 and H_1 are the width and height of original image respectively.

Linear normalization is insufficient to deal with the deformations such as slant, imbalance of gravity and stroke non-uniformity. In handwritten Chinese character recognition, the nonlinear normalization (NLN) based on stroke density equalization [5], [6], [7] is very efficient to deal with stroke density imbalance. The NLN method has various algorithms depending on the computation of stroke density. In previous studies [7], it was shown that the algorithm of Yamada [5] gives the best performance, the algorithm of Tsukumo [6] gives a little lower accuracy but is much less complicated in computation. In our experiments, we adopt the algorithm of Tsukumo. The scaling functions of NLN are computed from stroke density histograms in horizontal and vertical axes.

By ARAN, the size of normalized image (width W_2 and height H_2) is not fixed but adaptive to the aspect ratio of original image via aspect ratio mapping. The mapping functions are introduced in the following subsection.

2.2 Aspect Ratio Mapping Functions

Denote the aspect ratio of original image as $r_1 = W_1/H_1$ if $W_1<H_1$ and $r_1 = H_1/W_1$ if $H_1<W_1$ such that $r_1<1$. Needless to say, when $W_1=H_1$, the character image is exactly square-shaped and the normalized image remains a square. When $W_1<H_1$, the normalized image has fixed standard height H_2 whereas the width is adaptive to the aspect ratio $r_2 = W_2/H_2$. And when $H_1<W_1$, the normalized image has fixed standard width W_2 whereas the height is adaptive to the aspect ratio $r_2 = H_2/W_2$.

The mapping function of aspect ratio thresholding is as

$$r_2 = \begin{cases} r_1 & \text{if } r_1 < t_0 \\ 1 & \text{otherwise} \end{cases}$$

where t_0 is a threshold with $0<t_0<1$. We test four values of threshold $t_0=2/5$, 1/3, 1/4 and 1/5, and refer to the normalization strategies based on them as S2, S3, S4 and S5 respectively. The strategy that the normalized image is always square ($W_2=H_2$=constant) is referred to as S1. Some adaptive strategies are as follows.

- S6:

$$r_2 = \begin{cases} r_1 + 0.5 & \text{if } r_1 < 0.5 \\ 1 & \text{otherwise} \end{cases}$$

- S7:

$$r_2 = \begin{cases} 1.5r_1 + 0.25 & \text{if } r_1 < 0.5 \\ 1 & \text{otherwise} \end{cases}$$

- S8:

$$r_2 = \sqrt{\sin \frac{\pi}{2} r_1}$$

- S9:

$$r_2 = \begin{cases} 0.8\sin \frac{2\pi}{3} r_1 + 0.2 & \text{if } r_1 < 0.75 \\ 1 & \text{otherwise} \end{cases}$$

The curves of some mapping functions are plotted in Fig. 1.

In implementation of ARAN, we still use a square frame for normalized images. Either the horizontal axis or vertical axis of normalized image has full size. In the dimension of deficient size, the image is centered in the square frame. Fig. 2 shows some examples of conventional normalization (S1) and ARAN, where the images of upper row are transformed by linear normalization. The images of middle row have full square size, while the images of lower row have aspect ratio controlled by ARAN strategy S8. It is shown that ARAN considerably alleviates the distortion of conventional normalization.

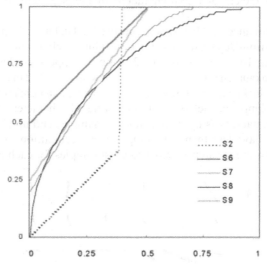

Fig. 1. Aspect ratio mapping functions

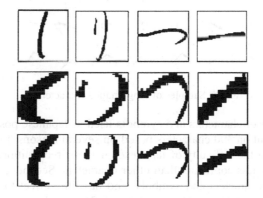

Fig. 2. Normalized images by conventional method and ARAN

3 Experimental Results

3.1 Database, Feature Extraction and Classification

We have test the ARAN method in multilingual character recognition as well as numeral recognition. The application background is Japanese mail address reading [11]. It is a problem of multilingual character recognition in that Kanji (Chinese characters), hiragana, katakana, Roman letters, Arabic numerals, and some symbols are used concurrently. Some characters naturally have elongated shape (Fig. 3 shows some samples). In recognition, all the segmented character images are normalized to square frame as for Chinese characters. By conventional normalization, the distortion of long-shaped characters and non-character image segments may lead to mis-recognition.

To test the performance of ARAN, we compiled a database composed of character images segmented from Japanese mail images. This database has 300 categories of characters, including 10 Arabic numerals, 24 English upper-case letters, 65 hiragana characters, one katakana character (the second row in Fig. 3. This character is often used as a delimiter in Japanese address while other katakana characters are not often used), and 200 Kanji characters. For multilingual character recognition (300 categories), each character has up to 400 training samples and up to 400 test samples (some classes may have less than 400 samples). In recognition of Arabic numerals alone, we use 1000 training samples and 1000 test samples for each class.

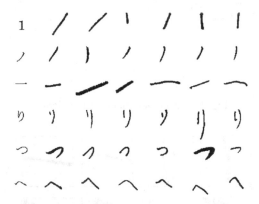

Fig. 3. Samples of long-shaped characters

The total number of characters and symbols used in Japanese postal mails is about 4000. We experiment on 300 characters based on several reasons. First, this character set well accounts for the difficulty of Japanese character recognition, where the Kanji characters are much less confusing than other characters. Second, ARAN is effective mainly to numeric characters and symbols, and has little effect to Kanji characters. Third, in lexicon-driven mail address reading [12], the character set to classify is dynamic depending on linguistic context. The maximum number of characters to classify is less than 500 and the average number is less than 100.

For multilingual character recognition, each sample is represented as a feature vector of 256 measurements. The character image is first scaled to 64x64 grid by linear or nonlinear normalization with aspect ratio controlled by 9 strategies S1~S9. The normalized image is smoothed by a connectivity-preserving procedure and the contour pixels are assigned to four directional planes corresponding to four orientations [13]. From a directional plane, 8x8 blurring masks are imposed and the convolution of a mask with the plane gives a measurement. The blurring mask is a Gaussian filter with the variance parameter determined by the sampling theorem.

For numeral recognition, each sample is represented by 64 feature measurements. The character image is scaled to size 32x32 by linear normalization. After normalization, smoothing and directional decomposition as for multilingual recognition, 4x4 measurements are extracted from each directional pane by blurring.

Two classification schemes are used in recognition. The Euclidean distance between the test pattern and the mean feature vector of each class is used as an instance of naive classifier. The modified quadratic discriminant function (MQDF2) proposed by Kimura et al. [14, 15] is used as a sophisticated classifier. MQDF2 was proposed to overcome the bias of covariance matrix estimated on limited samples by replacing the minor eigenvalues with a greater constant. The MQDF2 between a test pattern $\mathbf{x}$ and a class ω_i is computed by

$$g_2(\mathbf{x},\omega_i) = \sum_{j=1}^{k} \frac{1}{\lambda_{ij}}[\phi_{ij}^{T}(\mathbf{x}-\mu_i)]^2 + \sum_{j=k+1}^{n} \frac{1}{\delta}[\phi_{ij}^{T}(\mathbf{x}-\mu_i)]^2 + \sum_{j=1}^{k}\log\lambda_{ij} + \sum_{j=k+1}^{n}\log\delta$$

$$= \sum_{j=1}^{k} \frac{1}{\lambda_{ij}}[\phi_{ij}^{T}(\mathbf{x}-\mu_i)]^2 + \frac{1}{\delta}\{\|\mathbf{x}-\mu_i\|^2 - \sum_{j=1}^{k}[\phi_{ij}^{T}(\mathbf{x}-\mu_i)]^2\} + \sum_{j=1}^{k}\log\lambda_{ij} + (n-k)\log\delta$$

where μ_i denotes the mean vector of class χ, λ_{ij} denote the eigenvalues sorted in decreasing order, ϕ_{ij} are the corresponding eigenvectors, δ is the modified value of minor eigenvalues, n is the dimensionality of feature vector, and k is the number of principal eigenvectors. MQDF2 is also helpful to save memory space and computation compared to traditional quadratic discriminant function (QDF) because only the projections onto principal eigenvectors are computed. MQDF2 has produced promising results in handwritten numeral recognition as well as in Chinese character recognition [15, 16].

In our experiments, the number k is set to 40 for multilingual character recognition and 30 for numeral recognition. The parameter δ is class-independent and is tuned so as to give high accuracy. Empirically, the value of δ is proportional to the variance of training samples. In multilingual recognition by MQDF2, in order to reduce the computation cost, the Euclidean distance classifier is used to select 30 candidate classes then MQDF2 is then use to identify the unique class.

3.2 Recognition Results

In multilingual character recognition, we investigate the overall recognition rate as well as the correct rates of two disjoint subsets. The subset 1 contains 10 Arabic numerals, a katakana (as in the second row of Fig. 3) and a Kanji (the third row of Fig. 3), and the subset 2 contains the rest 288 characters. In the following, we refer to the subset 1 simply as numeric characters even though it is not purely numeric. The recognition rates from linear normalization are listed in Table 1, and the results of nonlinear normalization are listed in Table 2. The terms "Rate 1", "Rate 2", and "Overall" denote the recognition rates of subset 1, subset 2, and whole character set respectively. It is evident that nonlinear normalization improves the accuracy of non-numeric characters compared to linear normalization. However, for numeric characters, nonlinear normalization does not exhibit advantage. In comparison of two classifiers, MQDF2 is superior to Euclidean distance.

In comparison of the aspect ratio strategies, the ARAN strategies S6~S9 generally outperform the conventional normalization S1 and aspect ratio thresholding strategies S2~S5. The difference mainly lies in the accuracy of numeric characters, while for non-numeric characters, the difference of accuracy is not so evident. It is noteworthy that in the case of nonlinear normalization and MQDF2 classification, the aspect ratio threshold strategy S5 gives highest accuracy to numeric characters, but the difference from ARAN strategies is marginal. Overall, the ARAN strategy S8 provides the best performance, either for numeric characters or in overall recognition accuracy.

Table 1. Recognition accuracies from linear normalization

Strategy	Euclidean			MQDF2		
	Rate 1	Rate 2	Overall	Rate 1	Rate 2	Overall
S1	75.46	88.00	87.46	92.65	97.73	97.52
S2	71.75	87.72	87.04	92.35	97.62	97.39
S3	74.02	87.86	87.28	92.73	97.69	97.48
S4	75.06	87.97	87.42	93.13	97.72	97.52
S5	75.17	87.99	87.44	93.39	97.72	97.53
S6	75.71	88.04	87.51	92.98	97.72	97.52
S7	76.65	88.01	**87.58**	93.23	97.73	97.54
S8	**77.50**	87.96	87.51	**93.58**	97.75	**97.57**
S9	77.10	87.99	87.53	93.17	97.73	97.54

Table 2. Recognition accuracies from nonlinear normalization

Strategy	Euclidean			MQDF2		
	Rate 1	Rate 2	Overall	Rate 1	Rate 2	Overall
S1	75.17	92.80	92.05	93.73	98.51	98.30
S2	71.81	92.50	91.62	92.50	98.35	98.10
S3	74.75	92.66	91.90	93.19	98.46	98.24
S4	75.94	92.76	92.05	93.67	98.48	98.28
S5	76.25	92.78	92.08	**93.94**	98.50	98.30
S6	76.17	92.84	92.13	93.90	98.52	98.32
S7	75.96	92.82	92.10	93.69	98.52	98.31
S8	**76.44**	92.91	**92.21**	93.88	98.56	**98.36**
S9	75.98	92.89	92.17	93.83	98.53	98.33

Table 3. Accuracies of handwritten numeral recognition

Strategy	S1	S2	S3	S4	S5	S6	S7	S8	S9
Euclidean	88.52	86.89	89.83	90.05	89.68	89.85	**90.41**	90.32	90.37
MQDF2	98.69	98.10	98.44	98.60	98.69	98.82	**98.85**	98.77	98.76

The recognition accuracies of handwritten numeral recognition are given in Table 3. Again it is shown that either by Euclidean distance or MQDF2 classification, the ARAN strategies outperform conventional normalization and aspect ratio thresholding strategies. Among these strategies, the ARAN strategy S7 yields the best performance. Further investigating into the confusion of characters, we found that by ARAN, the substitution of "1" to "2, 3, 5" and the substitution of "9" to "3, 8" were considerably decreased.

4 Conclusion

The motivation behind ARAN is to alleviate the shape distortion caused by conventional normalization and to preserve useful geometric information. It was proven to be effective to improve recognition performance in experiments of multilingual character recognition and numeral recognition. In character field recognition integrating segmentation and classification, more geometric features are available, such as the relative size and relative position of character images. The integration of these features into normalization or classification will be beneficial to overall recognition.

References

1. A. Gudessen, Quantitative analysis of preprocessing techniques for the recognition of handprinted characters, Pattern Recognition 8 (1976) 219-227.
2. S. Mori, K. Yamamoto, M. Yasuda, Research on machine recognition of handprinted characters, IEEE Trans. Pattern. Anal. Machine. Intell. 6(4) (1984) 386-405.
3. R.G. Casey, Moment normalization of handprinted character, IBM J. Res. Dev. 14 (1970) 548-557.
4. G. Nagy, N. Tuong, Normalization techniques for handprinted numerals, Communi. ACM 13(8) (1970) 475-481.
5. H. Yamada, K. Yamamoto, T. Saito, A nonlinear normalization method for handprinted Kanji character recognition--line density equalization, Pattern Recognition 23(9) (1990) 1023-1029.
6. J. Tsukumo, H. Tanaka, Classification of handprinted Chinese characters using non-linear normalization and correlation methods, Proc. 9th ICPR, Roma, Italy, 1988, pp.168-171.
7. S.-W. Lee, J.-S. Park, Nonlinear shape normalization methods for the recognition of large-set handwritten characters, Pattern Recognition 27(7) (1994) 895-902.
8. J. Schüermann, et al., Document analysis--from pixels to contents, Proc. IEEE 80(7) (1992) 1101-1119.
9. G. Srikantan, D.-S. Lee, J.T. Favata, Comparison of normalization methods for character recognition, Proc. 3nd ICDAR, Montreal, Canada, 1995, pp.719-722.
10. S. Mori, H. Nishida, H. Yamada, Optical Character Recognition, John Wiley & Sons, 1999, Chapter 3: Normalization, pp.60-104.
11. H. Ogata, et al., A method for street number matching in Japanese address reading, Proc. 5th ICDAR, Bangalore, India, 1999, pp.321-324.
12. M. Koga, R. Mine, H. Sako, H. Fujisawa, Lexical search approach for character-string recognition, Document Analysis Systems: Theory and Practice, S.-W. Lee and Y. Nakano (Eds), Springer, 1999, pp. 115-129.
13. C.-L. Liu, Y-J. Liu, R-W. Dai, Preprocessing and statistical/structural feature extraction for handwritten numeral recognition, Progress of Handwriting Recognition, A.C. Downton and S. Impedovo (Eds.), World Scientific, 1997, pp.161-168.
14. F. Kimura, et al., Modified quadratic discriminant functions and the application to Chinese character recognition, IEEE Trans. Pattern Anal. Machine Intell. 9(1) (1987) 149-153.
15. F. Kimura, M. Shridhar, Handwritten numeral recognition based on multiple algorithms, Pattern Recognition 24(10) (1991) 969-981.
16. F. Kimura, et al., Improvement of handwritten Japanese character recognition using weighted direction code histogram, Pattern Recognition 30(8) (1997) 1329-1337.

A Neural-Network Dimension Reduction Method for Large-Set Pattern Classification*

Yijiang Jin[1], Shaoping Ma[2]

State Key Laboratory of Intelligence Technology and System
Department of Computer Science, Tsinghua University
Beijing 100084, China
[1]yijiangjin@sina.com, [2]msp@tsinghua.edu.cn

Abstract. High-dimensional data are often too complex to be classified. K-L transformation is an effective dimension reduction method. However its result is not satisfactory in large-set pattern classification. In this paper a novel nonlinear dimension reduction method is presented and analyzed. The transform is achieved through a multi-layer feed-forward neural network trained with K-L transformation result. Experimental results show that this method is more effective than K-L transformation being applied in large-set pattern classification such as Chinese character recognition.
Keywords. Dimension Reduction, Neural Network, Principal Components Analysis, Chinese Character Recognition

1 Introduction

It's usually a very complex task to analyze high-dimensional data set. Pattern recognition systems are often bothered by high-dimensional problems. Dimension reduction can always decrease the complexity of computing and speed up the recognition process. Furthermore, the subordination components may come from noise. So it may be beneficial for classification to wipe them off.

There are many methods to reduce the number of dimension, such as principal components analysis, scene analysis and feature cluster, etc.[1], but none of them are very suitable for large-set high-dimensional pattern classification.

A novel nonlinear transformation method for dimension reduction has been proposed in this paper. A point in high-dimensional space is mapping to a low dimension space by this transformation (called as NN transformation in the rest of this paper) using a neural network. The neural network was trained by back propagation method using results of Karhunen-Loeve transformation as teacher.

Chinese character recognition is a typical large-set high-dimensional pattern classification task. It has more than three thousand classes. The pattern vector has hundreds of dimensions. We have tested our dimension reduction method with

---

* Supported by the National Natural Science Foundation and the "863" National High Technology Foundation

T. Tan, Y. Shi, and W. Gao (Eds.): ICMI 2000, LNCS 1948, pp. 426-433, 2000.

Chinese character patterns. The experimental results show that it is effective in such cases.

2   Principal Components Analysis (K-L Transformation)[1]

Principal components analysis (PCA, also called K-L transformation) is an optimal orthogonal transform in the sense of minimal mean square error. It can remove the correlation among the variables.

Suppose X is an n-dimensional random vector, $X = [x_1, x_2, \cdots, x_n]^T$. Its mean is 0 and its covariance matrix is $C_X = E\{XX^T\}$. Its eigenvalues and eigenvectors are λ_i and $e_i, (i = 1, 2, \cdots, n)$ respectively.

Suppose $\lambda_1 > \lambda_2 > \cdots > \lambda_n$.

Let $\Phi = [e_1, e_2, \cdots, e_m]$

Then

$$Y = \Phi^T X \qquad (1)$$

defines Karhunen-Loeve transformation.

3   Architecture and Algorithm of NN Transformation

We hope that after dimension reduction transform, the patterns belonging to the same class should congregate together. At the same time, the neighbor patterns should still be close to each other after the transform. In order to meet this goal, we use a multi-layer feed-forward network to map a high-dimensional vector to a low-dimensional vector.

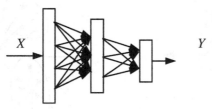

Fig. 1. Architecture of NN Transformation

The architecture of the network is shown as Fig. 1.

It has been proved that multi-layer perceptron with one hidden layer and nonlinear transformation function has the ability to approach any complex function[1]. So if only the right training set has been selected suitably, the multi-layer perceptron can achieve the transform we need.

We use the following algorithm to construct training set.

Assume a classification task with k classes, $C_1, C_2, \cdots, C_k$. Every pattern X is an n-dimensional vector, $X = [x_1, x_2, \cdots, x_n]^T$.

The mean vector (central point) of each class is $\overline{X}_1, \overline{X}_2, \cdots, \overline{X}_k$ respectively. The central point needs not to be a real pattern.

Vector X is transformed to vector Y after the dimension reduction operation.

$$Y = F(X) \tag{2}$$

where F is the transformation function, Y is an m-dimensional vector.

$$Y = [y_1, y_2, \cdots, y_m]^T \tag{3}$$

Generally we have $m \ll n$.

The training set is constructed in two steps. The first one is to calculate K-L transform matrix Φ from training pattern set S through PCA. Φ is an $n \times m$ matrix. The second step is to calculate the corresponding teacher vector T_X of every pattern $X \in S$ using the following formula:

$$T_X = \Phi^T \overline{X}_i \text{ if } X \in C_i, i = 1, 2, \cdots, k \tag{4}$$

Thus we map all patterns belonging to a same class to just one point which is the K-L transform result of the central point of this class.

Back propagation algorithm[1] is employed to train the network after the training set is constructed. Then the trained neural network is applied to carry out the NN transformation.

4   Analysis and Comparison with K-L Transformation

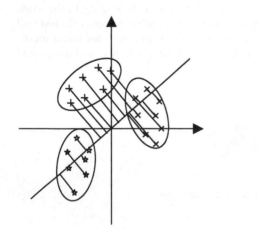

Fig. 2 Illustration of K-L transformation

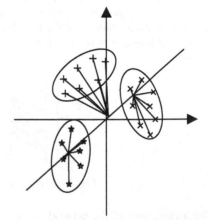

Fig. 3 Illustration of NN transformation

The difference between NN transformation and K-L transformation is shown in Fig. 2 and Fig. 3.

As a kind of global linear transformation, K-L transformation uses the same transform matrix for all patterns, despite the fact that they belong to different classes. When K-L transformation is used to reduce dimension, the subordinate components are omitted.

This can lead to information lost. So the classes may overlap each other after transform even though they might be dividual in the high-dimensional space, especially when the number of classes is very large.

On the contrary, NN transformation has the ability of local mapping. Patterns of different classes will be treated discriminatingly. In fact, it can firstly classify an input pattern and then map it to the corresponding point. All patterns belonging to a same class will be mapped to one point so that the classes will not overlap each other.

NN transformation and K-L transformation are still tightly related to each other. The space of NN transformation is the same with that of K-L transformation, so that it is orthogonal, too. The training set of NN transformation is constructed from the result of K-L transformation. So NN transformation is a quasi-continuous transform. This means that even if a pattern is mis-classified, it can still be mapped to a neighbor class.

5 Experiment and Result

The patterns in the experiment are features obtained from Chinese character samples[2]. There are 300 training patterns and 200 testing ones for each class.

The neural networks have 1 or 2 hidden layers. We have tested various numbers of neurons in hidden layers. The input layer has 256 neurons because the feature we used has 256 dimensions. The output layer has 2 or 4 neurons according to the object dimension number of dimension reduction. The training method is resilient back propagation.

The experimental results are shown in tables 1 to 4. The performance is measured by two parameters. One is the class-distributing radius after transform; the other is the overlapping degree between classes, which is measured by the number of classes covered by the distributing area of one class.

The results of 2-dimensional experiments are shown in table 1 and 2. The results of 4-dimensional ones are shown in table 3 and 4.

Table 1. class-distributing radius of 2-dimensional experiments

a	b	c	d	e	f	g	h
10	130	0.118	0.379	0.207	0.030	0.132	0.053
10	100	0.118	0.379	0.207	0.031	0.126	0.055
10	62	0.118	0.379	0.207	0.025	0.113	0.044
10	38	0.118	0.379	0.207	0.031	0.132	0.055
10	30	0.118	0.379	0.207	0.028	0.131	0.050
10	25	0.118	0.379	0.207	0.032	0.139	0.058
10	70x20	0.118	0.379	0.207	0.012	0.069	0.021
10	50x20	0.118	0.379	0.207	0.015	0.085	0.028
10	35x10	0.118	0.379	0.207	0.025	0.121	0.047

Table 1. (continued)

50	60x20	0.105	0.313	0.182	0.037	0.171	0.067
50	80x20	0.105	0.328	0.183	0.050	0.197	0.089
100	50x20	0.102	0.302	0.177	0.059	0.201	0.102
100	50x20	0.104	0.313	0.181	0.053	0.190	0.093
100	80x20	0.104	0.313	0.181	0.058	0.199	0.101

a: Class number
b: Neuron number in hidden layer
c: Average radius obtained by K-L transformation
d: Maximum radius obtained by K-L transformation
e: 90% pattern radius obtained by K-L transformation
f: Average radius obtained by NN transformation
g: Maximum radius obtained by NN transformation
h: 90% pattern radius obtained by NN transformation

Table 2. overlapping degree of 2-dimensional experiments

a	b	c	d	e	f
10	130	8.4	6.5	2.3	1.7
10	100	8.4	6.5	2.1	1.6
10	62	8.4	6.5	2	1.6
10	38	8.4	6.5	2.1	1.7
10	30	8.4	6.5	2.7	1.8
10	25	8.4	6.5	2.5	1.9
10	70x20	8.4	6.5	1.5	1.2
10	50x20	8.4	6.5	1.6	1.3
10	35x10	8.4	6.5	2.2	1.8
50	60x20	27.34	23.44	11.14	5.78
50	80x20	29.18	23.5	13.72	8.18
100	50x20	51.73	43.96	29.08	19.45
100	50x20	54.08	45.37	26.38	16.41
100	80x20	54.08	45.37	28.26	19

a: Class number
b: Neuron number in hidden layer
c: Class number covered by maximum radius obtained by K-L transformation
d: Class number covered by 1.5 * 90% pattern radius obtained by K-L transformation
e: Class number covered by maximum radius obtained by NN transformation
f: Class number covered by 1.5 * 90% pattern radius obtained by NN transformation

Table 3. class-distributing radius of 4-dimensional experiments

a	b	c	d	e	f	g	h
10	70x20	0.154	0.418	0.241	0.020	0.103	0.034
10	80x20	0.154	0.418	0.241	0.030	0.133	0.048
10	150x40	0.154	0.418	0.241	0.026	0.099	0.042
10	70x20	0.154	0.418	0.241	0.028	0.105	0.047

a: Class number
b: Neuron number in hidden layer
c: Average radius obtained by K-L transformation
d: Maximum radius obtained by K-L transformation
e: 90% pattern radius obtained by K-L transformation
f: Average radius obtained by NN transformation
g: Maximum radius obtained by NN transformation
h: 90% pattern radius obtained by NN transformation

Table 4. overlapping degree of 2-dimensional experiments

a	b	c	d	e	f
10	70x20	9	7.6	1.9	1.6
10	80x20	8	6.1	1.6	1.2
10	150x40	8	6.1	1.2	1.2
10	70x20	8	6.1	1.2	1.2

a: Class number
b: Neuron number in hidden layer
c: Class number covered by maximum radius obtained by K-L transformation
d: Class number covered by 1.5 * 90% pattern radius obtained by K-L transformation
e: Class number covered by maximum radius obtained by NN transformation
f: Class number covered by 1.5 * 90% pattern radius obtained by NN transformation

Fig.4 shows some examples of NN transformation compare with K-L transformation. × means the result point of NN transformation. + means the result point of K-L transformation. It's clear that the results of NN transformation are better than ones of K-L transformation. From the experimental result, it can be seen that the performance of NN transformation is much better than that of K-L transformation.

6 Summary

The dimension reduction method using NN transformation is a promising method for pattern classification. It combines the strong classification ability of neural network with the advantage of K-L transformation, which is a continuous transformation. We can expect that the speed of Chinese character recognition will be improved dramatically by using this method. The next phase of our research work is to apply this method into practical recognition task.

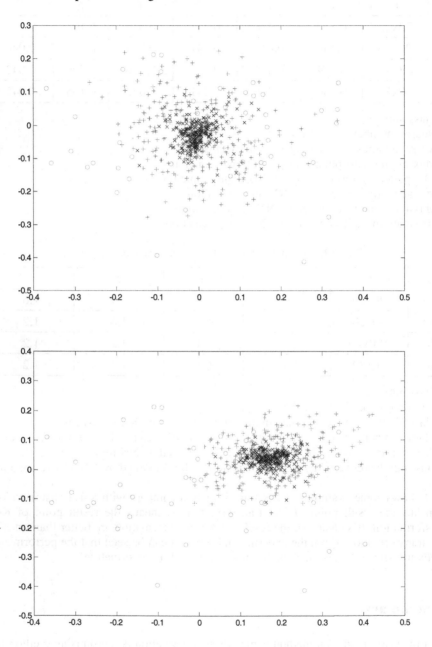

Fig. 4 Some results of the NN transformation

References

1. A. S. Pandya and R. B. Macy, *Pattern Recognition with Neural Networks in C++*, CRC Press, 1996
2. Ma Shaoping et al, *Handwritten Chinese characters recognizing based on fuzzy directional line element feature*, Journal of Tsinghua University (Sci. & Tech.), Vol.37 No.3, p42~45, March 1997

Local Subspace Classifier in Reproducing Kernel Hilbert Space

Dongfang Zou

Character Recognition Laboratory, Institute of Automation, Chinese Academy of Sciences,
P. O. Box 2728, Beijing 100080, China
Dfzou@hw.ia.ac.cn, Dfzou@yahoo.com

Abstract. Local Subspace Classifier(LSC) is a new classification technique, which is closely related to the subspace classification methods, and a heir of prototype classification methods. And it is superior to both of them. In this paper, a method of improving the performance of Local Subspace Classifier is presented. It is to avoid the intersection of the local subspaces representing the respective categories by mapping the original feature space into RKHS(Reproducing Kernel Hilbert Space).

1 Introduction

Local Subspace Classifier is a new classification technique[1]. It is closely related to the subspace classification methods, and an heir of prototype classification methods, such as the k_NN rule [2][3][4]. It fills the gap between the subspace and prototype principles of classification. From the domain of the prototype-based classifiers, LSC brings the benefits related to the local nature of the classification, while it simultaneously utilizes the capability of the subspace classifiers to produce generalizations from the training samples.

In fact, the local subspaces are the linear manifolds of the original feature space. Because the dimensionality of the feature space is usually finite and comparatively small, the local subspaces often have intersections. When an unlabeled sample is in or near one of the intersections, the distance between it and each of the corresponding two manifolds is zero sheerly or approximately. The difference between the distances is small, and we can not classify it.

The similar problem also exists in the conventional subspace classifier, which has been solved by mapping the feature space into the Hilbert Space[5]. Similarly, we map the feature space into RKHS to improve the performance of the Local Subspace Classifier.

The rest of this paper is organized as follows: In section 2, the principle of Local subspace Classifier is introduced; In Section 3, the concepts of general RKHS and the RHKS we use($F_N(\rho)$)are given; In Section 5, the algorithm of Local Subspace Classifier in RKHS is presented; Section 6 contains experimental results and conclusion.

T. Tan, Y. Shi, and W. Gao (Eds.): ICMI 2000, LNCS 1948, pp. 434-441, 2000.
© Springer-Verlag Berlin Heidelberg 2000

2 Local Subspace Classifier

A D-dimensional linear manifold L of the d-dimensional real space is defined by a matrix $U \in R^{d \times D}$ of rank D, and an offset vector $\mu \in R^d$, provided $D \le d$,

$$L_{U,\mu} = \{x \mid x = Uz + \mu, z \in R^D\}$$

The same method can alternatively be defined by $D+1$ prototypes provided that the set of prototypes is not degenerate. The prototypes forming the classifier are marked m_{ij}, where $j = 1,\ldots,c$ is the index of the category and $i = 1,\ldots,N_j$ indexes the prototypes in that class. The manifold dimension D may be different for each class. Therefore, the class-dependent dimensions are denoted D_j. When we are to classify vector x, the following is done for each category $j = 1,\ldots,c$:

Step 1. Find the $D_j + 1$ prototypes closest to x and denote them $m_{0j},\ldots,m_{D_jj}$

Step 2. Form a $d \times D_j$-dimensional basis of the vectors

$$\{m_{1j} - m_{0j},\ldots,m_{D_jj} - m_{0j}\}$$

Step 3. Orthogonize the basis to get the matrix $U_j = \begin{pmatrix} U_{1j} & \cdots & U_{D_jj} \end{pmatrix}$

Step 4. Find the projection of $x - m_{0j}$ on the manifold $L_{U_j,m_{0j}}$:

$$\hat{x}'_j = U_j U_j^T (x - m_{0j})$$

Step 5. Calculate the residue of x relative to the manifold $L_{U_j,m_{0j}}$:

$$\tilde{x}_j = x - \hat{x}_j = x - (m_{0j} + \hat{x}'_j) = (I - U_j U_j^T)(x - m_{0j})$$

The vector x is then classified according to minimal $\|\tilde{x}_j\|$ to the class j, i.e.

$$g_{LSC}(x) = \arg\min_{j=1,\ldots,c} \|\tilde{x}_j\|$$

In any case, the residue length from the input vector x to the linear manifold is equal to or smaller than the distance to the nearest prototype, i.e.

$$\|\tilde{x}_j\| \le \|x - m_{0j}\|$$

These entities are sketched in Fig.1.

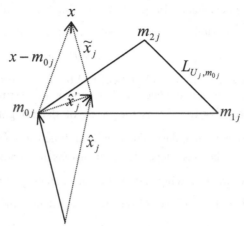

Fig. 1. The d-dimensional entities involved in the LSC classification when $D_j = 2$

3   Mapping the Feature Space into RKHS

3.1   About RKHS

Considering the Hilbert space $H(\chi)$ based on domain K and composed of functions(functionals) defined on set χ,

$\forall x \in \chi$, We design a functional F_x on H as follows:

$$F_x(f) = f(x) \ (\forall f \in H), \text{ then } F_x \text{ is linear.}$$

If F_x is continuous or bounded on H, then due to the Riesz representation theorem, there exists an only $K_x \in H$, satisfying:

$$f(x) = F_x(f) = < f, K_x >$$

Now+ we define a mapping $K : \chi \times \chi \to K$, satisfying $K(x, y) = < K_x, K_y >$ $(\forall x, y \in \chi)$, then:

$$K_x(y) = < K_x, K_y > = K(x, y)$$

$$K(x,\cdot) = K_x \in H$$

We call K the reproducing kernel on Hilbert Space $H(\chi)$, and correspondingly call $H(\chi)$ the Reproducing Kernel Hilbert Space.

3.2 Mapping the Feature Space into $F_N(\rho)$

$F_N(\rho)$ is an infinite dimensional RKHS, which was put forward by Rui.J.P.deFigueiredo, who constructed OI(Optimal Interpolative)-Net[6][7]. It is sketchily introduced as follows:(with details in [8])

Assume $x = (\xi_1 \quad \xi_2 \quad \cdots \quad \xi_N)^T \in R^N$

Define $f(x) = \sum_{k_1=0}^{\infty} \sum_{k_2=0}^{\infty} \cdots \sum_{k_N=0}^{\infty} f_{k_1 k_2 \cdots k_N} \dfrac{\xi_1^{k_1} \xi_2^{k_2} \cdots \xi_N^{k_N}}{k_1! \, k_2! \quad k_N!}$

$$\left(f_{k_1 k_2 \cdots k_N} \in R, k_1, k_2, \ldots, k_N = 0,1,\ldots,\infty \right) \quad (1)$$

Define $F_N = \{ f \mid f_{k_1 k_2 \cdots k_N} \in R, k_1, k_2, \ldots, k_N = 0,1,\ldots,\infty \}$, then F_N is a linear space.

Given a bounded sequence $\{ \rho_{k_1 k_2 \cdots k_N} \geq 0 \mid k_1, k_2, \ldots, k_N = 0,1,\ldots,\infty \}$,

$\forall f, g \in F_N$, We define the inner product of f and g :

$$< f, g > = \sum_{k_1=0}^{\infty} \sum_{k_2=0}^{\infty} \cdots \sum_{k_N=0}^{\infty} \frac{\rho_{k_1 k_2 \cdots k_N}}{k_1! k_2! \cdots k_N!} f_{k_1 k_2 \cdots k_N} g_{k_1 k_2 \cdots k_N} \quad (2)$$

Define $F_N(\rho) = \left\{ f \mid f \in F_N, \sum_{k_1=0}^{\infty} \sum_{k_2=0}^{\infty} \cdots \sum_{k_N=0}^{\infty} \frac{\rho_{k_1 k_2 \cdots k_N}}{k_1! k_2! \cdots k_N!} \left| f_{k_1 k_2 \cdots k_N} \right|^2 < \infty \right\}$,

then $F_N(\rho)$ is a subspace of F_N.

$\forall f, g \in F_N(\rho)$, Define:

$$< f, g >_{F_N(\rho)} = < f, g >_{F_N}, \quad \|f\| = \sqrt{< f, f >}$$

then $F_N(\rho)$ is a Hilbert space.

$\forall x = (\xi_1 \quad \xi_2 \quad \cdots \quad \xi_N)^T \in R^N$, Define the functional F_x on $F_N(\rho)$:

$$F_x(f) = f(x) \qquad (\forall f \in F_N(\rho))$$

then F_x is a bounded linear functional, which has been elucidated in [8].

So there exists $K_x \in F_N(\rho)$, satisfying:

$$F_x(f) = f(x) = < f, K_x > \tag{3}$$

and K_x is unique.

Observe (1), (2) and (3), and it is easy to ensure:

$$K_x(w) = \sum_{k_1=0}^{\infty} \sum_{k_2=0}^{\infty} \cdots \sum_{k_N=0}^{\infty} \frac{\xi_1^{k_1} \xi_2^{k_2} \cdots \xi_N^{k_N}}{\rho_{k_1 k_2 \cdots k_N}} \frac{\eta_1^{k_1}}{k_1!} \frac{\eta_2^{k_2}}{k_2!} \cdots \frac{\eta_N^{k_N}}{k_N!}$$

$$\left(w = \begin{pmatrix} \eta_1 & \eta_2 & \cdots & \eta_N \end{pmatrix}^T \in R^N \right)$$

More over, we can assert that $F_N(\rho)$ has a reproducing kernel, which is

$$K(x,w) = < K_x, K_w > = \sum_{k_1=0}^{\infty} \sum_{k_2=0}^{\infty} \cdots \sum_{k_N=0}^{\infty} \frac{(\xi_1\eta_1)^{k_1} (\xi_2\eta_2)^{k_2} \cdots (\xi_N\eta_N)^{k_N}}{k_1! k_2! \cdots k_N! \rho_{k_1 k_2 \cdots k_N}}$$

$$= K_x(w) = K_w(x)$$

$$K(x,\cdot) = K_x \in F_N(\rho)$$

From all the above, we can assert that $F_N(\rho)$ is an RKHS.

If when $\sum_{i=1}^{N} k_i = n$, all the $\rho_{k_1 k_2 \cdots k_N}$'s are equal, marked ρ_n

then: $K(x,w) = \sum_{n=0}^{\infty} \frac{(< x, w >)^n}{n! \rho_n} = \varphi(s)|_{s=<x,w>}$, $\varphi(s) = \sum_{n=0}^{\infty} \frac{s^n}{n! \rho_n}$

Especially, if all the ρ_n's are equal, marked ρ_0, then $\varphi(s) = \frac{1}{\rho_0} \exp(s)$

In this paper, we add a scaling factor $\alpha_0 > 0$, and set ρ_0 to be 1, then:

$$K(x,w) = \exp(\alpha_0 < x, w >)$$

4 Discussions on the Intersection of the Local Subspace

Theorem 1. For two sets of prototypes: $\{m_{0i}, \ldots, m_{D_i i}\}$, $\{m_{0j}, \ldots, m_{D_j j}\}$ $(i \neq j)$,

if $m_{0i}, \ldots, m_{D_i i}, m_{0j}, \ldots, m_{D_j j}$ are linearly independent, then the local

subspaces(manifolds): $L_{U_i, m_{0i}}$ and $L_{U_j, m_{0j}}$ have no intersection.

Proof. $L_{U_i, m_{0i}} = \{ x \mid x = m_{0i} + U_i z, z \in R^{D_i} \}$,

$$L_{U_j,m_{0j}} = \{x \mid x = m_{0j} + U_j z, z \in R^{D_j}\},$$

where

$$U_i = (m_{1i} - m_{0i}, \ldots, m_{D_i i} - m_{0i})\Lambda_i, U_j = (m_{1j} - m_{0j}, \ldots, m_{D_j j} - m_{0j})\Lambda_j,$$

and Λ_i, Λ_j are upper triangle matrixes, thus are invertible.

Assume $x \in L_{U_i,m_{0i}} \cap L_{U_j,m_{0j}}$, there exist $z_i \in L_{U_i,m_{0i}}$ and $z_j \in L_{U_j,m_{0j}}$, satisfying:

$$x = m_{0i} + U_i z_i = m_{0j} + U_j z_j$$

$$m_{0i} + (m_{1i} - m_{0i}, \ldots, m_{D_i i} - m_{0i})\Lambda_i z_i = m_{0j} + (m_{1j} - m_{0j}, \ldots, m_{D_j j} - m_{0j})\Lambda_j z_j$$

Because $m_{0i}, \ldots, m_{D_i i}, m_{0j}, \ldots, m_{D_j j}$ are linearly independent, it can be inferred from the above equation that neither z_i nor z_j exists. Thus the assumption is wrong. So $L_{U_i,m_{0i}}$ and $L_{U_j,m_{0j}}$ have no intersection.

Theorem 2. If the feature space is mapped into an RKHS with a strictly positive kernel, then for finite vectors in the original feature space: $\{x_1, x_2, \ldots, x_n \mid x_i \neq x_j (i \neq j)\}$, $\{K(x_1, \cdot), K(x_2, \cdot), \ldots, K(x_n, \cdot)\}$ is a set of linearly independent vectors in the RKHS.

Proof. Assume $a_i \in R(1 \leq i \leq n)$

$$\sum_{i=1}^n a_i K(x_i, \cdot) = 0 \Leftrightarrow < \sum_{i=1}^n a_i K(x_i, \cdot), \sum_{i=0}^n a_j K(x_i, \cdot) > = 0$$

$$\Leftrightarrow [a_1 \quad a_2 \quad \cdots \quad a_n] \cdot \begin{bmatrix} K(x_1, x_1) & K(x_1, x_2) & \cdots & K(x_1, x_n) \\ K(x_2, x_1) & K(x_2, x_2) & \cdots & K(x_2, x_n) \\ \cdots & \cdots & \cdots & \cdots \\ K(x_n, x_n) & K(x_n, x_2) & \cdots & K(x_n, x_n) \end{bmatrix} \cdot \begin{bmatrix} a_1 \\ a_2 \\ \vdots \\ a_n \end{bmatrix} = 0$$

It has been proved in [9] that the median matrix in the above equation is positive definite, thus $a_i = 0 (1 \leq i \leq n)$.

From theorem 1 and theorem 2, it can be inferred that the local subspaces, which are manifolds of $F_N(\rho)$, whose kernel is strictly positive, representing the respective categories, have no intersections between them.

5   The Improved Local Subspace Classifier Algorithm

When the feature space has been mapped into $F_N(\rho)$, the pattern vectors become infinite dimensional, and we can not use the LSC algorithm directly. But we can use it indirectly.

Assume that the prototypes forming the classifier are marked m_{ij}, where $j = 1, \ldots, c$ is the index of the category and $i = 1, \ldots, N_j$ index the prototypes in that category. The dimension of the manifolds representing the respective categories are denoted $D_j, j = 1, \ldots, c$.

Now we are to classify an unlabeled sample x. The improved algorithm is as follows:

1. Find the $D_j + 1$ prototypes closest to x and denote them $m_{0j}, \ldots, m_{D_j j}$

2. Define $\Phi = K(x, \cdot), \Phi_{ij} = K(m_{ij}, \cdot); \ A_j = \left(\Phi_{1j} - \Phi_{0j} \quad \cdots \quad \Phi_{D_j j} - \Phi_{0j} \right);$

$j = 1, \ldots, c; \ i = 1, \ldots, D_j$, then the distances between Φ and the corresponding

manifolds are: $d_j(x) = \left\| \left(I - A_j \left(A_j^T A_j \right)^{-1} A_j^T \right) \left(\Phi - \Phi_{0j} \right) \right\|$

$$d_j^2(x) = \left(\Phi^T - \Phi_{0j}^T \right) \left(I - A_j \left(A_j^T A_j \right)^{-1} A_j^T \right) \left(\Phi - \Phi_{0j} \right)$$

$$= \Phi^T \Phi - 2\Phi_{0j}^T \Phi + \Phi_{0j}^T \Phi_{0j} + 2\Phi_{0j}^T A_j \left(A_j^T A_j \right)^{-1} A_j \Phi$$

$$- \Phi^T A_j \left(A_j^T A_j \right)^{-1} A_j^T \Phi - \Phi_{0j}^T A_j \left(A_j^T A_j \right)^{-1} A_j^T \Phi_{0j}$$

where $\Phi^T \Phi = K(x, x), \Phi_{pj}^T \Phi_{qj} = K(m_{pj}, m_{qj}), \Phi^T \Phi_{pj} = K(x, m_{pj}),$

$0 \leq p, q \leq D_j$

Define:

$$h_j(x) = \Phi^T \Phi - 2\Phi_{0j}^T \Phi + 2\Phi_{0j}^T A_j \left(A_j^T A_j \right)^{-1} A_j^T \Phi - \Phi^T A_j \left(A_j^T A_j \right)^{-1} A_j^T \Phi$$

then:

$$g_{LSC}(x) = \arg\min_{j=1,\ldots,c} h_j(x)$$

6   Experimental Results and Conclusion

Our experiment was carried on to recognize the handwritten numerals. We had 10000 samples of every numeral. We took 8000 samples of each as training samples, and took the rest of each as test samples. The dimension of the original feature space was 64. After it was mapped into RKHS, we defined the dimension of every local subspace to be same, marked D. The results are shown in table 1.

The Subspace Classifier differs from the prototype classification methods in that the former separates the training samples with a conicoid bound, whereas the latter do with a hyperplane. The nonlinear degree of the former is even greater. The Local Subspace Classifier is superior to the conventional subspace classifier because it makes full use of the local nature of the classification. Considering that the subspaces generated by the conventional subspace classifier or Local Subspace Classifier usually have intersections, we map the feature space into RKHS to avoid this happening. Thus the Local Subspace Classifier comes into being. Compared with OI-Net, the classification bound of which is a hyperplane in RKHS, the Local Subspace Classifier in RKHS has the advantage of its classification bound is a conicoid in RKHS.

Classifier	error rate	Parameters	
LSC in RKHS	1.9%	$d = \infty$	$D = 12$
LSC	2.6%	$d = 64$	$D = 12$
LSC in RKHS	1.4%	$d = \infty$	$D = 24$
LSC	2.2%	$d = 64$	$D = 24$

Table 1.

References

1. Jorma Laaksonen: Local Subspace Classifier. Proceedings of ICANN's 97. Lausanne, Switzerland (October, 1997),637-642
2. Oja, E: Subspace Method of Pattern Recognition. Research Studies Press
3. R. O. Duda , P. E. Hart: Pattern Classification and Scene Analysis. New York:Wiley(1972)
4. K. Fikunaga: Introduction to Statistic Pattern Recognition. New York :Academic(1972)
5. Koji Tsuda: Subspace classifier in the Hilbert Space. Pattern Recognition Letters, Vol. 20(1999), 513-519
6. Sam-Kit Sin, Rui. J. P. deFigueiredo: An Evolution-Oriented Learning Algorithm for the Optimal Interpolative Net. IEEE Transactions on Neural Networks, Vol. 3(March,1992), 315-323
7. Sam-Kit Sin , Rui. J. P. deFigueiredo: Efficient Learning Procedures for Optimal Interpolative Net. Neural Networks, Vol. 6(1993), 99-113
8. Dongfang Zou: Research of OI-Net and RLS-OI Algorithm. Research Report of Institute of Automation(1999), CAS
9. Haykin, S.: Neural Networks: A comprehensive Foundation. IEEE Computer Society Press, Silver Spring MD(1995)

A Recognition System for Devnagri and English Handwritten Numerals

G S Lehal and Nivedan Bhatt

Department of Computer Science & Engineering, Punjabi University, Patiala, INDIA.
gslehal@mailcity.com

Abstract

A system is proposed to recognize handwritten numerals in both Devnagri (Hindi) and English. It is assumed at a time the numerals will be of one of the above two scripts and there are no mixed script numerals in an input string. A set of global and local features, which are derived from the right and left projection profiles of the numeral image, are used. During experiments it was found that the Devnagri numeral set had a much better recognition and rejection rate as compared to the English character set and so the input numeral is first tested by the Devnagri module. The correct recognition enables the system to be set to the appropriate context (Devnagri/English numeral set). Subsequent identification of the other numerals is carried out in that context only.

1. Introduction

Handwritten numeral recognition has been extensively studied for many years and a number of techniques have been proposed [1-7]. However, handwritten character recognition is still a difficult task in which human beings perform much better. The problem of automatic recognition of handwritten bilingual numerals is even more tough. Recently there has been a growing interest in script and language identification for developing international OCRs for multi-lingual scripts[8-12]. A relevant research work in the field of bilingual character recognition of Indian scripts is by Chaudhary and Pal[12].They have developed an OCR system to read two Indian language scripts: Bangla and Devnagri. A stoke feature based tree classifier is used to recognize the basic characters. For some characters additional heuristics are used. In a multilingual environment like India, there are many situations where a single document may contain handwritten numerals in two or more language scripts. As English and Hindi are the two most prominent languages used in India, an integrated approach towards the recognition of numerals of both Devnagri and English script is helpful in OCR development in India. Also such approach will have commercial applications in postal and banking environments. In this paper a bilingual OCR system for isolated handwritten numerals of Devnagri(Hindi) and Roman scripts has been presented. To the best of our knowledge, this is the first paper on recognition of bilingual handwritten numerals in Devnagri and English.

This paper is based on the work done by Kimura and Shridhar[3]. Kimura and Shridhar have presented two algorithms to achieve good classification rates of handwritten numerals. The first algorithm is a statistical one which used a modified discrimination function(MQDF) with features derived from chain codes of the character contour and the second one is the structural algorithm which uses the left and right profile features. We have modified the structural classifier used in the above mentioned paper for the recognition of numerals of both Devnagri and English scripts. The use of such a classifier has allowed a use of a variety of different feature descriptors, which more precisely describe the character and yet remain under the domain of same classifier. Also the use of a single classification scheme has enabled fast and flexible learning.

T. Tan, Y. Shi, and W. Gao (Eds.): ICMI 2000, LNCS 1948, pp. 442-449, 2000.

2. Data Processing for Feature Extraction

The input samples were taken in a specially designed form, which contained sets of guide boxes. The use of guide boxes for the samples allowed us to avoid the segmentation phase completely as the input was already in an isolated form. Some of the samples of handwritten Devnagri and English numerals are shown in fig.1. The form was processed by a scanner at 300dpi and gray-tone to two-tone was automatically performed by the scanner software. Each numeral occupied a field that measured 60Wx77H pixel units. Then a contour extraction algorithm was applied to obtain the closed contours of the numerals. As the input numerals can be of arbitrary size, the scaling factor had to be dynamically adjusted and then the normalization was applied to the numeral contour so that the enclosing frame measured 60Wx77H pixel units. The width varied from a minimum of 5 pixels to 60 pixels. This minimum width of 5-6pixel units(referred to as pen width) will be one of several feature descriptors used for the recognizing the numerals.

Fig. 1(a). Samples of the handwritten Devnagri numerals.

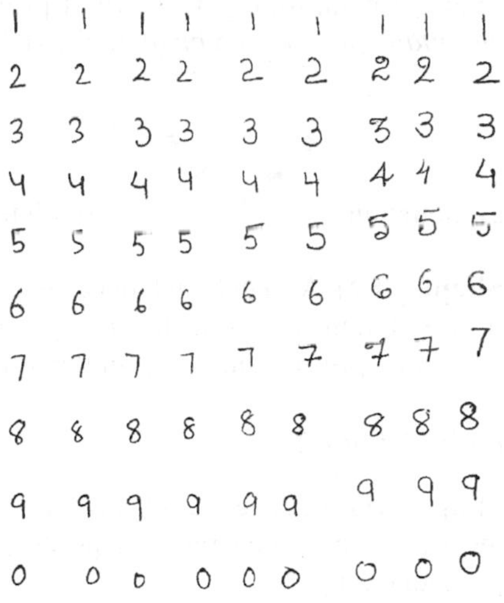

Fig. 1(b) Samples of the handwritten English numerals.

3. Feature Extraction

The input for the feature extraction phase is the normalized version of the candidate numeral's contour. The leftmost and rightmost foreground values are searched and the contour is split at those points to yield left are right profiles respectively. All the features required in our work are obtained from the shape variations in the profiles. The left profile lp[i] and the right profile rp[i] are nothing but the collection of distances from the left and right edges. i ranges from 1 to the length of the numeral. These are the global features. In addition to these there is an additional set of global features called the first differences in the left and right profiles i.e. ldiff[i] and rdiff[i] where i ranges from 2 to the length of the numeral.

In addition to these, we need local features to account for local variations in the shape of the contour. To extract these features, the numeral contour is visualized as a complex structure which is composed of individual shape descriptors(features).The various categories of such features are described below:

CATEGORY 1)

Shape_ 1.1) A sequential occurrence of a minimum, maximum and a minimum i.e. rpmin, rpmax, rpmin in the right profile generates shape descriptors of type *mountain, plateau* and *bump* in the right profile for a particular range R1.

(a) mountain   (b) plateau   (c) bump

Shape_ 1.2) A sequential occurrence of a minimum, maximum and a minimum i.e. Ipmin, Ipmax, lpmin in the left profile generates shape descriptors of type *mountain, plateau* and *bump* in the left profile. for a particular range R1.

Shape_ 1.3) A sequential occurrence of a maximum, minimum and a maximum i.e. rpmax, rpmin, rpmax in the right profile generates shape descriptors of type downward *mountain, plateau* and *bump* in the right profile for a particular range R2

a) mountain   b) plateau   c) bump

Shape_ 1.4) A sequential occurrence of a maximum, minimum and a maximum i.e. Ipmax, lpmin, lpmax in the left profile generates shape descriptors of type downward *mountain, plateau* and *bump* in the left profile for a particular range R2.

CATEGORY 2)

Shape_2.1) A sequential occurrence of a minimum and a maximum i.e. rpmin, rpmax in the right profile generates shape descriptors of type *ascender* in the right profile for a particular range R1.

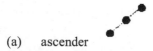

(a)   ascender

Shape_2.2) A sequential occurrence of a minimum and a maximum i.e. lpmin, lpmax in the left profile generates shape descriptors of type *ascender* in the left profile for a particular range R1.

Shape_2.3) A sequential occurrence of a maximum and a minimum i.e. rpmax, rpmin in the right profile generates shape descriptors of type *descender* in the right profile for a particular range R2.

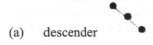

(a)   descender

Shape_2.4) A sequential occurrence of a maximum and a minimum i.e. lpmax, lpmin in the left profile generates shape descriptors of type *descender* in the left profile for a particular range R2.

CATEGORY 3)
Shape_3.1)Discontinuity(successive pixels are widely separated) in the right profile for a particular range R1.
Shape_3.2)Discontinuity(successive pixels are widely separated) in the left profile for a particular range R1.
Shape_3.3)Sharp discontinuity in the right profile for a particular range R1.
Shape_3.4)Sharp discontinuity in the left profile for a particular range R2.
Shape_3.5)A smooth(relatively less discontinues) right profile for a particular range R1
Shape_3.6)A smooth(relatively less discontinues)left profile for a particular range R2.

CATEGORY 4)
Shape_4.1) Occurs when the right profile is more or less a straight line i.e. the difference between rpmax and rpmin is a small value for a particular range R1.
Shape_4.2)Occurs when the left profile is more or less a straight line i.e. the difference between lpmax and lpmin is a small value for a particular range R2.
Shape_4.3)Occurs when the difference between rpmax and rpmin is greater than a specific value for a particular range R1.
Shape_4.4)Occurs when the difference between lpmax and lpmin is greater than a specific value for a particular range R2.

CATEGORY 5) Width Features:
Shape_5.1)These occur when the width of the contour w[i]=lp[il - rp[il is approximately equal to the pen-width for a particular range R1.
Shape_5.2)These occur when the contour width stays nearly the same for particular range R2.
Shape_5.3) These occur when the width at the particular point is greater than any other particular point.
Shape_5.4) These occur when the width of the contour approaches the standard width value for the normalized version (which is 60 in our case).

CATEGORY 6)
Shape_6.1) These occur when the ratio of length to width is less than a particular value.

In order to classify the members of each numeral subclass, distinct groups containing features describing the numerals have to be found. It is worth mentioning that in each group of shape features of all the numeral subclasses there are two kinds of features present. First are those features, which uniquely define that particular numeral. Second are those, which are used specifically to avoid the confusion with the other similar numerals of its own or other character set. Although they may seem trivial in the manner they describe the numeral but in some cases they act as a decisive key for differentiating the two confusing numerals. For example the feature (shape_2.1) in the range 70 to 75 helps differentiating the Devnagri 3 from the English 3. The feature (shape_5.4) in the range 1 to 25 helps to differentiate between the Devnagri 4 and the distorted sample of English 8. Also the features (shape_1.2) in the range 1 to 55 and (shape 2.2) in range 40 to 75 helps differentiating a Devnagri 5 from a distorted English 4 and so on.

All the possible features for each numeral subclass are grouped together. These form the classified character groups for the numerals. The recognition process uses a tree structure to establish the identity of the numeral candidate. The character groups were kept as distinct as possible to enhance the between -class variability and while minimizing the within - class variability. Role played by the context in conjunction with the recognition process is explained in the next section. Groups for all the numeral subclasses of both the character sets have been tabulated in the appendix.

A verbal explanation for the group of features for Devnagri 3 is given below (see Fig. 2).

I.   It has a bump feature (category 1.4) in its left profile in the range 40 to 75.
II.   It has a bump feature (category 1.1) in its right profile in the range 1 to 40 and 40 to 65.
III.   It has an ascender feature (category 2.1) in its right profile in the range 1 to 10.
IV.   It has a discontinuity (category 3.2) in its left profile in the ranges 1 to 25 and 40 to 75.
V.   It has an ascender feature (category 2.1) in the range of 70 to 75.

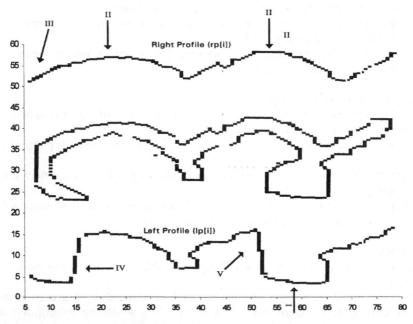

Fig 2: Projection profile of Devnagri 3

4. Identification of the Script of the Numerals

It is assumed that we are dealing with situations where at a time we shall be having the numeral candidates of a particular script only. For identification of script, no separate routine is used. During the recognition process when the first numeral of a particular guidebox set (Fig 3) is recognized correctly, the context (Devnagri/English) is set to the domain of that particular numeral's character set. Subsequent identification of the remaining numerals in that particular guidebox set is carried out in that context only which drastically reduces the search space and hence increases the performance of the system. It was observed that the Devnagri numeral set had a very good recognition and rejection rate, as compared to the English set. Also the Devnagri numeral set's recognition

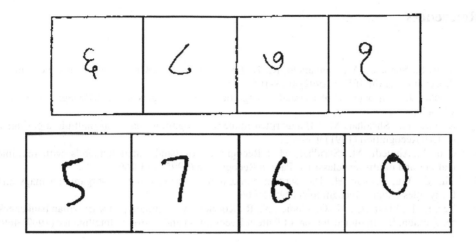

Fig. 3 : Guidebox containing (a) Devnagri and (b) English numerals.

module had good rejection rates for the numerals of the English character set. This property was exploited by adopting a *polling strategy* in which the input numeral is first tested by the Devnagri module. If the numeral is recognized then the context is set to Devnagri, else it is tested for English set and on recognition, the context is set to English. In case the numeral is rejected by both script sets, then the next numeral is tested, and this continues till one of the numeral is recognized. Subsequent identification of the other numerals in that particular guidebox set is carried out for character set of the recognized numeral. Numeral 0 is the same for both the character sets, thus in the case when the first numeral encountered in a particular guidebox set is a zero, subsequent numeral is checked before deciding the context

5. Results and Discussion

The performance of an OCR system is evaluated in terms of the parameters: recognition rate, confusion rate (indicating multiple membership of the numeral) and the rejection rate for a particular data set. In our case the results were tested on 1000 samples(in guideboxes) of both the Devnagri and English character set. For the Devnagri numeral set, a recognition rate of 89% and a confusion rate of 4.5% were obtained. The reason of the good performance for the Devnagri numerals is that these numerals are almost distinct from each other, thus increasing the inter-numeral subclass variability. Also the Devnagri numeral set's recognition module had good rejection rates for the numerals of the English character set. For the English numeral set we had a recognition rate of 78.4%, confusion rate of 18 % and rejection rate of 3.6%. The performance of the system can further be improved by including more features for English character set.

6. Conclusion

A recognition system that can read both Devnagri and English handwritten numerals has been described. The performance of the system for Devnagri character set is quite good but we are not so satisfied with the recognition rate for English character set. Work is currently being carried out for improving the performance for English character set.

References

1. Mori, S., Suen, C., Y., Yamamoto, K.: Historical review of OCR research and development. Proceedings of the IEEE (1992) 1029-1057.
2. Mantas, J. : An overviw of character recognition methodologies. Pattern Recognition (1986) 425-430.
3. Kimura, F., Shridhar, M. : Handwritten numeral recognition based on multiple algorithms, Pattern Recognition (1991) 976-983.
4. Cao, J., Ahmedi, M., Shridhar, M. : Recognition of handwritten numerals with multiple features and multistage classifier, Pattern Recognition (1995) 153-160.
5. Chi, Z., Wu, J., Yan, H.: Handwritten numeral recognition using self-organizing maps and fuzzy rules Pattern Recognition (1995) 59-66.
6. Lee, L., L., Lizarrage, M., G., Gomes, N., R., Koerich, A.: A prototype for Brazilian bankcheck recognition, International Journal of Pattern Recognition and Artificial Intelligence (1997) 469-487.
7. Stefano, C., D., Cioppa, A., D., Marcelli, A. : Handwritten numeral recognition by means of evolutionary algorithms, Proceedings ICDAR99, Bangalore, India, (1999) 804-807.
8. Spitz, A., L. :, Determination of script and language content of document images, IEEE Transactions on Pattern Analysis and Machine Intelligence (1997) 235-245.
9. Ding, J., Lam, L., Suen, C., Y.: Classification of Oriental and European scripts by using characteristic features, Proceedings ICDAR97, Ulm, Germany (1997) 1023-1027.
10. Lee, D., S., Nohl, C., R., Baird, H. S. : Language identification in complex, unoriented and degraded document images, Proceedings International Workshop on Document Analysis Systems, Malvern, Pennsylvania (1996) 76-98.
11. Pal, U., Chaudhuri, B., B. : Automatic separation of words in multi-lingual multi-script documents, Proceedings ICDAR97, Ulm, Germany (1997) 576-579.
12. Chaudhuri, B., B., Pal, U. : An OCR system to read two Indian language scripts : Bangla and Devnagri (Hindi), Proceedings ICDAR97, Ulm, Germany (1997) 1011-1015.

Appendix: Shape Features for Devnagri and English Numerals

Numeral	Shape features
Devnagri 1(१)	Shape_ 1.1 (1 to 30), Shape_1.4 (1 to 30), Shape_ 1 .3 (30 to 75), Shape_ 1.4 (30 to 75) Shape_2.3 (30 to 45), Shape_2.4 (30 to 45), Shape_2.1 (45 to 75), Shape_2.2 (45 to 75) Shape_5.1 (35 to 75), Shape_5.3 (10, 35 to 75) Shape_5.2 (35, 45 to 75)
Devnagri 2(२)	[Shape_3.4 (1 to 25) or Shape_3.2 (1 to 15)], Shape_3.2 (25 to 55), Shape_1.4(30 to 75) Shape_1.3 (30 to 75), Shape_2.1 (1 to 15), Shape_ 2.1(60 to 75), Shape_2.2 (60 to 75) Shape_ 5.1(60 to 75)
Devnagri 3(३)	Shape_ 1.1(1 to 40), Shape_ 1.1 (40 to 65), Shape_ (25 to 55), Shape_1.4 (40 to 75) Shape_2.1 (1 to 10), Shape_2.1 (70 to 75)*, [Shape_ 3.4(1 to 25) or Shape_3.2 (1 to 25)]
Devnagri 4(४)	Shape_ 1.3(25 to 55), Shape_1.2 (25 to 55), Shape_ 1.1(40 to 75), Shape_1.4 (40 to 75) Shape_ 2.1(40 to 65), Shape_2.4(40 to 65), [Shape_ 3.4(1 to 15) or Shape_3.3 (1 to 15)] Shape_ 5.3(15, 40), Shape_5.3 (65, 40), Shape_ 3.5(15 to 75), Shape_3.6 (15 to 75) Shape_ 5.4(1 to 10)*
Devnagri 5(५)	Shape_ 3.1(1 to 25), Shape_ 1.1 (1 to 40), Shape_ 2.2(1 to 40), Shape_2.2 (40 to 75) Shape_2.1(40 to 75), Shape_5.3 (25, 5) , Shape_1.2 (1 to 55)* , Shape_2.2 (40 to 75)* Shape_ 5.3(35, 55 to 75).
Devnagri 6(६)	Shape_ 1.4(1 to 40), Shape_ 1.4 (40 to 75), Shape_ 1.2(25 to 55), Shape_ 1.1 (40 to 75) Shape_2.4(1 to 10), Shape_3.1 (45 to 75), [Shape_ 3.1(1 to 25) or Shape_5.1 (1 to 15)] Shape_5.1 (70 to 75)
Devnagri 7(७)	Shape_5.2(15,25), Shape_5.2 (20, 25), Shape_5.2(35,40), Shape_5.2 (60, 65) Shape_5.2(65,75), Shape_2.2 (40 to 75), Shape_2.3(40 to75), Shape_3.5 (15 to 75)

	[Shape_3.3(1 to 25)or Shape_3.4 (1 to 25)], Shape_3.6(15 to 75), Shape_5.3 (25, 45 to 75) Shape_5.4(1 to 15)*
Devnagri **8(८)**	Shape_5.3(65,5to 35), Shape_2.3 (1 to 35), Shape_5.1(1 to 35), Shape_2.3 (1 to 35) Shape_2.4(1 to 65),Shape_3.5(1 to 30), Shape_3.3 (50 to 75),Shape_3.6(1 to 65)
Devnagri **9(९)**	Shape_5.3 (10, 35 to 75). Shape_5.1 (35 to 75), Shape_2.2 (30 to 75), Shape_2.1 (30 to 75) Shape_1.1 (1 to 25), Shape_1.4 (1 to 25), Shape_3.6 (1 to 75), Shape_3.5 (45 to 75)
0	Shape_2.1 (1 to 10), Shape_2.4 (1 to 10), Shape_2.3 (60 to 75), Shape_2.2 (60 to 75) Shape_3.5 (10 to 75), Shape_3.6 (10 to 75), Shape_5.2 (35, 40), Shape_5.2 (40, 45) Shape_5.3 (40, 15), Shape_5.3 (40, 65)
1	Shape_5.1 (15 to 65), Shape_3.5(1to65), Shape_3.6(15 to 65), Shape_4.1 (5 to 65) Shape_4.2 (15 to 65), Shape_5.2 (15,25 to65), Shape_5.2 (35, 40), Shape_5.2 (40, 45) Shape 6.1
2	Shape_2.4 (30 to 65), Shape_2.3 (15 to 65), Shape_5.1 (30 to 55), Shape_1.1 (1 to 40) [Shape_3.4 (1 to 25) or Shape_3.2 (1 to 25)] Shape_2.1 (1 to 10), Shape_2.3 (10 to 50) Shape_2.1 (50 to 75), Shape_2.2 (1 to 40)
3	Shape_1.1 (1 to 40), Shape_1.1 (40 to 75), Shape_1.3 (25 to 55), Shape_2.1 (1 to 15) [Shape_3.4 (1 to 25) or Shape_3.2 (1 to 25)], Shape_2.2 (1 to 40), Shape_2.4 (40, 75) Shape_2.3 (65 to 75)
4	Shape_3.1 (1 to 15), Shape_4.2 (55 to 75), Shape_4.1 (40 to 75), Shape_5.1 (50 to 75) Shape_3.2 (25 to 55), Shape_3.5 (50 to 75), Shape_5.3 (25, 50 to 75), Shape_5.2 (55, 60 to 75), Shape_3.6 (50 to 75)
5	Shape_2.3 (1 to 25), Shape_2.1 (25 to 60), Shape_2.3 (60 to 75), Shape_2.2 (5 to 40) [Shape_3.3 (1 to 15) or Shape_3.1 (1 to 15)], Shape_2.4 (40 to 75), Shape_1.1 (30 to 75) [Shape_3.2 (50 to 75) or Shape_5.1 (50 to 75)], Shape_1.2 (10 to 75), Shape_1.3 (1 to 50)
6	Shape_5.3 (65, 5 to 35), Shape_2.3 (1 to 35), Shape_2.3 (1 to 35), Shape_2.4 (1 to 65) Shape_2.2 (65 to 75),Shape_3.5 (1 to 30), Shape_3.6 (1 to 65), Shape_1.1(45 to 75)
7	[Shape_3.4 (1 to 25) or Shape_3.2 (1 to 25)], Shape_2.3 (5 to 75), Shape_2.4 (25 to 75) Shape_5.2 (25,30 to 75), Shape_2.2 (1 to 25), Shape_2.4(40 to 75) ,Shape_4.3 (25 to 75) Shape_4.4(25 to 75), Shape_5.1 (25 to 75)
8	Shape_1.1 (1 to 40), Shape_1.1 (40 to 75), Shape_1.4 (1 to 40), Shape_3.5 (40 to 75) Shape_3.6 (5 to 75), Shape_1.3 (25 to 55), Shape_1.2(25 to 55) ,Shape_5.3 (15,40) Shape_5.3(65,40)
9	Shape_1.4 (1 to 25), Shape_5.3 (15,45 to 75), Shape_5.2 (45,50 to 75), Shape_5.1 (45 to 75) Shape_3.2 (15 to 45) ,Shape_3.5 (35 to 75)

Note:

a) The notation Shape_(yl to y2) implies that the shape primitive is expected to occur in the range yl to y2.

b) The notation Shape_5.2(yl,y2 to y3) implies that the difference of the widths between the point yl and all of the points in the range y2 to y3 is a small value.

c) The notation Shape_5.3(yl,y2 to y3) implies that width at the point yl is greater than all the points lying in the range y2 to y3.

Deformation Transformation for Handwritten Chinese Character Shape Correction

Lianwen Jin[1*], Jiancheng Huang[2], Junxun Yin[1], Qianhua He[1]

[1]Department of Electronic and Communication Engineering,
South China University of Technology, Guangzhou, 510641, China
*eelwjin@scut.edu.cn
[2]China Research Center, Motorola (China) Electronics LTD, Shanghai,
200002, China

Abstract: In this paper, a novel 1-D deformation transformation based on trigonometric function for handwritten Chinese character shape correction is proposed. With a suitable selection of deformation parameters, the 1-D deformation transformation could deform a given handwritten Chinese character into 24 different handwriting styles. A deforming parameter controls the deformation degree for each style. The proposed deformation transformation could be applied as a non-linear shape correction method for Chinese character recognition. Our preliminary experiment has showed the effectiveness of the proposed approach.

1 Introduction

Object recognition using rigid model has been well established. Objects under such model only undergo translation, rotation, and simple affine transformation. However, in many practical situations, real world objects are not only translatable and rotatable, but also deformable. In such cases, a rigid model is less effective. Recently, modeling of objects as deformable elastic bodies has gained more attention in computer vision and pattern recognition. Many deformable models for image analysis have been established and successfully applied in object recognition, image matching and handwritten recognition[1-5,9].

Handwritten character recognition has been a challenging problem in the field of pattern recognition for many years. Handwritten Chinese character recognition is even more difficult due to the complex structure of Chinese characters, large vocabulary, many mutual similarity and great variability with different handwriting styles. From our point of view, handwritten Chinese character can be regarded as a kind of deformable object. Although there are large variations in the same category of character for different styles of handwriting, the basic topological structures of them are the same. Different handwriting variations can be treated as the distorted versions from one or several regular deformable models. In this paper, we will propose a new deformation transformation for handwritten Chinese character shape correction. By controlling the deformation parameters, the proposed 1-D deformation transform could deform a handwritten Chinese character into 24 different handwriting styles.

T. Tan, Y. Shi, and W. Gao (Eds.): ICMI 2000, LNCS 1948, pp. 450-457, 2000.

The proposed deformation transformation could be applied as a shape or position deforming correction method for handwritten Chinese character recognition.

2   Basic Idea

A binary handwritten Chinese character image C is consist of many black pixels, which could be represented by the following:

$$C = \{p_1, p_2, ..., p_n\} \tag{1}$$

where $p_i = (x_i, y_i)$ is the i th black pixel in character C, n is the total number of black pixles in C.

Deformation transformation (**DT**) determines the admissible deformation space, or equivalently the possible shapes that a deformable character can take. In theory, the deformation transformation can be any function which matches a 2D point to another 2D point, as it is used to approximate the displacement in a 2D plane. For a character image C, the deformation transformation $\mathcal{D}$ could be defined as:

$$\begin{cases} D(C) = \{D(p_1), D(p_2), ..., D(p_n)\} \\ D(p_i) = (x_i, y_i) + (D_x, D_y) = (x_i, y_i) + (f_x(x_i, y_i), f_y(x_i, y_i)) = p_i + D_i \end{cases} \tag{2}$$

where $D_i = (f_x(x_i, y_i), f_y(x_i, y_i))$ is called *displacement vector*, $f_x(x, y), f_y(x, y)$ are called *displacement functions*. Without loss of generality, we assume that the value of (x, y) are normalized in the unit square $[0, 1]^2$. We further assume that the deformation transformation meets the following constraints:

①. Displacement function are non-linear function.

②. Displacement functions are continuous and satisfy the following boundary conditions: $f_x(0, y) = f_x(1, y) = f_y(x, 0) = f_y(x, 1) = 0$.

③. Deformation transformation should preserve the topology structure of the character.

④. Deformation transformation should preserve the smoothness and connectivity of the character image.

In Eq.(2), displacement functions $f_x(x, y), f_y(x, y)$ are funcions of both x and y respectively, the transformation achieved is called the 2-D deformable transform. As a special case where $f_x(x, y) = f(x)$ and $f_y(x, y) = f(y)$, the transformation is called 1-D deformation transformation. In such case, the 1-D deformable transformation for character C is given by the following equation:

$$D(p_i) = (x_i, y_i) + (f(x_i), f(y_i)) \tag{3}$$

ie.

$$\begin{cases} D(x_i) = x_i + f(x_i) \\ D(y_i) = y_i + f(y_i) \end{cases} \tag{4}$$

3   1-D Deformation Transformation

The displacement function for 1-D DT is defined by the following trigonometric function:

$$f(x) = \eta \bullet x[\sin(\pi\beta x + \alpha)\cos(\pi\beta x + \alpha) + \gamma] \qquad (5)$$

where $\alpha, \beta, \gamma, \eta$ are constants. In the special case where $\alpha = 0, \beta = 1, \gamma = 0$, the displacement function becomes:

$$f(x) = \eta x \sin \pi x \cos \pi x \qquad (6)$$

Fig. 1 shows two curves of function '$x + f(x)$' with different parameters η.

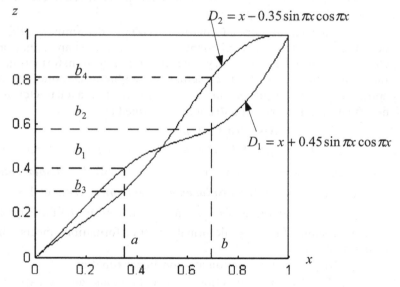

Fig. 1. *Two deformation curves with different deforming parameter*

Figure 1 demonstrates that the displacement function is a non-linear function, which maps a linear domain x into a non-linear domain z. The parameter η controls the degree of non-linear mapping, which is called **deforming parameter**. In Fig. 1, it could be seen that function D_1 compresses region $[a, b]$ of x-axis into $[b_1, b_2]$ of z-axis, while function D_2 expands region $[a, b]$ into $[b_3, b_4]$. For the region $[0, a]$ of x axis, function D_1 compresses it into the direction of zero point, while function D_2 expands it far away from zero point. Therefore, by selecting different x region and different deforming parameter η, different non-linear mapping effects could be achieved.

Considering region $[a, b]$ in Eq. (5), let:

$$\pi\beta x + \alpha\big|_{x=0} = a$$
$$\pi\beta x + \alpha\big|_{x=1} = b$$

such that:

$$\alpha = a, \ \beta = \frac{b-a}{\pi} \tag{7}$$

On the other hand, the displacement function $f(x)$ should satisfy the following boundary conditions:

$$f(0) = 0, \quad f(1) = 0$$

such that:

$$\gamma = -\sin b \cos b \tag{8}$$

From Eq.(5), (7), (8), and (4), the 1-D deformable transformation is finally defined as follows:

$$
\begin{aligned}
D(x_i) &= x_i + \eta x_i \big[\sin[(b-a)x_i + a]\cos[(b-a)x_i + a] - \sin b \cos b\big] \\
D(y_i) &= y_i + \eta y_i \big[\sin[(b-a)y_i + a]\cos[(b-a)y_i + a] - \sin b \cos b\big]
\end{aligned}
\tag{9}
$$

where $0 \le a < b \le 1$, η are constants. Note that the deformation varies with different $[a, b]$ and different deformation parameter η, as shown in Fig. 2.

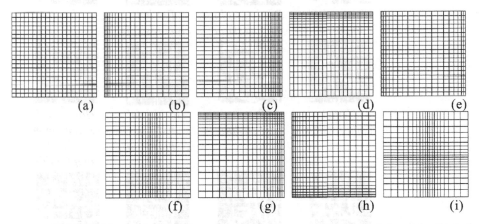

Fig. 2. *Different deformation efforts for a square grid. (a). Origin grid; (b).Left compression; (c).Right compression; (d).Top compression; (e). Horizontal expansion; (f).Horizontal compression; (g).Top right compression; (h).Bottom left compression; (i).Center compression*

4   24 Different Kinds of Deformation Styles for a Handwritten Chinese Character

In general, with suitable selection the value of a, b, η, we could perform 24 kinds of deformation processing for a given handwritten Chinese character, as shown in Fig. 3.

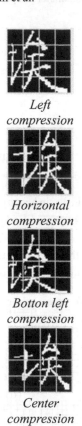

| Original Character | Left compression | Right compression | Top compression | Botton compression |

| Horizontal compression | Horizontal expansion | Vertical compression | Vertical expansion |

| Botton left compression | Bottom right compression | Top left compression | Top right compression |

| Center compression | Center expansion | Horizontal and Top compression | Horizontal expansion & Top compression |

| Right &Vertical compression | Left &Vertical compression | Horizontal and bottom compression | Horizontal expansion & bottom compression |

| Right compression & vertical expansion | Left compression & vertical expand | Horizontal expand &vertical compression | Horizontal compression &vertical expansion |

Fig. 3.   *24 kinds of deformation transformation for a given handwritten Chinese character '埃'*

5   The Deforming Parameter: A Control of Deforming Degree

Fig. 4 shows different 'Left compression' and 'Right compression' deforming effects with different deforming parameter η. In general, it is possible to transform a given handwritten character into hundreds of deformation copies by applying different deforming parameters with the 24 different deformation styles described above.

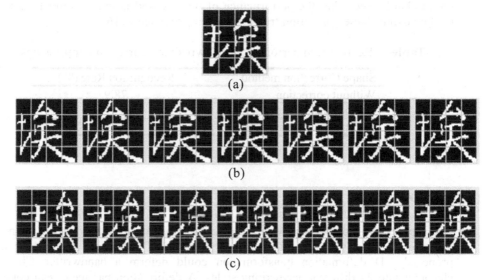

(a)

(b)

(c)

Fig. 4. *(a). Original character. (b). Different left compression effects with* η *values from 0.8 to 2.0 at a step of 0.2 (c). Different right compression effects with* η η values from 0.8 to 2.0 at a step of 0.2.*

6   Application   of   Deformation   Transform   in   Handwritten Chinese Character Recognition

It is known that one main difficulty in handwritten Chinese character recognition (HCCR) is the large variation and deformation in different handwriting styles. Many linear and non-linear normalization or shape correction approaches have been proposed for HCCR and these have been proved to be useful to improve the recognition rate in HCCR[6,7]. For the application of deformation transformation to HCCR, the proposed 1-D deformation transform could be applied as a normalization or shape correction method for handwritten Chinese character. It is expected that by applying the proposed deformation transformation to a deforming handwritten character, there is at least one regular sample could be found from the transforming copies generated by the different transformation shown in Fig. 3. In other word, the deformation transformation will try to correct the position deformation in the handwritten sample. Our experiment was conducted on 470 categories of handwritten Chinese characters. 40 sets of samples were used to train a minimum distance classifier, and another 10 sets of samples were used as testing data. The feature we used is fixed meshing directional cellular feature[8]. The experimental results of our method and the cosine shape correction approach proposed in [6] are shown in table 1. It can be seen that by using the deformation transformation, the recognition rate could be improved by 6.7%. It could also be seen that the performance of our method is much better than that of the cosine shape correction transformation proposed in [6].

Table 1. Recognition performance of two shape correction approaches

Shape Correction method	Recognition Rate (%)
Without correction	78.9
Cosine shape correction	82.9
Deformation transformation	85.6

7 Conclusion

In this paper, we proposed a new deformation transformation for handwritten Chinese character shape correction. With suitable selection of deforming parameters, the proposed 1-D deformation transformation could deform a handwritten Chinese character into 24 different handwriting styles. A deformation parameter can control the deforming degree for each style. The proposed deformation transformation could be applied as a non-linear shape correction method for Chinese character recognition. Experiment on 470 categories of Chinese characters showed that the proposed transformation is useful in HCCR. Extension of the deformation transformation could also be applied in other fields, such as shape matching, object recognition etc, which may merit our further study.

Acknowledgement

This paper is supported by the Natural Science Foundation of China, the Research Foundation from Motolora China Research Center, and Natural Science Foundation of Guangdong.

References

1.   Jain, Y. Zhong, S. Lakshmanan: Object matching using deformable templates. IEEE Trans. Pattern Anal. And Machine Intell, Vol 18, no. 3, (1996) 267-278

2. Yuille, P.W. Hallinan, D.S. Cohen: Feature extraction from faces using deformable templates. Int. Journey on Computer Vision, vol. 8, no. 2 (1992) 133-144
3. Terzopolous, J. Platt, A. Barr, K. Fleischer: Elastically deformable models. Computer Graphis, vol.21, no.4, (1987) 205-214
4. Tsang, C.K.Y., Fu-Lai Chung: Development of a structural deformable model for handwriting recognition. Proceedings. Fourteenth International Conference on, Pattern Recognition, 2 (1998) 1130 -1133
5. Lianwen Jin, Bingzhen Xu: Constrained handwritten Chinese character recognition using a deformable elastic matching model. ACTA Electronics Sinica, vol.25, no.5, (1997) 35-38
6. J.Guo, N.Sun, Y.Nemoto, M.Kimura, H.Echigo, R.Sato: Recognition of Handwritten Characters Using Pattern Transformation Method with Cosine Function- Vol. J76-D-II, No.4, (1993) 835-842
7. Dayin Gou, Xiaoqing Ding and Youshou Wu: A Handwritten Chinese Character Recognition Method based on Image Shape Correction. Proc. Of 1st National on Multimedia and Information Networks, Beijing, (1995) 254-259
8. Lianwen JIN, Gang Wei: Handwritten Chinese Character Recognition with Directional Decomposition Cellular Features, Journal of Circuit, System and Computer, Vol.8, No.4, (1998) 517-524
9. Henry S. Baird: Document Image Defect Models and Their uses, Proceedings of the third ICDAR, (1993) 488-492.

Offline Handwritten Chinese Character Recognition Using Optimal Sampling Features*

Rui ZHANG, Xiaoqing DING

Department of Electronic Engineering,
State Key Laboratory of Intelligent Technology and Systems
Tsinghua University, Beijing 100084, P.R.China
ray@ocrserv.ee.tsinghua.edu.cn

Abstract. For offline handwritten Chinese character recognition, stroke variation is the most difficult problem to be solved. A new method of optimal sampling features is proposed to compensate for the stroke variations and decrease the within-class pattern variability. In this method, we propose the concept of sampling features based on directional features that are widely used in offline Chinese character recognition. Optimal sampling features are then developed from sampling features by displacing the sampling positions under an optimal criterion. The algorithm for extracting optimal sampling features is proposed. The effectiveness of this method is widely tested using the Tsinghua University database (THCHR).

1. Introduction

Offline handwritten Chinese character recognition is one important branch of optical character recognition (OCR), which is used to translate human-readable characters to machine-readable codes, so as to allow direct processing of documents by computers. The difficulty in offline handwritten Chinese character recognition is mainly due to the stroke variations, therefore getting rid of the stroke variations and reducing the within-class pattern variability is important for the correct recognition. In previous researches, many normalization methods have been proposed to compensate for the variations. Linear normalization concerns with the linear variations. Nonlinear normalization is based on the density equalization and makes the strokes uniformly distributed. All linear and nonlinear normalization methods have limitations because they are class-independent. Wakahara proposed an adaptive normalization method. Adaptive normalization is class-dependent and normalizes an input pattern against each reference pattern by global and pointwise local affine transformation. This method greatly reduces the stroke variations, however its computational cost is very high.

---

* Supported by 863 Hi-tech Plan (project 863-306-ZT03-03-1) & National Natural Science Foundation of China (project 69972024)

T. Tan, Y. Shi, and W. Gao (Eds.): ICMI 2000, LNCS 1948, pp. 458–465, 2000.

In the design of an OCR system, feature extraction is essential. The directional features method, which utilizes directional factors of strokes, is considered a promising one and its efficiency has been clarified in previous works.

In this paper, we propose optimal sampling features which are developed from directional features and are adaptive to stroke variations. Optimal sampling features can reduce the within-class pattern variability effectively, so higher recognition accuracy can be achieved.

The paper is organized as follows. Section 2 describes the sampling features. Section 3 defines the optimal sampling features and gives the algorithm for the optimal sampling features extraction. Section 4 gives the coarse-to-fine strategy. Section 5 shows the experiment results, especially the improvement in accuracy when optimal sampling features are applied. Finally, section 6 outlines the conclusions of this study.

2. Sampling Features

Sampling features are based on directional features. The feature extraction procedure is as follows:

1. Normalization and smoothing operation are applied to the input binary image. The output is also a binary image (64 horizontal $\times$ 64 vertical).

$$f(m,n) \quad m,n=0,1,2; \cdot 64 \tag{1}$$

2. The chain coding is applied to the contour pixels of the character image. The direction of each contour pixel is quantified to one of four possible directions according to the chain codes, namely, horizontal, right-leaning, vertical and left-leaning. Then four directional feature images are generated from $f(m,n)$, each of them contains the contour pixels of different directional information of strokes. These four directional feature images are represented by:

$$f_i(m,n) \quad i = 0,1,2,3 \tag{2}$$

The two steps above-mentioned are the same as the extraction of directional features.

3. 2-D discrete cosine transform (DCT) is applied to each image, respectively. The corresponding frequency domain images are obtained:

$$F_i(p,q) = \mathrm{DCT}\left[f_i(m,n) \right] \tag{3}$$

4. The high frequency part of each frequency domain image is truncated using the following formula:

$$F_i'(p,q) = \begin{cases} F_i(p,q) & 0 \le p \le 8, 0 \le q \le 8 \\ 0 & p > 8, q > 8 \end{cases} \tag{4}$$

Fig. 1. The procedure in sampling features extraction

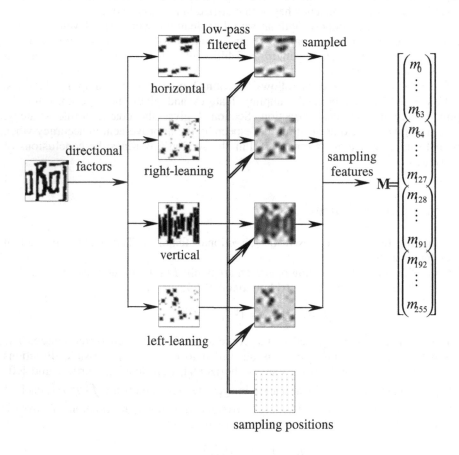

sampling positions

5. 2-D inverse discrete cosine transform (IDCT) is applied to each truncated frequency domain image, and the low-pass filtered images are obtained as follows:

$$f_i'(m,n) = \text{IDCT}\left[F_i'(p,q)\right] \tag{5}$$

6. Each low-pass filtered image is sampled at 64 sampling points. These sampling points are arranged into eight rows and eight columns. Each of the sampling points is denoted by (i,j). (i,j) refers to the row and column of its location in the arrangement. The values at these sampling points produce a feature vector of size $8 \times 8 \times 4 = 256$ (eight rows, eight columns and four directional images). This vector is defined as sampling features:

$$N = \left(n_0, n_1, \cdots, n_{254}, n_{255}\right)^T \tag{6}$$

In a low-pass filtered image, the positions of the sampling points can be flexible. The sampling positions play an important role in feature extraction. We define the sampling position matrix P to represent the sampling position:

$$P = \begin{pmatrix} p_{0,0} & p_{0,1} & \cdots & \cdots & p_{0,7} \\ p_{1,0} & \ddots & & & p_{1,7} \\ \vdots & & p_{i,j} & & \vdots \\ \vdots & & & \ddots & \vdots \\ p_{7,0} & p_{7,1} & \cdots & \cdots & p_{7,7} \end{pmatrix} \tag{7}$$

The size of P is 8×8. Each element is a vector $p_{i,j} = (x_{i,j}, y_{i,j})$ representing the coordinate of the sampling point (i, j). For certain low-pass filtered image, sampling features N depend on the sampling positions thus sampling features can be denoted by $N(P)$.

Fig. 1 demonstrates the procedure of sampling features extraction.

3. Optimal Sampling Features

Uniform Sampling Features

Regardless of feature extraction methods used, the extracted features must minimize the within-class pattern variability and maximize the between-class pattern variability to provide sufficient discriminative information among the different patterns.

When the sampling positions for feature extraction are uniform and rigid, the sampling position matrix P (Eqs. 7) is a constant matrix denoted by $P^{(0)}$ and the features are defined as uniform sampling features denoted by $N(P^{(0)})$. In handwritten Chinese characters, there are many variations between different handwritings. Since the sampling positions are fixed, uniform sampling features are easily affected by stroke variations and they can't efficiently decrease the within-class pattern variability.

Optimal Sampling Features and Feature Extraction

In order to decrease the within-class pattern variability and improve the discriminative ability, the sampling positions should be adaptable to stroke variations, i.e., the sampling positions are elastic and adaptive to each reference pattern. Sampling position matrix P is a variable matrix. For each reference pattern, sampling position matrix P should satisfy the following two conditions:

1. The Euclidean distance between sampling features of input pattern and the reference pattern should be minimized by displacing the sampling positions.
2. The smoothness of the displacement should be preserved.

The sampling position matrix satisfied above conditions is defined as the optimal sampling position matrix $P^{(l)}$, and the features are defined as optimal sampling features $N(P^{(l)})$. In the extraction of optimal sampling features, the optimal position matrix $P^{(l)}$ can be solved by optimization methods.

In general, an optimization method is comprised of a destination function and several constraint conditions. In this paper, the destination function is described as follows:

$$\min_{P} E = \min_{P} f(P, \overline{M}) = \min_{P} \left\| N(P) - \overline{M} \right\|^2 \tag{8}$$

where, $N(P)$ represents the sampling features of an input character. $\overline{M}$ is the reference pattern generated by averaging 1400 samples per class with uniform sampling features. $f(P, \overline{M})$ is the square of the Euclidean distance between sampling features and the reference pattern.

The restraint conditions are described by the following equations (Eqs. 9, 10):

$$p_{i,j}^{(l+1)} = p_{i,j}^{(l)} + \Delta p_{i,j} \tag{9}$$

$$p_{i+m,j+n}^{(l+1)} = p_{i+m,j+n}^{(l)} + T \cdot \Delta p_{i,j} \tag{10}$$

where

$$T = \begin{bmatrix} T_{00} & 0 \\ 0 & T_{11} \end{bmatrix} \tag{11}$$

$$T_{00} = T_{11} = w(m, n; i, j) \cdot (1 - k \cdot d(m, n; i, j))$$

$$w(m, n; i, j) = \begin{cases} 0 & |m - i|, |n - j| > W_T \\ 1 & |m - i|, |n - j| \le W_T \end{cases} \tag{12}$$

$$1 - k \cdot d(m, n; i, j) \ge 0$$
$$m, n \le M \tag{13}$$

$p_{i,j}^{(l)}$ is the (i, j) element in sampling position matrix P at the l th iteration. $\Delta p_{i,j}$ is the difference of $p_{i,j}^{(l)}$, namely, the horizontal and vertical displacements of the sampling point (i, j).

T is the smoothness matrix that denotes the displacement of sampling point $(i+m, j+n)$ with respect to sampling point (i, j). $d(m, n; i, j)$ is the Euclidean distance between sampling point (i, j) and its neighboring sampling point $(i+m, j+n)$. k is the smoothness parameter which determines the extent of displacement at sampling point $(i+m, j+n)$ with respect to the displacement at (i, j), and it is experimentally determined. $w(m, n; i, j)$ is the window function, which determines the neighboring region influenced by the sampling point (i, j), and it is determined by experiment.

By solving the above optimization method, the optimal sampling position matrix $P^{(l)}$ is determined, which minimizes the destination function $f(P, \overline{M})$ and satisfies the constraint conditions. Based on $P^{(l)}$, each low-pass filtered image $H_i'(m, n)$ is sampled. Thus optimal sampling features $N(P^{(l)})$ are obtained.

4. Coarse-to-Fine Strategy

Since Chinese character consists of many classes, a coarse-to-fine strategy is necessary to save computational costs. In this paper, the nearest-neighbor classification is employed. In coarse classification, 10 candidates are selected according to the Euclidean distances between uniform sampling features of the input character and all 3755 reference patterns.

In fine classification, optimal sampling features are extracted against each candidate pattern. The candidate with minimal destination function, namely the Euclidean distance between optimal sampling features and the candidate pattern, is output as the recognition result.

5. Experimental Results and Discussion

The performance of the recognition algorithm described in this paper is fully tested using the THCHR database. This database consists of 1500 sets and each set contains 3755 handwritten Chinese characters.

As mentioned above, the optimal sampling features are extracted by minimizing the destination function E (Eqs.8), which is defined as the Euclidean distance between the input and the reference pattern. As the number of iterations increasing, the distance between the input and each reference pattern decreases respectively. The experiment of optimal sampling features extraction is performed using the Chinese characters 哀. The top three candidates of coarse classification are "衰, 衷, 哀" and these candidates are similar in shape.

Fig. 2. Changes in destination functions

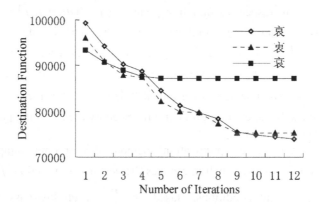

Fig. 2 shows a stable decrease of the destination function for each template with respect to the number of iterations. At the beginning of iteration, the destination function of the reference pattern 哀 is larger than that of 衷 or 衰, so the input 哀 is mis-recognized as 衷. As the number of iterations increases, the destination function of the reference pattern 哀 decreases faster than that of 衷 or 衰. And at the end of the iteration, the destination function of the reference pattern 哀 is smaller than that of 衷 or 衰, the input is recognized accurately. This result shows that optimal sampling features decrease the within-class pattern variability, and improve the discriminative ability.

Table 1. Recognition accuracy for optimal sampling features and uniform sampling features

Uniform sampling features	Optimal sampling features
90.09%	91.89%

Tab. 1 shows the recognition accuracy of optimal sampling features. The accuracy of uniform sampling features is also shown for comparison. This result proves the effectiveness of optimal sampling features.

6. Conclusions

In this paper, sampling features based on directional features are proposed. Optimal sampling features are adaptive to the stroke variations by displacing the sampling positions against each reference pattern while keeping smoothness. The optimal sampling positions are obtained by solving an optimal problem. Then optimal sampling features are extracted by sampling at the optimal sampling positions. In order to reduce the computational costs, a coarse-to-fine strategy is employed.

The performance of using optimal sampling features is tested on the handwritten Chinese character database (THCHR). Higher recognition accuracy is achieved as compared with using uniform sampling features.

Reference

1. Burr D. J.: A Dynamic Model for Image Registration. Computer Graphics and Image Processing, 1981, 15(1) 102-112
2. Wakahara T.: Adaptive Normalization of Handwritten Characters Using Global/Local Affine Transformation. IEEE Transactions on Pattern Analysis and Machine Intelligence, 1998, 20(12) 1332-1341
3. Trier O. D., Jain A. K., Taxt T.: Feature Extraction Methods for Character Recognition-a Survey. Pattern Recognition, 1996, 29(4) 641-662
4. Hildebrandt T. H., Liu W.: Optical Recognition of Handwritten Chinese Characters: Advances Since 1980. Pattern Recognition, 1993, 26(2) 205-225
5. Wang P. P., Shiau R. C.: Machine Recognition of Printed Chinese Character via Transformation Algorithm. Pattern Recognition, 1973, 3(5) 303-321
6. Mizukami Y.: A Handwritten Chinese Character Recognition System Using Hierarchical Displacement Extraction Based on Directional Features. Pattern Recognition Letters, 1998, 19 595-604
7. Bazaraa M. S., Shetty C. M.: Nonlinear Programming: Theory and Algorithms. New York, Wiley, 1979

A New Multi-classifier Combination Scheme and Its Application in Handwriting Chinese Character Recognition

Minqing Wu [1], Lianwen Jin [1], Kun Li [1], Junxun Yin [1], Jiancheng Huang [2]

[1]Department of Electronic and Communication Engineering,
South China University of Technology, Guangzhou, 510640, China
[2]China Research Center, Motorola (China) Electronics LTD,
Shanghai, 200002, China

Abstract. In this paper, a multi-classifier combination scheme based on weighting individual classifier's candidates is proposed. Four individual classifiers are constructed with four different feature extraction approaches. A confidence function is defined for the classifier integration through weighting the similarity of different individual classifier candidate outputs. Different integration methods are studied. Application of the multi-classifier to handwriting Chinese character recognition demonstrates that the recognition rate of the integrated system can be improved by 2% or so, showing the effectiveness of the proposed method.

1   Introduction

In recent years, classifier combination has been an important research direction in the field of pattern recognition, and has been successfully applied in many practical area, such as character recognition[1-5]. By the technology of integrating several classifiers together, the problem of attempting to find a high performance classifier becomes designing several better classifiers. In order to improve the overall recognition performance, different classification algorithms with different feature sets can be combined by a combination function, which integrates complementary characteristics of different classifiers together. The design of a combination function is one key factor for the multi-classifier combination.

In this paper, a multi-classifier combination scheme is proposed for handwriting Chinese character recognition. Four different individual classifiers are constructed with four different feature extraction approaches. A confidence function is defined for the classifier combination through weighting the similarity of different individual classifier candidate outputs. Application of the multi-classifier to recognizing 1034 categories of handwriting Chinese character shows that the recognition rate can be improved by 2% or so, demonstrating that the proposed approach is useful and effective.

T. Tan, Y. Shi, and W. Gao (Eds.): ICMI 2000, LNCS 1948, pp. 466-472, 2000.

2   A Multi-classifier Combination Scheme

Fig. 1 shows the block diagram of our proposed multi-classifier. The multi-classifier is composed of several individual classifier $C_1, C_2, \cdots, C_n$ and a decision combiner. An unknown Chinese character X is fed to each classifier C_i simultaneously. The m candidate recognition outputs of classifier C_i are represent as $C_{i1}, C_{i2}, \cdots, C_{im}$, while the similarity of each candidate are represented as $S_{i1}, S_{i2}, S_{i3}, \cdots S_{im}$. It is obviously that,

$$1 \geq S_{i1} \geq S_{i2} \geq S_{i3} \geq \cdots \geq S_{im} \geq 0$$

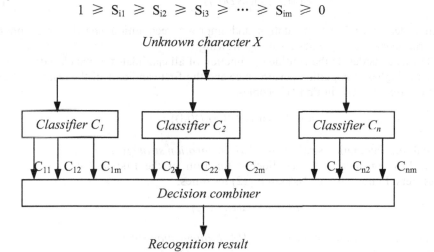

Fig.1. Diagram of multi-classifier system.

All recognition outputs from each individual classifier will be input into the decision combiner for integration decision. The decision combiner will yield the final recognition result. So that decision combiner is the key for the multi-classifier system. In designing decision combiner, we assume that:

(1) Each classifier is relatively independent and complementary, such that by appropriate intelligence integration strategy, higher overall performance can be achieved.

(2) The recognition performance for different classifiers is different, such that they make different contribution to the decision combiner. An individual classifier with the highest recognition rate should have the more contribution than the other individual classifiers. So we define a weighting coefficient for each classifier as $w_1, w_2, \cdots, w_n$ ($0 \leq w_i \leq 1$).

(3) In each classifier C_i, different candidates contribute to decision combiner differently. The first candidate should have the most contribution, while the mth has the least. Therefore, m candidate weighting coefficients are denoted as $p_1, p_2, \cdots, p_m$.

(4) The candidates may repeat in different classifier, i.e., the candidate output C_{ij} of classifier C_i is probably the rth candidate of classifier C_k, i.e., $C_{ij}=C_{kr}$. In general, we define a function $n(k,i,j)$ to represent the position where C_{ij} appears in classifier C_k, i.e., $C_{ij}=C_{kn(k,i,j)}$. If C_{ij} doesn't appear in C_k, then we let $n(k,i,j)=0$ and $p_0=0$.

According to these assumptions, a confidence function is defined for each candidate C_{ij}, as follows,

$$D(i,j)=w_i p_j s_{ij} + \sum_{k \neq i} w_k p_{n(k,i,j)} s_{kn(k,i,j)} \tag{1}$$

The integration strategy based on maximum confidence function principle for the multi-classifier combiner, which is called as "*ALL integration strategy*", is given by,
Integration system recognition result is C_{MN}, only if

$$D(M,N) = \max_{i,j} D(i,j) \tag{2}$$

In order to reduce computation and improve recognition speed, three simpler integration strategies are designed as following:

(1) Just considering the confidence function of all candidates of the classifier which have the highest recognition performance and the first candidate of the rest classifiers. In this case, $D\ (M,N)$ in Eq.(2) becomes,

$$D(M,N) = \max_{i,j} \{D(1,j), D(i,1)\} \tag{3}$$

We call this integration strategy as "*ONE integration strategy*".

(2) Just considering the confidence function of the first $m/2$ candidates of all classifiers. In this case, $D(M,N)$ in Eq.(2) becomes,

$$D(M,N) = \max_{i \leq m/2} D(i,j) \tag{4}$$

We call this integration strategy as   "*HALF integration strategy*"

(3)Especially, when $w_i=1$, $p_1=1$, $p_{j \neq 1}=0$ in Eq.(3), the integration strategy we got is voting integration strategy.

3   The Design of Four Individual Classifiers

3.1 Classifier Based on Contour Directional Feature[6]

Contour directional feature denotes contour pixels statistical characteristics of a Chinese character in horizontal, vertical, $+45°$, $-45°$ four direction. If a black pixel p in the handwriting character image is a contour point, it will be decomposed into one of the four directional strokes according to the different decomposition strategies. A set of elastic meshes is applied to the directional sub-patterns and the distribution of contour pixels in each mesh is computed as feature vectors. Two feature extraction methods are proposed:

1）Basic contour directional feature decomposition strategy;

2）Fuzzy contour directional feature decomposition strategy.

3.2 Classifier Based on Elastic Meshing Directional Decomposition Feature[7]

The elastic meshing directional decomposition feature extraction approach is shown in Fig. 2. According to the stroke statistical properties of Chinese character, a handwriting character is firstly decomposed into four directional sub-patterns, and then a set of elastic meshes is applied to each of the four sub-patterns respectively to extract the pixel distribution features. Two feature extraction methods are proposed[6]:

    1）OR feature extraction strategy;

    2）EDGE feature extraction strategy.

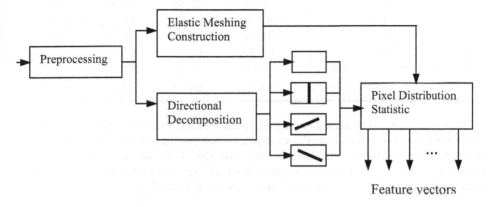

Fig. 2. Elastic meshing directional decomposition feature frame

According to the four feature extraction methods described above, the following four independent classifiers are used in our multi-classifier system:

    C_1: Classifier which uses fuzzy contour directional feature

    C_2: Classifier which uses EDGE elastic meshing directional decomposition feature

    C_3: Classifier which uses basic contour directional feature

    C_4: Classifier which uses OR elastic meshing directional decomposition feature

4　Experiments

4.1 The Complement of Four Individual Classifiers

Handwriting Chinese character database HCL2000 developed by BUPT is used for testing[8]. 1034 categories of handwriting Chinese characters are used in our experiments. Each category contains 60 different samples. We use 50 samples as training data to train the four classifiers described above, and the rest 10 sets of samples are used as testing data. The recognition results on the testing data with different classifier are shown in table 1.

Table 1. Recognition performance of individual classifier

Samples	C_1	C_2	C_3	C_4
1	76. 38	74. 68	75. 53	78. 94
2	99. 15	97. 45	98. 3	97. 87
3	99. 79	98. 72	99. 57	99. 57
4	99. 15	98. 94	98. 94	98. 51
5	99. 15	97. 45	98. 3	97. 87
6	95. 53	94. 89	93. 62	87. 02
7	76. 39	77. 02	72. 98	76. 81
8	85. 32	84. 26	84. 26	83. 83
9	97. 82	97. 66	97. 02	97. 66
10	87. 87	85. 11	85. 75	87. 02
Overall Recognition Rate	92. 96	91. 48	91. 34	91. 31
16 candidate Recognition rate	99. 47	99. 14	99. 28	99. 52

From table 1, it can be seen that the performance of the four individual classifiers is complementary with each other. For example, C_1 has the highest recognition rate, but there are several samples with lower recognition rate. However, the other classifier has higher corresponding recognition rate. So it is expected that the multi-classifier combination can improve the recognition rate by complementing.

4.2 Comparison of the Four Integration Strategies

The recognition performances of the four integration multi-classifiers are shown in table 2. It can be seen that the recognition rate of the multi-classifiers is higher than that of each individual classifier. The highest recognition rate is 95.160%, which is 2.2% higher that C_1. It can also be seen that the recognition rate is affected by the weights of classifier.

Table 2. Different integration strategies testing results

Classifiers Weights	{1, 1, 0, 0}	{1, 0. 95, 0. 7, 0. 6}	{1, 1, 0. 9, 0}	{1, 1, 0, 0. 8}
Voting	93. 21	93. 21	93. 21	93. 21
All	95. 16	94. 28	94. 19	94. 31
ONE	93. 02	94. 05	93. 82	94. 44
HALF	94. 63	94. 07	93. 88	94. 06

4.3 Performance against Different Candidate Character Weights

The recognition results of *ALL integrating strategy* for different candidate weights are shown in table 3. It can be seen that the recognition rate is also affected by the weights of candidates. When all the weights are chosen as constant '1', which means ignoring the influence of candidates, the recognition rate we got is the lowest. This demonstrates that considering not only the similarity of each candidate characters, but also the weights of different candidates' position is helpful in the design of multi-classifier combiner.

Table 3. Different candidates weights testing results

Candidates weights	Recognition rate
{ 1, 1, 1, 1, 1, 1, 1, 1 }	94. 95
{1, 0. 87, 0. 85, 0. 84, 0. 9, 0. 8, 0. 78, 0. 7}	95. 01
{1, 0. 86, 0. 85, 0. 84, 0. 9, 0. 8, 0. 78, 0. 7}	95. 02
{1, 0. 86, 0. 85, 0. 85, 0. 9, 0. 8, 0. 78, 0. 7}	95. 03
{1, 0. 86, 0. 85, 0. 85, 0. 86, 0. 8, 0. 78, 0. 7}	95. 14
{1, 0. 86, 0. 85, 0. 85, 0. 86, 0. 85, 0. 78, 0. 7}	95. 16

5   Conclusion

In this paper, a new multi-classifier combination scheme is proposed for handwriting Chinese character recognition. A confidence function is defined for the classifier integration through weighting the similarity of four different individual classifiers candidate outputs. Classifier weighting coefficients and candidate weighting coefficient are determined by experience. The result is not optimal, but application of the multi-classifier to handwriting Chinese character recognition demonstrates the effectiveness of the proposed method.

Acknowledgement

This paper is supported by the Natural Science Foundation of China, the Research Foundation from Motolora China Research Center, and Natural Science Foundation of Guangdong.

References

1. Jing Zheng, Xiaoqing Ding, YouShou Wu: Dynamic Combination of Multi-classifiers Based on Minimum Cost Criterion. Chinese J. Computers, 2 (1999) 182-187
2. Quhong Xiao, Ruwei Dai: A Metasynthetic Approach for Handwritten Chinese Character Recognition. ACTA Automatica Sinica, China, 5 (1997) 621-627
3. Xiangyun Ye, Feihu Qi, Guoxiao Zhu: An Integrated Method of Multiple Classifiers for Character Recognition. ACTA Electronica Sinica, China, 11 (1998) 15-19

4. Xiaofan Lin, Xiaoqing Ding, Youshou Wu: Combination of Independent Classifiers and its Application in Character recognition. Pattern Recognition and Artificial Intelligence, China, 4 (1998) 403-411

5. Anil K.Jain, Robert P.W.Duin, Jianchang Mao: Statistical Pattern Recognition: A Review. IEEE Transactions on Pattern Analysis and Machine Intelligence, Vol. 22, No. 1 (2000) 4-37

6. Yi-Hong Tseng, Chi-Chang Kuo, His-Jian Lee: Speeding Up Chinese Character Recognition in an Automatic Document Reading System. Pattern Recognition, Vol. 31. No. 11 (1998) 1601-1612

7. Lianwen JIN, Gang Wei: Handwritten Chinese Character Recognition with Directional Decomposition Cellular Features. Journal of Circuit, System and Computer, Vol.8. No. 4 (1998) 517-524

8. Jun Guo, Zhiqing Lin, Honggang Zhang: A New Database Model of Off-line Handwritten Chinese Characters and Its Applications. ACTA Electronica Sinica, China, 5 (2000) 115-116

Off-Line Handwritten Chinese Character Recognition with Nonlinear Pre-classification

Zhen Lixin[*], Dai Ruwei

AI Lab., Institute of Automation Chinese Academy of Science P.O. Box 2728, Beijing
100080 P. R. China

Abstract - In this paper, we describe a new Chinese character recognition system, in which neural networks are employed as a nonlinear pre-classifier to pre-classify similar Chinese characters, and an algorithm of clustering called Association Class Grouping algorithm (ACG) is hired to cluster similar Chinese characters. In our system, feature of contour direction is extracted to form a Bayesian classifier. Experiments have been conducted to recognize 3,755 Chinese Characters. The recognition rate is about 92%.

Key words: Chinese Character Recognition, Association Class Grouping, Neural Network, Feature of Contour Direction, Bayesian Classifier.

1. Introduction

Off-line handwritten Chinese character recognition is one of the most challenging problems in the literature of character recognition. Various approaches for handwritten character recognition have been developed. Great progresses have been made, for examples, in [1], [2], which use ETL9B, the public database of Chinese characters, as recognition set, very high recognition rate, above 99%, have been achieved. ETL9B is a comparatively good quality Chinese character set. In practice, such good quality cannot always be obtained. To put off-line Chinese character recognition into practical use, more effort is needed.

One of the difficulties of Chinese character recognition is the character set is very large. There are more than 6,000 commonly used Chinese characters. Clustering characters into cluster [3], [4] seems to be a promising direction to

T. Tan, Y. Shi, and W. Gao (Eds.): ICMI 2000, LNCS 1948, pp. 473-479, 2000.
© Springer-Verlag Berlin Heidelberg 2000

explore. Clustering transforms the large class problem to small one. But for so many similar Chinese characters, to obtain high hitting rates, the number of characters in one class would be still large. In [4], output characters are about 1,000 while hitting rate is 99.76%. And how to distinguish similar characters is still a tough task.

In this paper, a different strategy is proposed to establish a Chinese character recognition system, in which ACG algorithm is employed to cluster similar characters into classes. Neural networks as nonlinear pre-classifiers are constructed to separate similar characters in these classes. A Bayesian classifier makes the final decision.

2. Feature of Contour Direction

Transforming a character image into a feature vector is an important step of character recognition. Many efforts have been made to design effective methods to extract features. Better features provide more classification information. The Feature of Contour Direction (FCD), which is base on pixel orientation, is one of effective features.

2.1. Pixel Direction

For a character image, inner black pixels are set to white and outer ones remain which form the contour of character image. Each outer pixel is set to one or two types of direction elements, vertical, horizontal and two oblique lines slanted at $\pm 45°$, according to eight neighbors' structure. [2], [5] (Figure 1).

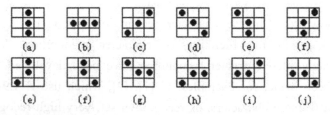

Figure 1. Pixel Direction. (a) – (d) are described by one type of direction element. (e) – (j) are described by two types of direction element.

2.2. Feature of Contour Direction

The procedure of feature extraction is shown in Figure 2. An input pattern is normalized in a 64×64 image. Then contour extraction and dot orientation is completed. The preprocessed image is divided into $k×k$ uniform rectangular zones. The FCD is calculated as follows:

$$p_i^j = (d_{i1}^j, d_{i2}^j, d_{i3}^j, d_{i4}^j) \tag{1}$$

$$s_j = \sum_{i=1}^{t} p_i^j \tag{2}$$

$$v = (s_1, s_1, \cdots, s_T) \tag{3}$$

p_i^j is the vector of four direction elements of i-th pixel in j-th sub-area, and s_j is that of j-th sub-area. v is the FCD of input pattern. T and t denote the number of sub-area and that of pixel in each sub-area respectively. Since each sub-area has four dimensions, the FCD vector for one character has $k \times k \times 4$ dimensions.

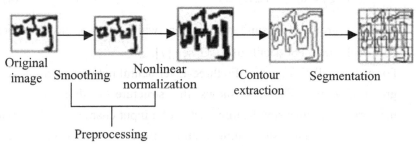

Original
image Smoothing   Nonlinear   Contour   Segmentation
                    normalization   extraction

Preprocessing

Figure 2. Procedure of Feature Extraction

3. Discrimination of Similar Chinese Characters

The difficulties of handwritten Chinese character recognition are mainly so many categories of Chinese characters, distortion of handwriting and similitude of Chinese characters. By experience, in the case of Chinese handwritten character, recognition failure caused by deformation can be corrected by different feature extractions, but that by similitude is hard to overcome. The hair like difference among similar Chinese characters can be flooded by noise. Figure 3 shows some examples of similar characters. Neural networks have strong learning and nonlinear simulation ability. Theoretically, MLP can approximate the Bayesian optimal discrimination function [6]. But MLP cannot deal with problems with large number of categories.

Figure 3. Examples of Similar Characters

In our study, discrimination of similar Chinese character is divided into two phrases: clustering similar characters by ACG algorithm and classification by neural networks.

- ACG algorithm clusters similar characters into small classes according to the association character matrix, which is formed by recognition results of training examples. Given N categories, pattern space C consists of c_1, c_2, $\cdots$ c_N. Suppose l similar groups are found, g_1, g_2, $\cdots$, g_l. Following conditions are satisfied:

$$g_1 \cup g_2 \cdots g_l \cup \tilde{g} = C$$

$$\forall_{i \neq j}(g_i \cap g_j = 0) \qquad 1 \leq i, j \leq l \qquad (4)$$

  $\tilde{g}$ is the group of dissimilar characters which are not easily confused with each other. For details of ACG, see [7].

- For each small class, a specific three-layer neural network is employed as group classifier. Figure 4 shows the structure of three-layer neural networks. τ-dimension feature vector of a input character is the input of network and n-dimension output vector represents n categories of a class. The number of notes of middle layer is determined by experiments.

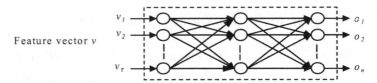

Figure 4. Structure of Neural Network

4. System Architecture

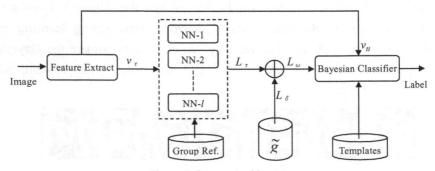

Figure 5. System Architecture

Figure 5 shows the architecture of our system. v_τ and v_u are features for neural networks and Bayesian classifier respectively; L_τ is the group of neural networks' outputs; L_g is the set of dissimilar characters and L_ω the candidate set for Bayesian classifier. The recognition procedure can be described as following:

Step 1: Feature extraction, v_τ and v_u generated.

Step 2: For all neural networks, use v_τ as input vector and output the most possible character (label). Form the vector L_τ. The most similar characters with the true character of the input sample have been removed.

Step 3: Candidate characters $L_\omega = L_\tau + L_g$.

Step 4: Classification, using City Block Distance with Deviation [2].

$$y = \min_{u \in L_\omega} \sum_{i=1}^{n} \max\{0, |v_{ui} - \mu_{ui}| - \lambda_{ui} \times \theta\} \tag{5}$$

Where μ_{ui} is the i-th dimension of template vector of u-th character's. θ is a constant. λ_{ui} denotes the standard deviation of i-th dimension.

5. Experiments

3755 most frequently used characters are selected as our research objects. For each character, 250 samples are picked up from 4-M Samples Library (4MSL), which was collected by Institute of Automation, Chinese Academy of Sciences [8], where 200 for training and 50 for testing. Some of them are illustrated in Figure 6. The samples listed in the same line belong to the same category.

Figure 6. Some samples in the database.

Among these characters, 1712 characters are clustered into 549 similar groups by ACG, while a dissimilar group, namely $\tilde{g}$, contains 2043 ones. The largest g_i ($i\in[1, 549]$) contains 50 characters and the smallest ones 2 characters. For each g_i, a specific neural network is constructed. The input feature of these neural networks is 4×4 FCD, 64 dimensions. The average recognition rate of these neural networks is 95.6%. Figure 7 shows some groups of g_i.

Figure 7. Illustration of g_i, The Largest group and several smallest ones

The Bayesian classifier only deals with characters in L_ω. In our system, $|L_\omega| = |L_\tau| + |L_g|$ = 2592. For the Bayesian classifier, the most confusing characters with respect to input character have been removed in step 2. 7×7 FCD, 196 dimensions, is assigned to Bayesian classifier. An experiment without NN pre-classification is also constructed as a contrast. System recognition rate with NN pre-classification is 91.7%, while that without NN pre-classification 85.4%.

6. Conclusion

Neural networks have been applied to character recognition, mainly to numeric recognition, and usually are used as fine classifier [9]. In this paper, neural networks are employed as nonlinear pre-classifiers to distinguish similar characters. This strategy avoids the problem of network selection (or hitting rate)[4] [9] and meanwhile harnesses strong classification ability of NN to deal with similarity of Chinese characters.

For traditional clustering methods, it is difficult to cluster so many Chinese characters into small groups. The ACG, which is different from traditional clustering methods, plays a very important role in the system. It is not designed for completeness but does find out similar characters.

Although system performance is improved highly, replacing Bayesian classifier with multi-classifier combination, the system can achieve higher recognition rate.

References

[1]   T. Wakabayashi, Y. Deng, S. Tsuruoka, F. Kimura and Y. Miyake, "Accuracy Improvement by Nonlinear Normalization and Feature Compression in Handwritten Chinese Character Recognition," *Trans. IEICE*, vol. J79-D-II, no. 1, pp. 45-52, 1996.

[2]   Nei Kato, Masato Suzuki, Shin'ichiro Omachi and Yoshiaki Nemoto, "A Handwritten Character Recognition System Using Directional Element Feature and Asymmetric Mahalanobis Distance," *IEEE Trans. PAMI*, vol. 21, no. 3, pp. 258-262, 1999.

[3]   K. P. Chan and Y.S. Cheung, "Clustering of Clusters," *Pattern Recognition,* vol. 25, no. 2, pp. 211-217, 1992.

[4]   Yuan Y. Tang, Lo-Ting Tu, Jiming Liu, Seong-Whan Lee, Win-Win Lin, and Ing-Shyh Shyu, "Offline Recognition of Chinese Handwriting by Multifeature and Multilevel Classification," *IEEE Trans PAMI*, vol. 20, no. 5, 1998.

[5]   Dai Ruwei, Hao Hongwei and Xiao Xuhong, Systems and integration of Chinese Character Recognition, Zhejiang Science and Technology Press, 1998.

[6]   D.W.Rucky, S.K.Rogers, M.Kabrisk, M.E.Oxley and B.W.Suter. The multi-layer perceptron as an approximation to a Bayers optimal discrimination function. IEEE Trans. Neural Network 1(4), 296—298 (December 1990).

[7]   Zhen Lixin and R.W. Dai, "Finding Similar Chinese Characters," Technical Report, AI Lab., Institute of Automation Chinese Academy of Science, 2000.

[8]   R. W. Dai, Y. J. Liu and L. Q. Zhang, "A new approach for feature extraction and feature selection of handwritten Chinese character recognition," *Frontiers in Handwriting Recognition,* pp. 479-489, Elsevier, Amsterdam, 1992.

[9]   H.W. Hao, X.H. xiao and R.W. Dai, "Handwritten Chinese Character Recognition By Metasynthetic Approach," *Pattern Recognition*, vol. 30, no. 8, pp. 1,321-1,328, 1997.

Detection of the Indicated Area
with an Indication Stick

Takuro Sakiyama, Masayuki Mukunoki, and Katsuo Ikeda

Graduate School of Informatics, Kyoto University
606–8501 Kyoto, Japan

Abstract. This paper describes a detection method of indication action
and the corresponding indicated area with a stick. In human communication, there exists ambiguity of indication action, so we need some useful
knowledge about indication action. We propose a detection method of
the indicated area based on the knowledge which we clarify by means
of observation of some lectures. We formulate a potential value of intentional indication action of an instant, and cast the value for printed
area using a weighted vote method. An experimental result shows that
85 percents of indication actions are correctly detected.

1 Introduction

In lectures and conferences it has become popular for a lecturer to use OHP
slides and presentation slides on PC, which are prepared beforehand. The slides
are projected on a screen and added explanation by the lecturer. The lecturer
not only shows each slide on the screen to participants of the lectures and conferences, but points to significant terms or where the lecturer explains with an
indication stick in order to emphasize them. Pointing gesture of the lecturer
helps participants to understand key points or relations between the terms.

Thus pointing gesture is so important that it is useful to investigate where
the lecturer points to with the indication stick. If we can detect an indicated area
on the screen by the lecturer, we can retrieve the lecture slide and lecturer's indication action anywhere. It means that we can make archives of those lectures
and conferences, which have hyper-links between the indicated areas and lecturer's explanations, and that in remote lectures we can send less information
to remote classrooms for remote students to understand the lectures [1,2].

There are some studies about detecting the position and movement of the
indication stick in three-dimensional space, most of which are in a field of human-
computer interface [3,4]. To utilize the indication stick as a 3-D mouse device,
it is necessary to detect precise position of the stick and gesture movement that
is related to a particular command like clicking a button. In this case, taking a
pause before making a gesture makes it easy to decide whether a person intends
to make a gesture or is only moving the stick.

But in the lecture, the lecturer uses the stick for students, not as a computer input device, so its position and movement have ambiguity. This ambiguity makes it more difficult to detect where the lecturer points to, because there

T. Tan, Y. Shi, and W. Gao (Eds.): ICMI 2000, LNCS 1948, pp. 480–487, 2000.

exists a gap between the indicated area and the tip of the stick (we call the tip *pointer*).

In our work we first observe how lecturers use the indication stick in some lectures. We derive useful knowledge from the observation, and propose a detection method of the indicated area with a vote method.

2   Indication Action with a Stick

2.1   The Ambiguity of Indication Action as Human Communication

It is important to detect the precise position of the pointer to utilize the indication stick as a computer mouse device. But in human communication, it is more important to recognize the area where the lecturer indicates. For this purpose, it is not sufficient to detect precise position of the stick, but to cope with the ambiguity of the position and movement of the stick.

If the lecturer uses the indication stick to explain the meaning of the lecture slide to students, the position and movement of the stick have ambiguity:

the position of the stick The lecturer indicates some areas on the screen with the stick to emphasize there, so the lecturer often indicates beside the area in order not to conceal the indicated area from students. This makes a little spatial gap between the pointer and the indicated area.

the movement of the stick It is difficult to decide clearly whether the lecturer intends to indicate somewhere or is just moving the stick, because there are some methods to indicate the area (Fig. 1), and indication action consists of a combination of them.

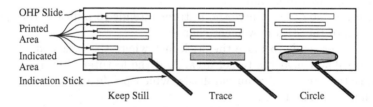

Fig. 1. There are some methods to indicate the area (described as gray rectangles) in the lecture slide with the indication stick

This means that it is difficult to detect when and where the lecturer indicates with the stick. Students can usually understand the indicated area, mainly by means of checking the lecturer's voice with terms near the pointer in the lecture slide. A speech recognition technology may help us; in particular extraction of registered keywords is now well reliable [5]. But the lecturer frequently uses demonstrative pronouns instead of keywords, so keywords extraction is useless.

Therefore we consider that there are some relations between the indication action and the position and movement of the stick.

2.2   Useful Knowledge about Indication Action

In order to clarify the relations between the indication action and the position and movement of the stick, we make a preliminary experiment to observe some lectures and trace the locus of the pointer (Fig. 2). The result shows us the followings.

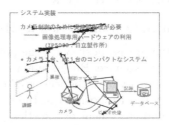

Fig. 2. Example of a lecture slide, on which the locus of the pointer is drawn

1. The lecturer indicates the area in where characters or figures are printed.
   Most of lecture slide consist of some simple sentences and figures, each of which are separated spatially and semantically. In a lecture, the lecturer indicates one of the areas and explains the meaning about that written there, or the lecturer never indicates where nothing is written. We call those areas, in which some sentences and figures exist, the *printed area.*
2. Keeping the stick still for a while usually means that the lecturer indicates the printed area intentionally.
3. Near the printed area where the lecturer intends to indicate, the speed and direction of the stick movement change.
   The lecturer tends to move the stick fast where the target area is far, and to move slower near the target area. And the lecturer often repeats tracing the printed area to give an emphasis, which produces changes of direction around the printed area.
4. The lecturer indicates below or beside the printed area instead of the center of the area, in order not to conceal the area from students.
5. The lecturer moves the stick along the printed area.
   The lecturer tends to trace over a long printed area or to circle around a square area.

We propose a detection method, which makes good use of this knowledge, of the indicated area with the stick

3   Detection Method of an Indicated Area

In this chapter we describe our detection method of the indicated area.

We face two problems: one is to decide whether or not a lecturer is pointing to some areas, the other is to detect which area the lecturer indicates.

We consider that the lecturer indicates each printed area, i.e. one of the printed areas is the indicated area. First, we define *a potential value of intentional indication action* of an instant, which is derived from the knowledge of indication action. For the latter, considering spatial and temporal continuity of indication action, we adopt a vote method. We cast the potential value of intentional indication action for the corresponding printed area, count the votes in a short term, and decide which area the lecturer indicates.

3.1   A Potential Value of Intentional Indication Action

We introduce a potential value of intentional indication action, which shows the possibility of intentional indication action by the lecturer of each instant.

In Sect. 2.2, we derive some knowledge about indication action, some of which show the characteristic pointer movement in the neighborhood of indicated areas. From 2 and 3, we can say that the speed of pointer movement near the indicated area is relatively low: in some cases the pointer stops, in other cases the speed decreases. It is also said that the pointer movement becomes so active that the direction of pointer movement changes frequently.

Based on those, we formulate the potential value of intentional indication action *pval* described below.

$$pval = \sum w_{pv_i} \cdot pval_i \quad \left(\sum w_{pv_i} = 1\right) . \tag{1}$$

a degree of keeping still

$$pval_1 = \frac{1}{|\boldsymbol{v}_t| + 1} . \tag{2}$$

a degree of slowdown (only if $|\boldsymbol{v}_{t-1}| \geq |\boldsymbol{v}_t|$)

$$pval_2 = 1 - \frac{1}{(|\boldsymbol{v}_{t-1}| - |\boldsymbol{v}_t|) + 1} . \tag{3}$$

a degree of turning back

$$pval_3 = \frac{1}{2}(1 - \cos\theta_t) . \tag{4}$$

where $\boldsymbol{v}_t$ is a pointer moving vector at time t, and θ_t is an angle between $\boldsymbol{v}_t$ and $\boldsymbol{v}_{t-1}$.

3.2   Detection with a Vote Method

We cast the potential value of intentional indication action for the corresponding printed area. But we need to consider the spatial ambiguity of indication action. From this point of view, we adjust the pointer position to fill the gap mentioned in 4, Sect. 2.2, and we adopt a weighted vote method to cope the ambiguity of indication action in human communication.

Pointer Position Adjustment It is mentioned in Sect. 2.2 that there exists a gap between the pointer and the indicated area. So we adjust the pointer position toward the adequate before voting, using the knowledge of 4 and 5 (Fig. 3).

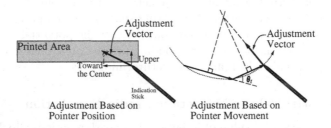

Fig. 3. Pointer position adjustment consists of two separate adjustment: one is based on the pointer position, the other on the pointer movement

adjustment based on the pointer position Adjust the pointer position a little upper and toward the center of the area. Most of lecture slides consist of horizontal printed areas, and the lecturer often indicates below or outside of each area.

adjustment based on the pointer movement Adjust the pointer position toward the center of the arc, if a locus of pointer movement draws the arc. This adjustment must be scaled in proportion to $\sin \theta_t$.

A Weighted Vote Method We adopt a weighted vote method to distribute the effect of the vote value. It is possible that the adjusted pointer position is out of each printed area. This is caused by the ambiguity of indication action: the indicated area may exist not just at the adjusted pointer position, but near there.

The weighted vote method is to cast spatially weighted vote values, which are larger near the center, and become smaller according to the distance from the center (Fig. 4). With this method we cast the value not for a printed area, but for the vote space $V(x, y)$. We also adjust the shape of weighted vote value in order to reflect the knowledge 5, mentioned in Sect. 2.2.

$$shape(x, y) = pval \cdot \exp(-\frac{x^2}{2s_x^2}) \exp(-\frac{y^2}{2s_y^2}) \ . \tag{5}$$

$$s_x = w_{\mathrm{dis}}(1 + w_{\mathrm{ratio}}(|dx_t| - |dy_t|)) \ . \tag{6}$$

$$s_y = w_{\mathrm{dis}}(1 + w_{\mathrm{ratio}}(|dy_t| - |dx_t|)) \ . \tag{7}$$

where $\boldsymbol{v}_t = (dx_t, dy_t)$.

shape(x,y)

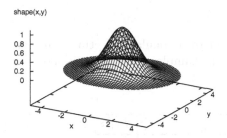

Fig. 4. The shape of weighted vote value is defined as $shape(x, y)$. This sample shows the shape where $(pval = 1,\ s_x^2 = 3,\ s_y^2 = 2)$

3.3   Detection of the Indicated Area

As described above, every time when the position of the indication stick is extracted, weighted vote values are cast for the vote plane $V(x, y)$. Considering the temporal continuity of indication action, we hold some vote planes in time order $(V(x, y, t))$, and judge which area the lecturer indicates from sum_n, the amount of votes in each printed area in a short term.

$$sum_n = \sum_{i=0}^{T-1} w_{\text{dec}}^i \sum_{area} V(x,\ y,\ t-i) \quad (0 < w_{\text{dec}} < 1)\ . \tag{8}$$

If sum_n is enough large, we decide the corresponding printed area is the indicated area by the lecturer.

4   Experiments and Discussions

In this chapter the results of testing our method are presented.

4.1   An Experimental Method

We carried out some simulated lectures using a few OHP slides. In these lectures, we took video of the screen on which the OHP slides were projected. Then we asked the lecturer to report when and where the lecturer intended to indicate. We use those reports as right answers of our experiments.

We extracted by hand the pointer position from video streams at a rate of 30 frames per a second for these experiments, so no error of pointer position extraction exists. We also divided the OHP slides into printed areas separately by hand.

4.2   Results

We applied our method to 5 simulated lectures, in which lecturers used 3–5 pieces of OHP slides. An example of the result is shown in Fig. 5, and the total results in Table 1.

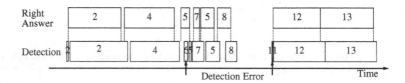

Fig. 5. An example of the result. Each box represents the area number of printed area, and the term of indication action

Table 1. This table means the result for the simulated lectures. Each row represents the total number of right answers, correct detection by our method, and excessive detection.

Indication Action	118
Correct Detection by Our Method	99
Excessive Detection Error	21

In Fig. 5, each box represents the area number of printed area, and the term in which the lecturer indicates the corresponding area. The upper part means the right answer, the lower shows the result detected by our method.

In Table 1, the total number of indication action means the number of right results reported by the lecturers. We regard that each indication action is detected correctly, if the corresponding printed area is detected at least once within the period of indication action, extended before and after 5 frames each other. We count other detection as excessive error if the detected area is never indicated by the lecturer on that time.

4.3   Discussions

In these experiments we can detect 99 indication actions and corresponding indicated areas correctly, which accounts for 85 percents of the total number of indication actions. But the examination of the results in detail, like Fig. 5, tells us that most of the detected periods of indication actions have time lags from those of the right answers. These delays are essential problems on the vote method. It is not detected until a sum of the votes exceeds an adequate threshold, and the voted value remains for a while after the pointer left from the area. Thus the timing when the indication action is detected is delayed, and once detected, it lasts for a while though the indication action is over.

Most of detection errors in Table 1 are instantaneous. On the other hand these are the cases that we cannot detect that the lecturer indicates some areas successively, or the lecturer moves the stick around a large square area. We need more examinations about those cases whether or not we can detect depending on the setting of parameters.

5   Conclusion

We proposed a detection method of indication action and the corresponding indicated area with an indication stick.

Appropriate knowledge about the indication action was required in order to cope with the ambiguity in human communication. So we examined the relations between the indication action and the position and movement of the indication stick by means of observation of some lectures. According to the examination, we proposed a detection method based on a potential value of intentional indication action and a vote method.

We implemented a prototype of our method and applied it to some simulated lectures. As the result 85 percent of the indication actions and corresponding indicated areas are detected correctly. In the future we examine about those that are not detected in this experiment.

References

1. N.Ohno, T.Sakiyama, M.Mukunoki, K.Ikeda, "Video Stream Selection according to Lecture Context in Remote Lecture," Proceedings of the 3rd Symposium on Intelligent Information Media, pp.31–38, 1999 (In Japanese). 480
2. H.Miyazaki, K.Kichiyoshi, Y.Kameda, M.Minoh, "A Real-time Method of Making Lecture Video Using Multiple Cameras," MIRU'98, Vol.I, pp.123–128, 1999 (In Japanese). 480
3. T.Ohashi, T.Yoshida, T.Ejima, "A recognition method of gesture drawn by a stick," MIRU'96, Vol.II, pp.49–54, 1996 (In Japanese). 480
4. K.Nakahodo, T.Ohashi, T.Yoshida, T. Ejima, "Stick gesture recognition using directional features," Technical Report of IEICE, PRMU97–273, pp.57–64, 1998 (In Japanese). 480
5. T.Kawahara, K.Ishizuka, S.Doshita, "Voice-operated Projector Using Utterance Verification and Its Application to Hyper-text Generation of Lectures," Transactions of IPSJ, Vol.40, No.4, pp.1491–1498, 1999 (In Japanese). 481

Heuristic Walkthroughs Evaluation of Pen-Based Chinese Word Edit System (PCWES) Usability

Zhiwei GUAN, Yang LI, Youdi CHEN, and Guozhong DAI

Intelligent Engineering Laboratory
Institute of Software, Chinese Academy of Sciences
P.O. Box 8718, Beijing 100080, China
gzw@imd.cims.edu.cn

Abstract. An evaluation experiment was conducted to compare four interface styles: Hanwang Fixed Window, Translucent Mobile Single Window, Translucent Mobile Double Window, and No Window. Subjects tested the four styles on a pen_based word edit system in free-form mode. The result shows that traditional interface style of the Hanwang Fix Window didn't satisfy the requirement of the nature interaction of the end user. And the TMDW mode are very suitable for user to perform the word processing task, in terms of the task performing time and mean accuracy rate. And the NW is the most nature and promising style, in terms of the subjective preference and the error rate of the performing time and accuracy rate of the experimental data. The experiment also shows that the moving window would influent the accuracy rate of the recognition. This experiment confirmed that a proper employment of interface style could improve the interactive efficiency of word editing systems. Also we give out some useful interface hints to improve the performance of the interface. Our tests for the first time give statistical support to the view that free pen_based interface is meaningful in the nature pen and gesture input of the word edit system.

1 Introduction

Word processing is a basic and important task type in our daily life. Officers are discussing with each other by writing their opinion on the paper note, which can illustrate their idea more clear. Teachers are teaching the students by writing down their curriculum on the blackboard, which can help the students apprehend the context more easily. Writing down the text and papers directly on the electric form by using computer becoming the main mode of working in the office and school. The type of writing programs varies from the very beginning of Notepad, WordPad, to the well-organized Microsoft Word. However, almost all of the text editing applications are developed based on the traditional input devices, keyboard, and mouse.

With the post-WIMP and Non-WIMP is coming forth [1]. The more user pay attention on is the freedom and natural of the usability. They wish the computer could work in a more humanistic mode. They want computer can hear, look, and even smell [2]. The new-fashioned interaction devices pen and voice, are just feasible. With the

T. Tan, Y. Shi, and W. Gao (Eds.): ICMI 2000, LNCS 1948, pp. 488–495, 2000.

usage of pen in word-processing program, end-user can write down the words just like he wrote on a paper.

Many former studies had focused on the comparison on the different interaction devices such as speech versus keyboard input, mouse versus speech input, speech versus gesture input, which include package sorting and Computer Aided Drafting, and in the communication environment [3][4][5][6][7]. Much research had been performed which announced that the pen+voice is the best mode among the diverse interaction mode. Pen is the main implement in the interaction task. However, there had little experimental evaluation on a practical interface style of typical pen-based systems, such as the word processing.

In our study, we evaluate the interface styles in which we employed the original pen as the main interaction implement in PCWES (Pen_based Chinese Word Edit System). Section 2 gives a brief description of PCWES and the interface styles. Section 3 gives a description of our experiment. Section 4 shows the results based on analyses of the experimental data. Following is the discussion, as well as the conclusion.

2   Word Edit Program

We proposed PCWES, which enabled the pen to perform the word input and gesture command in the task performance. To set pen to be usable, we employed a pen_based development C++ class (PenBuilder). There are some gestures used in the word edit program. They are insert, select, delete, and exchange gesture.

2.1   Overall System

In the PCWES, we separate the screen into 3 parts, which are the menu, the main word edit area, and the prompt section. The purpose to adopt menu is to suit the traditional desktop computer users. The purpose of adopting a prompt section is to provide user with a dynamic feedback, which can hint user what gesture can be used in the current performing state.

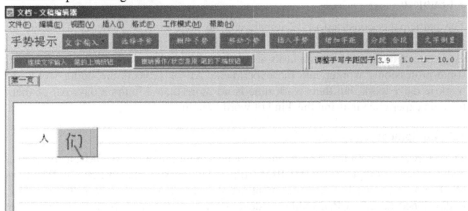

The PCWES provides many basic tasks, which include word input, section delete, section replace, section insert, word convert and section copy. And we set four typical interfaces to facility the user to fulfill their work. Below are the details of the interface style when using pen interaction.

2.2 Pen Input Interface

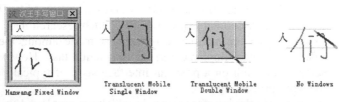

Hanwang Fixed Window   Translucent Mobile Single Window   Translucent Mobile Double Window   No Windows

Fig. 1. Four Interaction modes in the experiment

We provided four types of interface, in which user can write the word. The first is HFW (Hanwang Fixed Window), which is provided by hand-recognition program. The second is TMSW (Translucent Mobile Single Window). User can write down in a translucent mobile window. System uses time-interval as the criterion of the discrimination of word and gesture input. The third is TMDW (Translucent Mobile Double Window). This interface style is similar with the TMSW. However, because the single input window is always hop together with the moving of the pen, we adopt the double window strategy. The second window is somewhat smaller than the first window. The translucent window is aligning with the text line, moving smooth without intensive leap. By using double window, system separates the gesture and word input by analyzing the time interval and the trigger point of writing in the second box. The last one is NW (No Window). User can write down freely at any location on the screen. By analyzing the size of the bounding box of the strokes, and the distance between the bounding box, system can distinguish the gesture and word input.

3 Method

3.1 Subjects

The participants consisted of 12 female and 36 male college students. They are all right handed and between 20 and 30 years old. Considering that word compile process is one of the most common tasks at the daily life and work, we have not affiliated some professional worker into the experiment. All of subjects are doing word edit work in daily work, but their work time is no more than 4 hours a day. All of them have no experience in using this kind of word system.

3.2 Equipments

The hardware used in the experiment was a pen-input tablet (WACOM), a stylus pen, and a personal computer (PII350, IBM Corp.). All experiments were run on Windows 98. The software used in the experiment mainly include a word process system which is developed based on PenBuilder (a development platform support of pen usability developed by us), a gesture recognition system (developed by us), a handwriting recognition system (provided by Hanwang Corp.)

3.3   Task Design

3.3.1   Heuristic Walkthrough

Nielson and Mack defined inspection-based evaluation technique as a set of methods to evaluate user interface [8]. These techniques tend to require less formal training, can reduce the need for test users. Heuristic walkthroughs is one of the inspection-based evaluation techniques, which combines the benefits of heuristic evaluation, cognitive walkthroughs, and usability walkthroughs [9].

Heuristic walkthrough is a two-pass process. During Pass 1, evaluators explore tasks from the prioritized list. The list of user tasks guided Pass 1. During Pass 2, evaluators are guided by their task-oriented introduction to the system and the list of usability heuristics. Pass 2 allows evaluators to explore any aspect of the system they want while looking for usability problems.

Along with the evaluation principle of heuristic walkthrough, we define two Pass of the process.

3.3. 2   Pass 1:   Format Task-oriented Evaluation

In the experiment, subjects are instructed to perform a succession of task. Five sentences to be input, five sentences to be inserted, and five sentences to be exchanged.

The detail of experimental plan had been presented in the [10].

3.3.3   Pass 2:   Free-Form Evaluation

During the pass 2, evaluators can freely explore the operation aspects of the system. Subjects were given a section of words that should be input to the system. Different from the Pass 1 evaluation, subjects can freely set the succession of operation. They can input first, once they find there are some mistake in their text, they delete the mistake right now. And insert the correct word, and continue input. Or they can input at first, and once they find mistakes, they exchange the false by the correct words. Or they can input first, after all of the words are written, then they exchange the error word one by one, then insert some omitted word. All of their actions are guided by the knowledge gained during Pass 1.

Total operation time of subjects was counted. Once the end of the experiment, the tasks that evaluators performed will be compared with the tasks given, and the accuracy rate of the experiment tasks will be set.

After subjects finished their testing, they would give out their evaluation regarding four aspects of the system according their testing experience respectively.

CONTINUOUS: Will user know what they need to do next?

SUGGESTIVE: Will user notice that there is additional gesture he can use?

SIMPLISTIC: Will user fell it is easy to know how to use it?

FEEDBACK: If the users think input can lead the response of system immediately?

Ten scales were provided to user to measure the feeling of using regarding each of four aspects. The average of the scale is set as the subjective preference data.

3.4 Procedure

We had fulfilled the experiment of the Pass 1. And the evaluation had been discussed in the [10]. Continued with the experiment of Pass 1 evaluator performed their task in free-form mode. Subjects should writing down the text of the task in term of their instruction.

At the beginning of the experiment, system will first set an interface, which include the list of four-interaction styles. Subject can select coordinate list to begin the test. In the following steps, user will work in the accordant interface style. After finished their task, subject can use coordinate tools to end their test. Subjects' performance data in terms of their working time will be recorded by system. The additional item system would measure include how much degree the writing is familiar with writing plan. The result would be set as the mean accuracy rate. After subject finished their experimental task, they were questioned about their preferences about mode of interaction style just tested. They were asked to rank (on a scale of 1-10) each mode according to their satisfaction and desire to use.

4 Results

We had discussed the evaluation result of the Pass1 in detail in the [10]. We found that subjects can perform the basic input task correctly by employing the HFW mode. However, if the task evolved more complexity task, such as insert and exchange, the efficiency would be dropped. On the other way, working under TMDW and NW mode would perform more efficiency. With the composed task setting, TMDW is performing best. And the NW mode is the most satisfactory modes, followed by TMDW mode.

In the evaluation result of Pass2, we performed an ANOVA (analysis of variance) with repeated measures on the with-in subject factors, with the total time, perform accuracy and subjective performance as dependent measures in the free-form evaluation.

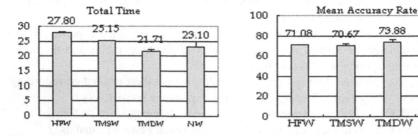

Fig. 2. The total performing time and the mean accuracy rate in terms of interaction mode.

4.1 Total Time of Free-Form Task Performing

With the total time of the free-form task performing, we found that users who working under the TMDW mode took the shortest time, with the following of the NW mode, TMSW mode, and HFW mode. There was significant difference exit in the four interface style (F(3,92)= 219.629, p<0.05). The result shows that the TMDW mode is very suitable for user to perform the word processing task. The error rates of the experimental result are also show in the figures. The result shows that the usage of TMDW mode is well proportioned, whose error rate is 0.57384. With the comparing with even TMDW mode, the experimental result of the NW mode is more fluctuant, whose error rate is 1.88129. However, the shortest time of all the experimental result is exit in the NW mode, which is 20.3 minutes. These results show that user worked under the NW mode is not stable enough.

4.2 Mean Accuracy Rate of Free-Form Task Performing

By fitting the ANOVA analysis in the accuracy rate of the experimental data, we found that there exit a significant difference among four interface styles, $F(3,92)=$ 28.38852, $p<0.05$. In the experimental data of the free-form task performing, we found that the accuracy rate of task performing under the TMDW mode is the highest, which is 73.88, followed by NW mode (72.39), HFW mode (71.08). The accuracy rate of task performing under the TMSW interface style is the lowest (70.67). It indicates that users who work under the TMDW would make fewer mistakes than other three styles in the experiment. This result may due to the prompt function of the second input area. Once the user set the tin of pen in the second area of the input window, the word recognizer would separate the two words, and apply the identification respectively. After eliminate the best performing mode of the TMDW, we found there also exit a significant difference among remain three interface styles, $F(2,69)=$ 12.3761, $P<0.05$. This result reveals that NW is also a reasonable interface style, in spite of the lack of the prompt window. The intuition of subjects would set a large space between the two linked words, which would provide some help to the word recognition process.

Furthermore, the error rate of the experimental data may reveal the stability of the word process under different interface styles. The error rate of the NW mode is highest (2.97471). This result show that the performance under the NW may fluctuated because of its free interface of the word input. The error rate of the TMDW is 2.375, which is less than that of NW. This data reveal that the continued translucent input window may help the user and the system to correctly separate the linked handwriting input, then correspondingly enhance the accuracy rate of the recognition.

4.3 Subjective Preference

In the subjective preference data of the experiment, there were significant difference among four interface styles for subjective evaluation, $F(3,92)=$ 198.077, $p<0.05$. The HFW mode is the last one user preferred (6.95), followed by TMSW mode (7.05), and followed by TMDW mode (7.52). After eliminate the NW mode, there were also significant difference between the remain three interface style, $F(2,69)=$ 70.0477, $p<0.05$. There also exit significant difference between the TMSW mode and HFW mode, $F(1,46)=$ 4.78909, $p<0.05$. The result means that a single and fixed window may not suit the preference of the users.

Regarding the error rate aspect, the NW mode is the highest (0.64493); the second is the TMDW mode (0.1721). The comparison of the error rates reveals that the NW is a promising interface style, although its stability still need to be improved. However, some of our testers said that they prefer working under the style of the NW mode, although the freedom of the operation would results more mistakes comparatively.

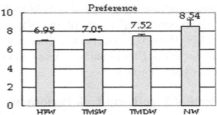

Fig. 3. The subjective preference in terms of interaction mode.

5   Discussion

1. The traditional interface style of the Hanwang Fix Window didn't satisfy the requirement of the nature interaction. When working in the fixed mode, user should separate their attention into two parts, one is focused on the writing window, another is focused on the main window, in which main paragraph is presented there. The experimental data of the FHW mode show that when people worked under this version of interface style, he can do it in a comparatively accuracy rate, but their efficiency is quite low, together with the very low preference rate of subjects. Furthermore, with the combination of gestures into the program, user should change their hand situation from the input area to the paragraph area at every turn. This action may cause a confusion of the recognition of the gesture and word input.
2. With the employment of the moving window, there was a translucent input window that is linked with the location of the pen subjects used. The handwriting in this area would be recognized by the embedded recognition engineer. By analyzed the experimental data of the four interface styles, the accuracy rate of the TMSW mode is lower than HFW mode, and TMDW mode is higher than the TMSW mode. These results show that the moved window would influent the accuracy rate of the recognition. Under the single moved window, users would be confused by the movement of the window. This would cause the accuracy rate of the recognition decline. However, with the second translucent input window encapsulate, the rate of the word process may be enhanced more.
3. The NW interface mode is a truly nature interaction mode. Under this interface style, user can write down at any location he wants. This interface mode is very similar with the nature writing mode in daily work. By referring the accuracy rate, we can say that the NW mode of the interface style is a promising style. It gets the highest selection in the subjects. We should pay more attention on the applying of this type interface style. However, there still has a technology nodus, which is how to separate two linked words, so that system can perform the recognition correctly. This experimental provided us a significant suggestion that we should provide some effective engineers to separate the linking word, set the proper writing area, to support the NW interface mode. By discussing with subjects, we found this free

input mode may cause the user automatically set more space in the words. The separated words will get more high recognition rate.

6   Conclusion

This paper presented an evaluation experiment based on a word process system. The experimental results showed that TMDW and NW interface styles may perform efficiency in text edit systems. The analyses show that the NW interface style is comparative more suitable in the pen_based word systems. Overall, we have provided information about how users would perform under each of the four interface styles, and the promising direction of interface style in order to suit the nature of interaction.

References

1. Andries. Van Dam, Post-WIMP User Interfaces, Communications of the ACM, Vol. 40, No. 2, (February), 1997, 63-67.
2. Yang L., Zhiwei G., Youdi C. and Guozhong D., Research on Gesture-based Human-Computer Interaction, Proceedings of Chinagraph'2000, China.
3. Gale L. Martin, The utility of speech input in user-computer interfaces, International Journal of Man-Machine Studies, 1989, No. 30, 355-375.
4. H. Ando, H. Kikuchi, and N. Hataoka, Agent-typed Multimodal Interface Using Speech, Pointing Gestures and CG, Symbiosis of Human and Artifact, 1995, 29-34. J.J.
5. Mariani, Speech in the Context of Human-machine communication, ISSN-93, 1993, Nov, 91-94
6. Steve, W. Patrick, H. & Myrtle, W., Flochat: Handwritten Notes Provide Access to Recorded Conversations, Human Factors in Computing Systems , 1994, April 24-28
7. Luc J. and Adam C., A Multimodal Computer-augmented Interface for Distributed Applications, Symbiosis of Human and Artifact, Elsevier Science B.V., 1995, 237-240.
8. Nielsen, J., & Mack, R., Usability inspection methods. 1994, New York: Wiley
9. Andrew Sears, Heuristic walkthroughs: Finding the problem without the Noise, International Journal of Human-Computer Interaction, 1997, 9(3), 213-234
10. Zhiwei Guan, Yang Li, Hongan Wang and Guozhong Dai, Experimental Evaluation of Pen-based Chinese Word Processing Usability, Asia Pacific CHI & ASEAN Ergonomics 2000

Design and Realization of 3D Space Coordinate Serial Input[*]

Liu Jingang[1]   Zheng XiangQi[1]   Li GuoJie[2]   Qian Yaoliang[2]

[1]Join Lab for Computer Application Capital Normal University
Beijing 100037, China
[2]National Research Center for Intelligent Computer Systems
Beijing 100080, China

Abstract. This paper is about design and realization of communication circuit, which connects three-dimension coordinate input device to computer. Communication protocol and the key technique of device driver are published in this paper. It is told how the applied software is developed on the basis of communication protocol and driver. One applied example is introduced. This paper provides essential technique condition for a broad application of the device.

Keywords. Space coordinate, Input device, Communication protocol

1   Introduction

Human-computer interface technique is developing now. Keyboard, mouse and joystick are the master peripheral equipment at present. We can manipulate computer on plane through these devices. But they can only offer two-dimension information, therefore they can not be used in 3D space. In order to manipulate computer in 3D space, a kind of device, which can input three-dimension coordinate at same time, is necessary. Since the late 80's some developed countries had begun to research this kind device, in the early 90's had produced some input devices such as spaceball, 3D joystick, 3D space probe etc. But for the inconvenient or the expensive, they are not applied widely.

---

[*] This project gets support of 863 high-tech, contract No: 863-306-08-01

T. Tan, Y. Shi, and W. Gao (Eds.): ICMI 2000, LNCS 1948, pp. 496-501, 2000.
© Springer-Verlag Berlin Heidelberg 2000

Great efforts are made to research a kind of 3D space coordinate input device, which is convenient, cheap and widely applied. It is important for the Virtual Reality (VR), more real human-computer interface and is a kind of key technique of VR. Manipulating computer in the 3D space is an inevitable trend and applied field is wide. This kind of device will provide a new computer manipulation environment to people. Under the support of 863 high tech, we had begun the study of applied VR technique-multidimensional space information input device since 1996. Now the project has been completed successfully, a kind of applied 3D space coordinate input device has been developed. While a little emitter in the hand is moved around in the font of screen, 3Dinformation of space is inputted into the computer. The device's response time is 6 milliseconds and spatial precision is 1.5 millimeter, so it completely satisfies the demand of input in 3D space. It is convenient, cheap and applied. At present we are working on manufacture it largely.

Though this kind of device can be manufactured, it needs the support of 3D applied software. Manipulating computer in 3D space, people will feel intuitionistic, quick, real and like in the computer.

2   Design of Serial Communicate Hardware

The kernel of the device is a singlechip, 89C2051. Because serial export voltage difference between the high and low is 5v, serial input logic "1" is from -5v to -15v and logic "0" is from +5v to +15v in computer, serial export signal from singlechip can not be input into computer directly and must be processed through hardware circuit.

Usual method that realizes serial communicate is to convert voltage from 0~5v to −12v~+12v through 1488, but 1488 needs ±12v power. This device's power is supplied with COM and need not outside power, showed in Figure 1. The method of power supply is that the two of three serial signal cables is set high level and another is set low. Because the energy of COM is limited, its' voltage is pull down from ±12v to ±5v when it supply power for the device. So 1488 can't be used in this device.

In order to realize serial communication in the case of the device supplied with ±5v power, we did plenty of research and experiments. Finally we decide to adopt 74HC4053 Analog Multiplexer which power is little, switch speed is quick, in Figure 2, $t_p \approx 18ns$. Serial communicate circuit can work correctly.

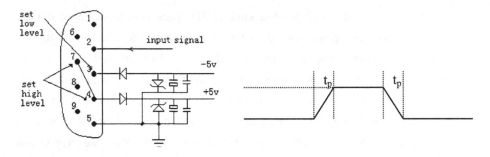

Figure 1   Figure 2

74HC4053 is 2-channel multiplexer, in Figure 3, principle showed in Figure 4.

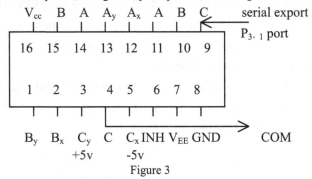

Figure 3

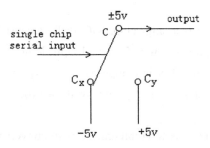

Figure 4

The serial output of singlechip is connected to control port of conversion switch C. when the output is high level 5V, namely logic "1", conversion switch C connects to negative power and the voltage of signal sent to computer COM is −5v. When the output is low level 0v, namely logic "0", conversion switch C connects to positive power and the voltage of signal sent to computer COM is +5v. Thereby serial output voltage is converted from 0 ~+5v to +5v~-5v, the voltage difference is 10v. The question about level conversion is resolved.

This serial communication circuit designed is simple, cheap and reliable. Tested by experiments, data signals are transmit correctly when the transfer cable is 5 meter long. This circuit is widely could be used in other serial communication circuit which need voltage change.

3   Communication Protocol

This device completes collection and transmission of space coordinate data under the control of singlechip. It transmits not only x, y and z coordinate but also the status of two key-presses. When x, y and z coordinate don't change, namely stop operation, no data is transmit. only when we operate the 3D device, namely x, y or z changes, the data is transmit to computer. In order to insure the data is correct, the communicate protocol between singlechip and computer is constituted. It is convenient for development of applied software that the protocol is published in this paper.

Serial port of singlechip works on mode 1, baud rate is 19.2Hz and 8 bits asynchronous communicate port is adopted. Frequency of crystal oscillation is 22.1184MHz. Data bits are transmit from TXD port. A frame data is 10 bits, the first bit is start bit, the middle 8 bits are data and the last bit is stop bit. Serial control register SCON is 01000000, namely 40H. Timer T_1 works on mode 2, TMOD=20H. While CPU run a command which function is to write data to sending buffer SBUF, transmitter start transmitting data. After data have been transmit, interrupt flag TI is set "1".

Capacity of a set of space coordinate data is as large as 4000. So every set of data includes 38 bits, thereinto every coordinate 12 bits, status of key-press two bits. It is made of five frames. The first two bits is the status of key-press, which are 01, left key is pressed, 10, right key is pressed, 11, both keys are pressed, 00, no key is pressed. The third and forth are 0, others are coordinate bits which include 12 bits, in Figure 5.

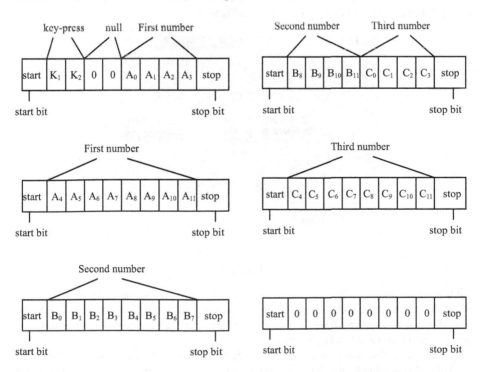

Figure 5

For avoiding data error, a special data frame 00H is sent between every set of data. Signal baud rate is 19.2KHz, it needs T = 60/19.2 = 3.125ms to finish sending a set of data. Computer COM works on interrupting mode, once data input, the computer responds at once. When a frame data collected is all 0, computer complete the collection of a set of integrated space coordinate.

According communicate protocol, six frames data received was converted to a set of space coordinate, namely $(x : A_{11}A_{10} \cdots A_0, y : B_{11}B_{10} \cdots B_0, z : C_{11}C_{10} \cdots C_0)$, A, B and C is binary code. Coordinate (x, y, z) is gotten and then object can move in 3D space according the coordinate. Because 3D coordinate owns more a dimension coordinate z than plane coordinate, more information can be communicated and you feel more real. The device has more functions than common mouse. This makes computer can finish some work that before can't.

4   3D Applied Software Example

The complement of this convenient, practicable and quick device widens the applied field of computer, promotes the development of computer applied software and makes the human-computer interface more perfect. 3D bowls applied software is an example the device is applied.

This software need 3D space coordinate, conventional mouse can not manipulate it because the speed of ball is measured according the change of coordinate z.

The software was developed in Visual C++. First to establish a 3D environment of bowls court and then control the bowls by the device, in Figure 6.

Figure 6

When we are holding a little emitter in the hand and moving it, the cursor moves with it. We place the cursor on the ball and press the left key-press, then the bowls will be locked by the key. If we push the emitter from the appropriate position and release the key at same time, the bowls would bump and bowl over bottles.

The speed and direction of the cast ball decide how many bottles are knocked down. The speed of ball is decided by the change of coordinate z. Convention mouse only moves on plane and can't control the movement speed of the ball.

Because the paper length is limited, only one applied example is introduced. 3D applied software has more wide use and foreground than the plane. The device will promote the progress of computer and applied software.

Reference

[1] Liangju Shi. "A Collection of Selected IC Application", Electronic industry publishing company, 1989.

[2] Limin He. "System Configure and Interface Technology", Beijing University of Aeronautics and Astronautics publishing company, 1990.

[3] Peter Norton, RobMcGregor. "Developing Windows 95/NT4 Application in MFC", Tshinghua publishing company, 1998.

[4] Teffrey Richter. "Windows 95/NT3.5 Advanced Program Technique", Tshinghua publishing company, 1997

[5] Zhonghua Shen. "Three-dimension Animation Software 3D Studio Release 3.0", Study garden publishing company, 1994.

[6] Qilin Fu etc. "Microcomputer Interface and Communicate", Electronic industry publishing company, 1992.

User's Vision Based Multi-resolution Rendering of 3D Models in Distributed Virtual Environment DVENET

Xiao-wu Chen   Qin-ping Zhao

Virtual Reality and Visualization Laboratory
Department of Computer Science and Engineering
Box6863#, Beijing University of Aeronautics & Astronautics
Beijing 100083
cxw@vrlab.buaa.edu.cn

Abstract. Funded by National High Technology 863 Project, the distributed virtual environment network DVENET has been focused at incorporating realistic 3D visual simulation into a joint exercise. Visual tracking allows visual simulation to measure visual field and visual angle relied on by multi-resolution rendering techniques including LOD (level of detail), morphing and LOL (level of light), which economize the computational cost. Based on vision perception of user, this paper describes our recent research in real-time multi-resolution rendering of 3D models in virtual environment. This method calculates the critical parameters, such as view-distance, imaging parameter, brightness parameter, and so on, which control the switches between neighbor multiple resolution models and are correlative to user's visual field and visual angle. It might also be utilized for kinds of simulators responding to the presence of the multi-resolution target but based on the special visions of infrared equipment or radar. **Keywords:** DVENET, user's vision, multi-resolution rendering, level of detail, morphing, level of light, critical parameter.

1 Introduction

DVENET is a distributed virtual environment network sponsored by National High Technology 863 Project [1]. It includes a wide-area network, a set of standards and toolkits, and a realistic three dimensional visual simulation system, all of which involve many exiting kinds of man-in-the-loop simulators and some virtual simulators controlled by different users, which are distributed from different locations. In an exercise, most of simulators can take parting in and interact in a 110 kilometers by 150 kilometers synthetic environment using multiple modalities including visual tracking.

Vision perception is the main way for people to get information, and the 3D visual simulation of simulator in DVENET is the most important way to make the users to be immersed in the virtual environment [2, 3]. To generate realistic scene, the visual simulation firstly measures user's visual field by tracking his eye-movements, which is one of the human-computer interactive modalities [4], and secondly calculate the visual angle with respect to the object size and view distance.

T. Tan, Y. Shi, and W. Gao (Eds.): ICMI 2000, LNCS 1948, pp. 502-509, 2000.

On the other hand, in order to reduce the polygon flow and the computational cost, thereal-time visual simulation utilizes multi-resolution rendering techniques including LOD (Level of Detail), morphing and LOL (Level of Light) to render many complex objects in the virtual world [5, 6, 7]. And the user's visual parameters being corrective to visual field and visual angle have been proposed as important factors on selecting the correct resolution models, which impact the fidelity of scene rendered and the frame rate of the simulation.

In this paper, we describe our recent progress in real-time rendering multi-resolution objects with 3D models based on user's vision including visual field and visual angle in virtual environment. We present a method used to calculate the critical parameters controlling the switches between neighbor multiple resolution models and being correlative to user's vision while these multi-resolution render techniques are utilized to keep the frame rate above the threshold in DVENET. Finally, it is proposed that the users imply not only human but also all kinds of infrared equipment or radar simulated in the distributed virtual environment DVENET.

2 Visual Field and Angle

There are two important tasks must be accomplished in order to generate realisticvisual scene. The first of these is to measure the visual field with respect to user's eye movements. Figure 1(a) shows the visual field of right eye while the head and the eyes are all hold still. Figure 1(b) shows the visual fields while the eyes moving with still head [8].

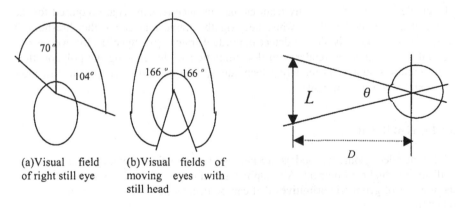

(a)Visual field of right still eye

(b)Visual fields of moving eyes with still head

Fig. 1. Visual fields

Fig. 2. Visual angle

The second task is to calculate the visual angle to estimate the size of object in virtual world. Figure 2 shows how to define visual angle θ :

$$\theta = 2 \cdot \arctan \frac{L}{2D} \qquad (1)$$

D is the view distance, and L is the length of a line object. When $D \gg L$, then:

$$\theta \approx \frac{L}{D} \qquad (2)$$

Table 1 shows several visual angles of the familiar objects observed by human.

Table 1. Visual angles of familiar objects

Object	Distance	Angle
sun	1.5×10^8 kilometers	30 minutes
moon	386000 kilometers	30 minutes
coin	80meters	1 minute
12 points word	0.4meters (reading distance)	13 minutes

The experimental data results that the visual angle should be more than 15 minutes with natural illumination.

3 Multi-resolution rendering of 3D Models

The challenge is to add as many features that enhance realism as are required for the intended use of the simulator while keeping the frame rate above the threshold at which the human eye is able to detect individual frames [9]. Specifically, we seek to expand the visual scene rendered in the simulation while reducing the polygon flow using multi-resolution 3D rendering techniques, such as LOD (level of detail), morphing and LOL (level of light).

3.1 Level of Detail

Highly detailed geometric models are necessary to satisfy a growing expectation for realism in virtual environment. All graphics systems have finite capabilities that affect the number of geometric primitives that can be display per frame at a specified frame rate [7].

To further improve the rendering performance of real-time visual simulation, it is a better way to define several geometric models of an object at various levels of detail. A higher resolution model is used when the object is close to the user's eyes and lower approximations are substituted as the object recedes. Figure 3 shows the illustration of multi-resolution rendering using LOD.

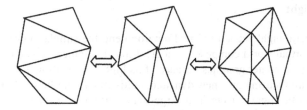

Fig. 3. Level of detail

Recently, we present a part LOD method based quad-tree. Figure 4 shows two near LODs, a lower resolution model named LODa and a higher resolution model named LODb divided into four parts. We can just do partly switching in the area titled LODb0, in which the mostly different representation between LODa and LODb appears [10].

In DVENET most of objects have multiple geometry levels of detail located in certain terrain,which is stored as a heap-sorted quad-tree.

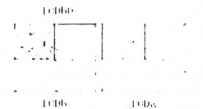

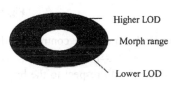

Fig. 4. Part LOD using quad-tree   **Fig. 5.** Morphing range

3.2 Morphing between LODs

One problem that arises when using LOD techniques is the visual discontinuity, or popping, that occurs as one level of detail is abruptly replaced with another [10].

Morphing, the smooth merging of one level of detail into another, can solve this problem. When the higher resolution level of detail switch in, the real-time visual simulation first display the morph vertices and then update the visual scene in gradual steps until the level of detail's real vertices are displayed. It is shown in Figure 5.

Morphing can refer to the level-of-detail attributes, such as the color, the normal, the switch in value, the switch out value, the transition distance, the texture coordinate, and the coordinates. Each vertex in a higher resolution level of detail is allocated a morph vertex, the simplest way to set up a morph between two consecutive levels of detail is to select both and create a morph vertex by copying the closest vertex in the lower resolution level of detail.

3.3 Level of Light

To render realistic night scene in virtual environment, the simulation of many illuminations in the virtual world composed of a great lot of complex geometric models is a key challenge in computer graphics because of its high computational cost [11, 12]. Therefore, we are focusing on a new dynamic multi-resolution illumination method for generating realistic night effects in a large-scale virtual environment. This method allows each illuminant can have several levels of light (LOL) shown in Figure 6, such as point light, local part light, local spotlight, entirely environmental light, and so on.

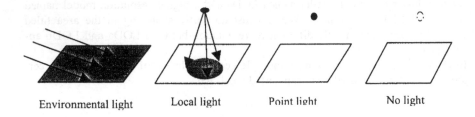

Environmental light   Local light   Point light   No light

Fig. 6. Level of light

The critical parameters controlling the switch between LOLs include imaging parameter ξ with respect to the image size of object mapped into screen, and brightness parameter η with respect to the brightness of object because of certain illumination. Figure 7 shows the projected line L in virtual world and its image line l in screen.

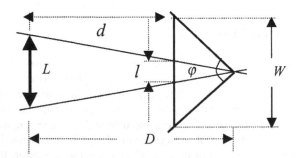

Fig. 7. The length of imaged line

For perspective projections the length of line l can be calculated as following:

$$l = L \cdot \frac{D-d}{D} = \frac{L \cdot W}{2 \cdot D \cdot tg\frac{\varphi}{2}} \qquad (3)$$

W is the size of screen, φ is angle with respect to view point and screen size, and d is the distance between line L and screen.
If there are ρ pixels in unit length, then line has ξ pixels:

$$\xi = \frac{L \cdot W \cdot \rho}{2 \cdot D \cdot tg\dfrac{\varphi}{2}} \qquad (4)$$

Figure 8 supposes a circle area illuminated by certain light. The triangles around the center of the circle and ring boundary are equably sampled to compute the brightness average U_{ave}, the maximum U_{max}, and the minimum U_{min} of this area.

According the requirements of simulation, the tunable imaging parameters and brightness parameters of certain illuminant are used to select a more economical level of light from its all of levels in some time during the simulation. This method also utilizes quad-tree to fleetly find objects illuminated by certain level of light, to load correct models at multiple LOD, and to calculate the critical parameters.

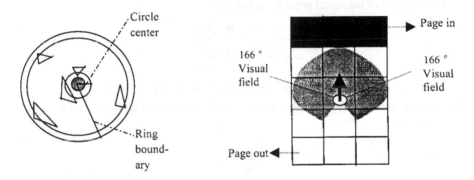

Fig. 8. A illuminated circle area   **Fig. 9.** Load correct terrain using visual field

4 User's Vision based Parameter Calculation

While using multi-resolution 3D rendering techniques including LOD, morphing and LOL to economize the computational cost, it is common to calculate the parameters controlling the switches between different resolution models. And most values of parameters are correlative to user's visual perception.

First, DVENET had already founded a 150km by 110km synthetic environment including kinds of objects using the actual terrain data from the National Basal Geographical Information Center. But it is very difficult to input all of the data (about 400Mbytes) portraying the virtual world into the computer at the same time. In order to increase the rendering performance, the data set has been divided into 4125 text files based on the two kilometers standard of the military "grid square" with each file containing data for four square kilometer.

If user's visual field is no more than a 6 kilometers by 6 kilometers square, 3 by 3 block files are loading into memory at any given time. It is shown as figure 9. During terrain paging, in order to update the active area of the simulator, files corresponding to a 1 by 3 km. strip of terrain are consecutively block-load into the data structure, and the opposite strip is freed.

Second, Because the distance between objects and user's eyes is the only parameter used to select correct LOD, according formula (2) we represent the visual-distance as follows:

$$D \approx \frac{L}{\theta} \tag{5}$$

Generically we suppose length L is static, then the visual angle θ of user is the only factor utilized to select the right LOD.

Third, the LOL technique relies on user's vision in two ways:

a) We measure the imaging parameter ξ using user's visual angle and variable length of line L as formula (4). And using formula (2), formula (4) can be written as a factorization of the visual angle θ as follows:

$$\xi = \frac{W \cdot \rho}{2 \cdot tg \dfrac{\varphi}{2}} \cdot \frac{L}{D} = \frac{W \cdot \rho \cdot \theta}{2 \cdot tg \dfrac{\varphi}{2}} \tag{6}$$

Then imaging parameter ξ is only corrective to user's visual angle θ.

b) The luminance contrast can be written as a function of brightness maximum and minimum:

$$K_{contrast} = \frac{U_{max} - U_{min}}{U_{max} + U_{min}} \tag{7}$$

Based on user's vision principle, the domain of luminance contrast $K_{contrast}$ should be:[0.01, 0.02].

So the brightness parameter η can be calculated after synthesis of the luminance contrast and the brightness average U_{ave}.

5 Conclusion and Future Work

Visual tracking allows visual simulation to calculate visual field and visual angle relied on by multi-resolution rendering of 3D models in virtual environment. Based on user's vision the multi-resolution rendering techniques includes LOD, morphing and LOL, which economize the computational cost and calculate their critical parameters controlling the switches between different resolution models and being correlative to user's visual perception.

DVENET had already developed kinds of man-in-the-loop simulators and virtual simulators controlled by different users, which are distributed from different locations.

The users imply not only human but also all kinds of infrared equipment or radar simulated in the distributed virtual environment. So the multi-resolution rendering method based on user's vision can be utilized for some simulators, such as tank, helicopter, and fighter plane, which can respond to the presence of the multi-resolution target based on the special visions of infrared equipment or radar.

References

1. Zhao Qinping, Shen Xukun, Xia Chunhe, and Wang Zhaoqi. DVENET: A Distributed Virtual Environment. Computer Research & Development. Vol.35, No.12, 1064-1068, Dec. 1998
2. Michael J. Zyda, Networking Large-Scale Virtual Environments, Proc. Of Computer Animation '96, Geneva, Switzerland, IEEE Computer Society Press, p1-4. June 1996.
3. Chen Xiaowu, He Hongmei, Duan Zuoyi. Generation of Virtual Vision Based on Multiple Platforms. Computer Research & Development. Vol.35, No.12, 1079-1083, Dec. 1998
4. L. Ngay, et al., A Generic Platform for Addressing the Multimodal Challenge. In Proc. CHI'95 Human Factors in Computing System. ACM New York, Denver, 1995.
5. Lindstrom P, Koller D, Ribarsky W. Real-time, Continuous Level of Detail Rendering of Height Field. Computer Graphics, 1996, 30: 109-117
6. Jonathan Cohen, Amitabh Varshney, et al. Simplification Envelopes. Computer Graphics Proceedings, Annual Conference Series 1996, p119-128.
7. Hugues Hoppe. Progressive Meshes. Computer Graphics Proceedings, Annual Conference Series 1996, p99-108.
8. Wang Cheng-wei et al. Virtual Reality Technology and harmonic Human-machine Simulation Environment. Computer Research & Development. Vol.34, No.1, pp1-12, Dec. 1997.
9. John S. Falby, Michael J. Zyda, David R. Pratt and Randy L. Mackey. NPSNET: Hierarchical Data Structures for Real-Time Three-Dimensional Visual Simulation. In computer & Graphics, Vol. 17, No. 1, pp.65-69. 1993.
10. Qinping Zhao, Xiaowu Chen. Realistic Three Dimensional Visual Simulation In Real-time Distributed Virtual Environment DVENET. International Conference on CAD/CG'99. Shanghai China, pp1136-1141. Dec. 1999.
11. Drettakis G., Sillion FX. Interactive Update of Global Illumination Using a Line-space Hierarchy. In Computer Graphics, SIGGRAPH'97 Proceedings, Annual Conference. Vol.31, pp57-64. 1997.
12. Wolfgang Heidrich, Hans-Peter Seidel. Realistic, Hardware-accelerated Shading and Lighting. In Computer Graphics, SIGGRAPH'99 Proceedings, Annual Conference. Vol.33, pp171-178. 1999.

Vision-Based Registration Using 3-D Fiducial for Augmented Reality

Ya Zhou, Yongtian Wang, Dayuan Yan, Tong Xu

(*Department of Opto-electronic Engineering, Beijing Institute of Technology,
Beijing 100081, China*)

Abstract. One of the key issues in the realization of Augmented Reality is the registration problem. Synthetically, vision-based registration can offer superior solutions. In several existing registration methods, 2-D objects or pictures are used as positioning fiducials, and at least two cameras and/or separate tracking devices are needed. A new registration method is proposed which uses a solid fiducial and a single color CCD camera. The new method significantly simplifies the registration system and eliminates tracker-to-camera errors.

Keywords. Augmented Reality, Registration, solid fiducial

1. Introduction

Augmented reality (AR)[1] is a technology in which a user's view of the real world is enhanced or augmented with additional information generated from a computer model. The enhancement may consist of virtual artifacts to be fitted into the environment, or a display of non-geometric information about existing real objects. With an AR system, the user can see the real world around him, with computer graphics superimposed or composited with the real world. Instead of replacing the real world as Virtual Reality (VR) does, AR supplement it. Ideally, it would seem to the user that the real and virtual objects coexist. The technology can be used as productivity aids in assembly, maintenance, training and logistics.

One way to implement AR is with an optical see-through Head-Mounted Display (HMD). This device places optical combiners in front of the user's eyes. The combiners let light in from the real world, and they also reflect light from miniature monitors inside the HMD

T. Tan, Y. Shi, and W. Gao (Eds.): ICMI 2000, LNCS 1948, pp. 510–517, 2000.

displaying computer graphic images.   The result is a combination of the real world and a virtual world drawn by the monitors.

A key issue to realize AR is the registration problem - the registration of the positions of the objects on which virtual information is overlaid.   Several techniques are developed, including

1) Knowledge-based 3-D registration[2].   A head-tracker is employed to determine the user's head position and orientation, and other 3D trackers are attached to key components in the real world, whose positions can then be monitored by the system.   The problems associated with this scheme are system latency and lack of accuracy due to coordinating errors among the tracking devices.

2) Image processing based 3-D registration[3].   The tracking information is abstracted from the digital image of the real world by identifying some special objects, and the position and orientation of user's head are registered accordingly.   The scheme adopts optical tracking and does not need any separate tracking subsystem, thus the whole system is simple as well as versatile. However, the image processing involves enormous computation, which often causes serious latency.

3) Vision-based 3-D registration[4].   Special marks (fiducials) are placed on real objects.   These marks are recognized in real time by the computer vision system, and the positions and orientations of the user's head as well as the objects of interest are calculated. Vision-based registration can provide superior overall performance, and it is especially suitable for use with optical see-through HMDs.

In vision-based registration systems developed previously, planar fiducial marks are used, and at least two cameras are required to obtain enough information.   The computer needs to compare the different image outputs from different cameras in order to complete the registration task.   The system setup and computation algorithm are complex, and extra errors are introduced.   A solid fiducial is proposed, which gives sufficient information for the registration of the user's head with only one camera and relatively simple computations.

2. Configuration of Registration System

In our registration scheme, a cube, of which the edge length is known, is placed in the AR working space and used as the 3-D fiducial object.   A corner point of the cube is used to designate the origin of the spacial coordinates of the real world, and the three perpendicular edges extended from the corner represent (the negative directions of) the three coordinate axes, which are painted with different saturate color such as red, green and blue.   A single color CCD camera with a fixed focal length is fitted on the HMD, whose optical axis represents the line of sight of the user.   During the operation of the AR system, the 3 colored edges of the cube are required to remain in the field of view of the camera.   Therefore, the working area of the registration device is confined in the first quadrant of the coordinate system defined by the extension of the three colored edges, as seen in Fig.1.

3. Algorithm of 3-D Registration

When the camera moves with the HMD, it images the cube fiducial from different positions, resulting in different projection patterns on the CCD for the three colored edges.   The position and orientation of the camera in the coordinate system defined above can then be uniquely determined.

In Fig.1, OX, OY and OZ are the three colored edges of the cube fiducial, which coincide with the three coordinate axes respectively. At a specific moment, the camera moves to a certain position, and λ is a plane passing through O and perpendicular to the optical axis of the camera.   If the distortion of the camera lens is considered negligible, the images of OX, OY and OZ are proportional and parallel to their projections on the plane λ, namely OX', OY' and OZ'.

By definition, $\triangle$XOY, $\triangle$YOZ and $\triangle$XOZ are all isosceles right-angled triangles, and under the condition of operation depicted above, their projections on the plane λ are triangles too.   Fig.2 illustrates $\triangle$XOY and its projection $\triangle$X'OY', in which γ is the angel between the XOY plane and the plane λ, and θ is the angle between OX and the line of intersection of the two planes.   We denote OX=OY=a, thus XY=r=$\sqrt{2}a$.   The lengths of their projections on the plane λ are signified by x', y' and r' respectively, and the ratios among them can be obtained by measuring the lengths of the images of OX, OY and OZ on the CCD, which can be written as

$$x': y': r'= m : n : l \tag{1}$$

Referring to Fig.2 and using simple geometry, we have

$$a^2 \cos^2 \theta + a^2 \sin^2 \theta \cos^2 \gamma = x'^2 \tag{2}$$

$$a^2 \sin^2 \theta + a^2 \cos^2 \theta \cos^2 \gamma = y'^2 \tag{3}$$

$$2a^2 - (a\sin\theta \sin\gamma - a\cos\theta \sin\gamma)^2 = r'^2 \tag{4}$$

Eqs. (2) and (3) are divided by (4) to give

$$\frac{\cos^2\theta + \sin^2\theta\cos^2\gamma}{2 - (\sin\theta\sin\gamma - \cos\theta\sin\gamma)^2} = \frac{m^2}{l^2} \tag{5}$$

$$\frac{\sin^2\theta + \cos^2\theta\cos^2\gamma}{2 - (\sin\theta\sin\gamma - \cos\theta\sin\gamma)^2} = \frac{n^2}{l^2} \tag{6}$$

If we defining $t = \cos\theta$, the value of t^2 can be reckoned out. Then we have, after simplification

$$\cos^2\gamma = \frac{m^2 - t^2(m^2 + n^2)}{n^2 - t^2(m^2 + n^2)} \tag{7}$$

Within the first quadrant working range of the registration system, we have $\gamma > 0$. Thus $\cos\gamma$ can be uniquely determined.

In a similar way, we can calculate $\cos\alpha$ and $\cos\beta$, where α is the angel between the YOZ plane and λ, and β is the angel between the XOZ plane and λ. $(\cos\alpha, \cos\beta, \cos\gamma)$ are in fact the direction cosines of the normal of the surface λ, which by definition is parallel to the optical axis of the camera. Hence, the orientation of the line of sight of the user in the XYZ coordinate system is found.

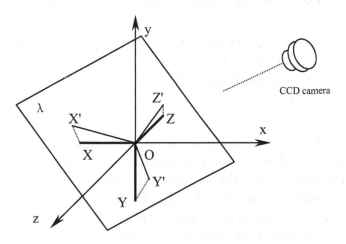

Fig.1   Coordinate axes and definition of plane λ

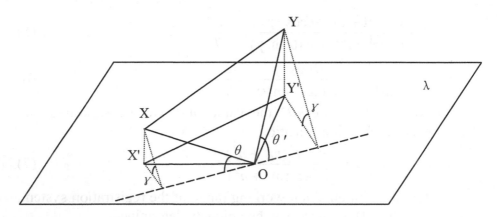

Fig.2 $\triangle$XYZ and its projection on plane λ

Now the projection lengths of OX, OY and OZ on the plane λ can be easily calculated.

$$OX'=\sqrt{OX^2-OX^2\sin\alpha\sin\theta_x}=\sqrt{a^2-a^2\sin\alpha\sin\theta_x}$$
$$OY'=\sqrt{OY^2-OY^2\sin\beta\sin\theta_y}=\sqrt{a^2-a^2\sin\beta\sin\theta_y}$$
$$OZ'=\sqrt{OZ^2-OZ^2\sin\gamma\sin\theta_z}=\sqrt{a^2-a^2\sin\gamma\sin\theta_z}$$

$$(8)$$

The lengths of the image of OX, OY and OZ on the image plane of the camera lens, denoted as (O"X"), (O"Y") and (O"Z"), are known. Thus the magnification of the lens is given by

$$\Omega=\frac{(O''X'')}{(OX')}\qquad(9)$$

Using the principles of geometrical optics, the position of the camera can also be determined.   For this purpose, we designate the principal point of the lens as the reference point for the camera and the user's head, which is indicated as C' in Fig.3.   C" is the center of the CCD image plain, and C is the intersection point of the optical axis with the plane λ.   The object and image distances are signified as l and l' respectively, and it is noted that the image distance approximates to the focal length of the lens since the object distance is large, namely $l'\approx f'$, which gives

$$l=\frac{l'}{\Omega}\approx\frac{f'}{\Omega}\qquad(10)$$

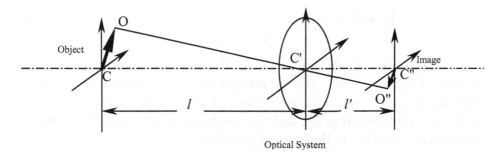

Fig.3   Camera lens

To find the coordinattes of the point $C(X_C, Y_C, Z_C)$, we refer to its image C" which is the centre of the CCD plane as shown in Fig.4. The angle $\phi_x = \angle C''O''X''$ can be obtained from the lengths of (O"C"), (O"X") and (C"X") using the cosin theorem.   On the conjugate plane λ , for a distortion-free camera lens we have $\angle COX' = \angle C''O''X'' = \phi_x$ and $(OC) = (O''C'')/\Omega$.   The formula for the x-coordinate of the point C is thus

$$X_C = (OX_C) = (OD)\frac{(OX')}{(OX)} = \frac{(O''C'') \cdot (O''X'') \cdot \cos\phi}{\Omega^2 \cdot a}$$

$$(11)$$

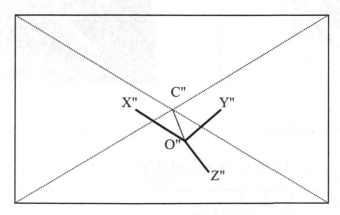

Fig.4   Image plane

Similarly, the values of Y_C and Z_C can be calculated.   The position of the reference point C' in the O-xyz coordinate system is then given by

$$X_{C'} = X_C + l\cos\alpha$$

$$Y_{C'} = Y_C + l \cos \beta \qquad (14)$$
$$Z_{C'} = Z_C + l \cos \gamma$$

The position and the orientation of the camera (and the user's head) are thus fully resolved.   All the objects in the real world which have fixed location relative to the solid fiducial can now be easily registered with the computer program, and the augmented information or virtual objects can be added to the desired positions.

4. Result of the Experiment

After some experiments, we can get the registration result as follow.   Fig.5 is the original image come from the camera, and Fig.6 is the manipulated image.   The result of registration is listed in Table 1.

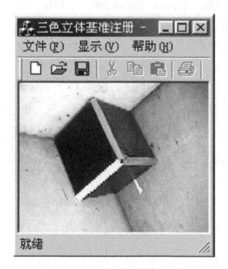

Fig. 8   Original Image

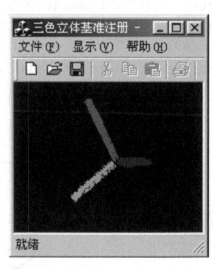

Fig. 9   Manipulated Image

Table 1   Result of Registration

	Actual value	Calculated value						
		1	2	3	4	5	6	Average
$\cos \alpha$		0.406763	0.404524	0.406151	0.407471	0.406151	0.406151	0.406202
$\cos \beta$		0.281797	0.281797	0.299425	0.308362	0.299425	0.303577	0.295731
$\cos \gamma$		0.798678	0.787183	0.806498	0.796618	0.806494	0.803868	0.799890
X-a(mm)	430	435.263	419.49	447.071	419.524	447.071	447.874	436.048
Y-a(mm)	410	418.667	418.363	420.393	401.877	420.393	407.662	414.559
Z-a(mm)	730	726.747	732.47	732.33	728.314	732.33	735.94	731.355

5. Conclusion

The registration system presented in this paper uses one solid fiducial and a single CCD camera. It completely eliminates the coordination errors involved in systems using more than one cameras and/or other tracking devices. As the edges of the solid fiducial are painted with different colors, it needs only picking up the pixels of high color saturation on the CCD to recognize the axes of the coordinate system. The algorithm requires little computation and can be easily realized in real time without latency. And through experimental verification, we can conclude that the algorithm is feasible.

6. Reference

[1] R. T. Azuma, "A survey of augmented reality", *Presence: Teleoperators and Virtual Environments*, Vol.4, 1997, pp.355- 385

[2] S. Feiner, B. MacIntyre, and D. Seligmann, "Knowledge-based augmented reality", *Communications of the ACM*, July 1993, pp.52-62.

[3] P. Milgram, S. Zhai, D. Drascic and J. J. Grodski, "Applications of augmented reality for human-robot communication", *Proceedings of IROS'93: International Conference on Intelligent Robots and Systems*, Yokohama Japan, July 1993, pp.1467-1476

[4] U. Neumann and Y. Cho, "A self-tracking augmented reality system", *Proceedings of the ACM Symposium on Virtual Reality Software and Technology*, 1996, pp.109-115

The Study on Stereoscopic Image Generating Algorithm Based on Image Transformation[1]

Sun Wei   Huang Xin-yuan   Qi Dong-xu

（CAD Research Center, North China University of Technology, Beijing,100041）
sunwei@ncut.edu.cn, hxy@ncut.edu.cn, qidongxu@ncut.edu.cn

Abstract. Nowadays, image-base modeling is one of hotspots in the study of Computer graphics. Along with the requirement of putting the VR technology into real-life environment being enhanced, the immersion of the scene becomes more and more important. Allow for the complexity of the scenes, we have employed the approaches based on Digital image Transformation to obtain stereoscopic image. In this paper, the stereoscopic image generation algorithm based on the Stereo Vision and Digital image Transformation is given. At same time, image and video processing examples are presented to discuss the algorithm's feasibility and practicability.

Keywords. Stereoscopic image, Image Transformation, Image-base modeling, VR, algorithm

1 Introduction

In the virtual environment, the optical interface provides the observant with the illusion of immersion. The technique of modeling virtual environment and algorithm of generating stereoscopic image are always two hotspots in the study of optical interface. Presently, compute graphics is the main approach used to generate the models of virtual environment. At the same time, some algorithms of generating stereoscopic image has been proposed, such as the Stereoscopic Ray-tracing Algorithm, which can run line by line, discussed by Adelson (1993)[1] and the Speedup Algorithm, which has acquired considerable evolvement in China [2]. Along with the requirement of putting the VR technology into real-life environment being enhanced, the complicacy of modeling has been aggravated. To solve these problems, we must take the advantage of compute optical approaches or combining them with Computer Graphics. Recently, owing to the development of modeling based on the digital images provides us with new approaches to simulate real environment, such as 3D face model building from photos and 3D models from panoramas [3], it makes it possible to acquire certain effect which may be inaccessible only using computer

---

[1] The research work of this paper is supported by National '863' program under grant No.863-306-ZT03-08-1.

T. Tan, Y. Shi, and W. Gao (Eds.): ICMI 2000, LNCS 1948, pp. 518–525, 2000.
© Springer-Verlag Berlin Heidelberg 2000

graphics. This method have comprehensive application prospect and broad implication[3].

In a series of studies, we have attempted to construct a three-dimensional model to create stereoscopic image by using 3DS MAX etc and get favorable application [4]. Allow for the complexity of scenes, we have employed the approaches based on Digital image Transformation to obtain stereoscopic image according to our prior experiments. Furthermore, we can generate stereoscopic image and stereoscopic video with the characters of higher reality by applying certain transforms on the static image or video section. In this paper, we will begin with the optical principles and mainly discuss validity and practicability of such stereoscopic image-generating algorithm combining with the conclusion drawn from the experiments.

2 Optical Principle and Stereoscopic Image Display

2.1 Two illuminating optical phenomenon — illusion and hallucination

We have known that the optical channel in the virtual environment is found on the people' physiological optical perceptivity. It is a commonplace for people to misread a image, whether stem from the physiological characters of human eyes or from the psychological characters. For such kind of mistakes seemed unavoidable, we could not just call them a mistake, but call them an illusion. For example, two parallel straight lines are displayed in Figure 1 (1). Although they are equal in length, it gives us an illusion that the lower one seemed a little longer than the upper. If we remove the additional inward and outward arrows at the end of the straight lines, such feeling will disappear. Therefore we can draw a conclusion that it is the additions that lead to the illusion. The same case is also given in Figure 1 (2). When dotted with many oblique short lines in different direction, a group of parallel lines seemed as if no longer run parallel with each other.

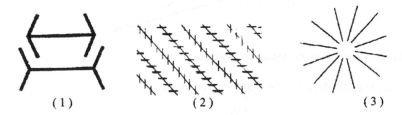

(1)   (2)   (3)

Fig. 1. Samples of illusion and hallucination

Hallucination engenders optical impact by making certain geometric texture and produces various motile senses, such as eradiation, flow, rotation and glint. Hallucination is a sort of virtual illusion arising when the eyes focus on the image. Some additional effects out of the original image has been added into the final image formed in the observer's mind. Consequently, there is a strong relationship existing between the optical effect of the hallucination and the distance from observer to the image. Figure 1(3) shows it. **Focus your eyes on the center of the blank area of the ring; keep on staring at it, then moving slowly towards the image**. You will see the flashing phenomenon if only your eyes are close enough to the image.

Some points can be given from previous description of different optical phenomenon:
- Optical effect can be changed using contrived methods
- Complying with general rules of ocular, the approaches could be found to satisfy some special requirements on the visual.

Well, is it possible to find a approach through which we can transform plane image into stereoscopic image? The answer is certain!

2.2 Observation geometry of stereoscopic vision

The depth perception formed by the overlapping area of the visual field is called Double Eyes Stereoscopic Perception. The optical system can get depth perception through four clues: lateral retinal image disparity, motion parallax, differential image size and texture gradient. There into, in the terms of distance measurement lateral retinal image disparity is more sensitive and reliable to some extent; motion parallax is the most powerful clue in the terms of depth perception.

Figure 2 describes the observation geometry of stereoscopic vision, stereoscopic image pair has been set on the point in distance D front of the observer's eyes, the interval between the eyes is defined by S. When it is observed, the character projects at point x_l, y_l in the left-view and x_r, y_r in the right-view is under the impression that it is at the point P.

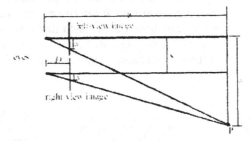

Fig. 2. Observation geometry of stereoscopic image pair

From Figure 2 the dependence of the depth can be given

$$z = \frac{DS}{x_l - x_r} \tag{1}$$

According to the basic principle of the stereoscopic photograph, if

$$DS = fd \tag{2}$$

Then, the scene, which is observed, is same as the real-life one. In this equation, f is the focus of camera; d is the distance between two cameras.

To rebuild the 3D scene exactly, following are needed:

- Observer's eyes must focus on infinitude distance away, but be able to see the correct focus of the stereoscopic image pair
- The line of sight must coincide with axis Z.

Thus, it is clear that we can construct the algorithm to satisfy the condition required, and we can program to make the generation of the stereoscopic image pair to match the geometry of stereoscopic photograph, to make the display of the stereoscopic image pair to match the observation geometry of stereoscopic vision. So that, the immersion can be get in this condition.

3   The study on the Algorithm Based on the Image Transformation

Image geometry transformation or space transformation is a kind of function that can make a mapping relationship between the original image and the transformed one. It can be express as following:

$$[x, y] = [X(u, v), Y(u, v)]$$ (3)

Where, $[x, y]$ is the location of the pixel in transformed image, $[u, v]$ is the homologous location in original image. X, Y is mapping function.

3.1   Stereo vision algorithm

If the depth image can be calculated based on the stereo image given, then, we can get a stereoscopic image pair from a brightness image and a depth image. In fact, this technique makes it possible for us to create stereoscopic images. So at first, let

$A_0 = \{a_{ij}\}, \quad i = 1,2,\cdots n, \quad j = 1,2,\cdots,m,$

$B_0 = \{b_{ij}\}, \quad i = 1,2,\cdots n, \quad j = j_0 + 1, j_0 + 2, \cdots, m$

$\tilde{A}_0 = \{a_{ij}\}, \quad i = 1,3,\cdots 2[\frac{n}{2}] + 1, \quad j = 1,2,\cdots,m$

(4)

$\tilde{B}_0 = \{b_{ij}\}, \quad i = 2,4,\cdots 2[\frac{n}{2}], \quad j = j_0, j_0 + 1, \cdots, m \qquad m, n \text{ are integer}$

$B = \tilde{B}_0 \cup S, \quad S = A_0 - B_0, \quad C = \tilde{A}_0 \cup B$

Where $\cup$ means a kind of algorithm that make the elements of the matrix arranging by the order of odd-numbered line and even-numbered line. Now the goal is how to generate the stereoscopic image pair based on the given image, make the processing as simple as possible. Here we fathom out the generating algorithm based on the transformation of image subset.

To create the left-view image, we can get subset from image A_0, and transform it to $\tilde{B}_0$. For this purpose, define the operator as follows:

$$T_{pq}f_{ij} = g_{p(i),q(j)}, \quad \forall i, j \in \{1,2,\cdots,n\} \tag{5}$$

where $p(i), q(j)$ represent new coordinates of the pixel (i, j) in original image, the new pixel's grayscale, $g_{p(i),q(j)}$, is depended on the neighboring points of (i, j). With different $T_{p,q}$, there are several algorithms can be given.

● **Algorithm based on affine transform.**
Define operator

$$E_{pg}f_{ij} = f_{i+p,j+q}, \quad i, j, p, q \in Z_+ \tag{6}$$

to create the left-view image, for A_0, B_0. Let

$$b_{ij} = E_{0j_0}a_{ij}, \quad \forall i, j \in \{1,2,\cdots,n\} \tag{7}$$

So that $\tilde{B}_0$ can be calculated, and its pixels on u line can be expressed as following:

$$(\tilde{B}_0)_u = (b_{ij})_u = (a_{2u+2,j_0+1}, a_{2u+2,j_0+2}, \cdots, a_{2v+2,n}, a_{2v+2,n-j_0}, a_{2v+2,n-j_0+1}, \cdots, a_{2v+2,j_0}) \tag{8}$$

To create the right-view image, same as the processing above, we can get $\tilde{B}_0$, and its pixels on v line can be expressed as following:

$$(\tilde{B}_0)_v = (b_{ij})_v = (a_{2v+1,1}, \cdots, a_{2v+1,1}, a_{2v+2,1}, a_{2v+2,2}, \cdots, a_{2v+2,n-j_0}) \tag{9}$$

● **Algorithm based on perceptive transform**
Using the algorithm above, in experiment, the left-view and right-view image can be generated quickly, and the vision effect is good. To make the vision effect better, we have a test using perceptive transform. This time, we define operator F_{pq} instead of E_{pq} as

$$F_{pq}f_{ij} = f_{p(i),q(j)} \tag{10}$$

where $p(i) = \alpha i + \beta$, $q(j) = \gamma j + \delta$.

Perceptive transform (also named mapping transform) could be easily represented as following:

$$\left[x', y', w' \right] = \left[u, v, w \right] \begin{bmatrix} a_{11} & a_{12} & a_{13} \\ a_{21} & a_{22} & a_{23} \\ a_{31} & a_{32} & a_{33} \end{bmatrix} \tag{11}$$

Where, $x = \dfrac{x'}{w'}, y = \dfrac{y'}{w'}, \ a_{13} \neq 0, a_{23} \neq 0, w = 1$.

The forward mapping function of the perspective transform could be given as:

$$x = \frac{x'}{w'} = \frac{a_{11}u + a_{21}v + a_{31}}{a_{13}u + a_{23}v + a_{33}}, \ y = \frac{y'}{w'} = \frac{a_{21}u + a_{22}v + a_{32}}{a_{13}u + a_{23}v + a_{33}} \tag{12}$$

4   Implementation And Experiments

The hardware platform of the implementation is Pentium III450 + 128MB RAM, and the programming language is VC++ 6.0. During the experiments, several image and video segment with different resolution and complexity had been transform into stereoscopic scenes, in addition, the generated video files had been compressed and compared with itself in vision effect.

Figure 3 (1)shows the original true color image in the size of 320×240. Figure 4 are stereoscopic image pair based on Figure 3 (1), and Figure 4(1) is the left-view image, Figure 4(2) is the right-view image. Figure 3 (2) shows the stereoscopic image generated from the stereoscopic image pair. The algorithm of this case is based on the Depth Estimation Using Stereo Lenses and Perspective Transformation.

Figure 5 shows the stereoscopic image generated from Figure 3 (1) using the algorithm based on Affine Transformation. The value of the parallax in the image shown in Figure 6(2) is triple than that of the image shown in Figure 5(1), and in the terms of vision effect, the image shown in Figure 5(2) is better. But it doesn't mean that the bigger the value of the parallax is the better the result is. Therefore, to get the best effect, the available value is needed.

(1) the original image(320×240,24bits)   （2）stereoscopic image 1

Fig. 3. The original image and stereoscopic image based on it

(1) left-view image   (2) right-view image

Fig. 4. Stereoscopic image pair based on Figure 3(1)

(1) stereoscopic image 2   (2) stereoscopic image 3

Fig. 5. Stereoscopic images generated using the algorithm based on Affine Transformation

As shown in Figure 6, the digital video had been tested in our Experiments. Figure 6(1) shows the original video in the size of 320×240. What is shown in Figure 6(2) is the stereoscopic video generated from the original one, uncompressed, the size of the .AVI file is 9,038KB, and Figure 6(3) shows the stereoscopic video compressed with Intel Indeo® Video compression standard, the compression rate is 0.56, the size of the .AVI file is 579KB. Compared with the video before compression the vision effect almost no reduce.

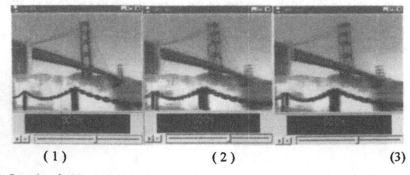

(1)   (2)   (3)

Fig. 6. Sample of video original, Figure6(1) is original video(320×240), Figure6(2) is the stereoscopic video which uncompressed , Figure6(3) is the stereoscopic video which compressed, the rate is 0.56.

The image shown in Figure 7 is more complex in scene and larger in size than the examples before, the original image is 800×600 pixel, true color image. The value of the parallax in the image shown in Figure 7(2) is triple than that of the image shown in Figure 7(1).

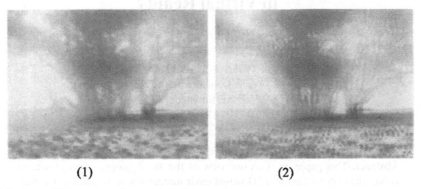

(1)   (2)

Fig. 7. Images with more complex scene and larger size

5 Summary

In this paper, the stereoscopic image generation algorithm based on the Stereo Vision and Digital image Transformation is given. At same time, image and video processing examples are presented to discuss the algorithm's feasibility and practicability. In further research, using Layer Stereo method to improve the vision effect of the stereoscopic image, combining with Panoramic Image Technology to make the virtual environment perfect, all of this, are valuable subjects to involve in.

Reference

1.Adelson S J, Hodge L F. Stereoscopic Ray-tracing. The Visual Computer, 1993,9(3): 127—144
2.Zhang Ping,Meng Xian-qi, Mo Rong, An Algorithm for Accelerating Stereoscopic Ray Tracing. CHINESE J. COMPUTER, 1999,22(7): 763—767.
3.Richard Szeliski, Image Based Modeling. MSR Course on Computer Graphics in Beijing. April. 2000.
4.Huang Xin-yuan, Sun Wei, Qi Dong-xu, The Generating and Composing System of Stereopsis Animation. Proceeding of Qingdao-HongKong International Computer Conference'99, 1999. 756-761
5.Wang Cheng-wei, Gao Wen, Wang Xing-ren, The theory, Implement techniques and Applications of VR. Tsinghua Univ. Press, Beijing (in Chinese), 1996.
6.S.Shah, J.K Aggarwal, Depth Estimation Using Stereo Fish-Eye Lenses, ICIP-94, Proc.IEEE Int. Conf. of Image Processing, 2:740-744,1994.
7.Qi Dong-xu, Fractal and its Generation on Computer, Academic Press, Beijing (in Chinese), 1994
8.Pratt W. K., Digital Image Processing, John Wiley &Sons, Inc, 1978

Jacob - An Animated Instruction Agent in Virtual Reality

Marc Evers[1], Anton Nijholt[1]

[1] University of Twente, Department of Computer Science
P.O. Box 217, 7500 AE Enschede, The Netherlands
{evers, anijholt}@cs.utwente.nl

Abstract. This paper gives an overview of the Jacob project. This project involves the construction of a 3D virtual environment where an animated human-like agent called Jacob gives instruction to the user. The project investigates virtual reality techniques and focuses on three issues: the software engineering aspects of building a virtual reality system, the integration of natural language interaction and other interaction modalities, and the use of agent technology. The Jacob agent complies with the H-Anim standard. It has been given a task model and an instruction model in order to teach the user a particular task. The results of the project can be generalised so that the agent can be used to instruct other tasks in other virtual environments.

1. Introduction

The Jacob project investigates the application of virtual reality techniques and involves the design and construction of an animated agent in a 3-dimensional virtual environment. The agent is called Jacob and provides instruction and assistance for tasks that the user has to learn to perform in a virtual environment. The user interacts with Jacob by performing actions as well as by using natural language. The use of a lifelike agent in an interactive learning environment has a strong positive impact on students, which is shown by an empirical study performed by Lester et al. Such an agent can increase both the learning performance and the student's motivation [8].

The Jacob project involves an integration of knowledge from different disciplines, like intelligent tutoring systems, virtual reality, intelligent agent technology, natural language processing, and agent visualisation and animation techniques. It is a pilot project of the VR-Valley Twente Foundation, which aims at establishing a regional knowledge centre on virtual reality in the Netherlands. Two other pilot projects we would like to mention are AneuRx and Digimap. AneuRx aims at creating a medical training system in virtual reality for a specific type of surgery. Digimap aims at creating a 3D map of the region of Twente, which can be used to visualise and navigate through proposed changes in the environment. In the future, Jacob could play a role in these projects. More information about VR-Valley Twente can be found at http://www.vr-valley.com/.

T. Tan, Y. Shi, and W. Gao (Eds.): ICMI 2000, LNCS 1948, pp. 526-533, 2000.

Jacob is also closely related to the Virtual Music Centre (VMC), an ongoing project of our research group [12]. The VMC is a model of an existing music theatre in virtual reality. It contains agents that provide information about performances, that can handle theatre bookings, and that assist the user in navigating through the building. Eventually, the Jacob agent will be integrated in the VMC.

We will first describe the objectives and the approach of the Jacob project; then we will give an overview of the current state of the Jacob system. Related work and future research will be discussed. Last, we present some conclusions.

2. Objectives

The research focus of the Jacob project is the use of virtual reality techniques and the design and implementation of virtual reality based systems. Important questions addrexssed in this project are:

- Can traditional software engineering and HCI technology be applied for designing and building a virtual reality system or does it require different technology?
- How can different interaction modalities like natural language, gestures, gaze, and manipulation of objects be integrated in a task oriented virtual reality system?
- How can agent technology be used in virtual reality systems?

To support the research activities, a prototype system is being developed that has to meet a number of requirements. First, the interaction between the user and Jacob should be multimodal. An important issue is how natural language dialogue and non-verbal actions are to be integrated and what knowledge of the virtual environment is needed for this purpose. Second, the Jacob agent should behave in an intelligent way, helping the user proactively and learning from the interaction. Third, visualisation of Jacob plays an important role, including natural animation of the body, generation of facial expressions, and synchronisation of lip movement and speech. Fourth, both the user and Jacob should be able to manipulate objects in the virtual environment, e.g. by using a dataglove. Fifth, there should be interaction between multiple agents in the virtual environment. Jacob should be able to communicate with other (artificial) agents and multiple users.

As the problem domain for the project, we have selected instruction of tasks in the virtual environment. This concerns tasks that consist of manipulating objects in the virtual environment, e.g. moving an object, pressing a button, or pulling a lever.

3. Approach

We use existing knowledge, theories, and frameworks from different areas like intelligent tutoring systems [2] [15], computer graphics [11], and multi agent technology [17].

Software engineering plays a prominent role in the Jacob project. We apply object oriented techniques [3], design patterns [4], and software architecture knowledge [14]. We are investigating how to design and build such a virtual reality system in a maintainable and adaptable way.

For the current version of the prototype, we have applied the following technology: the basic virtual environment and the Jacob agent have been defined using VRML 2.0 (Virtual Reality Markup Language). The 'intelligent' part of the system has been written using the Java 1.1 programming language. The Java part is linked to the VRML world through the external authoring interface (EAI). The system runs in a web browser with a VRML plug-in. In this way, the Jacob system is highly portable and can be executed on a regular (powerful) PC or workstation.

4. Current State of the Project

In the current version of the prototype, Jacob teaches the user the Towers of Hanoi game, a classic toy problem from the field of artificial intelligence. In this game, the user has to move a stack of blocks from one peg to another using an auxiliary peg. The user can move single blocks from peg to peg; it is not allowed to place a larger block on top of a smaller one. Figure 1 shows a screenshot of the Jacob agent moving a block in the Towers of Hanoi.

Fig. 1. Screenshot of the Jacob prototype

Figure 2 depicts the software architecture of the system. We have applied a layered architecture to separate the concerns of the 3D visualisation from the basic functionality of the system. The concrete 3D world layer consists of a hierarchical structure of VRML nodes, like transformation nodes, geometry nodes, and sensor nodes. The example node objects in the figure depict this.

The abstract 3D world contains objects representing e.g. blocks, pegs, and Jacob's physical manifestation. Each user is represented by an Avatar object. This abstract 3D world layer also provides simulation of physical properties. For the

Towers of Hanoi, we have implemented basic collision avoidance so that objects cannot be moved through each other. Furthermore, we have implemented a simple gravity variant that makes objects fall at a constant speed.

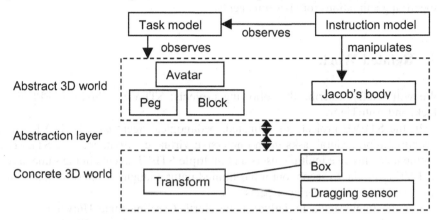

Fig. 2. Software architecture of the Jacob system

We have defined an interface layer between the abstract and concrete 3D world layers to make the system more robust to changes. This is necessary because standards like EAI are under development and are subject to change. This interface layer exposes only the essential properties of the nodes in the concrete 3D world, like the position of an object. All other details are hidden.

The task model and the instruction model together form Jacob's mind. They act as controllers that observe the abstract world and try to reach specific instruction objectives by manipulating Jacob's body.

The task model encapsulates knowledge about the task and the performance of the task: which objects are relevant for the task, how should the task be performed, and what errors can the user make. The task model also specifies why specific steps and actions should be taken. It observes the events and actions that happen in the world to keep track of the current state of the task.

The instruction model encapsulates instruction knowledge. This knowledge is generic and independent of the specific task. It includes adaptation of the instruction to the user, giving feedback on the user's actions, helping the user when he/she gets stuck, giving explanations, and showing the next step. An example of an instruction model is one where Jacob lets the user find his/her way in solving the Towers of Hanoi task while he occasionally gives feedback and answers questions from the user. For users that need more guidance, an instruction model can be defined where Jacob takes initiative and directs the user towards the solution of the task.

To keep track of the user's characteristics and his/her performance on the task, a student model is needed. Because of time limitations we have decided to leave this model out of the prototype.

The visualisation of Jacob's body has been created to comply with the H-Anim standard [10]. H-Anim is a standard for describing a humanoid body in terms of joints and segments. The use of this standard makes it relatively easy to plug in a different body for Jacob. A small number of simple animations like walking and jumping have been defined for Jacob. These animations can work with any agent body that complies with the H-Anim standard. Note that there is not yet a standard for specifying animations for H-Anim agents.

5. Related Work

We will briefly describe the following projects: STEVE, WhizLow, PPP Persona, Jack, and AutoTutor.

In the STEVE project, an animated, pedagogical agent called STEVE gives instruction in procedural tasks in an immersive virtual environment. The STEVE environment can involve multiple users and multiple STEVE agents in the same task [13]. STEVE is similar to Jacob, but it has limited natural language interaction and focuses less on software engineering aspects.

For the IntelliMedia Initiative of the North Carolina State University three animated pedagogical agents have been developed, of which the WhizLow project involves a 3D virtual environment [9]. The WhizLow agent is an animated, lifelike agent in a virtual environment that provides instruction about computer architecture. The virtual environment consists of a 3D representation of concepts like CPU and buses. The student specifies a program and the agent shows and explains how the program is carried out by the different components in the world. There is no direct interaction with or manipulation of objects in the virtual environment.

PPP Persona (Personalized Plan-based Presenter) is a project of the German Research Center for Artificial Intelligence (DFKI), where an animated agent (called PPP Persona) provides instruction and presents information to users using the PPP system [1]. An agent creates a hierarchical plan how to present certain multimedia information to the user, taking into account e.g. characteristics of the user. A PPP Persona is represented in 2D and does not offer multimodal interaction.

Jack is a software system for controlling and animating human-like figures in an interactive, 3D environment. It has been developed at the Center for Human Modeling and Simulation at the University of Pennsylvania. It provides collision detection and avoidance, realistic grasping of objects, realistic graphics and visualisation. A commercial version of the Jack software is currently available from Transom Technologies Inc. (part of Engineering Animation Inc.) This commercial version of Jack aims at helping product designers to test their products in a virtual environment [16]. It enables users to define a virtual environment populated by virtual humans. Constraints can be used to specify the interaction between the virtual humans and the environment. Tasks can be defined for the virtual humans. Finally, the performance of the virtual humans can be analysed. It is for example possible to determine whether the virtual human can see or reach certain objects.

A system based on Jack is SodaJack, an architecture for animated agents that search for objects and manipulate these objects [5]. This architecture uses three hierarchical planners to locate objects and manipulate them. These planners produce low-level motion directives that are executed by the underlying Jack system. Both

Jack and SodaJack have realistically animated agents in a 3D virtual environment, but they are not interactive like Jacob.

AutoTutor is an intelligent tutoring system developed by the Tutoring Research Group of the University of Memphis. AutoTutor uses natural language dialogues for tutoring. The dialogue is delivered using an animated agent, specifically a talking head [6].

6. Future Research

We are currently working on the natural language interaction part of the Jacob system. We will restrict ourselves to keyboard input, but the intention is to add speech recognition and speech synthesis later on in the project. Natural language interaction involves parsing and interpretation of the user's utterances using an appropriate knowledge representation of the virtual environment. Jacob will respond by producing utterances or performing actions. For generation of Jacob's utterances, annotated templates will be used. The natural language dialogue manager and the instruction model are closely related and will be integrated.

To determine the lexicon and grammar for the user utterances, we will perform Wizard of Oz experiments [7]. In a Wizard of Oz experiment, users work with the Jacob system in an experimental setting where Jacob and Jacob's responses are controlled by a human. The dialogues are logged and analysed. We have already constructed a distributed version of the Jacob system where Jacob can be controlled from a remote system. Wizard of Oz experiments can also be useful to find out what kind of actions and nonverbal behaviour to expect from users.

The animation for Jacob should be made more natural. Jacob should also show facial expressions, like anger and surprise, as a response to utterances or actions of the user.

The Towers of Hanoi game is quite restricted as an example task. As a result, the task and instruction models are currently restricted as well. We will extend and generalise the Jacob system to other tasks and analyse the impact on the task and instruction models. We are investigating formalisms for representing these models. We will integrate Jacob in the Virtual Music Centre mentioned in the introduction. There, Jacob can give instruction for tasks like navigating through the VMC (to teach the user what to do and to find where) and operating the lights of the theatre.

We will also investigate the use of a dataglove. A dataglove allows the user to manipulate objects in the virtual environment directly. It can also be used for pointing at objects and for making gestures.

7. Conclusions

Although the current version of the Jacob system is restricted, we think that the results can be generalised very well. The concept of instruction and assistance is very generic and can be applied to various tasks and situations.

Furthermore, the use of stable knowledge from several fields like intelligent tutoring systems makes the components of the system quite generic.

The use of software engineering knowledge and techniques improves the flexibility of the Jacob system. The concepts of such a virtual reality system map well onto an object oriented model. We do not yet have results on the applicability of agent technology and the issues of multimodal interaction.

More information about the project, including the latest results and the prototype, can be found at: http://www.cs.utwente.nl/~evers/jacob

References

1. André, E., Rist, T., Müller, J.: WebPersona: A Life-Like Presentation Agent for the World-Wide Web. Knowledge-based Systems **11**(1) (1998) 25-36
2. Bonar, J., Cunningham, R., Schultz, J.: An Object-Oriented Architecture for Intelligent Tutoring Systems. In: OOPSLA'86 Proceedings (1986) 269-276
3. Booch, G., Rumbaugh, J., Jacobson, I.: The Unified Modeling Language User Guide. Addison Wesley (1999)
4. Gamma, E., Helm, R., Johnson, R., Vlissides, J.: Design Patterns - Elements of Reusable Object-Oriented Software. Addison-Wesley (1995)
5. Geib, C., Levison, L., Moore, M.B.: SodaJack: an architecture for agents that search for and manipulate objects. Technical Report MS-CIS-94-16 / CINC-LAB 265. Dept. of Computer and Information Science, University of Pennsylvania (1994)
6. Graesser, A.C., Wiemer-Hastings, K., Wiemer-Hastings P., Kreuz, R.: AutoTutor: A simulation of a human tutor. Journal of Cognitive Systems Research **1** (1999) 35-51
7. Hoeven, G.F. v.d., Andernach, J.A., Burgt, S.P. v.d., Kruijff, G-J.M., Nijholt, A., Schaake, J., De Jong, F.M.G.: SCHISMA: A Natural Language Accessible Theatre Information and Booking System. In: Proc. First International Workshop on Applications of Natural Language to Data Bases (1995) 271-285
8. Lester, J.C., Converse, S.A., Kahler, S.E., Barlow, S.T., Stone, B.A., Bhogal, R.S.: The persona effect: affective impact of animated pedagogical agents. In: CHI'97 Proceedings (1997) 359-366
9. Lester, J.C., Zettlemoyer, L.S., Grégoire, J.P., Bares, W.H.: Explanatory Lifelike Avatars: Performing User-Centered Tasks in 3D Learning Environments. In: Agents'99, Proceedings of the Third International Conference on Autonomous Agents, ACM Press (1999) 24-31
10. Humanoid Animation Working Group: Specification for a Standard Humanoid Version 1.1. http://ece.uwaterloo.ca/~h-anim/spec1.1/ (August 1999)
11. Lin, M., Manocha, D., Cohen, J., Gottschalk, S.: Collision Detection: Algorithms and Applications. In: Laumond, J.P., Overmars, M., Peters, A.K. (eds.): Algorithms for Robotics Motion and Manipulation (Proc. of 1996 Workshop on the Algorithmic Foundations of Robotics) (1996) 129-142
12. Nijholt, A., Hulstijn, J.: Multimodal Interactions with Agents in Virtual Worlds. In: Kasabov, N. (ed.): Future Directions for Intelligent Information Systems and Information Science, Physica-Verlag: Studies in Fuzziness and Soft Computing (2000)
13. Rickel, J., Johnson, W.L.: Animated Agents for Procedural Training in Virtual Reality: Perception, Cognition, and Motor Control. Applied Artificial Intelligence **13** (1999) 343-382
14. Shaw, M., Garlan, D.: Software Architecture - Perspectives on an Emerging Discipline. Prentice Hall (1996)
15. Tekinerdogan, B., Krammer, H.P.M.: Design of a Modular Composable Tutoring Shell for Imperative Programming Languages. In: Proceedings of the International Conference on Computers in Education (1995) 356-363

16. Transom Technologies: Jack Software. http://www.transom.com/Public/transomjack.html
17. Weiss, G. (ed.): Multiagent Systems. The MIT Press (1999)
18. Zhang, D.M., Alem, L., Yacef, K.: Using Multi-Agent Approach for the Design of an Intelligent Learning Environment. In: Wobcke, W., Pagnucco, M., Zhang, C. (eds.): Agents and Multi-agent Systems. Lecture Notes in Artificial Intelligence, Vol. 1441. Springer-Verlag, Berlin Heidelberg New York (1998) 220-230

Penbuilder: Platform for the Development of Pen-Based User Interface*

Yang LI, Zhiwei GUAN, Youdi CHEN and Guozhong DAI

Intelligent Engineering Laboratory, Institute of Software, Chinese Academy of Sciences
P.O.Box 8718, Beijing 100080, China
ly@imd.cims.edu.cn

Abstract. Pen-based user interfaces is widely used in mobile computing environment, which provides natural, efficient interaction. It has substantial differences from any previous interface and it is difficult to be implemented. A well-designed platform will improve the development of pen-based user interface. In this paper, the architecture of Pen-Book-Page-Paper is presented, which is extended from the metaphor of Pen-Paper. Penbuilder is a development platform based on this architecture. Pen-based user interface can be constructed from three layers of Penbuilder: modal-primitive layer, task-primitive layer and task layer. Each layer provides different extent supports for flexibility and reusability. The components of the platform are discussed in detail and some properties of the platform are argued. Penbuilder is also compliant for the development of distributed user interface. In the section of discussion, we analyze the platform through task tree. The platform provides full supports for extensibility and new interactive devices can be easily added.

1 Introduction

With the development of computer network and mobile devices, the computing mode evolves from traditional centralized mode to distributed mode. And the metaphor of human–computer interaction (HCI) shifts beyond desktop-based metaphor. The dominant WIMP (windows, icons, menus and pointer) interface (i.e., Graphical User Interface, GUI) is a desktop-based paradigm of user interface, which employs two major interactive devices, mouse and keyboard. Both of them are obviously cumbersome in mobile computing environment. Human has used pen for writing for thousands of years. Using pen as interactive device is a natural way for human. Recently a lot of powerful pen devices emerged (such as *Wacom Intuous Pen*) and they provide plenty of interactive information, such as pressure, position and orientation. As a result, pen is more powerful than mouse and it is able to naturally express interactive intention of human.

---

* This paper is supported by the key project of China 863 Advanced Technology Plan (No. 863-306-03-01, 863-511-942-012). We gratefully acknowledge their support.

T. Tan, Y. Shi, and W. Gao (Eds.): ICMI 2000, LNCS 1948, pp. 534–541, 2000.

Pen-based user interface has substantial differences from GUI. Firstly, it provides multimodal interactions. The modality mentioned here is referred as human's perceptual modality, such as the sense of pressure and orientation. Pen is capable of expressing these senses and transferring them to the computer simultaneously. Consequently, there are multiple input streams in pen-based user interface. There are continuous and discrete interactive behaviors in pen-based interaction but GUI only cares about the user's discrete behaviors. Alterations of pressure and orientation are continuous from the user's point of view, but button is discrete because the user only cares about whether button is down or up. In fact, Pen-based user interface is a kind of non-WIMP user interface [1]. Pen-gesture is an efficient way for expressing interactive intention [2]. The interaction in pen-based user interface is based on the user's existing skills and it doesn't need much cognitive effort of users. It releases user from the frustrating mode switching and gives a broader communication bandwidth between human and computer.

Pen-base user interface is more complex than GUI. Consequently, it is difficult to be constructed. A well-designed developing tool will greatly improve the development of user interface [3]. Traditional tools, such as MFC, Motif, are GUI-oriented developing toolkits and they are not suitable for the development of pen-based user interface. Especially, the multimodal, continuous and real-time aspects of pen-based user interface are not easy to be achieved by these toolkits. In addition, pen-based user interface is a kind of distributed user interface (DUI) because it is usually applied in the mobile computing environments.

In this paper, the architecture of Pen-Book-Page-Paper is introduced and it is also suitable for the development of DUI. Platform "Penbuilder" based on this architecture is presented. We will introduce the major components of Penbuilder in detail. The reusability, extensibility and flexibility are stressed in the design and implementation of Penbuilder. The development can be performed from three layers of platform, the layer of modal primitive, task primitive and task. Each of these layers provides different extent supports for reusability and flexibility. A simple CFG-based task specification is given which specifies the pen-based interaction in word processing and this specification can be analyzed and executed by the task framework provided by the platform. We will introduce them in the following sections.

2   Architecture of Penbuilder

Penbuilder is implemented as a C++ library that can be embedded into a C++ development environment, such as Microsoft Visual C++. Penbuilder acquires better independency of interactive devices and window system. The architecture of Penbuilder is based on multi-agent model which models interactive system as a set of agents cooperating with each other. The metaphor of Pen-Paper [4] is employed for pen-based interaction. In order to meet the special requirements of pen-based user interface, it is enhanced as *Pen-Book-Page-Paper*, which derives from the structure of the normal book. A normal book contains multiple pages and it usually has a catalog to index each page. Each page is actually a piece of paper that can be marked by pen. In our architecture, *pen, book, page* and *paper* are implemented as class *UIPenAgent, UI-Book, UIPage* and *UIPaper* respectively. The architecture is shown in figure 1. *UIClient* acting for an interactive agent is the aggregation of *UIPage* and *UIPaper*. *UIPaper* is responsible for processing the interactive information, invoking semantic actions and it can be attached to a window for producing feedback. *UIPage* is the communicating agent for *UIPaper* so that the details of communicating among agents can be masked by this component. *UIBook* contains a *UITaskTrigger* object for integrating and dispatching the interactive primitives. *Name service* of *UIBook* has the similar functions as the catalog of a book. *UIPenAgent* is the manager of the pen device which encapsulates device-dependent information. Although Penbuilder is designed for pen-based interaction, it is supported for adopting various devices, such as voice, keyboard and mouse. *UIBook* is capable of managing multiple modalities. *UIBook* also can manage other *UIBook* objects so that there is recursive characteristic in the architecture. Consequently, a *UIBook* object can manage a set of *UIClient* objects in a domain. The topology of the architecture is like an inverted tree.

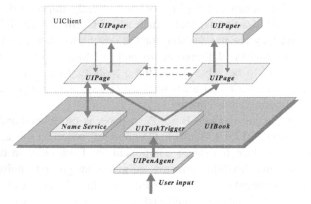

Fig. 1.   Architecture of Penbuilder

UIBook is the router and name server for the agents of interactive system and provides a communicating layer for a set of cooperating agents. It provides facility for initializing task trigger, adding modal agents and *UIClient* objects. The communication between *UIClient* objects can be inter-process, local and remote. *UIClient* object can be distributed on different machines. Because *UIPage* masks the details of communication, the implementation of *UIPaper* is transparent to the way of communication. The structure for DUI based on this architecture is shown in figure 2. The communication is fulfilled by establishing a virtual path between *UIPage* objects and it depends on the package redirection of each node in this virtual path. So this architecture supports multi-tier structure and it is compliant for DUI.

The architecture abstracts the interactive system as the virtual book and gives a clearly layer-based structure for a complex pen-based interactive system. The interactive agents can be dynamically imported and removed without many side effects to others. It is suitable to be applied in the complex and mutable environment, such as internet. The details of the architecture will be discussed through following sections.

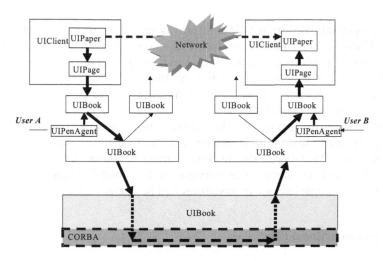

Fig. 2.   Penbuilder for DUI

3   Abstracting and Dispatching Interactive Events

3.1 Modal Primitive

In order to achieve modal-independency and multimodal integration, it is necessary to abstract interactive events from different modalities and unify them based on fundamental interactive semantics. By analyzing and synthesizing most of interactive devices and human's usual behaviors, five varieties of interactive primitives are concluded: *position, orientation, pressure, button* and *word*. Pen is able to directly produce four kinds of primitives except *word* primitive which is usually generated by keyboard or voice modality. A sequence of interactive events are transformed and packaged into a modal primitive object which is an instance of class *UIModalPrimitive*. *UIModalPrimitive* defines the format of the communicating among agents. By inheriting *UIModalPrimitive*, some extensions for modal primitives can be done. Some new interactive primitives can be added whenever new interactive devices emerge. This brings much extensibility to PenBuilder. There is time stamp in *UIModalPrimitive* in order for temporal-based integration. The type of modality is kept in *UIModalPrimitive* for special processing.

3.2 Managing Interactive Devices

UIModalAgent is an abstract class for the manager of interactive device which should be inherited for particular devices. The manager is responsible for initializing modal devices. It provides facilities for configuring the devices and transforming interactive events into device-independent modal primitives. One of major functions of the manager is to fulfill the device-dependent feedback.

UIPenAgent inherited from *UIModalAgent* is special for managing pen device. At present, *UIPenAgent* is implemented based on an open industry standard WinTab [5], which provides a universal interface for accessing digitizer devices. The instance of this class receives events from pen, transforms them into abstract modal primitive and sends it to *UITaskTrigger*. *UIPenAgent* accomplishes pen-dependent visual feedback, such as position and orientation.

Multiple interactive devices can be imported into platform. A modal agent class should be provided. If we want to employ speech interaction, a class *UISpeechAgent* inherited from *UIModalAgent* should be implemented. *UISpeechAgent* will recognize user's voices and transform voice information into modal primitive *"word"*. The integration of multiple interactive modalities will be discussed later.

3.3 Dispatching Modal Primitives

There are multiple cooperating agents in an interactive system and they communicate with each other to fulfill an application task. Each agent is equipped with an interactive control and can be suspended or resumed. *UITaskTrigger* is a component for dispatching modal primitives to the desired agent (the instance of *UIClient*) [6]. As shown in figure 3, *UITaskTrigger* firstly buffers modal primitives and then filters them under a certain policy. *UITaskTrigger* maintains two lists, one for *UIClient* objects and the other for modal agents. *UIClient* objects and modal agents can be added and removed dynamically. Reliability of system can be improved by multimodal interaction because the disability of any modality can be overcome by other modalities. Simple lexical-based integration is performed in order to reduce the traffic of communication. Finally, modal primitives are routed to *UIClient* object. In order to accomplish the processing correctly, consulting with *knowledge base* is necessary. *UITaskTrigger* is configurable. By modifying the reasoning rules of *knowledge base*, the policy of filtering, integrating and dispatching of *UITaskTrigger* can be adapted.

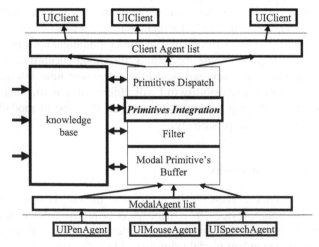

Fig. 3.   Structure of UITaskTrigger

4   Constructing Task Primitive: Pen Gesture

A sequence of modal primitives composes a task primitive which is related to special application task and a sequence of task primitives will construct an application task. For pen-based user interface, task primitives are implemented as pen gestures. The process of constructing pen gestures usually follows a common framework. *UITask-PrimitiveConstructor* implements this framework as shown in figure 4. *Modal Primitive Analyzer* is a component for analyzing modal primitives and it is dependent of special gestures. Penbuilder provides an interface class *UIPrimitiveAnalyzer* for this component and the developers can inherit and implement this interface. In fact, this component is a pattern recognizer for pen gestures.

In order to analyze modal primitives, it is necessary to buffer them firstly. The format of the buffer employs digital-ink, which stores the information of position, pressure and orientation. It is just like the ink track of the normal pen. In the construction, the context information will improve the accuracy of recognizing and enhance the capability of capturing user's intention. *UIPrimitiveAnalyzer* is a lexical analyzer from the point of view of the compiler. A sequence of modal primitives will construct a *"token"* which will be processed in syntax analysis. The syntax analysis is fulfilled in *UIPaper* which will be introduced in section 5.

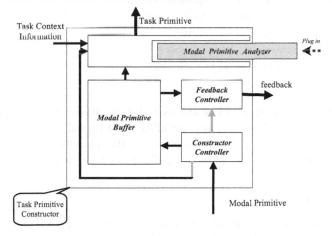

Fig. 4.   Constructor of Task Primitives

5   Interactive Agent

An application task is often composed of a set of interactive tasks and each interactive task can be implemented by an interactive agent, which is implemented as a *UIClient* object. The major part of *UIClient*, *UIPaper*, aggregates three components: interactive controller, *UITaskPrimitiveConstructor* and *application model link* (AML). Interactive controller receives modal primitives and redirects them to *UITaskPrimitiveConstructer* for constructing task primitives. AML is a connector for application model in order to acquire better semantic-independency.

Interactive controller is the most complex part in user interface. In order to ease the developers' burden and improve the reusability, it is an efficient way to specify task by task grammar, which is easy to clarify the relationship of control of interaction. *UICommonFramework* employed by *UIPaper* implements a common framework for interactive controller. A semantic-directed syntax analyzer is implemented here. The syntax-based and semantic-based multimodal integration can be fulfilled in the process of analyzing. The framework is instantiated by a task specification. If the developers use *UIPaper* without *UICommonFramework*, much flexibility can be acquired but they must construct interactive controller from the scratch. Contrarily, the developers just need to provide task specification. In this situation, it's not easy to fulfill some special requirements because of some constraints of the framework. A specification for pen-based word processing is given below and it is based on CFG (Context Free Grammar). Task primitives (i.e., terminal symbols of grammar) are described as *input, insertGesture, selectGesture, deleteGesture, moveGesture, copyGesture* and *interval*. The semantic actions are embraced into a pair of brackets. The rules of grammar are shown as following.

$Word\,Processor \rightarrow input\{add\quad word\} \cdot Word\,Processor | Edit \cdot Word\,Processor | \varepsilon$

$Edit \rightarrow Insert | Copy | Move | Delete$

$Insert \rightarrow insertGesture\{Get\quad insert\quad position\,s\} \cdot In$

$Copy \rightarrow Select \cdot copyGesture \cdot moveGesture\{copy\quad words\}$

$Move \rightarrow Select \cdot moveGestur\,e\{move\quad words\}$

$Delete \rightarrow Select \cdot deleteGest\,ure\{delete\quad words\}$

$Select \rightarrow selectGest\,ure\{Get\,\&\,highlight\quad selected\quad word\}$

$In \rightarrow input\{Insert\quad word\} | int\,erval$

The specification grammar will be enhanced for specifying more complex interaction in the future. Hybrid automaton is a well-defined formal tool for specifying the system with continuous and discrete alteration [3]. By submitting the specification to *UICommonFramework*, the interactive controller can be automatically constructed.

6   Discussion

Penbuilder provides a set of fundamental interactive styles (i.e., pen gestures) and the facilities for constructing interaction but it doesn't confine the implementation of various interactive styles. The reusability, extensibility and flexibility of Penbuilder are embodied from three layers of the platform as shown in figure 5. *UIPaper* is designed to fulfill an interactive task and it employs *UICommonFramework* for synthesizing task primitives. Task primitives are gestures maintained by *UIPenGesture*. These gestures are generated by *UITaskPrimitiveConstructor* from a sequence of modal primitives. Modal primitives are maintained by *UIModalPrimitive* and they are generated by *UIPenAgent* from low-level events of pen device. The part of "*modal primitives*" in figure 5 is semantic-independent layer, which is also independent of particular interactive modalities. The developer can develop pen-based user interface based on this layer and much flexibility can be acquired but it's too complex. The middle part of figure 5 is task-primitive layer, which includes a set of pen gestures and the related recognizer. This layer is special to the application semantics. This layer can be reused based on general interactive semantics. The toppest part of figure 5 is task layer, which includes a set of pre-implemented interactive components. Each component implements a certain interactive task and it complies with a certain component model, such as CORBAR or COM. This layer provides the task-level reusability.

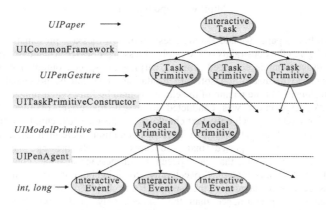

Fig. 5.   Task tree

7   Conclusion

Penbuilder employs the architecture of Pen-Book-Page-Paper which is designed for pen-based user interface in mobile computing environment. The major components of the platform are introduced here in detail. Penbuilder provides much flexibility, extensibility and reusability to the developers. Layer-based multimodal integration is employed in this platform. Penbuilder is compliant for the development of DUI. Instances of pen-based word processing [2], conceptual CAD design and geography information processing have been implemented based on Penbuilder. The developers don't need to care about the details of fundamental interactive devices and many universal gestures can be reused. In the future work, high-level specification facility for pen-based interaction will be imported into Penbuilder for more efficient development.

References

1. Mark Green and Robert Jacob, 1991, Software Architectures and Metaphors for Non-WIMP User Interfaces, Computer Graphics, Vol. 25, No. 3, (July) 229-235.
2. Yang LI, Zhiwei GUAN, Youdi CHEN and Guozhong DAI, 2000, Research on Gesture-based Human-Computer Interaction, Proceedings of Chinagraph2000, China.
3. Myers, 1996, User interface software technology. ACM Computer Survey, Vol. 28. No. 1. (March), 189-191.
4. Andre Meyer, 1995, PEN COMPUTING: A Technology Overview and a Vision, ACM SIGCHI bulletin.
5. Rich Poyner, 1996, Wintab Interface Specification 1.1, LCS/Telegraphics, http://www.wintab.com.
6. Zhiwei GUAN, Hongan WANG, Zhiming NIE and Guozhong DAI, 1999, Agent-Based Multimodal Scheduling and integrating, CAD/CG 99, Shanghai, China.

Information Presentation for a Wearable Messenger Device

Jacques Terken[1], Liesbet Verhelst[1]

[1]IPO, Center for User-System Interaction, Eindhoven University of Technology
P.O. Box 513, 5600 MB Eindhoven, The Netherlands
j.m.b.terken@tue.nl
http://www.ipo.tue.nl/ipo

Abstract. In this paper we address the question of how message presentation can be optimized given the constraints of a wearable messenger device. The device delivers short messages and is equipped with an earplug for speech output and a small screen. Due to hardware and software limitations, text can be presented on the screen in chunks, but the pacing of the chunks cannot be fully synchronized with the speech. An experiment was conducted to assess the effect of five different presentation modes: speech only, text chunks only, speech + text chunks, full text at once, speech + full text at once. The latter two conditions were included as control conditions. Subjects reported main concepts in each message right after presentation. It was found that the "speech + full message at once" condition did give better performance than the "speech only" condition but not better than the "full text message at once" condition; the "speech + text chunks" condition gave a (non-significant) better performance than the "speech only" and "text chunks only" conditions. The results are interpreted as evidence that, even if full synchronization of the speech and text cannot be achieved, combined presentation of speech and text is superior than either alone. The findings are discussed in relation to cross-modal compensation effects.

1   Introduction

In the EU-sponsored COMRIS project a mixed reality is created by coupling the real physical and social world of human actors with a virtual world inhabited by software agents who represent the interests of the human actors. In the physical world, full-fledged social interaction requires physical proximity. Thus, the selection of whom to talk to and what activities to engage in is determined to a large extent by the proximity of human beings and objects in the physical vicinity (physical proximity). Finding human beings to communicate with and activities to engage in that are not in the immediate vicinity requires active effort and search. Since physical barriers are absent in the virtual world, the software agents inhabiting the virtual world are able to navigate through the agent space and select other agents to communicate with on the basis of interests only (interest proximity). The idea is that an agent inhabiting the virtual space, representing the interests of a human actor, looks for fellow agents with matching interests, representing other human actors and possible activities, and gives suggestions to the actor about whom to meet and what activities to engage in.

In order to enable the human actors to receive suggestions from the software agents wherever and whenever relevant and suitable, the messages of the agents are broadcast to wearable devices which the users carry with them. In this way the users may receive messages regardless of whatever activity they happen to be engaged in. An adaptive filtering mechanism filters out the most relevant messages, in order to prevent information overload. A control mechanism enables the users to accept messages, or to queue them

T. Tan, Y. Shi, and W. Gao (Eds.): ICMI 2000, LNCS 1948, pp. 542-548, 2000.
© Springer-Verlag Berlin Heidelberg 2000

when they prefer their current activities not to be interrupted by messages from the virtual world.

The primary output modality for the messages of the agents is speech. This will enable the user to keep his attention directed to the environment, rather than having to switch attention to a screen. In fact, the metaphor that comes to mind is of a parrot that is sitting on one's shoulder, or of a guardian angel, whispering messages in one's ear. In addition, the wearable is equipped with a small display that mainly serves to assist the user in controlling the wearable, by displaying the semantics of five buttons that implement the wearable's limited functionality.

Since there is no upper limit on the number of messages that can be delivered by the agents, the spoken messages are generated by means of speech synthesis, the quality of which may cause problems with intelligibility. A possible solution is to use the screen also for displaying the content of the messages. However, since the display is rather small, it cannot display full messages at once. This gives rise to the question that is the focus of the current research: does it make sense to combine speech and text presentation of the messages, given the constraints imposed by the size of the display and the quality of the synthetic speech.

1.1   Text and Speech Presentation

There are clear differences between text and speech presentation. For text, the speed of processing is determined by the receiver, while for speech the speed is determined by the sender. This makes text more user-centered than speech. However, many of the advantages of text disappear with small-sized displays. If the message is larger than what can be shown at one stroke, the display needs to be refreshed a number of times to show the whole message.

There are several ways to refresh a display and show the full message. One way is to scroll the text from right to left across the display. This technique has been called "leading". Another possibility to refresh the display is by overwriting the contents of the display by the next part of the message. This method has been developed in experimental psychology and is known as RSVP, for Rapid Serial Visual Presentation. It has been shown that RSVP gives better performance than Leading [3]. Furthermore, it produces optimal readability levels at reading rates of 200-300 words per minute, if display size is 12 characters or more and successive displays include an average of about 2-3 words, with syntactic groupings preserved as much as possible, resulting in a refreshment rate of about 600 milliseconds [1].

From the previous observations, we conclude that there are quite adequate ways to present text messages even with small screens. In addition, text presentation does not suffer from the problems with intelligibility that are associated with synthetic speech. On the other hand, using speech output rather than text output makes sense because speech impinges itself on the user and draws his attention, so that the user does not have to initiate activities like watching a screen, but instead may continue to monitor the visual environment while listening to the speech. Thus, there is no reason to make an a priori choice for text or speech.

Rather than being forced to choose, we may consider combining speech and text presentation, the more so as this has clear potential advantages. In the first place, there is the issue of individual differences. Some users may prefer text presentation as they find it easier or more pleasant to pick up information from written text, while others may prefer speech presentation as they find it easier to pick up information from the auditory modality. Secondly, the two modalities may enhance each other, the more so in situations where the implementation details of text and speech presentation are such that either by itself may impair the comprehension of the message by the user. In such cases we may expect that the redundancy created by presenting the same information simultaneously in two modalities may facilitate message comprehension. However, we would expect that such a beneficial effect would arise only if the information in the two modalities is synchronized. When the information is presented a-synchronously, interference may arise instead of a beneficial effect. In the context of the COMRIS project such synchronization cannot be guaranteed. Thus, we ask whether we find a beneficial effect of multi-modal presentation of text and speech over uni-modal presentation if they are presented synchronously, and an interference effect in the absence of full synchronization. An experiment addressing this question is described in the following sections.

2   Method

Materials. Thirty-two messages were constructed which were representative of the messages that the wearable device will present in the actual application. An example is (in translation) "*Peter Moens has been working on pattern recognition for several years, and he will give a presentation on this topic February on 17th*". Each message contained between three and four concepts from the following set: name, affiliation, topic, type of meeting, date, time, location. The concepts in the example are: name: *Peter Moens*, topic: *pattern recognition*, type of meeting: *presentation*; date: *February 17th*. The messages were converted to speech by means of a state-of-the-art diphone-based speech synthesizer. Since the text messages in the final application will be generated from data structures by means of a language-speech generation module [2], the utterances were manually corrected with respect to phoneme errors and prosody (location of emphasis, boundaries and pause). This is reasonable since the system we aim for has full knowledge of the message properties, and the data-to-speech conversion will be error-free in terms of phoneme properties and prosody, which count among the major problems in unrestricted text-to-speech conversion.

For the "text chunks" condition the RSVP method was chosen, since this method is superior to scrolling across a text box. The messages were divided in chunks of at most twelve characters leaving words intact and leaving syntactic grouping intact as much as possible, which gives optimal performance with RSVP. Durations of the speech utterances were measured and divided by the number of text chunks. On the basis of this calculation, the update rate for the presentation of text chunks was fixed at 720 milliseconds. This was done for two reasons: first, to get equal exposure intervals in the text and speech conditions; second, to get as close to synchronization of the text and speech conditions as possible.

We anticipated that subjects might develop special strategies to cope with the expected interference in case the text and speech would get out of synchronization. More concretely, we expected that they might neglect the speech message altogether and attend only to the text message. In order to prevent them from adopting such strategies, sixteen catch trails were constructed in which there was a discrepancy between the information in the text message and the speech message, and where they had to report the contents of the speech message. In this way they were forced to attend to the speech messages at all trials.

Subjects. Eighty subjects were recruited from various academic and vocational backgrounds. Ages ranged from 18 to 25 years. All were native speakers of the Dutch language. No subject participating in the speech conditions reported hearing problems. Subjects were paid for participation.

Procedure. A between-subjects design was chosen to keep the experiment within reasonable limits: in a within-subjects design, balancing the order effects would get quite complicated, in particular given the controls needed in connection with expected practice effects for synthetic speech. Subjects took the experiment in groups of between three and eight persons. They were explained about the task and worked through ten practice trials to get acquainted with the task and to make sure that they understood the instruction correctly. Texts were presented on a screen by means of a video projector. Font size was chosen such that reading would cause no difficulties at the given distance and illumination. Speech was presented through headphones at comfortable loudness levels. In the text + speech conditions, onsets of text and speech presentation were synchronized. Exposure time for the text messages was set equal to the duration of the speech utterance. Administration of the trials was controlled by a computer.

Subjects wrote their responses on an answer sheet. In order to prevent the act of writing down the responses to interfere with retaining the information in working memory, inspiration was taken from experiments by Sperling [4], who showed that conventional recall measures may underestimate actual recall considerably due to the interference between the recall processes and the reporting activity. Thus, subjects needed to write down only two of the concepts in each message. The concepts to be reported were specified immediately after the message was presented. It was assumed that this would constitute an accurate measure of the amount of information that was actually extracted from the messages. After subjects had written down their response to a trial, the experimenter initiated the next trial. Total duration of the experiment was about thirty minutes.

Subjects' responses were counted as correct if a concept was reported accurately and wrong otherwise. Thus, for each message subjects would score "2" (both concepts reported correctly), "1" (one concept reported correctly, the other concept missed or reported incorrectly or partially) or "0" (both concepts reported missed or reported incorrectly or partially). For the catch trials, subjects were requested to report the content of the speech messages.

3   Results and Discussion

Results are shown in Figure 1. For sake of comparison, the results for the "speech only" condition are shown both in the "text chunks" and "full text at once" panels.

A one-way analysis of variance was conducted with conditions as a fixed factor and subjects as a random factor (analyses were conducted with the percentage score for each subject, computed by dividing the total number of concepts reported correctly by the total number of concepts to be reported). The effect of conditions was found to be significant ($F_{4,75}=9.12$, $p \le .005$). Tukey post-hoc analysis for the relevant comparisons showed that the mean for "speech only" differed significantly from those for the "speech+full text" and "full text" conditions. None of the differences between conditions in the "text chunks" pane reached significance. An analysis with arcsine transformed data gave similar results.

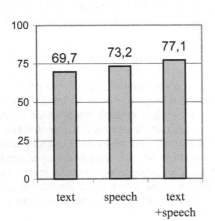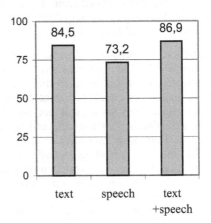

Figure 1. Percentages of concepts reported correctly. Left panel: "text only", "speech only" and "speech+text" conditions for "Text chunks" condition. Right panel: "text only", "speech only" and "speech + text" condition for "Full text at once" condition

Both for the "full text at once" and "text chunks" conditions presenting messages simultaneously in the visual and auditory modality results in better comprehension and recall than presenting the messages only by speech or only as text. Even though for the "full text" case the difference between the "text+speech" and "text only" condition is rather small and not significant, and for the "text chunks" condition none of the differences reaches significance, this finding provides evidence for the beneficial effect of multi-modal presentation.

The fact that for the "full text at once" comparisons the "speech+text" condition is only slightly (and non-significantly) superior to the "text only" condition may relate to the fact that the exposure time for the text in the "text only" condition was set equal to the duration of the speech utterance. Since reading is usually faster than listening, the consequence would be that subjects have relatively more time to read the sentence in the "only text" condition. It seems reasonable to assume that therefore they achieve quite a high level of performance in the "text only" condition, and that adding speech does not result in a further improvement in comprehension and recall.

The fact that in the "text chunks" case the multi-modal condition performs better than either uni-modal condition is contrary to our original expectation, which was based on the fact that text and speech in the multi-modal presentation were not synchronised. We predicted that the lack of synchronisation would give rise to interference between the two modalities, so that the combined presentation would result in poorer perfomance than either modality by itself. The natural way of coping with this interference, by means of attending to just one channel was prevented in the current set-up by the presence of catch trials, by which we forced subjects to attend to both modalities simultaneously,

In retrospect we can understand this pattern of results. Our original expectation was based on the observation that the text and speech would not be synchronized, since the Comris application did allow only a fixed refreshment rate for the screen rather than a dynamic rate dependent on the speech. However, the fixed refreshment rate for the text chunks was calculated by reference to the duration of the utterances: we determined the refreshment rate such that the total exposure time for the sequence of chunks would be approximately equal to the duration of the utterance. Although the fixed exposure interval for each chunk would lead to a-synchronicity between the onset of presentation of the text chunks and the onset of the corresponding fragments in the speech utterance, at the same time it would give considerable overlap. That is, even if the onset of a particular text chunk would occur 100 milliseconds before or after the onset of the corresponding speech fragment, with an exposure interval of 700 milliseconds there would be an overlap of 600 milliseconds. Thus, most of the time text chunks would at least partially overlap with the corresponding portion of the speech signal, thereby reducing the interference between modalities. Therefore, even though the "text chunks only" condition by itself resulted in a relatively poor performance, it is plausible to assume that the subject in the "text+speech" condition might take up some information from the text chunks to contribute to the uptake of information from the speech signal, especially given the fact that the synthetic speech, which is by itself already problematic, contains low-redundancy materials such as proper names.

Looking at the uni-modal conditions, we see that text does not give consistently better performance than speech, even though text does not suffer from the problems that are associated with synthetic speech. In the "text chunks" case, the "speech only" condition gives a better perfomance than "text only", even if the difference is not significant (the latter may be due to the rather large inter-subject variation within conditions). This may be understood in relation to the properties of the "text chunks" condition. As we mentioned above, the text chunks condition left the words intact, but not always the syntactic grouping. Thus, syntactic constituents might be spread across subsequent screens, so that subjects had to impose the grouping themselves by integrating information across subsequent screens, and apparently, the quick succession of text fragments makes this a difficult task.

The results may be understood in terms of a cross-modal uncertainty reduction effect. Due to the way these presentation modes are implemented, subjects face difficulties with the "text chunks only" and the "speech only" conditions. Combining the two presentation modes will increase the redundancy of the total information so that subjects can compensate for problems in processing the input from one modality by combining it with information from the other modality. The question as to the locus of this effect remains to be answered. Either information from one modality may be used to compensate for problems arising when processing information in the other modality. Or the resulting memory representations may be more stable due to dual coding. The fact that the "speech+full text" condition was not really superior to the "full text only" condition seems to argue in support of the former interpretation and against a "dual coding" explanation.

4   Conclusion

In accordance with our hypothesis, in the "full text at once" condition we find that speech+text performs better than speech only. However, contrary to our expectation we do not find that speech+text performs better than text only. The fact that exposure time for text was the same as for speech, while the information intake is usually faster for text than for speech, may explain this lack of facilitation from multi-modal presentation. Thus, it appears that readers had excess time in the text-only condition for processing and memorizing the text, so that adding speech did not further improve the performance.

Our expectation that combined presentation of speech and text in the "speech+text chunks" condition may create interference due to a-synchrony effects is rejected. Instead, the "speech+text chunks" presentation results in better performance than the "speech only" and "text chunks only" conditions by themselves. Here, we observe a beneficial effect of multi-modal presentation over either modality by itself. The fact that we do not find interference may be explained by the choice of the refreshment rate, which leads to considerable overlap between text and speech, even if they are not fully synchronised. And this appears to be sufficient for creating a facilitative effect of multi-modal presentation. Impediments that arise while processing information from one modality can be compensated for by information that is received in the other modality.

References

1. Cocklin, T.G., Ward, N.J., Chen, H.C. & Juola,, J.F.: Factors influencing readability of rapidly presented text segments. Memory and Cognition 12, (1984). 431-442
2. Geldof, S. & Van de Velde, W.: Context-sensitive hypertext generation. In: Working notes of the AAAI'97 Spring Symposium on Natural Language Processing for the Web, Stanford University Stanford, CA (1997) 54-61
3. Granaas, M.M., McKay, T.D., Laham, R.D., Hurt, L.D. & Juola, J.F.: Reading moving text on a CRT screen. Human Factors 26 (1984) 97-104
4. Sperling; G.: The information available in brief visual presentations. Psychol. Monogr. 74 (1960) no. 11

Usability of Browser-Based Pen-Touch/Speech User Interfaces for Form-Based Applications in Mobile Environment

Atsuhiko Kai[1], Takahiro Nakano[2], and Seiichi Nakagawa[2]

[1] *Faculty of Engineering, Shizuoka University, 3-5-1 Johoku, Hamamatsu-shi, Shizuoka, 432-8561, Japan*
kai@sys.eng.shizuoka.ac.jp
[2] *Faculty of Engineering, Toyohashi University of Technology, 1-1 Hibarigaoka, Tempaku-chou, Toyohashi-shi, Aichi, 441-8580, Japan*
{nakano,nakagawa}@slp.tutics.tut.ac.jp

Abstract. This paper describes a speech interface system for the information retrieval services on the WWW and the experimental result of a usability evaluation for the form-based information retrieval tasks. We have presented a general speech interface system which can be basically applied to many menu-based information retrieval services on the WWW. The system enables additional speech input capability for a general WWW browser. A usability evaluation experiment of the speech-enabled system for several existing menu-based information retrieval services is conducted and the results are compared with the case of a conventional system with pen-touch input mode. We also investigated the difference in the effect of usability for different operating conditions.

1 Introduction

Recently, many information retrieval services are deployed on the WWW and they used to provide easy access interfaces based on a graphical user interface(GUI). However, for some kind of the information retrieval tasks and under certain operating conditions, it may be insufficient for a small notebook PC or Personal Digital Assistant(PDA) (or wearable computer) users to access such services efficiently and satisfactory. Speech input modality will be one of promising alternatives to the GUIs which can be accessed only by keyboard or mouse[1]. However, since the existing forms on the WWW are not originally intended to be filled in by a spoken input, further investigation will be needed whether the spoken input is enabled for the WWW services and used as an alternative of the keyboard or mouse operation.

We have investigated a method to achieve a general speech interface which is task-independent and can be applied to most form-based information retrieval services on the WWW. So far, some researchers have presented the speech interface systems for the WWW services[2],[3],[4],[5]. We have reported a prototype system[7] which is basically task-independent and it can provide an additional speech input modality through the WWW browser with a GUI function. The user-side system works on a general WWW browser with a Java runtime support and a speech-input server.

T. Tan, Y. Shi, and W. Gao (Eds.): ICMI 2000, LNCS 1948, pp. 549–556, 2000.

Although the integrated use of several modality will provide an additional usability for an interface system, this study concentrates on a comparison of the usability of the speech alone and pen-touch alone input modes for simple menu-based tasks for the WWW information retrieval services. While the pen-touch input mode will be inherently excel for menu-based manipulations[8], the operating conditions such as the user's posture and the GUI display resolution will affect the usability of the direct manipulation by pen-touch input. Therefore, this study evaluates and compares the usability of the system with speech and conventional pen-touch input modes under the different operating conditions.

2   Form-Based Speech Interface System

2.1   Requirements of Speech Interface for Form-Based Services

Many information retrieval services which exist on the WWW have a form-based user-interface and they often have a menu-based GUI as shown in Figure 1. We focus on the form-based GUI function of the HTML documents on the WWW and intend to add a speech-interface function to a general WWW browser for allowing the users an alternative input modality which can be simultaneously or selectively used with other input modalities.

Since most of the operations for a form-based Web page are a fixed-form task, a speech interface system will reduce the difficulty of such operations by interacting with users by spoken dialogue in addition to a menu-based GUI. One example is to give a spoken dialogue interface for the forms in the HTML documents. In practice, some factors should be treated to automatically provide a speech interface for a menu-based GUI:

– Extraction of keywords contained in a menu list
– Generation of a grammar and lexicon for a speech recognizer
– Extraction of names (or labels) for input items or categories
– Generation of system prompts to inquiry and reply utterances
– Dialog control in the task of fulfilling the pages with multiple input items

Provided that a form-based Web page is described by the HTML, the first and second issues can be mostly handled by parsing the document based on HTML tags. Figure 2 shows a part of HTML description which is used for presenting a menu-based GUI as shown in Figure 1. Obviously, all menu items(keywords) can be extracted automatically since they are enumerated between two HTML tags(<select> and </select>) and preceded by a <OPTION> tag. A Web page with a menu-based GUI may be provided using other methods or languages such as Java and Javascript other than some HTML tags provided for presenting menu-based GUI functions. Essentially, in this case, it is impossible to extract each menu item or keyword only by parsing HTML document in terms of tags itself.

We adopt the same strategy employed in a voice-operating WWW browser system which has been developed before[6] for allowing a user to utter a voice

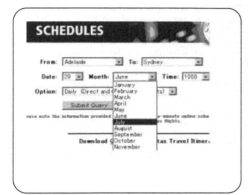

```
<SELECT NAME="CFrom1"size=1>
<OPTION>Adelaide
<OPTION>Brisbane
<OPTION>Cairns
<OPTION>Canberra
<OPTION>Darwin
<OPTION>Hobart
<OPTION>Melbourne
<OPTION>Perth
<OPTION>Sydney
         :
</SELECT>
```

Fig. 2. Example of HTML description for
Fig. 1. Example of form-based Web page menu-based GUI

command for jumping to a desired link without using a keyboard and/or mouse. The extracted keywords are decomposed by a morphological analyzer, and any keyword fragment of an allowed sequence of morpheme units and other common phrases are described in a context-free grammar.

The remaining three issues may not be sufficiently handled by a parser since the HTML specifies only a predefined set of tags which is not sufficient to describe a semantic structure for the form-based pages. Next section shows some examples and describes our approach for processing these issues.

2.2 Information Extraction from Form-Based Pages

Although the HTML does not provide strict rules or tags for describing a semantic data structure for a document, the HTML description for a display structure also partially contains such semantic information. We investigated the HTML descriptions for several form-based Web pages on the Internet. As a result, the extraction of names for input items or categories, which is the third issue described in the previous section, is mostly (there exist 71.4% out of 21 input items in 7 Web pages) handled by using an HTML form tag as a clue for the location of those names in the document. However, there are also many cases that the names cannot be extracted from an HTML document and another strategy is needed. These include (1)some items share the item category names (e.g., an item name "departure:" for both month and day items), (2)item or category names are represented by images, (3)the form-input function is described by other means such as the use of Javascript language, (4)the name and list of a form are separately located as an element of an HTML table structure and the correspondence of them is not clear. Although a system may not be able to extract the item names for the above cases, our system adds a graphical mark for each menu part (as described in the next section) in the target HTML document and we can know which item can be inputted by speech at a point of time in the form-filling task.

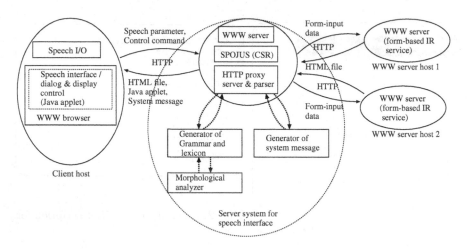

Fig. 3. Configuration of prototype system

The item or category information extracted from an HTML document for an information retrieval service often contains a query-style sentence and a request-style sentence as well as item names. The item names and such sentences are used for generating a system utterance produced by a text-to-speech system. For example, we can automatically generate a system message "Please input a departure date." for a menu item preceded by a keyword "departure date:".

3   Prototype of Speech Interface System

3.1   System Configuration

Figure 3 illustrates the components and configuration of a developed prototype speech interface system[7]. The user-side system consists of a WWW browser in which the Java applet can be run and a speech I/O server for providing speech detection, feature extraction and speech synthesis functions. Other main processing is performed on a hub host and some tasks such as speech recognition and morphological analysis may be distributed to the other hosts. The main role of the hub host is the function as a proxy server, applet server and communication center for the other tasks.

This system employs a continuous speech recognizer(CSR) SPOJUS[9],[10] which is based on a client-server architecture. The SPOJUS uses a set of syllable-unit based HMMs as an acoustic model and a language model described by a context-free grammar. The CSR process is based on a frame-synchronous algorithm and allows the concurrent processing of speech analysis and CSR for a speech input. During the search process, it permits an interrupting request from a client system to allow a flexible speech input interface and to be used in a multi-modal interface system. A Japanese morphological analysis server[11] is also employed in the system for dynamically extracting keyword fragments and

```
System:Please input the departure
       date.
User:  July.
System:''''(no prompt generated
       from system)
User:  16th.
System:Please input the departure
       city.
User:  Tokyo.(a fragment of
       ''Tokyo Haneda'')
System:Please input the arrival
       city.
User:  Uh, Nagoya.
System:Please input the desired
       departure time.
User:  Back (for correcting the
       arrival city)
System:Please input the arrival
       city.
       :
```

Fig. 4. Example of WWW browser display with a speech interface system

Fig. 5. A spoken dialog example

generating a lexicon and a grammar. While this system is intended to be used for Japanese Web pages, also the pages or keywords written in other languages can be accessed by spoken input if the system allows the speech input by the item number other than a keyword to be selected.

3.2   Examples

Figure 4 shows an example of the access to a public information service Web page through our speech-enabled browser interface system. The display window is separated by three frame regions. In the top frame of this browser window, a Java applet is running for providing speech interface and browser control functions. The middle frame is the main frame and displays the filtered output of a target Web page which originally has a menu-based GUI. The bottom frame displays a prompt message and a keyword candidate list for a focused menu item which is indicated by two opposite arrows surrounding a menu in the middle frame. All the candidate lists for each menu are dynamically updated only once each time a Web page is accessed. The grammar and vocabulary for the speech recognizer are switched each time one item is selected by mouse or voice and the focus moves to another item.

Some common keyword commands such as 'back' and 'next' (for selecting an item which should be corrected or modified) are provided. Figure 5 shows an example of spoken dialog for a airline timetable information service provided on the WWW. A system prompt is not provided in some cases where such information cannot be extracted from the HTML source. Even for such a case, the candidate list and fixed messages are produced and can be proceeded by user's spoken input.

4   Experiments

We implemented the user-side system on a notebook PC in which mainly a general WWW browser (Netscape Navigator 4.61) and a speech I/O server are running. We used a notebook PC equipped with a 300MHz-Celeron CPU and a touch-panel 10.4 inch LCD display. The server-side system of this speech-enabled interface was implemented on a PC-UNIX system using a client-server architecture. We used a 300MHz-PentiumII desk-top PC for the server-side system. Although this experiment employed only a single PC for running all of the server-side components(CSR, WWW proxy server, HTML morphological analyzer and response generator), they can be distributed on several hosts over the network. As mentioned in the previous section, the hub host has to produce a processed (Java-controllable graphical markers inserted) Web page and to generate a lexicon and grammar for the speech-enabled system dynamically in response to the user's request of browsing a desired WWW service. It took about 13 seconds in average to complete the pre-processing of the Web pages used in the following experiments, while the pre-processing is performed only once each time a Web page is accessed.

We compare two modes of user interfaces, which are pen-touch alone and speech input alone modes in our pen-touch/speech-enabled Web browser system for existing menu-based WWW information retrieval services. Although the integrated use of two input modes is already implemented and it will provide different usability, we don't mention it here since also another usability factor due to the GUI design will arise and it is out of the focus of this study.

Since this study also concentrates on the effect of mobile computing environment for the use of pen-touch and speech-enabled interfaces, we examine the effect of two kinds of user's conditions: the size of fonts and the display area (by changing the display resolution in two levels: 800×600 for *small-font* and 640×480 for *large-font*), and the using-styles of the subjects (by forcing users to keep *sitting* with the notebook PC on the desk and *standing* postures with the notebook PC holded in one hand). To compare the effect of two different modes of interfaces and four experimental conditions, each of 6 male subjects performed 6 different tasks in total under the above-mentioned different conditions. The subjects were divided into two groups. First, a subject group(A,B,C) used the system in a condition: *sitting-posture* and *large-font* condition(**Sitting-Large**). On the other hand, another subject group(D,E,F) used the system in the opposite condition: *standing-posture* and *small-font* condition(**Standing-Small**). Second, each of two groups used the system in the opposite condition. In advance to the experiment, all the subjects are allowed to use the systems for about 20 minutes.

The information retrieval tasks are all based on the existing Japanese public Web pages in the Internet. They include an airline timetable information service(**AT**), a train timetable information service(**TT**), and a recipe information service for daily Japanese cooking(**CR**). The numbers of menu items included in a Web page of each task are 5, 4 and 3, respectively. Some menu items include a large number of choices in them and their choice's lists are displayed by the

Table 1. Results of menu-based information retrieval tasks

Posture-Font	Sitting-Large						Standing-Small					
Input mode	Speech			Pen-touch			Speech			Pen-touch		
Subject	Task	Time	Point	Task	Time	Point	Task	Time	Point	Task	Time	Point
A	AT	201	3	TT	25	4	AT	182	3	TT	32	4
B	CR	123	2	AT	30	5	CR	85	4	AT	35	4
C	TT	149	4	CR	18	4	TT	101	4	CR	16	4
D	AT	154	4	TT	26	5	AT	160	4	TT	30	4
E	CR	71	5	AT	38	4	CR	70	5	AT	40	3
F	TT	115	4	CR	19	4	TT	108	4	CR	25	4
Average			3.6			4.3			4.0			3.8

pull-down menu with a scroll bar according to the behavior of the browser's GUI function. Since we want to exclude the effect of Web page's design, we force users to simply select or utter a keyword for every menu items in a sequential manner. Of course, since speech-enabled input mode sometimes causes input errors, two voice commands, "back" and "next", are provided for users. In case of speech-input mode, the subjects are instructed never to use the mouse and keyboard.

Table 1 shows the subjective and objective evaluation results for different tasks and conditions. The term *Time* indicates the elapsed time needed for completing a task. This figure includes the time for both the pre-processing of a target Web page and the search to the query. The term *Point* is the subjective score of the system's usability in 5-rank figure(larger figure as the degree of satisfaction becomes higher, and smaller figure as the degree of dissatisfaction becomes higher) which are asked to the subjects by a questionnaire form.

We observe the following things from the table. As for the speech-input mode, the change of posture and font-size conditions causes very little change for the subjective measure for the menu-based tasks. On the other hand, as for the pen-input mode, their condition's change affects the subjective measure, that is, the standing-posture and small-font condition degrades the usability. In particular, the difference appeared for the **AT** and **TT** tasks which have relatively many choices in a menu. However, the subjective measure is comparable with the speech input mode even for such a degrading condition. Consequently, as for the sitting-posture and large-font condition, the subjective measure is greater for the pen-touch input mode than the speech input mode.

The time figure consistently indicates that the pen-touch input mode is more advantageous than the speech input mode. This is partially because the speech input mode required about 25% of extra processing time due to the pre-processing of the Web page and the speech recognition, and also involved the redrawing of the display for indicating a focused menu item whenever one item is entered. These figures may be improved and thus further investigation is needed. However, the figure also shows that the change of the posture and font size significantly affects the time as for pen-touch input mode, that is, the time increases in case of the standing-posture and small-font condition. This suggests us to consider the fact that the time figure may be significantly changed in an adversed condition for pen-touch input mode and possibly also for the other direct-manipulation interfaces[8].

5   Conclusions

We presented a client-server based speech interface system for the form-based information retrieval services and showed some evaluation results of several information retrieval tasks for the existing menu-based Internet information retrieval services. This system is basically based on a method which is independent of the target tasks and applicable to most form-based Web pages. The experiment showed that our speech-enabled system provides a comparable usability with the case of a pen-touch input mode when the operating environment such as the posture and the font-size affects the usability. Also, the experiment showed the difference in the effect of efficiency degradation for both the speech and the pen-touch input modes.

References

1. P. R. Cohen: "Natural Language Techniques for Multimodal Interaction," *Trans. IEICE*, J77-D-II, 8, pp.1403–1416, 1994.   549
2. Alex Rudnicky, et. al.: "*Speechwear:* A mobile speech systems," *Proc. of ICSLP'96*, Philadelphia, 1996.   549
3. R. Lau, G. Flammia, C. Pao, and V. Zue: "WEBGALAXY - Integrating spoken language and hypertext navigation," *Proc. of EUROSPEECH'97*, pp.883-886, 1997.   549
4. Sunil Issar: "A speech interface for forms on WWW," *Proc. of EUROSPEECH'97*, pp.1343–1346, 1997.   549
5. K. Kondo, and C. T. Hemphill: "A WWW browser using speech recognition and its evaluation," *Proc. IEICE*, Vol.J81-D-II, No.2, pp.257–267, 1998.(in Japanese) 549
6. A. Kai, T. Nakano, and S. Nakagawa: "A voice-operating WWW browsing system using a continuous speech recognition server –SPOJUS–," *IPSJ SIG Notes*, SLP20-14, 1998.(in Japanese)   550
7. Atsuhiko Kai, Takahiro Nakano and Seiichi Nakagawa: "A speech interface system for information retrieval tasks on the WWW", Proc. of International Workshop Speech and Computer (SPECOM'99), pp.141–144, 1999.   549, 552
8. Martin G. L.: "The utility of speech input in use-computer interfaces," *International Journal of Man-machine Studies*, 30, 4, pp.355–375, 1989.   550, 556
9. Atsuhiko Kai and Seiichi Nakagawa: "Investigation on Unknown Word Processing and Strategies for Spontaneous Speech Understanding", *Proc. of EUROSPEECH'95*, Madrid, Spain, pp.2095–2098, 1995.   552
10. http://www.slp.tutics.tut.ac.jp/SPOJUS/   552
11. http://cactus.aist-nara.ac.jp/lab/nlt/chasen.html   552

A Lock Opening and Closing System with the Image Base Using a Cellular Phone through the Internet

Daisuke Shimada[1], Kunihito Kato[1], and Kazuhiko Yamamoto[1]

[1] Dept. of Information Science, Gifu University
1-1 Yanagido, Gifu, 501-1193, Japan
daisuke@yam.info.gifu-u.ac.jp
{kkato, yamamoto}@info.gifu-u.ac.jp

Abstract. Now a day, we use cellular phones to display color images, characters, and to connect to the Internet. As an Internet applications for future generations, we built a system in which electric equipment was connected to a web server and controlled by a cellular phone. More specifically we constructed a system to open and close a door using a cellular phone. One of the most important problems is that nobody knows the condition of a room, if somebody is not present in the room. In order to solve this problem, a pan-tilt camera is set in the room. The web server can control this camera, and the user can select various directions from which the image can be taken. An automatic lock system is built with human detection. Movement and color information is used in aiding human detection. By experimenting with this automatic lock system, the security of this system was confirmed.

1 Introduction

Recently, cellular phones[1] have become extremely popular. As the cause, anywhere can be used and it is because the cellular phone became small Due to progress of hard technology. Internet applications using the cellular phone have been researched in many fields.

Originally, the cellular phone was used to communicate by voice and characters, but recently it has been used to display high speed color images, and has the capacity to send packet information.

Many research projects have been proposed using a remote to control [2] various pieces of electrical equipment using the cellular phone through the internet, but in many cases, when controlled from a remote location, there is a problem recognizing the condition of the equipment and the status of the room. In order to solve this problem, we set a camera in the room and took images of the room to observe the condition of them. The camera images relayed much useful information.

In this paper, we proposed and built an automatic door-lock system as an example of a system to control the electric equipment. We also detected the human with the image processing [3] by camera images.

T. Tan, Y. Shi, and W. Gao (Eds.): ICMI 2000, LNCS 1948, pp. 557-563, 2000.
© Springer-Verlag Berlin Heidelberg 2000

When somebody is present, the room doesn't need to be locked. The current sy s-
tem locks the door every time, which isn't dependent on the condition of the room.
If we don't want to lock the door when some body exists in the room, the system
needs various sensors, such as infrared rays sensors, supersonic sensors, etc. This
equipment alone is not enough to completely recognize the condition of the room.
Therefore, by using a camera, we built a securely system that locks the door when
nobody exists in the room, as well as a system that doesn't lock the door as much as
possible when somebody exists in the room.

2 Lock opening and closing system

2.1 Outline of the lock opening and closing system

A "physica l key" is usually used as a general key. But it is a burden to carry the key
that is only effective in front of the door.

The key used characteristics of the human body, such as a face or a fingerprint
scan [4][5] is becoming popularized. In this system the user does not have the burden
of carrying the physical key, but it cannot use only near too.

Our system used the cellular phone to be able to carry at all times and use from
anywhere. This solves the distance-proximity problem. It is an effective area in the
position where the electric wave reaches.

Table 1. Merit and Demerit(Key opening and closing system)

	Merit	Demerit
Physical Key	Device is simple and cheap.	Only front of the device.
Human characteristic	Device is simple and cheap.	Must be near the device.
This system	All over.	Problems with security.

2.2 Construction of the lock opening and closing system

A diaglam of the lock opening and closing system is shown in Figure 1.

When a user wants to close the door, he sends a "clo se command" via cellular
phone. The web server receives this command through the internet and sends the
command to the lock device by using the RS-232C. Then, the door is locked by the
command.

The user can watch the logs and the image at current condition of the lock system
(opened or closed).

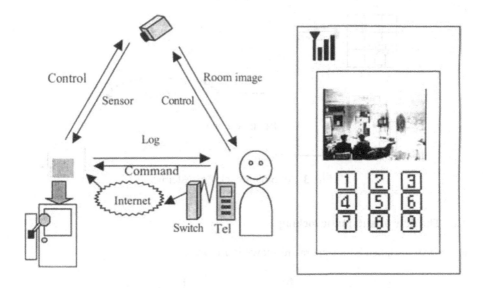

Fig. 1. The lock opening and closing system **Fig. 2.** User Interface(A cellular phone)

2.3 Cellular phone user interface

Our system includes a cellular phone user interface for confirmation. This interface is image-based. Its Figure of the user interface is shown in Figure 2. This image is taken by a pan-tilt camera, and connected and control through a PC. This camera takes a picture of the room from nine directions. The user has the ability to view an image from any direction by pushing a button.

3 Image based automatic locking system

3.1 The outline of the automatic locking system

At present there are many systems automatic locking doors, but these systems lock if somebody exists or does not exist in the room. Our system locks when nobody exists in the room, and does not necessarily lock when somebody exists in the room. This system uses a camera to observe the status of the room.

 We used image processing for recognition the human who is in the room.

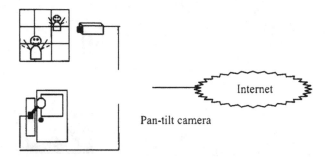

Pan-tilt camera

Fig. 3. Automatic locking system

3.2 The flow of automatic locking

Room images taken by a camera are shown in Figure 4.

Fig. 4. Room images from nine directions of the room

The size of these images are 320*240 pixels. This camera is set to take images from nine directions. 3 sheets of images are produced every 0.5 seconds. Our system recognizes somebody by using image processing of these images.

3.3 Image processing

When somebody exists, it supposes that a part of the recognized skin color moves. Therefore, our system detects within the skin color area and any movement associated. First, the moving area is found by using three images from each direction. The moving area is found by subtracting the differences between three images. The skin color area is found by using the number of R-B in RGB table, doing the labeling processing [6][7] to find these areas. This method takes the histogram in a vertical direction and marks the recognized part over a threshold. At last, it decides whether somebody exists by the combination of the logic products from the two areas. This flowing figure is shown in the Figure 5.

If image processing can't find somebody existing in the room all directions, it decides that nobody exists.

If it detects someone existing in one direction, it is detected locally, It also detects by same method from the first direction, when nobody exists in the frame.

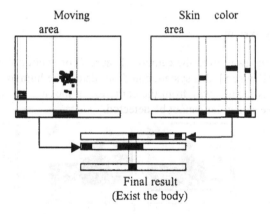

Fig. 5. The flow of image processing

3.4 Confirmation work

The lock closes by one confirmation work when it passes 30 seconds. As it is a very short time, this time alone is not enough to lock the door. So, the necessary time to confirm the work should be longer. Therefore, the confirmation work time was increased. This method has been used in order to increase the amount of time, and locks after five confirmation tests. (Figure 6).

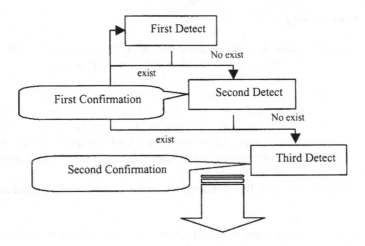

Fig. 6. Confirmation work

3.5 Time change system

If someone is facing away from the camera, [Figure 7] or is not moving at all within the viewing area [Figure 8], this system could not detect the human. In a situation in which the person is facing away from the camera, but moving, the moving area can be detected, but the skin color cannot be detected.

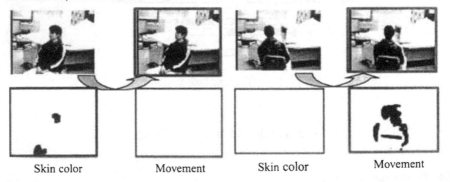

Fig. 7. Only moving area   **Fig. 8.** Only skin color area

So, this system keeps watching the position where higher possibility to exist any-body. Therefore, it cannot detect on its direction when somebody faced the back. Then it supports by only using the moving area.

In other words, it becomes higher reliability using the weight to detect the moving area and skin color area.

4   Experiment

4.1   Experiment 1

When nobody is in a room, it is the most important problem in security to lock the door securely. Therefore, this experiment confirms it.
<Place> In the room (Our laboratory)
<Number of date> 300 sets (One set represents one detection.)
This result is shown in the table 2.

Table 2. Result of experiment 1

	Detection of human	Locked
Nobody	0%	100%

4.2 Experiment 2

If somebody is facing away from camera, previous method cannot detect him. Therefore, this experiment estimates that the time change system is effective.
<Place> In the room (Our laboratory)
<Number of date> 100 sets (One set represents one detection.)
This result is shown in the Figure 9 and table 3.

Table 3. Result of experiment 2:"con=1" means that the confirmation is one time. "con=5" means that the confirmation is five time.

	Detection of human	Locked	
		con=1	con=5
No time change system	56%	44%	33%
Time change system	81%	19%	8%

HandTalker: A Multimodal Dialog System Using Sign Language and 3-D Virtual Human

Wen Gao[1,2], Jiyong Ma[1,2], Shiguan Shan[1], Xilin Chen[2], Wei Zeng[2], Hongming Zhang[2], Jie Yan[3], Jiangqin Wu[2]

[1]Institute of Computing Technology,
Chinese Academy of Sciences,100080, China
[2] Harbin Institute of Technology,Harbin,China,
[3]Microsoft Research China
{wgao,jyma}@ict.ac.cn

Abstract. In this paper, we describe *HandTalker:* a system we designed for making friendly communication reality between deaf people and normal hearing society. The system consists of GTS (Gesture/Sign language To Spoken language) part and STG (Spoken language To Gesture/Sign language) part. GTS is based on the technology of sign language recognition, and STG is based on 3D virtual human synthesis. Integration of the sign language recognition and 3D virtual human techniques greatly improves the system performance. The computer interface for deaf people is data-glove, camera and computer display, and the interface for hearing-abled is microphone, keyboard, and display. *HandTalker* now can support no domain limited and continuously communication between deaf and hearing-abled Chinese people.

1   Introduction

Gestures are part of everyday natural human communication – they are used as an accompaniment to spoken language as well as an expressive medium. Sign language, as a kind of structured gesture, is one of the most natural means of exchanging information for most deaf people. Researches of machine sign language recognition began in the 90's. To date, there are two ways to collect gesture data in sign language recognition, one is the vision-based approach, and the other is device-based. In this paper we report *HandTalker*: a translation system for direct communication between deaf people and hearing society in China. One critical technique involved in the system is Chinese Sign Language recognition (CSL). CLS recognition includes feature extraction and recognition approach two aspects. The other critical technique involved in the system is 3D virtual human synthesis. It includes automatic extraction feature points used for automatic generation of 3D face mesh, 3D virtual human modeling and gesture synthesis.

The organization of this paper is as follows: we begin with CSL recognition, then proceed to a discussion of human facial feature extraction for 3D virtual face synthesis for a given person and describe 3D virtual face synthesis for the given

T. Tan, Y. Shi, and W. Gao (Eds.): ICMI 2000, LNCS 1948, pp. 564-571, 2000.
© Springer-Verlag Berlin Heidelberg 2000

person. We then demonstrate the system setup. Finally, we give summary and conclusion.

2 Chinese Sign Language Recognition and Syntheses

Chinese Sign Language (CSL) is the primary mode of communication for most deaf people in China. CSL consists of about 5500 elementary vocabularies including postures and hand gestures. With the evolvement of CSL, the up to date CSL can express almost all meanings as natural spoken language can. Therefore, the task of CSL recognition is challenging and meaningful. Critical techniques involved in CLS recognition include feature extraction and recognition approach. Sign position independent feature extraction is very important to practical applications because it is not necessary to restrict a singer to a certain position when the signer is gesturing. The more detailed approach to extraction of sign position independent feature can be found in our previous work [1]. For large vocabulary recognition, Ho-Sub Yoon recently [2] reported that a sign vocabulary consisting of 1,300 alphabetical gestures were recognized using HMMs. And C.Vogler and D.Metaxas[3] pointed that the major challenge for sign language recognition is to develop approaches that will scale well with increasing vocabulary size.

In baseline system for sign language recognition, the data collected by the gesture-input devices are fed to the feature extraction module; then, the feature vectors are input to the fast match models. The fast match finds a list of candidates from a word dictionary. The Bigram model uses word transition probabilities to assign, *a priori*, a probability to each word based on its context. The system combines the fast match score to obtain a list candidates, each of which is then subject to a detailed match, the decoder controls the search for the most likely word sequence using Viterbi search algorithm. When the word sequence is output from the decoder, each word drives the speech synthesis module and 3D-virtual human gesture animation module to produce the speech and gesture synchronously.

Considering the scalability problems in sign language recognition, we propose the following approach. For each sign, a continuous HMM is trained using training data. For each stream such as left hand shape, right hand shape, left hand position, right hand position, left hand orientation and right hand orientation, the HMM parameters of all signs are clustered. This greatly reduces the amount parameters of signs, therefore reduces the computation load greatly without loss noticeable performance.

The integrated system was implemented on PIII-450. The first experiment was on a dialog between a sign language teacher and a hearing-abled. The database of gestures consists of 220 words and 80 sentences. Each sentence consists of 2 to 15 words. No intentional pauses were placed between signs within a sentence. Within 80 sentences, the deletion (D), insertion (I), and substitution (S) errors are D=3, S=2,I=2 respectively. The word recognition rate is 98.2%. The second experiment was carried on large vocabulary signs. The recognition rate of 5177 isolated signs is 94.8% in real time. For continuous sign recognition, the word correct rate is 91.4% for 200 sentences. A sign language teacher performed the signs. The average system response time is about 2 seconds delay after a sentence was performed.

Our early synthesis system of CSL, speech and the corresponding facial expression driven by the text was reported in [4]. A Chinese sentence is input to the text parser module, which divides the sentence into basic words. The parser algorithm is the Bi-directional maximum matching with backtracking approach. After the words are matched, each word in the sentence is then input to the sign language synthesis module and speech synthesis module. For the time being, the word library consists of above 3000 words.

3   Human Facial Feature Extraction for a Given Person's 3D Virtual Face Synthesis

In order to extract the facial features in frontal face images, the first step is face detection. We use our previous approach [5] to localize the face in an input image, which can give the location of the face and its approximate size. The second step is feature extraction of the frontal face. Deformable template [6] is an effective method to extract the location and shape of such facial salient organs as eyes, mouth and chin, but it does have some defects. Since it adopts the optimizing algorithm to minimize the cost function, it is highly dependent on the initial parameters and subject to trapping in local minimal. Furthermore, it is too time-consuming. To solve these problems, a coarse-to-fine feature extraction algorithm is proposed. Fig.1 illustrates some results of our feature extraction system. The size of the original images is 256x256, 256 gray-level while the typical size of face is 130x140. On a Pentium III 450 MHz computer, the time needed from detecting the face to detecting feature points is about 2 seconds.

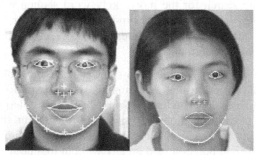

Fig. 1. Results of detecting facial features

Thirteen feature points are defined in profile model. Color information is used to segment the facial image. A threshold operation is carried out for the profile image, and it produces a binary image. Extracting the profile feature points includes two main parts: extraction of profile outline and extraction of feature points. On the profile image with blue background, the face region is detected and the outline is located. The feature points can be obtained by utilizing geometric relation among feature points.

4 Image-Based 3D Face Model Automatic Generation and Virtual Face Synthesis for a Given Person

Feature points are defined and automatically extracted as mentioned in the preceding section. In order to rectify the possible errors of automatic feature extraction, an interactive mechanism for the modification of feature points is introduced. According to these features points, the general 3D face model is adjusted to fit a given person's characteristics. Then multi-direction texture mapping is utilized to enhance the reality of the synthetic face. Then expression synthesis and lip-motion synthesis are realized.

4.1 Transforms from the General 3D Face Model to the 3D Face Model of a Given Person's

Global transform and local transform are designed in order to change the general 3D neutral face model to the given person's 3D neutral face model [7]. Global transform is used to adjust the global contour of the face and the position of the organs on the face. It is accomplished by scaling the coordinate values of each vertex of the model. The scaling factor can be calculated using the coordinates of the feature points before and after transform. While local transform aims at adjusting the shape of the organs such as the eyes, eyebrows, mouth, nose and the chin to fit the given person. Fig.2 illustrates the adjusting procedure of model.

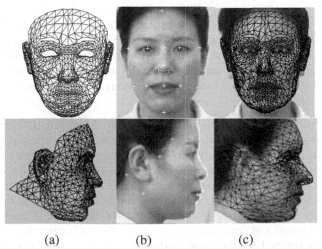

(a)   (b)   (c)

Fig. 2. Modification from the general face model to the given person's face model (a) General face model (b) Feature points extracted (c) Given person's face model

4.2 Multi-direction Texture Mapping

Texture mapping provides a valuable technique for further enhancing the reality of the synthetic human face. A texture mapping is constructed from two face images, one frontal face image and one profile, to a 3D-face surface. For each Bézier surface patch of face surfaces, the corresponding texture image is determined by mapping the boundary curve of the Bézier patch to face image. Which image to be chosen depends on the whole direction of the Bézier patch. When the angle of directional vector is less than 30 degree, frontal face image is used. Otherwise profile image is used. Fig.3 illustrates the result of one given person's synthetic face.

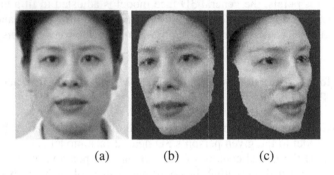

(a)   (b)   (c)

Fig. 3. Given person's synthesizing face (a) Input front image of given person (b) and (c) Synthetic face

4.3 Expression and Lip Motion Synthesis

In our system, the deformations caused by the facial expression are described by action units (AUs)[8]. AUs describe the contraction of several bunches of facial muscle. The expansion and contraction of each muscle vector reflect the changes of positions of facial feature points and propagate the movement to the related areas of the mesh. The expressions are driven by the motions of facial muscle vectors. (see reference[7] for details).

In the text-driven lip motion synthesis system, the first step is to divide the text into words. The second step is to convert the word into phonemes. The last step is to drive the movement of the 3D lip. We only consider some typical lip shapes in the process of pronunciation, such as a big, or a small of mouth and a round or a flat of lip shape, etc. In Chinese, one word corresponds to one syllable. Tongue's position and the lip shape do not change during pronunciation for a single vowel. But the lip shape of a compound vowel changes during pronunciation. We define eight basic lip-shapes corresponding to vowel and consonant, for any syllable in Chinese, the corresponding lip-motion is constructed by the combination of lip shapes in the eight basic lip shapes.

4.4 Synchronously Driving: Speech, Gesture, Lip Movement

Conversion text generated by the sign language recognition to speech is important to communication between deaf people and hearing society, while text to gestures and lip motions are useful for telecommunication between deaf and deaf. To synchronously drive speech, gesture and lip motion, we align the display time of gesture and lip motion to the time of playing speech so that human perceives it comfortably. The underlying assumption of this approach is that the display speed of gesture and lip motion is faster than that of playing speech. Fortunately, this requirement is usually satisfied.

5 System Setup

The system architecture of *HandTaler*: a translation system for direct communication between deaf people and hearing society based on sign language recognition and 3D virtual human synthesis is demonstrated in Fig.4.

The integrated system is implemented on two personal computers. One computer is for interacting with a person with hearing ability and the other for a deaf. These two computers are connected with LAN with two different interfaces. The agent B is responsible for the interaction of a person with normal hearing, the person faces 3-D virtual human face of the deaf at the other side as shown in Fig 5.

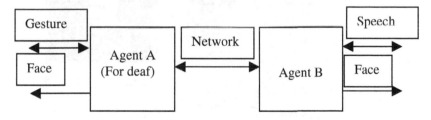

Fig. 4. The architecture of *HandTalker*

The agent A is responsible for the task of a deaf; the deaf faces to a virtual graphic human. The person with hearing ability can input or speaker a sentence to the computer. The agent *B* transfer the sentence to the agent *A*. When the agent *A* *has* received the sentence, the gesture synthesis module synthesizes the gestures corresponding to the sentence and the virtual graphic human performs the gestures for the deaf. The deaf can understand the meaning of the sentence via the synthesis gestures. If the deaf wants to "talk " with the person with hearing ability, he can perform gestures to the computer. The gesture recognition module recognizes the gestures and transfers the text to the agent *B*. When the agent *B* has received the text, it drives the realistic 3-D virtual human to speak the text accompanying speech and lip motion as shown in Fig.6.

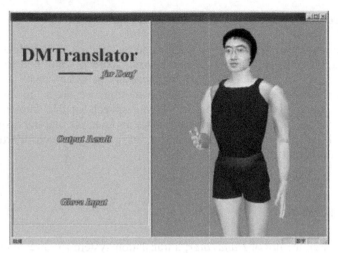

Fig. 5. The user interface for deaf people

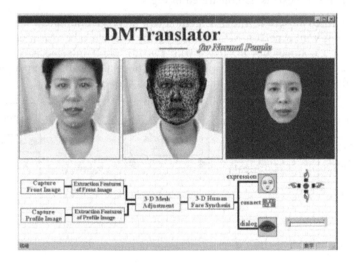

Fig. 6. The user interface for hearing-abled

6 Conclusion

We have described *HandTalker:* a system we designed for free communication between deaf people and normal hearing society based on sign language recognition and 3D virtual human synthesis.

In the system, CSL recognition has been developed using HMMs based recognizers. We have developed a new gesture feature extraction approach and a fast match approach to speed up the decoding process. The performance of these techniques ware evaluated using the CSL recognition system. Experimental results show that these techniques are capable of improving both the recognition performance and speed. The signer position independent feature extraction is very important to sign language recognition system in practical applications.

As to 3D virtual human synthesis, in order to obtain the facial feature of a given person automatically, we have developed algorithms for the feature extraction in both frontal face image and profile. In the aspect of feature extraction in the frontal face image, we have proposed a coarse-to-fine facial feature detection strategy based on the facial configuration prior, facial gray distribution characteristics and deformable template. Experiments have indicated that the strategy has good performance in both the accuracy and speed of the extraction. And we have introduced a new method for extraction of profile feature points. It combines profile detecting and locating of feature points. Experiments have shown that the method is reliable in various lighting conditions. Based on these features, 3D-face model of a given person can be generated. And multi-direction texture mapping greatly enhances the reality of the synthetic face. In order to make the synthetic face realistic, corresponding expression and lip-motion synthesis algorithms have been developed. The synthetic human face greatly improves the vitality and usability of the system for deaf people.

REFERENCES

1. Jiyong Ma,Wen Gao,Jiangqin Wu and Chunli Wang, A Continuous Chinese Sign Language recognition system, pp428-433, 28-31 March, FG'2000, Grenoble, France.

2. Ho-Sub Yoon. Jung Soh. Byung-Woo Min. Hyun Seung Yang. Recognition of alphabetical hand gestures using hidden Markov model. IEICE Transactions on Fundamentals of Electronics Communications & Computer Sciences, vol.E82-A, no.7, July 1999, pp.1358-66.

3. Vogler C., Metaxas D. Parallel hidden Markov models for American Sign Language recognition. Proceedings of the Seventh IEEE International Conference on Computer Vision. IEEE Comput. Soc. Part vol.1, 1999, pp.116-22 vol.1. Los Alamitos, CA, USA

4. Wen Gao,Yibo Song, Baocai Yin, Jie Yan and Ying Liu. Synthesis of sign language ,sound and corresponding facial expression driven by text in multimodal interface. In *Proceedings of the First International Conference on Multimodal Interface*, 1996, 244-248.

5. Wen Gao, Mingbao Liu. A hierarchical approach to human face detection in a complex background. In *Proceedings of the First International Conference on Multimodal Interface*,1996,289-292.

6. Yuille, A.L., Hallinan, P.W. & Cohen, D.S.Feature extraction from faces using deformable templates. *International Journal of Computer Vision*, 1992,8, 99-111.

7. Wen Gao, Jie Yan, Baocai Yin, Yibo Song, An individual facial image synthesis system for virtual human, In *Proceedings of the Second International Conference on Multimodal Interface* ,1999, 20-25

8. P. Ekman and W. V. Friesen. Facial action coding system. *Consulting Psychologists Press Inc.*, California, 1978

A Vision-Based Method for Recognizing Non-manual Information in Japanese Sign Language

Ming Xu[1], Bisser Raytchev[1], Katsuhiko Sakaue[1], Osamu Hasegawa[1],
Atsuko Koizumi[2], Masaru Takeuchi[2], and Hirohiko Sagawa[2]

[1] Electrotechnical Laboratory (ETL), Umezono 1-1-4 Tsukuba City, Ibaraki, 305-8568,
Japan
{xum, bisser, sakaue, hasegawa}@etl.go.jp
[2] RWCP Multi-modal Functions Hitachi Laboratory 1-280, Higashi-koigakubo,
Kokubunji-shi, Tokyo 185-8601, Japan
{koizumi, mtakeuch, h-sagawa}@crl.hitachi.co.jp

Abstract. This paper describes a vision-based method for recognizing the non-manual information in Japanese Sign Language (JSL). This new modality information provides grammatical constraints useful for JSL word segmentation and interpretation. Our attention is focused on head motion, the most dominant non-manual information in JSL. We designed an interactive color-modeling scheme for robust face detection. Two video cameras are vertically arranged to take the frontal and profile image of the JSL user, and head motions are classified into eleven patterns. Moment-based feature and statistical motion feature are adopted to represent these motion patterns. Classification of the motion features is performed with linear discrimant analysis method. Initial experimental results show that the method has good recognition rate and can be realized in real-time.

1 Introduction

Japanese Sign Language (JSL) is the usual method of communication used by the hearing impaired people in Japan. One characteristics of JSL is that it is a sequence of hand gesture combined with non-manual actions, such as head motion, face expression and body posture. Almost existing researches of JSL focused on the manual information but neglected the role of non-manual information, because their roles in JSL were not well investigated. Our preceding research [1] has found that non-manual action has explicit correlation relationship with hand action, thus provides meaningful constraints useful for JSL word segmentation and interpretation. For instance, head nodding usually occurs in the break between words or at the start or the end of a sentence. The data-glove based JSL translating system we are developing requires the user to return his/her hands to the starting position after each hand action, due to the difficulty of accurate segmentation.

Therefore, it is a natural thinking that non-manual information provides promising solution to these problems. For this purpose, we propose a vision-based method that can recognize human head motion in real-time. Although in this paper, our attention focused only on head motion, the most dominant non-manual information in JSL, the method can provides a common basis that can be expanded to copy with other non-

T. Tan, Y. Shi, and W. Gao (Eds.): ICMI 2000, LNCS 1948, pp. 572-581, 2000.
© Springer-Verlag Berlin Heidelberg 2000

manual information, such as gaze and facial expressions. The first part of the method is a face detector which adopted an interactive color-modeling scheme, it can detect face robustly even in environment with complex lighting and texture. After head is extracted, moment-based feature and statistical motion feature are used to represent the motion characteristics of head. Classification of motion features is performed with linear discriminant analysis. In the following sections, we will clarify the role of non-manual information, describe our method and show the experimental results.

2   Non-manual Information

In order to clarify the role of non-manual information in JSL, we constructed a multimodal JSL database in our previous study. The database consists of videos of JSL, their corresponding Japanese text, and labels of non-manual information. The videos are taken with video cameras from frontal and profile directions of the hearing impaired subject while he/she "says" the sample sentence. Each sentence is repeated five times, three subjects produced a total of 4000 sample video sequences. The JSL interpreter and the hearing impaired people make labels off-line for each video sequence to indicate the place where non-manual action occurs. Non-manual information is classified into several categories such as head, jaw, mouth, posture, eyebrow and gaze. Each category is further labeled according to its direction, shape or motion patterns. The number and type of labels in a sample sequence differs with the JSL content, so we use the occurrence frequency of the label to show the relationship of each label with JSL. The labels with higher occurrence frequency obtained from the database are listed in table 1.

Table 1. Major non-manual information

Head Motion	Facial Expression	Body Posture
Head nodding	Mouth shape	Body forward
Jaw up/down	Eye's open/close	Body up/down
Head shaking	Gaze direction	
Head slanting	Eye brows	

It is found that head motion, such as head nodding, the up and down motion of jaw has the highest occurrence frequency. It occurs in almost the 4000 samples of the database. Next to head motion, the facial expressions, such as shape of mouth, gaze, close and open of eyes etc. also have high occurrence frequency. Besides occurrence frequency, grammatical analysis also proves the functions of non-manual information. Head motion often occurs at the break between words or at the start/end of a sentence. Mouth shape, eye and gaze have special grammatical meaning for some particular JSL expressions.

We found that due to the habit of different users, even the same pattern of head motion may have different spatial and temporal scales. Some motions are clear and recognizable, while others are subtle. Moreover, the characteristics of the same motion pattern may change with the content of the sentence, even with the same user.

So it is convenient to describe head motion with several sub-patterns. Here, we define 11 sub-patterns as shown in table 2.

Table 2. Sub-patterns of head motion

ND	Nod Down	SL	Head Slant Left
NU	Nod Up	SR	Head Slant Right
JD	Jaw Down	MF	Head Move Forward
JU	Jaw Up	MD	Head Move Down
TL	Head Turn Left	MU	Head Move Up
TR	Head Turn Right		

{ND, NU} correspond to the regular nodding and are relatively large in motion scale. {JD, JU} are typical motion in JSL, their motion scale are usually small.

3   Head Detection

Before doing recognition of head motion patterns, head region needs to be detected at first. Since the JSL translation system is supposed to operate in practical circumstance, the head detector should be able to cope with complex background and variations in lighting. View-based method [2] is the most reliable approach to find human face from an image. This method has to search the whole image and use face templates at different scales, computation is heavy and therefore not possible to be implemented in real-time with a common PC. Skin color [3] is an often-used cue for detecting human face from image, real-time performance is easily achieved even in a low-end computer. But due to possible variations of circumstance lighting and personal difference in skin color, a general skin color model is not appropriate.

Here we propose an interactive color modeling method, which construct a most suitable color model for each specific user at each specific period. We assume only one subject is viewed at one time and the camera is fixed. HSV color space is adopted for color representation. From the collected facial samples, we found that the hue component (H) of skin color concentrates on a specific narrow band, even for different kind of skin colors under different lighting conditions (indoors or outdoors). By comparing the hue histogram with input image, skin-like pixels are extracted as C(x). Because only hue component is used, non-skin color pixels with similar hue component are also extracted in C(x).

Since the camera is fixed, motion cue is useful to remove this ambiguity. For each input image, sequential image subtraction is performed to extract pixels with intensity change, as shown in (1) and (2).

$$B(\mathbf{x},t) = T(|I(\mathbf{x},t-1) - I(\mathbf{x},t)| - S)$$ (1)

$$T(x) = \begin{cases} 1, & x \geq 0 \\ 0, & x < 0 \end{cases}$$ (2)

Where S is the fixed threshold corresponding the camera noise level, T is a thresholding function. Each time, when the number of the total pixels in image $B(x)$ is over a pre-determined value, we consider that there is a new subject entering the camera's viewing field, then color model adaptation is started. We use the logic AND operation between $C(x)$ and $B(x)$ to produce an image of the moving skin-like pixels. From the gravity center of the resultant image, a small rectangle is determined. Within the rectangle, each skin-like pixel is sampled to get the value of its corresponding S and V component. The sampling process sampled 100 input frames to construct the histograms of S and V components. Then an adapted skin color model corresponding to the specific subject and the specific circumstance is constructed. Furthermore, using the heuristic knowledge about human head position, it can stably sample the face region no matter in what way the subject comes into the camera's viewing field.

Using the adapted skin color model, extracted skin color pixels are grouped into blobs. The blobs with too small area or with unreasonable shape are removed [4]. In case of only one subject, we can simply take the largest blob as face candidate. Using the continuity of human motion, detected face region is tracked over frames.

This head detector has been tested under different circumstance, and with different subjects. It works well even under the complex lighting condition.

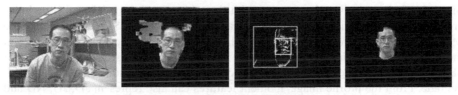

Fig. 1. Face extraction through interactive skin color modeling. The first one from the left is one of the images used for color modeling. The second from the left is extracted skin color regions using general skin color model. The third from the left is the result of the AND operation between $B(x)$ and $C(x)$, the small rectangle is the region to do skin color sampling. The right one is the extracted result using the adapted color model.

4   Recognition of Head Motion Patterns

From the output of head detector described in the previous section, some head motion patterns, i.e., SL, SR, MF, MD, MU can be directly estimated from the blob's moment properties: gravity center, long axis angle and the size of the blob's bounding rectangle. For blob $I(x,y)$, its spatial moment is defined as follows,

$$M_{pq} = \sum_x \sum_y x^p y^q I(x, y) \tag{3}$$

Where p and q is the order of moment. Then the center of the blob is calculated in (4),

$$x_c = \frac{M_{10}}{M_{00}}; \quad y_c = \frac{M_{01}}{M_{00}} \tag{4}$$

The central moment of blob $I(x,y)$ is as follows,

$$\mu_{pq} = \sum_x \sum_y (x - x_c)^p (y - y_c)^q I(x,y)$$ (5)

The long axis angle of the blob is calculated as follows,

$$\theta = (1/2)\tan^{-1}\{2\mu_{11}/(\mu_{20} - \mu_{02})\}$$ (6)

Fig. 2. Moment-based feature of facial region. The position of gravity center, the long-axis angle and the bounding rectangle are shown.

Since the facial area is well extracted with the adapted color model, and little holes are buried by morphological filtering, the angle θ reflects the SL and SR well. The MF, MU and MD can clearly reflected by the change of gravity center of the blob and the size of its bounding rectangle. Although the moment-based features describe planar motion very well, when motion is combined with rotation, the accuracy will degrade. From the profile image, moment features can also represent JD, JU, ND and NU. But the accuracy is not satisfactory for these relatively subtle patterns.

Therefore, we need other feature to represent these motion patterns. Many methods have been proposed about human motion recognition [5][6]. As our target is restricted to human head, it is natural to use the specific model knowledge. For instance, head motion can be explicitly known if facial features, such as pupils or mouth corner or the central point of jaw can be detected and tracked. But in practical situation, the feature extraction process is sensitive to noise, such as the variation of circumstantial lighting condition and head posture. Using factorization analysis, it is possible to reconstruct 3-D motion from image sequence, but the computation cost is high. Direct motion extraction method, like optical flow, requires heavy computational, and the results are very noisy in region with little texture, such as human face. Here, we adopt a feature statistically extracted from subtracted image sequence to represent these head motion patterns, and use the linear discriminant analysis for recognition.

4.1   Motion Features

Relative motion feature [7] was successfully used in gesture recognition. Here, we use this feature and adapt it to our task of head motion recognition. Let $I(i, j, t-1)$, $I(i, j, t)$ and $I(i, j, t+1)$ be three image frames obtained correspondingly at time t-1, t and t+1. From the three images, two binary images $B(i, j, t)$ and $B(i, j, t+1)$ are obtained by sequential subtraction followed by thresholding operation as shown similarly in formula (1). Therefore, in the binary image $B(i, j, t)$ and $B(i,j,t+1)$, a value "0" at coordinates (i,j) means that there has been no change at the location for the corresponding time interval, while "1" means that change has occurred. This representation only shows the position of change, but has no idea of the temporal corresponding relation. Therefore, a feature F is designed to statistically reflect the motion properties from two binary images. For an arbitrary point $p_1(t) = B(i,j,t)$, we select its spatial reference points $p_2(t)$ at the eight adjacent positions as shown in fig.3, l is the spatial parameter between $p_1(t)$ and $p_2(t)$. Similarly, the corresponding point pair $(p_1(t+1)\ p_2(t+1))$ are selected in $B(t+1)$.

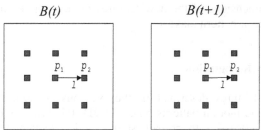

Fig. 3. Pixel structure for extracting statistical motion feature

The 4-tuples $(p_1(t)\ p_2(t)\ p_1(t+1)\ p_2(t+1))$ reflect the spatial-temporal change in one direction. Searching this 4-tuples in all the eight directions and accumulate them over all pixels in the operational region, we extract feature F as in (7):

$$F(u_1,u_2,v_1,v_2;l,k,t) =$$

$$\frac{1}{Z}\sum_{i,j}\Gamma\{[p_1(t),p_2(l,k,t),p_1(t+1),p_2(l,k,t+1)]\ Equ\ [u_1,u_2,v_1,v_2]\} \tag{7}$$

u_1,u_2,v_1,v_2 takes values 0 or 1, k takes value 0 to 7, which corresponds to the eight reference points p_2. Equ is a Boolean operator. Z in (3) is the normalization factor that takes the average value of the number of "1"s in the frames $B(t)$ and $B(t+1)$, function Γ is defined as follows,

$$\Gamma\{x_1,x_2,...,x_N\} = \begin{cases} 1 : x_1 = x_2 =...= x_N = 1, \\ 0 : otherwise, \end{cases} \tag{8}$$

Therefore, for each specific value l, feature F is a vector with length $(2^4-1)*8 = 120$, the combination (0 0 0 0) is eliminated, since no changed occurred. F is a statistical feature that reflects the overall spatial-temporal change of three consecutive images. It is also feasible to integrate F with different scale parameter l into a longer feature vector to better reflect motion with different spatial scales.

As discussed in the previous section, the dimensions of the motion feature F is 120 times the number of scales incorporated. We need to transform this high-dimensional feature into low-dimensional one. Here, we use linear discriminant analysis to transform the original motion feature, as shown in (9):

$$Y=XA \tag{9}$$

Where X is the given $n \times d$ pattern matrix, Y is the transformed $n \times m$ pattern matrix, and A is the $d \times m$ matrix of linear transformation. Here, n is the dimension of the original feature, d is the number of features, m is the dimensions of the transformed features. Discrimination performance of the new features is evaluated by the Fisher criterion, then optimal matrix A is obtained by finding the eigenvectors of $s_w^{-1} s_b$ (the product of the inverse of the within-class scatter matrix s_w, and the between-class scatter class s_b). The jth column of A is the eigenvector corresponds to the jth largest eigenvalues. The dimensions of the new features m takes the minimum of (K-1, d), here K is the number of pattern classes.

4.2   Head Motion Recognition

Using the motion features described in the previous section, we experiment to recognize the six head motion patterns ND, NU, JD, JU, TL, and TR. The first stage of the recognition process is to prepare sample video sequence for each of these motion patterns. Motion feature extraction operation are only performed in the region of extracted head, therefore remove the interference of other moving body parts. From the learning samples, transformation matrix A in (9) can be obtained. According to (9), each original motion feature is projected into the low-dimensional discriminant space. For a sequence of image frames, their motion feature forms a cluster of points in the discriminant space. Fig.4 shows the practical feature distribution of four patterns in the discriminant space.

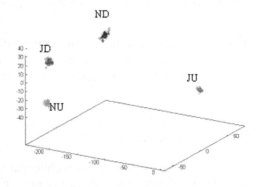

Fig. 4. Distribution of four sub-patterns in the discriminant space

In the second stage, for each obtained image frame, the extracted motion feature is projected into the discriminant space as one point. The distances between the point and the average of each learned pattern classes are calculated, and the point is classified into the class with the nearest distance. Thus, each time a new frame is inputted, a new classification result is outputted as a symbol. By analyzing these symbols, we can know what motion is performed. Since different motion patterns have different durations, and even the same pattern may have different duration due to different subjects or different JSL contents. Instead of using DP and HMM to recognize these temporal variation patterns, we use a simple approach by dynamically checking the number of each output symbols. From the JSL sample database, we can find out the statistical maximum M and minimum N of the temporal duration for the sub-pattern classes. Then we dynamically keep a set of M output symbols start from the latest one, if the number of symbols of one class is the largest and also over N, then we can decide that the motion pattern corresponding to the symbol has been performed.

5   Experiments

Our experimental system consists of two vertically arranged fixed-viewpoint video cameras. The breast-top shots of the subjects are captured from the frontal and profile directions. Images from two cameras are multiplexed synchronically with a multiple viewer (FOR.A MV-40E) and inputted into computer for processing. This configuration has the advantage of extracting the head motion patterns from the most appropriate viewing direction, i.e., extracting {ND, NU, JD, JU} from the profile view and {TL, TR, SL, SR, MF, MU, MD} from the frontal view.

First, we test the performance of face detector. Ten test sequences were taken from in the window side of a room, from morning to night respectively. The lighting consists of artificial light and sunlight, and changes slowly from time to time. The background is a complex clustering. For all of these ten test sequences, face is correctly extracted. The successful extraction of facial region makes the motion recognition of SL, SR, MF, MU and MD well, so long as the head does not have vigorous motion or have non-planar rotation.

Next, we test the performance for recognizing ND, NU, JD, JU, TL and TR, using the statistical motion feature. The learning data were taken from two subjects, there are twenty samples for each of the six patterns, and each sample has ten frames. In the process of extracting motion feature, two scales are incorporated. Its parameter l takes a small value 1 and 2, due to the small motion of these patterns. We selected two sets of test samples from the multimodal JSL database. The recognition results are shown in table 3.

Table 3. Recognition Results

	SAMPLE DESCRIPTION	CORRECT	FALSE
Set1	20 samples (20frames/sample, 42 patterns)	78.5%	10%
Set2	10 samples (200frames/sample, 100 patterns)	86%	20%

Correct rate shows the number of patterns corrected recognized, while false rate shows the number of non-patterns being falsely recognized as patterns. Analyzing the wrongly recognized sequence, we found that most of the non-recognition occurs when the motion duration is too short. Although it is possible to adjust the minimum duration value N in the recognition stage to cope with the motion with very short duration, on the contrast, other motion patterns would be falsely recognized. How to make a good balance is a problem needs further investigation.

Online experiments are also performed. The computer is Pentium-II 450MHz PC, and the extracted head region is about the size of 80*60 pixels. The system throughput is 15frame/second. When we enlarge the value l to 3 pixel and 6 pixels, the system throughput increases to 20frames/second.

6   Conclusions and Future Works

In this paper, we proposed a vision-based method to recognize the head motion patterns in JSL. Utilization of the robust head detector and statistical motion feature makes the proposed method robust to variations in the practical circumstances. Initial experiments have shown the efficiency of the method. At present stage, we used two fixed scales in the feature extraction process. Selection of optimal scale parameters is a direction to further improves the performance of motion recognition.

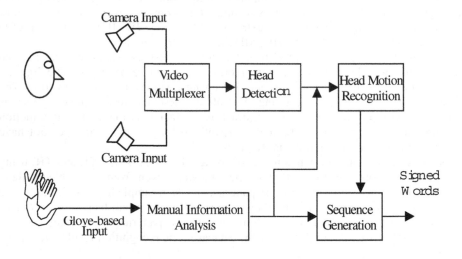

Fig. 5. Prototype for integrating non-manual and manual information

Non-manual information makes sense only when they are considered concurrently with manual information extracted from data-gloves. So it is important to integrate the proposed vision-based non-manual recognition system with the data-glove based translation system. The integration should be bi-directional, i.e., both manual and non-manual information could help its counterpart to perform better. Based on the architecture of current recognition system [8], Fig.4 shows a prototype for system integration. The output of the vision system is a sequence of symbols corresponds to different head motion patterns, which is synchronized with data-glove translation system. The output is utilized at the stage of sequence generation as a powerful grammatical constraint. Similarly, the manual information could also be useful in classifying meaningful non-manual information. The interactive use of information from two different modalities is supposed to improve the performance of JSL translation system as a whole. Therefore, integrated experiments are needed to show the performance of the multimodal system.

References

1. M.Takeuchi and H.Sagawa, "Sign Language Recognition Method Based On Spatial Information and Multimodal Sign Language Database", RWCP Symposium, June 9-10, 1998, pp.193-198.
2. M-H. Yang, N. Ahuja, and D. Kriegman, "Face Detection Using Mixtures of Linear Subspaces", International Conference on Face and Gesture, March 2000.
3. W.Skarbek and A.Koschan, "Colour Image Segmentation, A Survey", Technical Report of Technical University of Berlin, October 1994.
4. M.Xu and T.Akatsuka, "Multi-module method for detection of human face from complex background", Proceeding of SPIE Conference on Applications of Digital Image Processing, Vol. 3460, pp.793-802, 1998.
5. J.K.Aggarwal and Q. Cai, "Human Motion Analysis: A Review", Computer Vision and Image Understanding, Vol.73, No.3, March pp.428-440, 1999.
6. Mubarak Shah and Ramesh Jain(Eds.), "Motion-Based Recognition", Kluwer Academic Publishers 1998.
7. Bisser Raytchev, Osamu Hassegawa, Nobuyuki Otsu, "User-independent online gesture recognition by relative motion extraction", Pattern Recognition Letters, Jan. 2000, PP.69-82.
8. H.Sagawa and M.Takeuchi, "A Method for Recognizing a Sequence of Sign Language", Words Represented in a Japanese Sign Language Sentence", International Conference on Face and Gesture, March 2000.

A Parallel Multistream Model for Integration of Sign Language Recognition and Lip Motion

Jiyong Ma[1,2], Wen Gao [1,2] and Rui Wang[2]

[1]Institute of Computing Technology,
Chinese Academy of Sciences,100080, China
[2] Harbin Institute of Technology,Harbin,China
{jyma, wgao,rwang}@ict.ac.cn

Abstract. The parallel multistream model is proposed for integration sign language recognition and lip motion. The different time scales existing in sign language and lip motion can be tackled well using this approach. Primary experimental results have shown that this approach is efficient for integration of sign language recognition and lip motion. The promising results indicated that parallel multistream model can be a good solution in the framework of multimodal data fusion. An approach to recognize sign language with scalability with the size of vocabulary and a fast approach to locate lip corners are also proposed in this paper.

1 Introduction

Sign language is one of the natural means of exchanging information for the hearing impaired. The facial expression and lip motion, which accompany sign language, are less important than hand gestures in sign language, but to understand some hand gestures they may play important roles. This paper will address the problem of integration of sign language and lip movement. Automatic lip-reading is complementary to both speech recognition and hand gesture recognition. Some gestures can not be reliably recognized by using unimodal of sign language. Therefore, an approach, which combines hand gesture features and lip motion features, is proposed to recognize sign language.

It was seen that most researches on sign language recognition were taken on small test vocabulary. For large vocabulary recognition, Ho-Sub Yoon recently [1] reported that a sign vocabulary consisting of 1,300 alphabetical gestures were recognized using HMMs. And C.Vogler and D.Metaxas[2] pointed that the major challenge to sign language recognition is how to develop approaches that scale well with increasing vocabulary size.

The main difficulty in incorporating the visual information into a speech recognition system and hand gesture recognition is to find a robust and accurate approach for tracking the lip movements in real time and extracting important features. It is widely accepted that the information of lip motions is not enough for lip-reading [3]. It also requires the information of the teeth and the chin. There are

T. Tan, Y. Shi, and W. Gao (Eds.): ICMI 2000, LNCS 1948, pp. 582–589, 2000.
© Springer-Verlag Berlin Heidelberg 2000

two classes for lip-reading: the model-based approach and the image-based approach. For the case of model-based approach, a geometrical model of the lip contours is applied to the input image of the speaker's lip. The disadvantage of this approach is its time consuming during the matching of the lip model [3][4]. For the case of the image-based approach, the interest regions are segmented. The advantage of this approach is its fast speed. The system architecture of combining sign language recognition and lip-reading is shown in Fig.1.

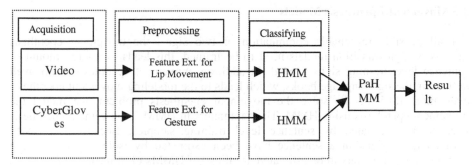

Fig. 1. The system architecture of integration of sign language recognition and lip reading

2 Sign Language Recognition

2. 1 Feature Extraction Approach

A hand gesture has a start position that corresponds to a spatial unit, transitional segment, and a stop position that also corresponds to a spatial unit. Therefore, a hand gesture can be viewed as a trajectory in the high dimensional space.

The basic spatial units in hand gestures should include the following six data streams,(right handshape, right hand position, right hand orientation, left handshape, left hand position, left hand orientation). To get invariant features to singer position, the relative angle between a unit directional vector of axis of the receiver at the left hand and a unit directional vector of axis of the receiver at the right hand and the distance between the receiver at the left hand and the receiver at the right hand are used as features [5].

2.2 Stream State Tying

For streams such as different handshapes, different hand positions and different hand orientations, the number of different patterns in each stream is usually not larger than 256. But the number of their combinations in the whole gesture space is very large. As C.Vogler and D.Metaxas[2] pointed that the number of possible combinations of different patterns in streams after enforcing linguistic constraints is approximately $5.5*10^8$. It seems impossible to obtain small number of spatial basic units in the whole gesture space. Therefore, the basic units should be found in each of streams. The basic idea is as the following, first, a whole spatial vector is formed using all stream vectors. For each hand gesture, its HMM is trained with training samples. After all HMMs have been trained, the observation probability densities in each data stream of all signs are tied with a few probability densities. The advantage of this approach is that the computation time is greatly reduced, as the state observation probability density in whole gesture space is the product of the state observation probability densities in six streams.

2.3 Movement Epenthesis Models

For all possible reasonable combination of two signs, the movement epenthisis models or sign transition models need to be trained, this leads to a large mount of sentence level training signs needed. In addition, movement epenthesis is not well defined in the sign language books, which leads to the modeling movement epenthesis even more difficult. To solve the movement epenthesis problem, we model the movement epenthesis using HMM. The parameters in HMM of movement epenthesis need to be estimated by sentence level training samples. Because the HMM parameters of signs in a sentence have been estimated by isolated sign training samples, the parameters to be estimated are those of HMMs of movement epenthesis. To estimate these parameters, the HMMs of signs in the sentence and HMMs for movement epenthesis are linked as a whole HMM. For example, a sentence consists of two signs such as u and v. The sentence level training samples are used to train the parameters in HMM of the movement epenthesis, during training the parameters of signs in the sentence are fixed, only parameters in the sign transition HMMs need to be estimated.

2.4 Fast Search Algorithm

The sign can be classified into one-handed class and two-handed class. In the case of one-handed, the left hand is motionless; the right hand conveys the information of a sign. To reduce the computation load, the right hand shape and position information is used firstly to prune unlikely words.

3 Lip-Reading

3.1 Face Location

The face detection and tracking module is based on template matching and skin color information [6][7]. Firstly skin-like regions is segmented, the input image is searched for pixels with skin-color and the largest connected region is considered as the region of the face [6]. A box containing the face is output to the module of lip localization.

3.2 Lip Location

For the IQ space, the color distributions of lips are estimated using the samples in lip areas. The log likelihood of the Gaussion distribution of lip color is used as a measure for labeling the possibility of a pixel belongs to the lip areas. Using this measure, the original image is transformed to an image in which area with color near to that of the lip is enhanced (See Fig.2 (b)). For different images, the lip colors are different. To get robust and consistent threshold for binarizing the image, the following approach is proposed. Firstly, a one-dimensional normal distribution is estimated by the log likelihood values of the image. Suppose that the mean value is μ and the standard deviation is σ. If the log likelihood value of a pixel is greater than $\mu+1.65\sigma$, the intensity value of the pixel is set to 255, otherwise set to 0. The threshold $\mu+1.65\sigma$ is got by experiments on thousands of images of different people under different lighting conditions. The threshold accounts for about 5% of pixels selected as candidates in lip regions. Secondly, after the candidates of lip pixels are selected using the above approach, to remove salt and pepper noise, morphological opening is applied to the resulting image (See Fig.2(c),(d)). The morphological operation removes the spurious pixels while still preserving the integrity of the lip area. Thirdly, the center of the lip area is estimated by the center of gravity of the resulting image. Then, a large 80×100 box containing the center is selected. A fast algorithm, which permits to detect the outline of the mouth, they are the top of the upper lip, the bottom of the lower lip and two corners.

The search algorithm for finding horizontal positions of lip corners is as the following: 1) Compute the sums of intensity values of pixels on each vertical line. The sum denoted as $Count_x[j]$,j is between j_1 and j_2; 2) Count on how many segments in the $Count_x[*]$, and calculate the start point and end point of each segment. If the values of consecutive elements in $Count_x[*]$ are greater than zero, the elements belong to the same segment. 3) Find the longest segment, the start point of the segment is set to the horizontal position of left lip corner, the end point of the segment is set to the horizontal position of the right lip corner.

The search algorithm for finding the top of the upper lip and the bottom of the lower lip is more complex due to the movements of the upper lip and the lower lip. The prior knowledge is that the contours of the two lips are convex and near to each other. The algorithm is as the following: 1)Compute the sums of intensity values of pixels on each horizontal line. The sum denoted as $Count_y[i]$,i is between i_1 and i_2 ;2)Join two segments into one segment if the distance between the end point of the preceding segment and the start point of the following segment is less than 3 pixels. The total number of segments is denoted as $Tseg$. The sum of $Count[*]$ in each segment is denoted as Sum[],the start position and the end position of a segment are denoted as $Start[*]$ and $End[*]$, respectively; 3)Find the segment that has the largest $Sum[*]$. The largest value is denoted as $Pmax$, the segment index is denoted as $Fmark$; 4)If the number of segments is greater than one, find the segment that has the second largest $Sum[*]$.The second largest value is denoted as $Smax$, the segment index is denoted as $Smark;$5)If $Tseg > 1$ and $Smark>Fmark$ and $Fmark>=1$, compute the $Ratio4=Sum[Fmark-1]/Pmax$;6)Compute the $Ratio1=Smax/Pmax$;7)Compute the

Ratio2=Length of the segment *Fmark*/distance of two lip corners;8)Compute the least distance between the pixels in segments *Fmark* and *Smark*. The distance is denoted as *Dis*;9)Compute the *Ratio3*=*Dis*/distance of two lip corners;10)If(*Ratio1*>0.15 and *Fmark*>*Smark* and ((*Ratio2*<0.44) or (*Ratio3*<0.15))and *Ratio3*<0.45 and *Dis*<15), then the top of the upper lip is set to *Start*[*Smark*],the bottom of the lower lip is set to *End*[*Fmark*] else if *Ratio1*>1 and *Fmark*>*Smark* ,then the top of the upper lip is set to *Start*[*Smark*],the bottom of the lower lip is set to *End*[*Smark*] ,else the top of the upper lip is set to *Start*[*Fmark*],the bottom of the lower lip is set to *End*[*Fmark*].

The located box is shown in the Fig.2.

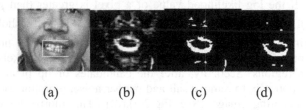

(a)   (b)   (c)   (d)

Fig. 2. (a)The original color face image and the lip region. (b) The enhanced face image using lip color information. (c) The binarized face image. (d) The image after morphological opening

3.3 Feature Extraction

Subsequent processing is restricted to the segmented lip area, which is a square and its side length equals to the distance of the lip corners. The center position of the square is estimated by the average of positions of two lip corners and by the average of positions of the top of the upper lip and the bottom of the lower lip. The original RGB image is converted into an intensity image. For illumination invariance, dividing the average intensity value of the segmented lip area, each pixel intensity value of is normalized. The segmented area is further segmented into 5×5 sub-areas. The average intensity value of each sub-area is used as feature. They consist of a vector with 25 dimensions.

3.4 Recognition Approach

Hidden Markov Models (HMMs) have been used successfully in continuous speech recognition, handwriting recognition, etc. An HMM is a doubly stochastic state machine that has a Markov distribution associated with the transitions across various state, and a probability density function that models the output for every state. A key assumption in stochastic gesture processing is that the lip motion signal is stationary over a short time interval. In the case of lip-reading, we used phrase HMMs for each phrase to be recognized. The models are trained used the Baum-Welch algorithm. Viterbi algorithm is used to compute the likelihood for each HMM of having generated the observed sequence. The model with the highest likelihood is chosen as the recognized phrase. The structure of HMM is left to right, the number of states of each phrase is 10.

4 Integration

The experiments for sign language recognition have shown that a sign (or a phrase) can be modeled with 3 or 5 states in HMM. A phrase can be modeled with the number of states relating to how many phonemes existing in the phrase in lip-reading. Therefore, different time scales exist in sign and lip movement. This leads to integration become even more difficult. The mathematical formalism of multistream model proposed by Bourlard is more suitable for this task [9][10]. For the two gesture and lip motion streams, the mathematical formalism of Parallel Multistream model, which is little bit modification version of multistream model, is proposed to integrate the two modalities. Assume that the'i input stream is X(i) and the model for a pattern P is composed of I parallel stream models P(i) that are forced to recombine their scores at the final anchor point. Each stream model P(i) is composed of sequential models P(i,j) (possibly with different topologies). The resulting statistical model is illustrated in Fig.3.

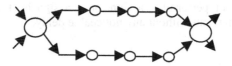

Fig. 3. Parallel multistream diagram

The topology structure of the parallel multistream model is a little bit different from that of the traditional multistream model. The basic idea of the parallel multistream model can be summarized as follows:(1) Extract appropriate feature vectors for each stream;(2) Normalize feature vectors of each stream;(3) Train independent recognizers for each stream. The fusion is taken at the final state of each sign model. The transition at sign boundary is determined by the fusion score at that boundary, including the language model. This enables the linguistics model to be used as early as possible. One characteristic of the search is that all streams are synchronous at sign boundary. It is ensured that the search is synchronous at sign level. It permits that the number of states in HMMs of different streams is different. As Viterbi decoding is taken in each stream, a lot of memory resources are needed. Since the decoding is taken synchronously for the two streams, the computation load is higher.

5 Experiments

5.1 Sign Language Recognition

We have chosen an isolated sign recognition task to demonstrate the advantage of the search approach. The number of states in HMM of each sign is between 1-7, which is determined by an adaptive approach. The HMM structure for each sign is left to right without skip. The hardware environment is Pentium III 450Hz.For the case of isolated sign recognition, 5177 signs in Chinese sign language were used as evaluation vocabularies, each sign was performed 5 times by a sign language teacher. 4 times were used for training and one for test. Using the approach of cross validation test, the test times for each word is 5. The number of states in HMM is 3. For different numbers of tree structured networks, the off-line recognition rates are listed in the Table 1.

Table 1. The recognition rates

128	256	512	1024
93.4	94.2%	94.8%	95.2%

The algorithm is about 10 times faster than a flat search one through all words in the dictionary when the number of the tree-structured networks is 512. The algorithm is about 10 times faster than a flat search one through all words in the dictionary when the number of the tree-structured networks is 512. For online test, because the property of Viterbi decoding is time synchronous and the recognition time is less than that of that needed for performing a sign, once a sign is stopped, the system will report the recognition result without any noticeable delay.

5.2 Lip-Reading

Experiments were performed using a database, which consists of color image sequences of ten phrases. Each phrase was spoken 5 times by a male speaker. Four was used for training, one for test. Each image sequence consists of twenty frames. The ten Chinese phrases (two words) not correctly recognized by sign language recognizer are /jueqi/, /shengchan/, /kuoda/, /laoban/, /zhuanshun/, /sikao/,/kuoda/, /kunnan/,/koudai/,/kaifa/. Among the ten phrases, only one word was not recognized correctly. If only lip shape such as lip width or height was used as feature, no phrase can be recognized correctly, this result shows that the lip shape has little discriminant power to lip-reading. In fact, the lips, teeth, tongue and lip shape all affect the recognition performance. The tracking result is shown in the Fig.5.

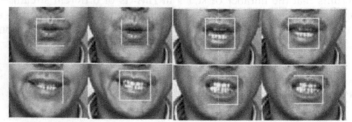

Fig.5. The tracking results of the Chinese phrase /jueqi/

5.3 Integration

For the ten Chinese phrases not correctly recognized by sign language recognizer, the parallel multistream model is used to integrate the gesture and the lip motion modalities. Off-line test was conducted on the ten phrases; only one word was not recognized correctly. This shows that lip-reading is really complementary to sign language recognition.

6 Conclusion

In this paper, we have discussed integration of sign language and lip motion using the parallel multistream model. Primary results suggest that the approach can indeed provide a new formalism for integration mulitimoal information. The preliminary work will be extended in the future.

References

1. Ho-Sub Yoon. Jung Soh. Byung-Woo Min. Hyun Seung Yang. Recognition of alphabetical hand gestures using hidden Markov model. IEICE Transactions on Fundamentals of Electronics Communications & Computer Sciences, vol.E82-A, no.7, July 1999, pp.1358-66. Publisher: Inst. Electron. Inf. & Commun. Eng, Japan.

2. Vogler C. Metaxas D. Parallel hidden Markov models for American Sign Language recognition. Proceedings of the Seventh IEEE International Conference on Computer Vision. IEEE Comput. Soc. Part vol.1, 1999, pp.116-22 vol.1. Los Alamitos, CA, USA

3. Luettin,J.,Thacker,N.A. and Beet,S.W.Speechreading using shape and intensity information,Proc.4th ICSLP Confference,Philadephia,PA,USA,1996.

4. Michael Vogt.Fast matching of a dynamic lip model to color video sequence under regular illumination conditions.In speechreading by human and machines, NATO ASI series F.,Vol 150,D.G.Stork, M.E.Hennecke (eds), pages199-407,1996.

5. Jiyong Ma,Wen Gao,Jiangqin Wu and Chunli Wang, A Continuous Chinese Sign Language recognition system, pp428-433, 28-31 March,FG'2000, Grenoble, France.

6. J.Matas.Color-based Object Recognition. Ph.D thesis, University of Surry,1995.

7. Karin Sobottka and Ioannis Pitas.Face Localization and facial feature extraction based on color shape and color information. ICIP,1996.

8. Uwe Meier,Rainer Stifelhagen,Jie Yang and Alex Waibel. Towards unrestricted lip reading.ICMI,1999.

9. Bourlard,H., Non-stationary multi-channel(multi-stream)processing towards robust and adaptive ASR,Proc.Tampere workshop on robust methods for speech recognition in aderse conditions,1995, pp.1-10.

10. Mirghafori,N.,A multi-band approach to automatic speech recognition, PhD dissertation, University of California at Berkely, Dec 1998.Reprinted as ICSI Technical report ,ICSI TR-99-04.

Multi-modal Navigation for Interactive Wheelchair[*]

Xueen Li   Tieniu Tan   Xiaojian Zhao

National Laboratory of Pattern Recognition (NLPR), Institute of Automation,

Chinese Academy of Sciences, Beijing, P.R. China

{lixe, tnt, xjzhao}@nlpr.ia.ac.cn

Abstract. In this paper ongoing work on an intelligent interactive wheelchair is presented. Also described is a new approach that integrates multimodal information for the navigational control of the wheelchair using a neuro-fuzzy network. Experiments show that the multimodal approach is capable of navigational control for the wheelchair, and of facilitating task and information sharing and trading between human and machine.

Keywords: intelligent wheelchair, neuro-fuzzy network,

multimodal integration, multimodal navigation

1. Introduction

Mobile robotics technology is starting to be massively applied in industrial applications. However, a large number of handicapped people are still waiting for the solutions that robotics can offer to their problems and enlarge their life space [1]. Thus, the research and development on intelligent wheelchair becomes a focus of the mobile robotics community. However, up to now, human-robot interfaces are one of the major bottlenecks for using intelligent wheelchair in real environments [2]. One possibility to overcome these drawbacks is using multiple modalities in the communication between humans and robots, which is a quite natural concept for human communications [3]. In this case, we distinguish different perceptive channels by which the system is connected with the environment.

In previous systems that enable natural human-computer interaction, the task of matching the interpretation of the visually observed scene and the verbal instruction

---

[*] This work is founded by research grants from the NSFC (Grant No. 59825105) and the 863 Program (Grant No. 863-512-98-20-3).

T. Tan, Y. Shi, and W. Gao (Eds.): ICMI 2000, LNCS 1948, pp. 590–598, 2000.

of the user is simplified. Recognized scenes are visualized on a screen and are directly accessible via a communication panel [4, 5] or some textual information about them is shown on a display and can directly be referred to by speech [6]. In the work presented here, the visually observed scene, the verbal instruction of the user, and ultrasonic sensors data are all integrated to a unified representation for navigational control of the interactive wheelchair. In addition, our system has to deal with uncertainties on all modalities, while previous systems are only concerned with uncertainty in one modality.

The outline of this paper is as follows: In Section 2, we will briefly introduce the intelligent wheelchair, a multimodal interactive wheelchair developed at the NLPR that has been a testbed for our research. In addition, some important aspects of ultrasonic, vision, and speech processing components in our system will be also described in this Section. Afterwards, we will propose a new approach that integrates those three modalities using neuro-fuzzy network (Section 3). Finally, we will give some experimental results of the implemented interactive wheelchair (Section 4) showing the robustness of our approach and a short conclusion (Section 5).

2. NLPR Interactive Wheelchair and Its Components

In order to make the multimodal integration and interactive scheme to be well understood, we will first briefly introduce the architecture of the NLPR wheelchair developed in our laboratory, and then we will describe its modality components of ultrasonic, vision and speech.

2.1 Architecture of the NLPR Interactive Wheelchair

The NLPR wheelchair is a multimodal interactive robotized wheelchair (cf. Fig. 1). It is driven by two rear wheels and controlled by a joystick or a portable computer according to the switcher's state. So far, the wheelchair is equipped with 5 ultrasonic sensors mounted at the left, front-left, front, front-right and right of the wheelchair. The wheelchair is also equipped with a color CCD camera mounted on the upper platform of the wheelchair. The camera could be controlled to zoom, tilt and pan by the command information passed from the portable computer's serial port (cf. Fig. 2). Moreover, the wheelchair has equipped some speech devices (microphone and speakers). Thus, there are multiple interaction modalities between the user and the robotized wheelchair, such as ultrasonic, vision, speech, etc.

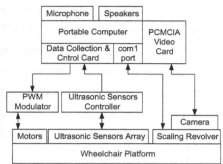

Fig. 1: NLPR interactive wheelchair   Fig. 2: Hardware diagram of the NLPR
interactive wheelchair

2.2 System Components

Integrated Multiple Ultrasonic Sensors Detection. In order to reduce the influence of noise, we filtered the ultrasonic readings with a digital low pass filter before the ultrasonic sensors' data fusion. To further simplify data fusion, we quantize the ultrasonic sensors' readings into four intervals: VN(Very Near), N(Near), F(Far) and VF(Very Far). Obstacle detection is mainly to detect the distribution of obstacles relative to the mobile robot, such as their orientation and distance. If we combine the results of left, front-left, front, front-right and right ultrasonic sensors' data processing to an obstacle distribution vector $[fd_1\ fd_2\ fd_3\ fd_4\ fd_5]$ according to the sequence of L-FL-F-FR-R, there will be a total of 1024 types of obstacle distribution patterns ($[pd_1^i\ pd_2^i\ pd_3^i\ pd_4^i\ pd_5^i]$, i=1,2,$\cdots$,1024). We just choose 16 typical types of obstacle distribution pattern by optimal training. Then we calculate the Euclidian distances between the detected obstacle distribution vector $[fd_1\ fd_2\ fd_3\ fd_4\ fd_5]$ and the typical obstacle distribution pattern $[pd_1^i\ pd_2^i\ pd_3^i\ pd_4^i\ pd_5^i]$, and take it as the degrees of similarity [7].

Vision Processing. The vision process components is based on a hybrid approach that integrates structural and holistic knowledge about the real world in semantic network formalism.

Holistic component: A current color camera image is used to match with index images that correspond to various viewpoints in the specified environment (c.f. Fig.3). The matching process is similar to content based image retrieval [8]. Then the holistic information of the scene around the wheelchair is acquired. In order to reduce the computing time, the index image database is a classified hierarchical directory structure, which has multiple resolutions. In the process of retrieval, we chose a group of index images according to previous scenes and the necessary resolution.

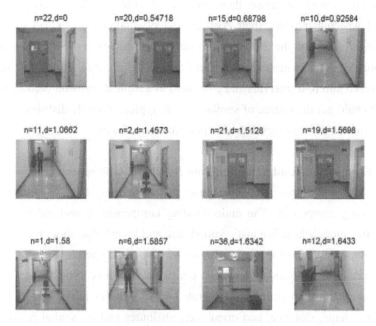

Figure 3: The result of image retrieval (n: the nth image in the image database; d: the distance between the index image (n=22) and n^{th} image)

Region component: A color camera image is segmented into homogeneously colored regions which are characterized by some shape features. Then the color and contour information of the segmented regions is acquired.

The visual abstraction hierarchy of a scene hypothesis consists of several results generated by the above vision components. Some components directly hypothesize scene classes or subclasses, and others generate results that are related to a scene class, such as the color of a region or rectangle. In our system, we distinguish the following visual evidences:

HOL_SCENE (holistic classification results): scene classes or subclasses generated by the holistic component using the method.

REGION_COLOR: color information associated with a segmented region.

CONTOUR_CLASS (contours found in a scene region): ellipses, parallel lines, rectangular closed contours, and so on.

All types of evidences are modeled as nodes in the Bayesian network. The conditional probabilities are estimated using a labeled test set of 180 scenes' images and the recognition results on this test set.

Random component: Because there may exist some random obstacles and the robotized wheelchair needs their position and orientation, we extract the edges of obstacles and calculate the distance of the closest obstacles in the left, front and right sub-regions of the area around the wheelchair [9]. Afterwards, the processing results are quantized into four intervals and combined to a triple dimension vector $[d_2, d_3, d_4]$. Then we could get the degree of similarity with typical obstacle distribution patterns, which is similar to the processing of ultrasonic sensors' data.

Speech Recognition, Understanding and Synthesis. The speech component of our system consists of three subcomponents: the recognizer, the synthesizer and the understanding component. The understanding component is realized in a semantic network that models both linguistic knowledge and knowledge about the construction task. An important aspect in the interaction of vision and speech is extracting scene's descriptions from spoken utterances. On a higher level in the understanding component, feature structures are generated specifying an intended scene by type, color, size, shape, distance, and orientation attributes and by spatial relations using reference scenes. Verbal descriptions of an intended scene class consist of some specified attributes that are partly defined on word level, such as "blue," "wide," or "door," and partly defined on simple grammatical attachments, such as "The front-left blue door." Currently, we distinguish six types of features mentioned in order to specify an intended scene class: type, color, size, shape, distance, and orientation information. All feature types are interpreted as random variables and are modeled as nodes in the neuro-fuzzy network. Conditional probabilities connecting these features with an scene class are estimated from experimented results.

3. Neuro-fuzzy Network Based Multimodal Integration

3.1 Multimodal Integration Scheme

The interactive wheelchair has multiple modalities to naturally communicate with human. Thus, the human speaker could verbally specify the navigation tasks or path by describing properties of them without knowing the exact terms. Then the system interprets mostly vague descriptions by multiple modalities.

In this paper, we integrate multimodal processing on several levels of abstraction. An approach to reduce uncertainty is needed for a consistent interpretation of the input data that takes into account the several layers and the dependencies between them. Therefore, we use neuro-fuzzy networks as a decision tool that can handle uncertainties, erroneous data, and vague meanings in a very comfortable way.

3.2 Neuro-fuzzy Network Based Multimodal Navigation

The neuro-fuzzy network is shown in Figure 4. From the view of function, it is equivalent to combination of a Bayesian network and a fuzzy inference system in series.

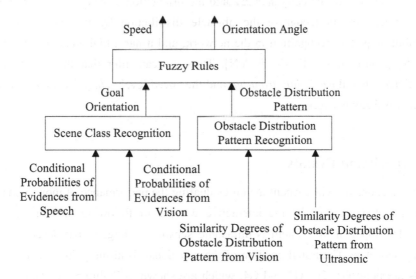

Figure 4: The neuro-fuzzy network

The inputs of the neuro-fuzzy network consist of fuzzified similarity degrees of obstacle distribution pattern generated by ultrasonic and vision components, and the conditional probabilities of all types of evidence generated by vision and speech components. In order to reduce the computing load, we decouple the scene class recognition and obstacle distribution pattern recognition. Their outputs are all used as the inputs of last layer of the neuro-fuzzy network and inferred to output the orientation angle and speed of wheelchair at next time according to the predefined fuzzy rules. The format of fuzzy rule is shown in the following:

If goal is on the front-left of the wheelchair, and the obstacle distribution pattern is [N, F, VF, F, N], then the moving orientation of wheelchair is front-left and the speed level is high.

Thus the network could integrate ultrasonic, vision and speech components and realize multimodal navigational control for the interactive wheelchair. For example, if a scene is recognized (see Fig. 3), the HOL_SCEN_CLASS is instantiated with the entry *scene22* (This is the label of index image, which is corresponding to one scene.) and the REGION_COLOR with the entry *blue*. At same time if the verbal description of goal is *"Enter the front-left blue door,"* the entry *blue, door,* and *front-left* are separately instantiated in the COLOR, TYPE and ORIENTATION evidence. Then the layer of scene class recognition in the network infers that the recognized scene has a high probability for the entry *scene22* and the orientation of goal is *front-left*. On the other side, we can recognize the obstacle distribution by the part of obstacle distribution pattern recognition in the network, and it has a high degree of similarity for the pattern of [N, F, VF, F, VN]. Thus we can infer that the orientation of wheelchair should be *slightly front-left* and the speed level is *high* by the last layer of the neuro-fuzzy network.

4. Experiment Results

In this section, we will concentrate on evaluating the performance of monomodal and multimodal navigation for the interactive wheelchair following different planning path. The input of the system is ultrasonic sensors reading, verbal description and image sequence. The initial position of the wheelchair is at the point S, and the four goals are point G1, G2, G3 and G4, which are shown in Figure 5. At the point O1, there exists a random obstacle (an ordinary chair).

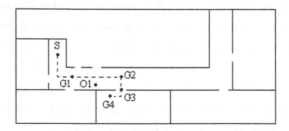

Figure 5: Partial sketch map of our laboratory

We do navigation experiments following four planning paths under the following conditions: (a) vision modality; (b) ultrasonic modality; (c) integration of vision and ultrasonic modalities; (d) integration of vision, ultrasonic, and speech modalities. In addition, same experiment is done 10 times for each case. Then we summarize all the experimental results and list them in Table 1.

Table 1: The rates of arriving goals successfully for the wheelchair under different navigation modality

Navigation Modality	Successful Rate (%)			
	Goal 1 (corridor A)	Goal 2 (corridor B)	Goal 3 (door)	Goal 4 (table)
Vision	100	70	60	50
Ultrasonic	100	90	10	0
Vision + Ultrasonic	100	100	80	80
Vision + Ultrasonic + Speech	100	100	90	90

From the table we can see that ultrasonic modality much adapt to obstacle avoidance, while vision modality is good at localization and goal seeking. The integration of vision and ultrasonic modalities much increases the successful rate of goal-seeking because it can reduce the influence of noise in ultrasonic sensor readings and the uncertainty of vision. The necessary verbal description is helpful to reduce the uncertainties brought by erroneous, vague, or incomplete data of vision and ultrasonic. Thus speech modality greatly improves navigational performance of the wheelchair in a clutter environment and make the wheelchair has the ability of natural interaction with human beings.

5. Conclusion

In this paper we presented a new approach to integrating ultrasonic, vision and speech modalities and it realized multimodal navigational control for the interactive wheelchair. As the wheelchair works in real environments, we often have to deal with erroneous, vague, or incomplete data. In order to find the most plausible interpretation of the scene and avoid random obstacles, the processing results of the vision, ultrasonic and speech components are connected and recognized by neuro-fuzzy networks. Then the neuro-fuzzy network infers to output the orientation and speed of the wheelchair according to the information of goal and obstacles related to the current location of the wheelchair.

In order to evaluate the effectiveness of our approach, we have conducted a series of navigation experiments under the conditions of monomodal and multimodal navigation and following four different planning paths. Experiment results show that the multimodal approach is capable of navigation for the robotized wheelchair, and the interactive wheelchair maybe adapted to people with severe physical handicaps.

References

1. F.Matia, R.Sanz and E.A.Puente, "Increasing Intelligence in Autonomous Wheelchair", Journal of Intelligent and Robotic Systems, Vol.22, pages 211-232, 1998.
2 Sven Wachsmuth, Gudrun Socher, Hans Brandt-Pook, et. al. Integration of Vision and Speech Understanding Using Bayesian Networks. Videre: Journal of Computer Vision Research, Volume 1, Number 4, pages 61–83, The MIT Press, 2000.
3 Hisato Kobayashi. Editorial: Special Issue on Human–Robot Interaction. Robotics and Autonomous Systems Volume 31 pages 129–130, May, 2000.
4. R. K. Srihari and D. T. Burhans. Visual Semantics: Extracting Visual Information from Text Accompanying Pictures. In Proceedings of AAAI-94, pages 793–798, Seattle, WA, 1994.
5. J. K. Tsotsos et al. The PLAYBOT Project. In J. Aronis (Ed.), IJCAI '95 Workshop on AI Applications for Disabled People, Montreal, 1995.
6. K. Nagao and Jun Rekimoto. Ubiquitous Talker: Spoken Language Interaction with Real World Scenes. In International Joint Conference on Artificial Intelligence, pages 1284–1290, 1995.
7. X.E. Li, X.J.Zhao and T.N. Tan. Ultrasonic Sensor Fusion Based Navigation for an Intelligent Wheelchair. SIRS'2000, UK, 2000.
8. X.J.Zhao, X.E. Li, and T.N. Tan. A Novel Landmark Tree Based Self-Localisation Method for An Intelligent Wheelchair. SIRS'2000, UK, 2000.
9. X.J.Zhao, X.E. Li, and T.N. Tan. A Novel Triangle Based Active Obstacle Avoidance Strategy for An Intelligent Wheelchair. submitted to ICARCV'2000, Singapore, 2000.

A Fast Sign Word Recognition Method for Chinese Sign Language[1]

Jiangqin WU[1] Wen GAO[1,2]

[1](Computer Science Department, Harbin Institute of Technology,150001)
[2](Institute of Computing Technology, Chinese Academy of Sciences, Beijing 100080)
wjq@vilab.hit.edu.cn wgao@ict.ac.cn

Abstract. Sign language is the language used by the deaf, which is a comparatively steadier expressive system composed of signs corresponding to postures and motions assisted by facial expression. The objective of sign language recognition research is to "see" the language of deaf. The integration of sign language recognition and sign language synthesis jointly comprise a "human-computer sign language interpreter", which facilitates the interaction between deaf and their surroundings. Considering the speed and performance of the recognition system, Cyberglove is selected as gesture input device in our sign language recognition system, Semi-Continuous Dynamic Gaussian Mixture Model (SCDGMM) is used as recognition technique, and a search scheme based on relative entropy is proposed and is applied to SCDGMM-based sign word recognition. Comparing with SCDGMM recognizer without searching scheme, the recognition time of SCDGMM recognizer with searching scheme reduces almost 15 times.

1   Introduction

Sign language is the language used by the deaf, which is a comparatively steadier expressive system composed of signs corresponding to postures and motions assisted by facial expression. And it is the language communicated by motion/vision.

The objective of sign language recognition research is to "see" the language of deaf. The integration of sign language recognition and sign language synthesis jointly comprises a "human-computer sign language interpreter", which facilitates the interaction between deaf and their surroundings.

Chinese Sign Language (CSL) uses approximately 5,500 gestures for common words and finger-spelling for communicating obscure words[1]. This paper aims at isolated sign word recognition.

Hidden Markov Model(HMM)[2] with standard topology, which has been widely used in the domain of dynamic gesture recognition, has strong capability to model space-time variation of sign language signal.Liang[3]'s Taiwanese Sign Language recognition system, Starner and Pentland[4]'s American Sign Language recognition

---

[1] The paper is partly sponsored by national 863 project (contact number: 863-306-ZT03-01-2)

T. Tan, Y. Shi, and W. Gao (Eds.): ICMI 2000, LNCS 1948, pp. 599-606, 2000.

system, Vogler and Metaxas[5]'s American Sign Language recognition system and Kirsti Grobel and Marcell Assan[6]'s Sign Language of Netherlands recognition system use HMM as recognition technique.

However, just the generalization of HMM topology leads to the difficulty to model sign language signal and the training and recognizing's computation complexity. Especially, for continuous HMM, the speed of training and recognition is comparatively slow, because of the need of computing large amount of emission probability and estimating too many model parameters. Aiming at the above problems, in this paper SCDGMM[7] is used to model sign language signal, and a search scheme based on relative entropy is proposed and is applied to SCDGMM-based sign word recognition. Comparing with SCDGMM recognizer without searching scheme, the recognition time of SCDGMM recognizer with searching scheme reduces almost 15 times.

SCDGMM is given in section two. SCDGMM-based sign word recognition is described in section three. In section four, the search scheme based on relative entropy is proposed and is applied to SCDGMM-based sign word recognition. The conclusion is drawn in the last section.

2   SCDGMM

In SCDGMM, sign language signal $X(t)(t=1,\cdots,T)$, the frame at moment t is modeled by a time-varing mixture probability density function with M components, i.e.

$$P(X(t))=\sum_{j=1}^{M}\pi_j p_j(t)q_j(X(t))\tag{1}$$

where t is time variable, π_j is Mixture Proportions, $\pi_j \geq 0, \sum_{j=1}^{M}\pi_j=1$, $q_j(X(t))$ is N-Gaussian mixture density, defined as

$$q_j(X(t))=\sum_{n=1}^{N}c_{jn}N_{\infty}(X(t),\mu_n,U_n)\tag{2}$$

c_{jn} is mixture coefficients, N_{∞} is Gaussian density with mean vector μ_n and covariance matrix U_n for the nth mixture component, i.e.

$$N_{\infty}(X(t),\mu_n,U_n)=\frac{1}{(2\pi)^{\frac{p}{2}}\sqrt{|U_n|}}exp\left(-\frac{1}{2}(X(t)-\mu_n)^T U_n^{-1}(X(t)-\mu_n)\right)\tag{3}$$

where $X(t)=(X_1(t),...,X_p(t))$ is p dimension feature vector.

$p_j(t)$ is a probability density of time variable, and is selected as Gaussian distribution, i.e.

$$p_j(t) = \frac{1}{(2\pi)^{\frac{1}{2}}\sigma_j} exp\left(-\frac{(t-\tau_j)^2}{2\sigma_j^2}\right) \tag{4}$$

here, τ_j, σ_j is mean and deviation respectively.

For model $\lambda = (\pi, c, U, \mu, \tau, \sigma)$, two basic problems must be solved, they are:

2.1 Evaluation

Assuming statistical independence of observations, the probability of the observation sequence $X = X(1)X(2)\cdots X(T)$, given model parameter λ, is

$$P(X/\lambda) = \prod_{t=1}^{T} P(X(t)/\lambda) = \prod_{t=1}^{T}\sum_{j=1}^{M}\pi_j p_j(t)\sum_{n=1}^{N} c_{jn} N_\infty(X(t),\mu_n,U_n) \tag{5}$$

where $U_n = diag(u_{n1}^2, u_{n2}^2, \cdots, u_{np}^2)$, $N_\infty(X(t),\mu_n,U_n)$ is approximated by one-order one-dimensional equivalent probability density, i.e.

$$N_\infty(X(t),\mu_n,U_n) = V_n(X(t))(u_{n1}u_{n2}...u_{np})^{-p^{-1}}, V_n(X(t)) = (1+d_h(X(t))p^{-1})^{-1} \tag{6}$$

$$d_h(X(t)) = \frac{1}{2}\|Y(t)\|^2, \quad Y(t) = \left(\frac{X_1(t)-\mu_{n1}}{u_{n1}},...,\frac{X_p(t)-\mu_{np}}{u_{np}}\right)$$

$p_j(t)$ is approximated by one-order one-dimensional equivalent probability density, i.e.

$$p_j(t) = w_j(t)\sigma_j^{-1}, \quad w_j(t) = \left[1+\frac{(t-\tau_j)^2}{2\sigma_j^2}\right]^{-1} \tag{7}$$

2.2 Model Parameters Estimation

For the given K groups of training data $x_k(t)$, $t=1,...,t(k)$, where $t(k)$ is the total frame number of the kth group training data, the reestimation formulas for $\lambda = (\pi, c, U, \mu, \tau, \sigma)$ are

$$\bar{\pi}_j = \gamma_j / \sum_{j=1}^{M}\gamma_j \tag{8}$$

$$\bar{c}_{jn} = \gamma_{jn}/\gamma_j \tag{9}$$

$$\bar{\tau}_j = \sum_{k=1}^{K} \sum_{t=1}^{t(k)} \gamma_j(t,k) w_j(t)t / \sum_{k=1}^{K} \sum_{t=1}^{t(k)} \gamma_j(t,k) w_j(t) \qquad (10)$$

$$\bar{\sigma}_j^2 = \sum_{k=1}^{K} \sum_{t=1}^{t(k)} \gamma_j(t,k) w_j(t)(t-\tau_j)^2 / \sum_{k=1}^{K} \sum_{t=1}^{t(k)} \gamma_j(t,k) \qquad (11)$$

$$\bar{\mu}_n = \sum_{j=1}^{M} \sum_{k=1}^{K} \sum_{t=1}^{t(k)} \gamma_{jn}(t,k) v_n(x_k(t)) x_k(t) (\sum_{j=1}^{M} \sum_{k=1}^{K} \sum_{t=1}^{t(k)} \gamma_{jn}(t,k) v_n(x_k(t)))^{-1} \qquad (12)$$

$$\bar{u}_{nl} = \sum_{j=1}^{M} \sum_{k=1}^{K} \sum_{t=1}^{t(k)} \gamma_{jn}(t,k) v_n(x_k(t))(x_{kl}(t)-\mu_{nl})^2 p^{-1} (\sum_{j=1}^{M} \sum_{k=1}^{K} \sum_{t=1}^{t(k)} \gamma_{jn}(t,k))^{-1} \qquad (13)$$

here,

$$\gamma_j(t,k) = \pi_j p_j(t) q_j(x_k(t)) / P(x_k(t)) \qquad (14a)$$

$$\gamma_j = \sum_{k=1}^{K} \sum_{t=1}^{t(k)} \gamma_j(t,k) \qquad (14b)$$

$$\gamma_{jn}(t,k) = \pi_j c_{jn} p_j(t) N (x_k(t), \mu_n, U_n) / P(x_k(t)) \qquad (14c)$$

$$\gamma_{jn} = \sum_{k=1}^{K} \sum_{t=1}^{t(k)} \gamma_{jn}(t,k) \qquad (14d)$$

3　SCDGMM-Based Sign Word Recognition

3.1　The Use of SCDGMM for Sign Word Recognition

Two CyberGlove with 18 sensors are used as gesture input device with 38400 baud rate of sampling, which can feature two bend sensors on each finger, four abduction sensors, plus sensors measuring thumb crossover, palm arch, wrist flexion and wrist abduction.

Assume we have a vocabulary of V sign words to be recognized. We have a training set of L tokens of each word and an independent testing set. To do sign word recognition, we perform the following steps:

1. First an SCDGMM for each sign word in vocabulary is built by equations (8)~(13), giving model $\lambda_v (1 \leq v \leq V)$.

2. For each unknown word in the test set, characterized by observation sequence $O=O_1\cdots O_T$, and for each word model $\lambda_v\,(1\le v\le V)$, calculate $P(O/\lambda_v)$ by equation (5).

3. The word whose model probability is highest is chosen, i.e.

$$v^{*}=arg\max_{1\le v\le V}\ \frac{1}{T}\sum_{t=1}^{T}logP(O_t/\lambda_v) \tag{15}$$

3.2 Experiment Result

Each sign word in a 274-sign vocabulary is gesticulated 10 times by the signer wearing datagloves, 8 of which are training samples, the remaining of which are testing samples. Each sign word is modeled and recognized by SCDGMM. By adjusting time mixture M and Gaussian mixture N of the model, different experimental results are obtained (as Fig. 1 is shown).

Considering the speed and performance of recognition, $M=2$, $N=1$ is selected, accuracy of 98.2% is reached, but the average recognition time per sign word is 0.61 seconds..

All the experiment is done on Pentium II 300 PC(64M RAM).

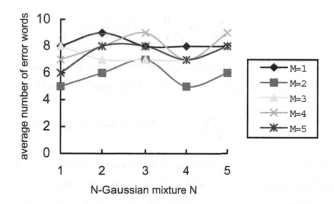

Fig. 1. The relation between error words number and model parameters M,N

By analysis of the experiment result shown in Fig.1, it is found that the result of some codebook for the same sample is very approximated, i.e., the distance between them in a given measure space is very close. The measure space is just relative entropy. Thus, a search scheme based on relative entropy is proposed and is applied to sign word recognition system based on SCDGMM to increase recognition speed.

4   Search Scheme Based on Relative Entropy

4.1   Relative Entropy

The relative entropy of two class probability[8] is the discrimination measurement for the two class. The larger the relative entropy, the better the two class discriminate. Assume that $\Omega_1, \Omega_2, ..., \Omega_N$ are N different class, the set composed of the i-th class training samples is denoted as X_i and the prior probability of each class is equal, the relative entropy of the i-th class relative to the j-th class（also referred to as information divergency）is defined as

$$H_{ij} = \sum_{x \in X_i} P(\Omega_i|x) log \frac{P(\Omega_i|x)}{P(\Omega_j|x)} \tag{16}$$

where

$$P(\Omega_i|x) = P(x|\Omega_i)P(\Omega_i)/P(x) \quad P(x) = \sum_{l=1}^{N} P(x|\Omega_l)P(\Omega_l)$$

thus

$$H_{ij} = \sum_{x \in X_i} \frac{P(x|\Omega_i)P(\Omega_i)}{P(x)} log \frac{P(x|\Omega_i)}{P(x|\Omega_j)} \tag{17}$$

The above equations for relative entropy are simplified and introduced into DGMM, and then the relative distance $D(\lambda_i, \lambda_j)$ from model λ_i to model λ_j is defined as

$$D(\lambda_i, \lambda_j) = \frac{1}{K} \sum_{O_k \in O} \frac{1}{T_k} [logP(O_k|\lambda_i) - logP(O_k|\lambda_j)] \tag{18}$$

where $O = \{O_1, O_2, ..., O_K\}$ is training sample set for model λ_i. The relative distance $D(\lambda_j, \lambda_i)$ from model λ_j to model λ_i is defined in the same way.

Considering the symmetry of distance, the distance between model λ_i and model λ_j can be defined as

$$D_s(\lambda_i, \lambda_j) = \frac{D(\lambda_i, \lambda_j) + D(\lambda_j, \lambda_i)}{2} \tag{19}$$

4.2   Search Scheme

In order to reduce the model matching time and increase recognition speed, a search scheme based on relative entropy is proposed. Decision tree is hierarchical structure. Comparing with linear searching algorithm, the searching complexity of a-tree is

$log_a N$, which makes large vocabulary sign language recognition in real time become possible.

Distance matrix D is firstly defined as following

$$D \overset{def}{=} \begin{bmatrix} D_{11} & D_{12} & ... & D_{1N} \\ D_{21} & D_{22} & ... & D_{2N} \\ ... & ... & ... & ... \\ D_{N1} & D_{N2} & ... & D_{NN} \end{bmatrix} \tag{20}$$

where $D_{ij}=D(\lambda_i,\lambda_j)$. Let $D_{ii}=0$, N is the number of sign words in a vocabulary.

According to the following process, a -tree based on D is then constructed:
1. Initialize the root by sign word vocabulary and label it as an active node;
2. An active node is selected and if the node can't be classified , it is labeled as a dead node, go to step 4;
3. According to the nearest-neighbor rule, the current active node is classified into a subset and all are labeled as active nodes. The current node is labeled as dead node, return step 2.
4. If all the nodes are dead nodes, end; otherwise, return step 2.

The leaf node of the a -tree corresponds to each sign word in the vocabulary, while each non-deaf node corresponds to a sign words set, which is called search node.

Finally, using training samples of all sign words in each search node except root a DGMM codebook corresponding to the search node is trained and is stored in search node codebook base. The recognizing module with search scheme is shown in Fig.2, where SNC stands for search node codebook base, SWC stands for sign word codebook base.

The recognizing speed not only relates with decision tree, but also is concerned with the structure of the tree. What kind of structure is optimal?

Theorem 1 For a -decision tree with N ($N>>a$, $a\geq 2$) leaf nodes, the computation amount traversing the entire tree is minimal when $a=3$. Here assume that the computation amount for each node is equal.

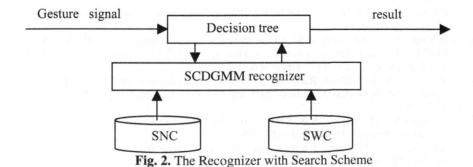

Fig. 2. The Recognizer with Search Scheme

4.3   Experiment Result

According to the above theorem, the average recognizing speed is maximal when $a=3$. In the experiment, the model parameters are selected: searching node codebook $M=2$, $N=1$, leaf node codebook(sign word codebook) $M=2$, $N=1$, the other testing environment is the same as in section 3. The result is shown in table 1.From the experiment result it is shown that the recognition rate with search scheme is equal to that without search scheme, but the recognizing speed has improved almost 15 times.

Table 1. Comparison of Experiment Result

Whether search scheme is adopted	y	n
Average recognizing time（s）	0.04	0.61
Recognition rate	97.4%	98.2%

5   Conclusion

To implement real-time sign word recognition, SCDGMM is used to model sign word, and the idea of relative entropy is introduced into search scheme and is applied to SCDGMM-based sign word recognition system. From the experiment result, it is shown that the fast sign word recognition method for CSL not only ensures the high recognition performance but greatly shortens the recognition time.

References

1.   China deaf association, Chinese Sign Language, Huaxia publishing company, Beijing,1991:i -xi.
2.   L.R.Rabiner and B.H.Juang, An Introduction to Hidden Markov Models, IEEE ASSP Mag., 1986;3(1):4-16.
3.   R. Liang & M. Ouhyoung, A Sign Language Recognition System Using Hidden Markov Model and Context Sensitive Search, in: Proc. of the ACM Symposium on VR software and Technology, Hongkong, 1996: 59-66.
4.   T. Starner, & A. Pentland, Real-time American Sign Language Recognition From Video Using   Hidden Markov Models, in: MIT Media Lab Perceptual Computing Section, TR-375, 1996.
5.   C.Vogler and D.Metaxas, ASL Recognition Based On a Coupling Between HMMs and 3D Motion Analysis, in: ICCV,Bombay,1998.
6.   K.Grobel and M.Assan. Isolated sign language recognition using hidden markov models.SMC'97:162-167.
7.   Ma Jiyong. Research on speaker recognition algorithms. Harbin Institute of Technology's dissertation for the Doctor Degree. 1999: 38-41
8.   T.Cover, J.Thomas. Elements of Information Theory. John Wiley & Sons, Inc. New York. 1991: 90-95

Automatic Keystone Correction for Camera-Assisted Presentation Interfaces

Rahul Sukthankar[1,2], Robert G. Stockton[1], and Matthew D. Mullin[1,*]

[1] Just Research, 4616 Henry Street, Pittsburgh, PA 15213, U.S.A.
{rahuls,rgs,mdm}@justresearch.com
[2] The Robotics Institute, Carnegie Mellon University, Pittsburgh, PA 15213, U.S.A.
rahuls@cs.cmu.edu
http://www.cs.cmu.edu/~rahuls

Abstract. Projection systems have become the ubiquitous infrastructure for presentation technology. However, unless the projector is precisely aligned to the presentation screen, the resulting image suffers from perspective (keystone) distortions requiring manual optical or digital correction. This tedious process must be repeated whenever the projector or screen is moved and is increasingly relevant given the emerging trend towards highly-portable LCD projection systems. This paper presents a presentation interface that pre-warps the image to be projected in such a way that the distortions induced by the projector-screen geometry precisely negate the warping. An uncalibrated, low resolution digital camera is used to infer the projector-screen geometry and to automatically determine the pre-warping parameters. This vision-based system is augmented with a natural interface that enables the user to interactively refine the suggested rectification. Arbitrary distortions due to projector placement are negated, allowing the projector (and camera) to be placed *anywhere* in the presentation room — for instance, at the side rather than the center of the room. Our solution works with existing projector hardware, and could easily be incorporated into the next generation of LCD projector systems.

1 Introduction

Projection systems are the ubiquitous infrastructure for presentation technology, and portable LCD projectors have become increasingly popular for multimedia presentations. However, unless the projector is carefully aligned to the projection surface (screen), the resulting image on the screen appears distorted, or *keystoned*[1]. Keystoning is undesirable, not only because viewers find this warping to be distracting, but also because the distortion is detrimental to the interpretation of visual information such as graphs, bar-charts and technical drawings.

---

[*] Rahul Sukthankar is now with Carnegie Mellon University and Compaq Cambridge Research Lab. Robert Stockton and Matthew Mullin are now with WhizBang! Labs.
[1] In related work, "keystoning" refers specifically to a symmetric, trapezoidal distortion caused by projector pitch misalignment. Here, the term refers to the broader class of distortions caused by *any* misalignments in projector position or orientation.

T. Tan, Y. Shi, and W. Gao (Eds.): ICMI 2000, LNCS 1948, pp. 607–614, 2000.
© Springer-Verlag Berlin Heidelberg 2000

Keystoning can be prevented by aligning the projection system's optical axis so that it is perpendicular to the screen, and ensuring that the image is not rotated with respect to the screen. For fixed projectors that can be mounted from the ceiling and carefully aligned once, these constraints are surmountable; however, portable projectors require alignment at the start of each presentation session. This manual process tedious; it can also be impractical to align a portable projector in a manner that eliminates all keystoning effects since optimal alignment may place the projector in an awkward position (such as in the middle of the audience). This motivates the need for a better presentation interface: one that allows arbitrary placement of the projector.

Sophisticated projectors now offer *manual* keystone correction that allow the user to counter the limited class of distortions caused by *vertical* misalignment. Unfortunately, this form of keystone correction only addresses projector pitch[2] and cannot compensate for distortions due to projector roll or yaw. More importantly, requiring the user to directly adjust the projector optics is undesirable from a user-interface standpoint: precise correction requires tweaking several coupled parameters, whose effect on the projected image is non-intuitive.

This paper presents a fully-automatic method for keystone correction. The two key concepts are: (1) a digital camera is used to observe the projected image; (2) the image to be displayed is pre-warped so that the distortions induced by the projection system will exactly undo the keystone distortion. The result is that an arbitrarily mounted projector (in an unknown orientation) displays a perfectly aligned and rectilinear image (see Figure 1).

2   Vision-Based Keystone Correction

The hardware requirements are modest: a standard computer, a low-resolution digital camera (e.g., Logitech QuickCam), and a standard LCD projector. The projector can be mounted anywhere, as long as the image falls entirely within the projection screen area (e.g., the projector could be placed near the edge of the room, even at an angle). The camera must be mounted such that the projection screen is within its field of view (but the camera can be placed anywhere in the room, and need not be level).

Our method for keystone correction is summarized as follows. (1) Determine the mapping between points in the computer display and the corresponding points in the camera image. (2) Identify the quadrilateral corresponding to the boundaries of the projection screen in the camera image. From this, compute a possible mapping between the projection screen and the camera image frame. (3) Infer a possible mapping from the computer display (source image frame) to the projection screen based upon the mappings computed in the previous two steps. (4) Determine an optimal placement for the corrected image on the projection screen. This is the largest rectangle that is completely contained within the projection of the computer display (i.e., the keystoned quadrilateral in Figure 1).

---

[2] Furthermore, current commercial keystone-correction systems are restricted to rectifying small misalignments in pitch, typically no more than $\pm 12°$.

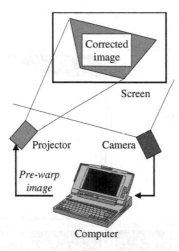

Fig. 1. An illustration of the apparatus. The computer is connected to a projector, and the projected image is observed by the camera. The positions, orientations and optical parameters of camera and projector are unknown. Due to projector misalignment, the rectangular screen appears as a distorted quadrilateral (shown shaded). However, by properly pre-warping the source image, the projected image appears rectilinear (shown by the white rectangle enclosed by the keystoned quadrilateral). The pre-warping parameters are automatically determined by the projector-camera system calibration.

(5) Pre-warp each application image to correct for keystoning. These steps are described below in greater detail, and results are presented in Figure 2.

2.1 Projector-Camera System Calibration

The goal of this step is to determine a mapping between points in the source image and the corresponding points in the camera image (see Figure 3). Surprisingly, although the positions, orientations and optical parameters of the camera and projector are unknown, this mapping (referred to as T in the remainder of this paper) can still be inferred.

First, we note that the mappings from the source image frame to the projected image frame, and from the projected image frame to the camera image frame are each perspective transforms. When these two transforms are composed (i.e., the projection of the source image is viewed through the camera), the resulting mapping, while not necessarily a perspective transform, can be expressed as a projective transform:

$$(x, y) = \left(\frac{p_1 X + p_2 Y + p_3}{p_7 X + p_8 Y + p_9}, \frac{p_4 X + p_5 Y + p_6}{p_7 X + p_8 Y + p_9} \right),$$

where (x, y) is a point in the source image frame, (X, Y) is the corresponding point in the camera image frame and the parameters $p_1 \ldots p_9$ are the unknowns

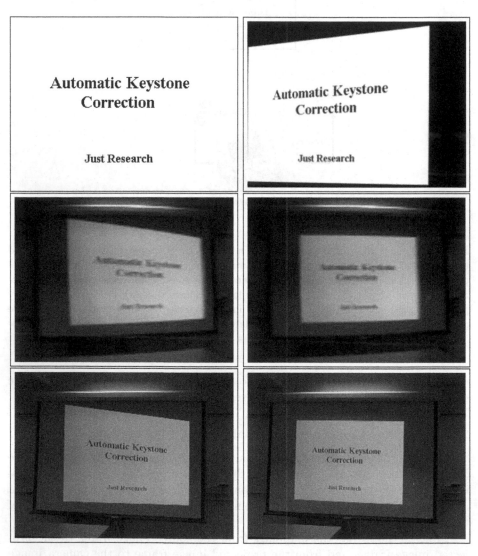

Fig. 2. A rectangular source image (top left) appears distorted when projected from an off-center projector (middle & bottom left). Using our method, the source image is pre-warped (top right). The resulting projected image is rectilinear and perfectly aligned to the projection screen (middle & bottom right). The middle row of images was captured with the low-quality camera used in our apparatus, and the bottom row was captured with a high-quality camera to illustrate how the presentation would appear to an audience member in the room.

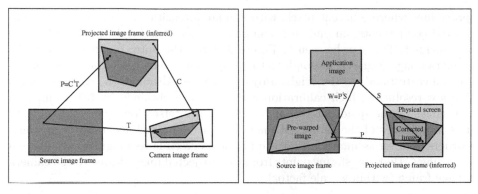

Fig. 3. Left: the relationships between the three frames of reference corresponding to the computer display (source image frame), camera (camera image frame) and physical screen (projected image frame). Note that the last can only be indirectly observed by our system, through the camera. T is obtained using the calibration method; C is obtained by locating the screen within the camera image. Finally, P, the mapping responsible for the keystoning distortion, is derived mathematically as T composed with the inverse of C, or $P = C^{-1}T$. **Right:** the application image can be appropriately distorted (pre-warped), using the mapping W so that it appears rectilinear after projection through a misaligned projector (modeled by the mapping P).

to be determined. Although there are 9 unknowns in this equation, there are only 8 degrees of freedom ($\sum_i p_i = 1$). Four point correspondences (where each point provides two constraints) are therefore necessary. Fortunately, in our system, these point correspondences can be automatically obtained by projecting a known rectangle into the environment, and observing the locations of its corners through the camera. Given four points, a unique solution for the parameters is obtained using standard linear algebra techniques.[3]

2.2   Identification of Projector Screen Boundaries

The general problem of keystone correction requires pre-warping a source image such that, when projected, its edges will be aligned parallel to the boundaries of the projection screen. In the typical case, where the projection screen is rectangular, this implies that the corrected image will also be rectangular. More precisely, the edges of the corrected image should converge towards the same vanishing points as the corresponding edges of the physical screen.

To do this, we must accurately locate the boundaries of the physical projection screen, which is a challenging computer-vision task. Since presentation rooms are typically darkened, the system first projects a bright white image onto the screen. By assuming that the projection surface is uniformly light in color, it is possible to extract the boundaries of the projection screen, even outside the boundaries of the projected image. This is done by a region-growing

---

[3] When greater than four points are available, a least-squares solution is used.

procedure where adjacent pixels with similar intensities are grouped into connected components, and all such components near the bright projected image are merged. This can be seen in Figure 3 (left): the outer quadrilateral in the camera image frame corresponds to the projection screen while the inner quadrilateral corresponds to the bright projected image. Although the camera images are low-resolution, high calibration accuracies are obtained by fitting lines to the edges of the quadrilateral, and computing intersections (corresponding to the coordinates of the corners of the screen) to sub-pixel accuracy. From these corners, and an assumption that the physical screen is rectangular, we derive a mapping C that transforms points from the projected image frame to the camera image frame (within a scale factor).

2.3   Inferring Mapping P

As shown in Figure 3 (left), the mapping (P) from the source image frame to the projected image frame may be computed from the two known mappings T and C derived above, by composing T with the inverse of C: $P = C^{-1}T$. In our system, P models the configuration of the projector setup that is directly responsible for the keystoning distortion. Our task is therefore to pre-warp the application image, so that it is rectilinear (distortion-free) after being mapped through P.

2.4   Determining Optimal Placement for Corrected Image

The computer display (source image frame) projects to an arbitrary quadrilateral (projected computer display) in the projected image frame as shown in Figure 3 (right). Since the pre-warped image can only be displayed within the bounds of the computer display, the corrected image must lie within the bounds of this quadrilateral, and (for best viewing) should be as large as possible. This is equivalent to finding the largest rectangle with appropriate aspect ratio within the projected computer display; a heuristic optimization is used to compute the dimensions of this rectangle.

Now, given the desired size and location of the corrected image, we compute a mapping S that scales and shifts the application image to lie within this area. We can compute the necessary pre-warping transformation W from the two known transformations S and P by the formula $W = P^{-1}S$, as shown in Figure 3 (right).

2.5   Interactive Refinement

The transformation, W, computed above is a candidate model for the correction required to undo projector misalignment. However, since W may be inaccurate due to errors in the calibration process, the user is given the opportunity to interactively refine the warping parameters. The rectangle bounding the source image is warped using the candidate W and projected onto the screen. If the calibration process were successful, the projected image would appear to be a rectangle aligned with the screen. The user, through a GUI, can directly manipulate the corners of the rectangle in order to fine-tune the alignment; this feedback is used by the system to refine W. Users of the presentation system have found this interface to be far superior to directly adjusting the eight degrees of freedom in W (e.g., using slider bars) since it allows the same level of control (the four corners of the rectangle provide eight parameters, uniquely constraining W) while being very intuitive. In fact, several users report that this interface alone (without the camera) enables them to perform excellent manual keystone-correction, indicating that such an interface would significantly improve current (manual) keystone-correction projectors.

2.6   Image Warping for Keystone Correction

Once a correct W has been determined, the pre-warped source image is created as follows. For each pixel in the pre-warped image, we find the corresponding point in the application image by applying the inverse mapping, W^{-1}. Since the computed point is real-valued, it will not typically correspond to a single (integer) pixel in the application image. Therefore, the four pixels in the application image that are closest to this point are blended using bilinear interpolation [1]. Note that the image warping computations are well-suited to exploit the accelerated 3-D graphics hardware available in typical computers. The application image is embedded in a virtual region of black pixels to ensure that only pixels within the corrected image are illuminated. The resulting image, when projected through the misaligned projector, appears undistorted.

3   Conclusion

Our system provides a better interface for presentation systems by allowing users to place portable projectors anywhere in the room, and automatically rectifying the resulting distortions. In particular, projectors may be placed at the side of the room, where they will not interfere with the audience. Recent related work in the area of teleconferencing [6] also demonstrates the benefits of distortion-correcting projector-based interfaces.

The availability of low-cost cameras and advances in processing power are making perceptual interfaces accessible to a general audience. Our automatic keystone correction system is one component in a larger *camera-assisted presentation interface*, prototyped at Just Research. Unlike typical perceptual interfaces, the camera-assisted presentation interface observes the presentation rather than the user, becoming a member of the audience. The user's interactions with the audience (e.g., laser pointer gestures) are observed and interpreted by the camera-assisted presentation system, allowing the user to control the presentation in a natural manner.

Although the system as described in this paper works with existing projector hardware, it could easily take advantage of newer projector technology. For instance, if the projector's position, location or optical characteristics could be computer-controlled, then the mapping W could be translated into commands to adjust these directly. Similarly, by integrating the camera hardware into the projector (e.g., the Proxima Cyclops [5]), the calibration problem is significantly simplified (T is the identity). We anticipate that newer generations of projectors will incorporate the automatic vision-based keystone-correction techniques described here.

If sufficient computational resources are available on the user's computer *during* the course of a presentation, then the keystone correction system could perform automatic online recalibration. This would have important benefits. For instance, if the camera or projector were accidentally moved during the course of the presentation, the system would automatically detect the disturbance by observing the discrepancy between expected and observed camera images. The pre-warp transformation would then be automatically adjusted so that the projected image could remain steady in spite of the motion. In an extreme scenario, one could even envision mounting the camera and/or projector on a moving platform (e.g., such as the user's body) with no adverse effect in the projected image.

The automatic keystone correction system described in this paper has been combined with other vision-based presentation interfaces (such as laser pointer presentation control) to create a *camera-assisted presentation interface*. The complete system has been deployed since December 1999, receiving very favorable response and attracting considerable commercial interest.

4   Acknowledgments

Thanks to Rich Caruana and Terence Sim, with whom the initial ideas for vision-based keystone correction were discussed, and to Gita Sukthankar for valuable feedback on the paper. Provisional patent applications for the inventions stemming from this work have been filed by Just Research [2, 3, 4].

References

[1] J. Foley, A. van Dam, S. Feiner, and J. Hughes. *Computer Graphics: Principles and Practice.* Addison Wesley, 1993. 613
[2] M. Mullin, R. Sukthankar, and R. Stockton. Calibration method for projector-camera system. Provisional U.S. Patent Filing, 1999. 614
[3] R. Sukthankar, R. Stockton, and M. Mullin. Automatic keystone correction. Provisional U.S. Patent Filing, 1999. 614
[4] R. Sukthankar, R. Stockton, M. Mullin, and M. Kantrowitz. Vision-based coupling between pointer actions and projected images. Provisional U.S. Patent Filing, 1999. 614
[5] G. Van Horn. Proxima's new Ovation+ projection panels do up multimedia. *Byte*, January 1995. <http://www.byte.com/art/9501/sec12/art9.htm>. 614
[6] R. Yang, M. Brown, B. Seales, and H. Fuchs. Geometrically correct imagery for teleconferencing. In *Proceedings of ACM Multimedia*, 1999. 613

A Multimodal User Interface for Geoscientific Data Investigation

Chris Harding[1], Ioannis Kakadiaris[1,2,3], and R. Bowen Loftin[1]

[1] Virtual Environments Research Institute, Univ. of Houston, Houston, TX 77023, USA
{charding, bowen}@uh.edu
[2] Department of Computer Science, University of Houston, Houston, TX 77023, USA
ioannisk@uh.edu
http://www.cs.uh.edu/~ioannisk/ICME00/

Abstract: In this paper, we report on our ongoing research into multimodal investigation of geoscientific data. Our system integrates three-dimensional, interactive computer graphics, touch (haptics) and real-time sound synthesis into a multimodal interface. We present applications of multimodal investigations of geoscientific data that pertain to surface meshes on which several typical properties were mapped and to geophysical volume data. Finally, we report on the preliminary results of a psychological study, which is being conducted to increase our understanding of the recognition of audio value in an absolute sense.

1. Introduction

In recent years, the resource industry has recognized 3D visualization and modeling of geoscientific data as an important part in exploration and exploitation of natural resources in order to increase efficiency. Several projects have demonstrated that the use of Virtual Environments has the potential to improve productivity and lower costs in areas such as petroleum exploration or mining exploration ([4],[5],[6],[9]). The fields of haptic force-feedback devices and real-time sound synthesis have matured sufficiently over the last years to allow research into the integration of touch and sound into (visual) Virtual Environments. Although both technologies have been used in a geoscientific context ([7],[1],[8],[2]), to the best of our knowledge, no research has been performed on the integration of visual, haptic and audio presentations into a multimodal, immersive system.

Most existing Virtual Environments focus entirely on improving the user's comprehension of geoscientific data by using interactive 3D (stereographic) techniques alone and largely ignore other senses. As real-world geoscientific data become routinely multidimensional (i.e., dozens of properties are sampled simultaneously for one point in space) the use of haptic force-feedback and real-time sound synthesis may allow the interpreter simultaneous access to more data dimensions during

[3] This work was supported in part by NSF Career Award IIS-9989482.

T. Tan, Y. Shi, and W. Gao (Eds.): ICMI 2000, LNCS 1948, pp. 615-623, 2000.
© Springer-Verlag Berlin Heidelberg 2000

his/her work. For most humans the visual sense is the most important input channel, and almost all computer applications focus heavily on it for human-computer interaction. Very few applications use the potential to feel and interact via touch and to analyze data via hearing.

The aims of our research are the following: 1) to provide a multimodal interface which includes these two underutilized channels in order to give the user the ability to work with multiple, overlaying properties, and 2) to establish methods of successfully mapping various properties of geoscientific data from their scientific domain into the domain of touch and sound. Presenting various aspects of data simultaneously through touch and sound could lead to enormous benefits (when done correctly) or greatly confuse the users (when done incorrectly).

2. Integration of Graphics, Haptics and Sound

In the following sections, we describe the application of our multimodal interface to two task-specific examples of geoscientific data investigation. Our system is being developed on a SGI Octane™ workstation with a Desktop PHANToM attached. The Desktop PHANToM provides a working volume of about 15 cm^3. Rendering of the 3D stereo images employs SGI's Performer™ toolkit, while 3D interaction is performed via the Desktop PHANToM, which uses Sensable's GHOST™ API. Several important lessons learned about geoscientific Virtual Environments from earlier projects ([5], [6]) have been incorporated into the graphical part of the system.

2.1 Digitizing of Line Segments on a 3D Surface

Surface-based data in the form of tessellated meshes is a data type common across the geosciences. Similarly, digitizing line segments on top of surfaces is used in many different geoscientific tasks, ranging from fault modeling to pipeline planning.

Currently, these tasks are being performed on 2D data (maps), however a 3D environment gives the user much better access to and understanding of the surface's morphology. In addition, the task of digitizing may involve the awareness of other variables projected on the surface. The following examples explore line digitizing on 3D surfaces using datasets from two different geoscientific domains.

The first surface-based dataset uses a high-resolution elevation model of a seafloor (using the original side-scan sonar data) with a maps of the residual mantle-boguer-anomaly (RMBA gravity map), which was used as a visual (texture) map, and a map of the age of the oceanic crust (calculated from magnetic data). The user changes the surface's position in space (navigates) by "grabbing" the surface with the PHANToM's stylus. Holding down the stylus button will attach the surface to the stylus. Grabbing the surface makes it is very easy to look at it from different angles and distances and visually investigate the change of curvature. Our application offers a special surface coloring, which shows the magnitude of the slope and its direction as a change of color (dip and azimuth coloring, Fig. 3). For this particular type of geological surfaces, a good understanding of the morphology is vital in order to model the underlying tectonic structure of the seafloor. The cyan cone in the lower right corner of Fig. 3 represents the users "fingertip".

Fig. 4 depicts the seafloor dataset after digitizing a main fault (yellow line). The gravity is visually mapped with a Blue-Green-Red color map, the magnetic age is mapped into the audio domain, and the change of slope was expressed as a friction map. When the tip of PHANToM's stylus touches the surface, the force-feedback allows the user to explore the sometimes-delicate surface features in conjunction with the visual input from 3D graphics. The user is also able to perceive inflection points by feeling an increase in friction (resistance). Line segments are laid down by clicking the PHANToM's stylus button. The application drapes a line between the current point on the surface and the last digitized points (yellow line in Fig. 4). In addition, the sonification gives the user access to the magnetic age of the surface.

For this abstract sounds (i.e., musical notes) are produced in real-time through a MIDI (Musical Instrument Digital Interface) system. MIDI allows easy synthesis of simple musical notes with different predefined instruments. In this sonification, a sound is defined by pitch, type of instrument (timbre), and duration. Although this approach could be used to sonify three independent data variables, we opted for a complete overlap of the sound properties: a single data variable is described by a fixed combination of pitch, instrument and duration. For a low data value we use the combination low pitch, bass type instrument and long duration, while for a high data value we use the combination high pitch, soprano type instrument and short duration, which can be easily understood by most people.

The second dataset explored with this application uses remote sensing data, which combines Digital Elevation Models (DEM) with three channels of multi-spectral satellite images mapped on it. Data classification is based on mapping a combination of spectral channels (e.g., through Boolean or arithmetic operations) on terrain (elevation) models. Working with both terrain data (such as elevation and slope) and different combinations of channel data is visually demanding and would require the use of several simultaneous views of the data or sophisticated color mappings. To investigate the use of touch and sound with remote sensing data, we used our interface to feel the DEM with the PHANToM and to translate the different terrain attributes, channels and their combinations into visual (texture) maps, sound maps and friction maps. Surface friction is perceived when the user's virtual "finger" touches or slides over the surface. By varying the value of *dynamic* friction, the user's movements can be more or less impeded, while high *static* friction value is noticeable even by simple touch. This setup is useful for digitizing features onto a part of the surface where a certain slope value is required (these slope values would be mapped to no friction, resulting in very smooth digitizing) and other slope values should be avoided (these slope values would be mapped to high friction values, making digitizing very difficult). Mapping the change of slope to audio allows the user to hear the inflection points.

2.2 Multimodal Interaction with Volume Data

Volume data play an increasingly important role in many parts of the geosciences dealing with subsurface data (e.g., seismic exploration). Currently, the interactive, visual exploration and modeling of large amounts of volume data presents a technical challenge. As with other types of geoscientific data, a volume dataset may contain several, simultaneously overlapping attributes. For the multimodal exploration of volume data, we have developed a volume renderer based on SGI's Performer toolkit and have added haptics and sound to it. We use a multi-attribute dataset from mining exploration. Fig. 5 depicts volume rendering of an inversion of a geoelectric survey (chargeability). The coloring being used is Blue - Green - Red. The opacity of the lower volume-density (Blue) values has been reduced to grant insight into the interior of the volume.

Haptic volume rendering is concerned with reproducing physical phenomena such as attraction, repulsion and viscosity by projecting a force to the user's hand while he/she moves through the volume. We have implemented a rudimentary form of repulsion - the PHANToM's tip is gently forced into a certain direction, depending on the value (density) of the volume the tip is currently traversing. The repulsion is tied to the opacity, i.e., the more solid a part of the volume looks, the more it repulses the tip which makes it feel more solid as well. Finally, the audio aspect of our interface allows the user to listen for a certain density value without the need to look at a readout of the value.

3. Psychological Study: Recognition of Absolute Audio-Signals

We are currently conducting a study to gain insight into the user's ability to recognize and remember absolute audio signals. We use the term *absolute* signal (as opposed to a *relative* signal) to refer to a signal that is evaluated outside the context of previous or following signals. Each signal will be defined first as corresponding to a certain number. The user has to remember the signal's properties well enough to be able to recognize it later, and connect it to the definition given earlier.

In our study, we define an audio signal by three simple parameters. These parameters are tied to the way (musical) tones can be synthesized through a MIDI interface and are the following: pitch (frequency), instrument (timbre) and tempo (tone repeat frequency). It is well known that the human hearing system is quite capable of discriminating even small changes in audio signals in terms of pitch or tempo. However, the absolute recognition of signals is considered much more difficult and usually dependent on musical abilities and training (the "perfect pitch" phenomena is an extreme example of recognition of absolute audio signals).

Thus, most sonifications are based on the relative recognition of audio signals (i.e., convey data by a change in the audio signal's makeup) and do not require the user to recognize and remember absolute audio signals. For example, in the context of multi-sensory data investigation the relative recognition would be useful to convey the rise in elevation or sudden drop of gravity while the user moves the stylus within the dataset. The absolute recognition of audio values would be useful in a "ballpark" setup. In this setup, the user partitions a certain property of a multi-dimensional dataset into a small (3 –7) number of bins, so that the user is able to perceive a rough idea of the current value (i.e., the value of the current stylus-tip position).

The idea of the ballpark scenario opens up several interesting questions: How should the audio signals (using pitch, tempo and instrument) be composed to allow for a short training and easy recognition? Should we use only single parameters (e.g., only pitch or only instrument) or some kind of combination? The psychological study aims to answer these questions by reducing the ballpark scenario to a simple "game" in which the test subject is guided through seven conditions and asked to recognize a series of audio signals from a set of five different sounds. For each condition, the make-up of the audio signal varies as follows:

1. Pitch only (single note C in five different octaves: C2, C3, C4, C5, C6, played by a grand piano),
2. Instrument only (single note C4 played by an church organ, a grand piano, a pan flute, a muted trumped and a cembalo),
3. Tempo (3 clearly separate notes (C4 played by grand piano) with speed increasing in five stages from overall 1.5 sec. to 0.3 sec.,
4. Combination of pitch and instrument (single notes),
5. Combination of pitch and tempo (played by grand piano),
6. Combination of tempo and instrument (playing a C4), and
7. Combination of pitch, instrument and tempo.

Each condition starts with a *training phase,* in which the subject is being made familiar with the current set of audio signals and a *recognition phase,* in which the subject is being asked to identify a random series of 15 audio signals (which is different for each condition but the same for each subject), and indicate his or her confidence in the choice.

Fig. 6 depicts the step in which the already trained subject plays the audio signal (PLAY), chooses a number (1 – 5) that corresponds to the audio signal and indicates his or her confidence (0-100). We use a questionnaire to collect information about each subject's age, gender, musical background and listening habits and professional background. We have begun testing subjects (which must not be music majors), and statistically evaluate the accuracy and the speed of audio signal recognition. The successful design of a set of audio signals for the ballpark scenario will most likely involve a lot of personal choices. However, the results of this study would determine the importance of each audio parameter and their combinations. This knowledge could provide us with a set of guidelines that would enable us to design a good first approximation of audio signals that the user would then be able to refine according to his or her personal taste.

Currently, we have collected data from 12 subjects (10 male, 2 female), which ranged in age from 21 to 55 (average 33). Table 1 depicts the average percentage of correct recognition of an audio signal and the average subjective confidence.

Table 1. Preliminary results of our study.

Condition	Success Ratio (%)		Confidence (%)	
	Mean	Std Dev	Mean	Std Dev
1. Pitch (P)	68.33	11.39	61.44	11.34
2. Instrument (I)	74.44	27.59	69.92	25.86
3. Tempo (T)	78.89	10.00	73.21	10.64
4. Pitch + Instrument	90.56	11.48	89.02	12.58
5. Pitch + Tempo	80.00	13.33	73.41	14.34
6. Instrument + Tempo	93.89	6.11	91.84	7.22
7. Pitch + Instrument + Tempo	95.00	5.83	93.14	6.69

The following diagrams (Fig. 1 and Fig. 2) present a graphical comparison of the success ratio and the subjective confidence across all seven conditions.

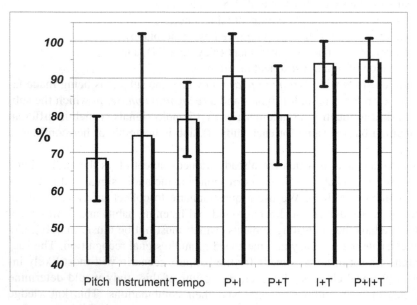

Fig. 1 Success ratio of recognition of an absolute audio value.

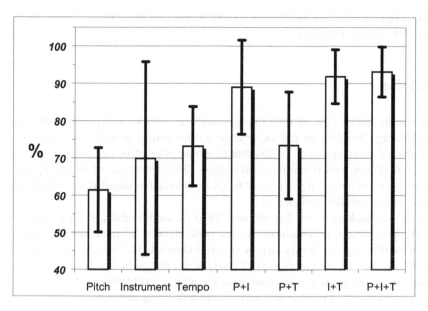

Fig. 2 Subjective confidence in the recognition of the audio value.

These preliminary results point to several interesting trends. It appears that of the three simple conditions (pitch, instrument, tempo) tempo is the most successful, while pitch is the least successful. This agrees with the generally accepted assumption that temporal variations are handled well by the human hearing system. Despite the fact that anecdotal evidence seems to suggest that the instruments are easy to distinguish, the lower success rate and high standard deviation of the Instrument condition indicates a wide range of fluctuations.

Conditions 4 - 7 are all combinations of two or three simple conditions. Our initial hypothesis was that these combinations should be easier to recognize and therefore more successful. Of the conditions that combine two sound properties, the presence of the instrument parameter seems to make the recognition more successful. For example, condition 4 (Pitch + Instrument) and condition 6 (Instrument + Tempo) are significantly more successful than condition 5 (Pitch + Tempo). Although condition 7 (Pitch + Instrument + Tempo) seems to be the most successful, it is not significantly different from condition 6 (Instrument + Tempo). This may indicate that pitch does not contribute much to the success rate within a combined condition. In general, the subjective confidence seems to be highly correlated to the actual success ratio, which may suggest that the subjects know fairly well how well they are able to recognize a certain audio value.

Note that we might have to account for a potentially high training effect, which could make a comparison between conditions invalid. However, even with a training effect present, there seems to be overall high success ratio and low standard deviation of the best conditions. This seems to indicate that it could be possible to train a subject to recognize a certain kind of audio signal, but that the training time may be longer than we initially expected.

4. Conclusion

Research into the integration of multimodal interfaces into Virtual Environments is still in its infancy. Building on our experience with previous projects, we have created a number of prototype applications that combine interactive computer graphics, haptic force-feedback, and scientific sonification. Demonstrations of these applications to various geoscientists have already resulted in valuable suggestions, which are being incorporated into the applications.

This early feedback suggests that the new field of multimodal data investigation promises to open up new opportunities for the presentation and modeling of geoscientific data by utilizing previously largely ignored channels for human-computer interaction. The results of the psychological study will help us in the design of audio signals that can be recognized in an absolute sense after a short training. We hope that our exploration into the use of touch and sound will bring more geoscientists into contact with this new technology and will help to open up new areas of potential applications in the geosciences.

References

1. Aviles, W., Ranta, J.: Haptic Interaction with Geoscientific Data. Proceedings of the Fourth Phantoms User Group (PUG) Meeting (1999)
2. Barrass, S., Zehner, B.: Responsive Sonification of Well Logs. Proceedings of The Sixth International Conference on Auditory Display (ICAD) (2000) 72-80
3. Chen, D., personal communication.
4. Fröhlich, B., Barrass S., Zehner B., Plate J., Goebel, M.: Exploring Geo-Scientific Data in Virtual Environments. Proceedings of 10th IEEE Visualization Conference (1999) 169-173
5. Harding, C., Loftin, B.: Interactive Visualization of Geoscientific data within a virtual Environment. Proceedings of SPIE Vol. 3960: Visual Data Exploration and Analysis (2000) 246-257
6. Harding, C., Loftin, B., Anderson A.:Visualization and Modeling of Geoscientific Data on the Interactive Workbench. The Leading Edge, Vol. 5, (2000) 77-83
7. McLaughlin, J. and Orenstein B.: Haptic Rendering of 3D Seismic Data. The Second Phantoms User Group (PUG) Meeting (1997)
8. Saue, S., Fjeld O.: A Platform for Audiovisual Seismic Interpretation. Proceedings of the 4. International Conference on Auditory Display (ICAD) (1997) 47-56.
9. Winkler, C., Bosquet F., Cavin X., Paul J.: Design and implementation of an immersive geoscience toolkit. Proceedings of 10th IEEE Visualization (1999) 429-432.

Appendix: Figures

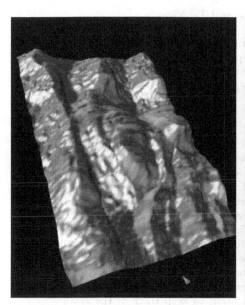

Fig. 3. Depiction of the seafloor data with curvature coloration.

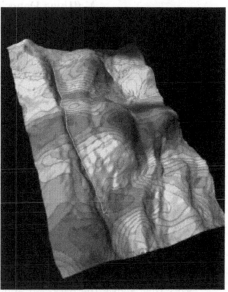

Fig. 4. Depiction of gravity data texture-mapped onto the seafloor.

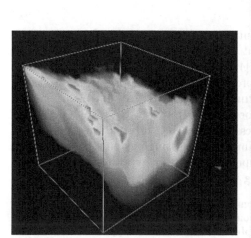

Fig. 5. Volume rendering of chargeability data.

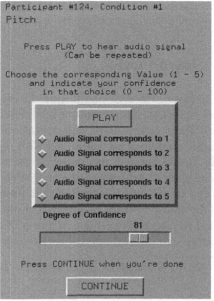

Fig. 6. Depiction of a widget used for playing an audio signal and relating it to signals learned during the training phase.

Partial Information in Multimodal Dialogue

Matthias Denecke and Jie Yang*

Human Computer Interaction Institute
School of Computer Science
Carnegie Mellon University
{denecke,yang+}@cs.cmu.edu

Abstract. Much research has been directed towards developing multimodal interfaces in the past twenty years. Many current multimodal systems, however, can only handle multimodal inputs at the sentence level. The move towards multimodal dialogue significantly increases the complexity of the system as the representations of the input now range over time and input modality. We developed a framework to address this problem consisting of three parts. First, we propose to use *multidimensional feature structures*, a straightforward extension of typed feature structures, as a uniform representational formalism in which the semantic content stemming from all input modalities can be expressed. Second, we extend the feature structure formalism by an *object-oriented framework* that allows the back-end application to keep track of the state of the representation under discussion. And third, we propose an *informational characterization of dialogue states* through a constraint logic program whose constraint system consists of the multidimensional feature structures. The multimodal dialogue manager uses the characterization of dialogue states to decide on an appropriate strategy.

1 Introduction

In the past, research on multimodal input processing systems has focused on how complementary information in different modalities can be combined to arrive at a more informative representation [4, 16]). For example, representations of deictic anaphora are combined with representations of the appropriate gestures. The fusion algorithms employed use symbolic [10], statistical and neuronal [15] techniques to achieve a reduction of error rate [3]. However, the results presented so far are limited to systems that process one sentence in isolation.

In another strain of research, multimodal processing systems have been applied to interactive error correction of speech recognizer hypothesis, where interactions extend over several turns during which a user may select one of multiple

* We are indebted to Laura Mayfield Tomokiyo for helpful comments on an earlier draft of this paper. Furthermore, we would like to thank our colleagues at the Interactive Systems Laboratories for helpful discussions and support. This research is supported by the Defense Advanced Research Projects Agency under contract number DAAD17-99-C-0061. Any opinions, findings and conclusions or recommendations expressed in this material are those of the authors and do not necessarily reflect the views of the DARPA or any other party.

T. Tan, Y. Shi, and W. Gao (Eds.): ICMI 2000, LNCS 1948, pp. 624-633, 2000.

input modalities for each turn [14]. No concurrent multimodal input takes place. In contrast to the multimodal systems described above, no deep understanding of the input is necessary since the interactions between the user and the computer take place on the surface level of the spoken words.

At the same time, researchers in the area of spoken dialogue systems address the question of how human-computer interaction through spoken language can be extended to natural dialogues with a machine. Dialogue managers hold a partial representation of the task to be performed and integrate complementary information over time. How this is done is prescribed in a *dialogue strategy*. In order to increase flexibility, many dialogue managers allow for scripted dialogue strategies that can easily be updated ([11, 12]).

In this paper, we propose a framework for multimodal dialogue systems that ties together different aspects of the three approaches described above. In the following, we refer to a *multimodal dialogue systems* as a system fulfilling the following two conditions. First, it should allow a user to perform one of a set of predefined tasks, and second, it should allow the user to communicate his or her intentions over a sequence of turns through possibly different communication channels.

The implementation of a multimodal dialogue system faces a set of challenges. Since the system is supposed to execute actions during and at the end of the interaction with the user, deep understanding of the input is necessary. The need for interactions that potentially range over a number of turns and modalities is not addressed in the work cited above, as interactions with the cited systems can range either over different turns or different input modalities but not both.

Furthermore, the input from different modalities may directly affect the information represented in the dialogue, which, in turn, may affect any sort of display presenting this type of information. Since the form of presentation may vary from object to object, we advocate an object-oriented methodology.

As the input channels are not entirely reliable, robust dialogue strategies are required. Since the switch of modalities has been shown to be effective, it is of interest to investigate dialogue strategies which actively suggest the switch of input channels when communication breakdown occurs. In order to implement this strategy, the system needs to keep track of the input channel associated with each piece of information in the discourse. Furthermore, the system needs to be capable to reason about these channels and their reliability.

The framework we develop in this paper to address these problems consists of three parts. We propose *multidimensional feature structures* as a representational vehicle to capture different aspects of the input provided by different channels. Multidimensional feature structures are a straightforward generalization of typed feature structures. Multidimensional feature structures allow to introduce as many different partial orders as is necessary for the task at hand. Moreover, we propose an object-oriented extension to the formalism of typed feature structures that allows the back-end application to be notified of changes in the representations in an object-oriented manner. Finally, in order to handle the growth of the state space, we propose to characterize the multidimensional representation through a constraint-logic program [8] where the constraints are formed by the multidimensional feature structures and where the logic program makes assertions as to which state the system is in.

2 Multidimensional Feature Structures

In many multimodal and spoken dialogue systems, variants of slot/filler representations are employed to represent the partial information provided by the different input sources. Abella and Gorin [1] provide an algebraic framework for partial representations. In multimodal systems, the multimodal integration then uses some scheme of combination to arrive at an augmented representation that integrates information either across modalities or across sentences.

Typed feature structures [2] have been proposed in the past as a representational vehicle to integrate information across modalities [10] and across time [6]. Typed feature structures can be considered as acyclic graphs whose nodes are annotated with type information and whose arcs are labeled with features.

In order to be able to represent different aspects of multimodal input, we propose to enrich the structure from which the type information is drawn. More specifically, we propose to annotate the nodes of feature structures with n-dimensional vectors $\mathbf{v} \in \mathbf{V}$ rather than with types. The vectors are drawn from the cross product of n possibly distinct sets $\mathbf{V} = P_1 \times \ldots \times P_n$, where each of the P_i is endowed with a partial order $\sqsubseteq_i$. This allows us to represent multimodal aspects of information in the atoms of the representations. Figure 1 shows an example of a multidimensional feature structure.

$$
\begin{bmatrix} t \\ F \begin{bmatrix} u \\ G \begin{bmatrix} v \\ H \quad w \end{bmatrix} \end{bmatrix} \end{bmatrix}
\qquad
\begin{bmatrix} (t_1, \ldots, t_n) \\ F \\ G \begin{bmatrix} (u_1, \ldots, u_n) \\ (v_1, \ldots, v_n) \\ H (w_1, \ldots, w_n) \end{bmatrix} \end{bmatrix}
$$

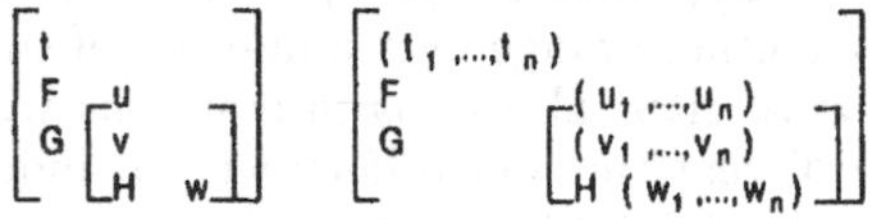

Fig. 1. A typed feature structure and a multidimensional feature structure. Symbols in capital letters denote features, symbols in small letters denote types and indexed symbols denote elements drawn from a partial order.

Note that both the definitions of unification and subsumption of typed feature structures [2] require only that the information associated with the nodes be drawn from a finite meet semilattice. This is a partial order in which all subsets of the elements have a unique greatest lower bound, and if for any subset of elements an upper bound exists, the least upper bound is equally unique. In multidimensional feature structures, the partial orders $\sqsubseteq_i$ of the elements v_i impose a natural partial order on the vectors in V according to

$$
\mathbf{v} \sqsubseteq \mathbf{w} :\Leftrightarrow v_i \sqsubseteq_i w_i \qquad \forall 1 \leq i \leq n
$$

Thus, standard unification and subsumption generalize in a straightforward manner to multidimensional feature structures.

The Ontology of Classes. The ontology of classes figures most importantly among the knowledge sources. It is used to represent inheritance relations between descriptions of objects, actions, states and properties of objects, actions and states in the domain at hand. The class hierarchy is the equivalent to the type hierarchy for typed feature structures, extended by a simple methodology to attach methods to the types (see section 3). We also use the class hierarchy to express linguistic information such as speech act types and the like. Figure 2 details two extracts of a domain model.

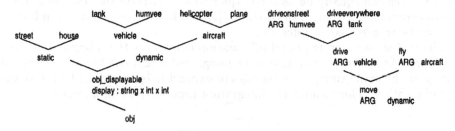

Fig. 2. An extract of an object-oriented domain model for a military application.

Spatial Partial Orders. A crucial feature in multimodal systems is the integration of pointing and moving gestures such as drawing arrows, circles and points. The informational content of a gesture is twofold. First, it communicates a certain semantic content. For example, in some contexts, the gesture of an arrow can be interpreted as a movement while the gesture of a circle can be interpreted as a state. Thus, the informational content conveyed by a gesture can partly be expressed through the concepts introduced in the class hierarchy.

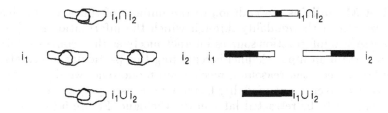

Fig. 3. Partial orders consisting of regions and temporal intervals.

In addition to the taxonomical information, a gesture conveys spatial information as well. This information is represented by sets of two- or three-dimensional points. As the power set of $\mathbb{R}^n$ forms a lattice under union and intersection, spatial information can be represented in multidimensional feature structures as well. An example for a spatial partial order is shown in figure 3.

Interestingly, it is the combination of semantic and spatial information that provides a substantial gain in the representations of multimodal inputs. For example, the system can infer from the domain model shown in figure 2 that streets and houses cannot be the argument of a movement action. If there are movable and unmovable objects in spatial proximity to a movement gesture, the system can infer based on the domain knowledge which objects to move.

Temporal Partial Orders. Temporal information can be used in two separate ways. First, temporal annotations of the hypotheses from different recognizers constrain the multimodal integration ([4, 16]). Second, temporal expressions in the utterance can equally be used be exploited to constrain the database access and to coordinate actions in the domain ([7]). In both cases, temporal information can be represented in intervals.

Figure 4 shows how temporal information constrains the integration across modalities. As the representations to be integrated might not have been created exactly at the same time, the intervals are expanded by a certain factor at the time of creation. This ensures the integration process to be monotonic.

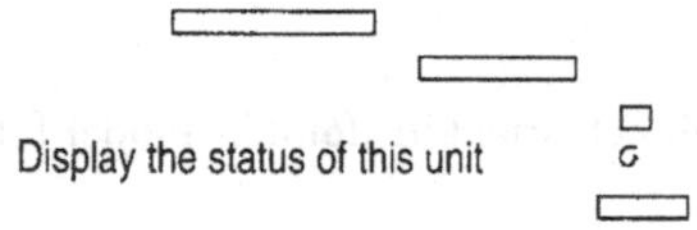

Display the status of this unit

Fig. 4. Overlapping intervals of representations in different modalities. The dark intervals show the creation time of the information while the light intervals show the time during which combination with other modalities is acceptable.

Query Counting. In order to adequately handle communicative breakdowns, the dialogue manager needs to keep track of the number of times a value for a feature has been queried. If this value surpassed certain threshold, alternative strategies can be pursued.

Record of Modalities. In addition to the number of times a value has been queried, we record the modality through which the information has been conveyed. These two information sources interact nicely as they convey information as to how certain an input channel is regarding this particular information.

In order to keep the reasoning mechanism monotonic, we do not represent the fact that a given input modality has not been used yet, as this information may be required to be retracted later in the dialogue. Figure 5 (a) presents the partial order used for the input modalities.

Confidence levels. In a similar manner, confidence level annotations can also be represented in multidimensional feature structures. Their partial order is shown in figure 5 (b).

3   Object-Oriented Descriptions and Changes

Multimodal dialogue systems need to address the fact that visualizations of objects may be subject to change at any time in the dialogue, caused either by external events or by user commands. As has been shown in many graphical user interfaces in the last decade or so, object-oriented frameworks greatly reduce complexity of this task. In multimodal dialogue processing, additional complexity is due to the fact that there is not a one-to-one relationship between descriptions

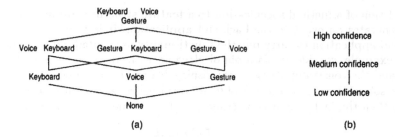

Fig. 5. (a) The partial order used to represent different input modalities. (b) Confidence levels of the input modalities.

(see, e.g. the input "*The units in here* + \<circle\>") and the objects themselves. For this reason, we develop an extension to the type hierarchy that allows a specification of methods as state constraints. Every time the status of the state constraint changes, the back-end application is notified about the constraint and all necessary parameters and can execute the relevant action. This is comparable in function to the handlers proposed by Rudnicky and Wu [13].

3.1   Method Specification

The domain model employed in the dialogue system uses a simple class hierarchy (see section 2). Class specifications may contain variables (whose type is a class from the ontology) and methods (whose arguments are classes from the ontology). In addition, class specifications may be related through multiple inheritance. While in conventional object-oriented design, objects in the domain correspond to classes, actions of the objects correspond to methods, and properties correspond to variables, we chose to model each of objects, actions and properties of objects and actions by classes. First, this allows us to uniformly express mappings from noun phrases, verbal phrases and adjuncts to classes. Second, any constituent of a spoken utterance may be underspecified.

3.2   Method Invocation

The type inference procedure for feature structures [2] can be generalized in a straightforward manner to include method specifications. More specifically, the method specifications form a partial order (defined by the subsumption ordering of the argument constraints) over which the type inference procedure can be defined. If the type of a feature structure is at least as specific as the type for which the method is defined and all constraints on the arguments are satisfied, the method specification is added to the feature structure. It can be seen easily that this extension is monotonic, increasing and idempotent as is required for inference procedures. Moreover, as the number of method specifications is finite, the type inference procedure halts.

A method specification does not implement any particular behavior of the class it belongs to. Rather, it detects that the informational content of a representation is specific enough for a procedure to be invoked. For this reason, the addition of a method specification to a feature structure through type inference generates an event to the back-end application. It is then the task of the back-end application to carry out the functionality associated with the method. As an example, consider a class obj_displayable with an associated method display() and the constraint string < obj_displayable.name,int < obj_displayable.x,int < obj_displayable.y (read: the variable obj_displayable.name contains more information than the fact that it is a string, i.e. it is instantiated). We thus have

$$
typeinf\left(\begin{bmatrix} obj\_tank \\ \text{NAME} & \text{"Bravo-1"} \\ \text{X} & 153 \\ \text{Y} & 529 \end{bmatrix}\right) = \begin{bmatrix} obj\_tank \\ \text{NAME} & \text{"Bravo-1"} \\ \text{X} & 153 \\ \text{Y} & 529 \\ display(n,x,y) \\ \quad n = \text{"Bravo-1"}, x = 153, y = 529 \end{bmatrix}
$$

As soon as the position and the name of the object become known to the dialogue system, the type inference adds the instantiated method signature to the feature structure and sends and event to the back-end application. The event contains a unique identifier of the representation, along with its name and coordinates as declared in the method specification. Should a description of an object refer ambiguously, an event is generated for each retrieved object that verifies the constraint. Not only does this approach provide a declarative way of specifying behavior and abstract over the form of the dialogue, it also decouples the natural language understanding component from the application itself in a natural way.

In this way, the method invocation interacts nicely with another characteristic of our approach to object-oriented design. While traditionally an instance of a class is an object, in dialogue processing an instance of a class can only be a (possibly incomplete) description of an object. Necessary information for object instantiation may be missing and can only be acquired through dialogue. Since descriptions of objects do not need to refer uniquely to objects, procedural method invocations become more complicated. For this reason, we chose the declarative approach to method invocation over a procedural one.

4   Informational Characterization of Dialogue States

Traditionally, one approach to describing dialogue is to explicitly model dialogue states and transitions. Here, all possible states all well as transitions through the state space need to be anticipated and specified during system design. The difficulties of the specification are aggravated as soon as behavioral patterns need to be replicated for each of the states. For example, when a misunderstanding occurs, it is a common dialogue strategy to repeat confirmation questions for any given filler only a few times. In finite state based dialogue managers, the uncertainty of the information is thus modeled by the state the dialogue manager is in.

Recent dialogue managers allow more flexible interaction through the specification of *dialogue goals* [6] or *forms* [12] which, when filled out entirely, adequately represent the users' intention. It is the duty of the dialogue manager to determine through interaction with the user which form to choose and how to fill its slots with values. This approach of information-based dialogue management gives up on the notion of an explicit state of the dialogue system. At the same time, it is very useful to make an assertion pertaining to the state of the dialogue manager, e.g., *the dialogue manager is in a state where conversational breakdown has occurred*. The information contained in these abstract states is then used to select appropriate dialogue strategies.

4.1 Constraint Logic Programming

In order to abstract away the concrete information available in the discourse and to arrive at a characterization of the dialogue state, we use a constraint logic program [8] to determine abstract states. The constraints are of the form

$$i : c \sqsubseteq_i x \quad i : c \text{ is compatible to } x \quad i : c \text{ unify } x$$

where i identifies the partial order P_i, $c \in P_i$ is an element from the partial order, and x is a variable instantiated with a multidimensional typed feature structure.

4.2 Characterization of States

The dialogue state is characterized by a set of five variables $s_1, \ldots, s_5$. These variables express the confidence in the representation of the current turn, the confidence of the overall dialogue, the speech act of the current utterance, whether or not the intention of the user could be determined uniquely, and whether or not referring expressions in the current utterance have unique or ambiguous referents, respectively. The values of the s_i range over one of the partial orders and are determined by the constraint logic program. Figure 6 lists the possible values for the five state variables. Details on a unimodal veriant of this approach as well as the determination of the users' intention are described in more detail in [5].

Variable	Meaning	Values taken from
s_1	modality confidence	confidence order × input modality order
s_2	dialogue confidence	confidence order
s_3	speech act type	class hierarchy
s_4	users intention	{none,unique,ambiguous}
s_5	reference of referring expressions	{none,unique,ambiguous}

Fig. 6. Values of the Dialogue State Variables

The integration of the multimodal information is achieved through additional clauses where the parameters are constrained so as to ensure the combination of appropriate representations.

4.3   Specification of Strategies

Additional clauses rely on the characterization of the informational state to decide the next action of the dialogue system. For example, if the confidence in the current utterance is medium, but the confidence in the overall dialogue is high, the system decides to ask for confirmation. Then, appropriate clauses determine the semantic content of the confirmation question, select the template, generate the clarification question and pass it on to the output module. The dialogue state variables decouple thus a concrete application specific dialogue state from a dialogue strategy that can be formulated in an application independent fashion.

5   Conclusion and Future Work

We have argued that a framework for multimodal dialogue systems not only needs to address the integration of information from different input streams, but also needs to be capable of representing and reasoning about input sources, input reliability and dialogue states. We have presented a framework for multimodal dialogue systems consisting of three central aspects addressing these requirements. First, we showed how multidimensional feature structures, a generalization of typed feature structures, can be used as a unified representational formalism for representing information stemming from different input sources. Second, we introduced an object-oriented extension to the feature structures that allows applications to receive notifications of state changes in the representations, to be employed for example for decentralized updates of displays. Finally, we demonstrated how the clauses of a constraint logic program over the multidimensional feature structures can be used to informationally characterize the informational content. These more abstract dialogue states are then used to determine appropriate dialogue strategies.

The work closest to ours is probably the the work by Johnston et al [9, 10]. In this work, multimodal input is represented in the standard types of typed feature structures and combined on a sentence level. The difference between this work and ours consists in the fact that the former encodes spatial information in the feature structures directly and relies on procedures external to the logic to perform the integration of multimodal input (e.g., intersection algorithms taking lists of types representing the coordinates of the points). In our work, however, this property is built in through the different dimensions in the feature structures and the combination with constraint logic programming. In addition, we propose object-oriented extension enabling the back-end application to track the state of the multimodal discourse and mechanisms to integrate information beyond the sentence level. Finally, the multimodal structures also allow to represent information that is necessary for guiding a multimodal dialogue; thus, the proposed representations enable the interaction to extend over the sentence level.

Future work includes the addition of logic to allow the dialogue system to determine an appropriate modality for the information being queried.

References

1. A. Abella and A.L. Gorin. Construct Algebra: Analytical Dialog Management. In *Proceedings of the 37th Annual Meeting of the Association for Computational Linguistics*, 1999.
2. B. Carpenter. *The Logic of Typed Feature Structures.* Cambridge Tracts in Theoretical Computer Science, Cambridge University Press, 1992.
3. P.R. Cohen, M. Johnston, D. McGee, S.L. Oviatt, J. Clow, and I. Smith. The Efficiency of Multimodal Interaction: A Case Study. In *Proceedings of the International Conference on Spoken Language Processing,Sydney*, pages 249–252, 1998. Available at http://www.cse.ogi.edu.
4. P.R. Cohen, M. Johnston, D. McGee, S.L. Oviatt, J. Pittman, I. Smith, L. Chen, and J. Clow. Quickset: Multimodal Interaction for Distributed Applications. In *Proceedings of the 37th Annual Meeting of the Association for Computational Linguistics*, 1997.
5. M. Denecke. Informational Characterization of Dialogue States. In *Proceedings of the International Conference on Speech and Language Processing, Beijing, China*, 2000.
6. M. Denecke and A.H. Waibel. Dialogue Strategies Guiding Users to their Communicative Goals. In *Proceedings of Eurospeech, Rhodos, Greece*, 1997. Available at http://www.is.cs.cmu.edu.
7. G. Ferguson and J.F. Allen. TRIPS: An Integrated Intelligent Problem-Solving Assistant. In *Proceedings of AAAI/IAAI*, pages 567–572, 1998.
8. J. Jaffar and J. L. Lassez. Constraint Logic Programming. In *Proceedings 14th ACM Symposium on Principles of Programming Languages, Munich*, pages 111–119, 1987.
9. M. Johnston. Unification-Based Multimodal Parsing. In *Proceedings of the 17th International Conference on Computational Linguistics and the 36th Annual Meeting of the Association for Computational Linguistics (COLING-ACL 98), Montreal, Canada.* Association for Computational Linguistics Press, 1998. Available at http://www.cse.ogi.edu.
10. M. Johnston, P.R. Cohen, D. McGee, J.A. Pittman S.L. Oviatt, and I. Smith. Unification-Based Multimodal Integration. In *Proceedings of the 35th Annual Meeting of the Association for Computational Linguistics.* Association for Computational Linguistics Press, 1997. Available at http://www.cse.ogi.edu.
11. E. Levin and R. Pieraccini. Spoken Language Dialogue: From Theory to Practice. In *Proceedings of the Workshop on Automatic Speech Recognition and Understanding*, 1999.
12. K.A. Papineni, S. Roukos, and R.T. Ward. Free-Flow Dialogue Management Using Forms. In *Proceedings of EUROSPEECH 99, Budapest, Ungarn*, 1999.
13. A. Rudnicky and X. Wu. An agenda-based Dialog Management Architecture for Spoken Language Systems. In *Proceedings of the Workshop on Automatic Speech Recognition and Understanding*, 1999.
14. B. Suhm, B. Myers, and A.H. Waibel. Model-Based and Empirical Evaluation of Multimodal Interactive Error Correction. In *Proceedings of the CHI 99, Pittsburgh, PA*, 1999. Available at http://www.is.cs.cmu.edu.
15. M.T. Vo. *A Framework and Toolkit for the Construction of Multimodal Learning Interfaces.* PhD thesis, School of Computer Science, Carnegie Mellon University, 1998. Available at http://www.is.cs.cmu.edu.
16. M.T. Vo and C. Wood. Building an Application Framework for Speech and Pen Input Integration in Multimodal Learning Interfaces. In *Proceedings of the International Conference on Acoustics, Speech and Signal Processing*, 1996. Available at http://www.is.cs.cmu.edu.

Multimodal Interface Techniques in Content-Based Multimedia Retrieval

Jinchang Ren[1,2], Rongchun Zhao[1], David Dagan Feng[2,3], and Wan-chi Siu[2]

[1] Department of Computer Science and Engineering, Northwestern Polytechnic University,
Xi'an, China 710072
{npurjc,rczhao}@nwpu.edu.cn
[2] Center of Multimedia Signal Processing, Dept. of Electronic and Information Engineering,
The Hong Kong Polytechnic University, Hong Kong
{enfeng,enwcsiu}@en.polyu.edu.hk
[3] Department of Computer Science, The University of Sydney, Australia

Abstract. Multimodal interfaces (MI) can well improve the interactivity between users and computers through cooperation of different interactive devices and methods to exchange information and understand their requirements or response. As a hotspot in information processing, content-based retrieval (CBR) of multimedia has intrinsic demand for multimodal interface techniques to suit for input / output of multiple media types. In this paper, different MI techniques in CBR of multimedia are introduced, which are classified into three classes, namely traditional CUI/GUI, multimedia UI and intelligent multimodal UI. The analysis and comparison of these MI techniques with corresponded media retrieval ways are also given. It is hoped that the investigation in this paper can much promote the work both in MI and CBR of multimedia for efficient and effective information interaction between human and machines.

1 Introduction

With the development of multimedia, virtual reality and scientific visualization, traditional graphic user interface (GUI) becomes more and more insufficient for efficient and effective information interactions. New efficient techniques, multimodal interface (MI), which can suit for user's properties, are strongly emphasized as an important research topic in computer science. MI can catch user's intention through cooperatively full usage of multiple feeling / action modalities including speech, gesture and vision.

The thought of MI was first presented by Negroponte of MIT Media Lab from the concept of "Conversional Computer", in which people could communicate with computer in natural ways using speech, gesture, facial expression, focusing or body languages, just as they did in daily lives. The concept of MI derived from the MediaRoom project by MIT's AMG (Architecture Machine Group), where the key person, Bolt, was still the top leader in the research of human-machine interactions.

From 1990's, more and more attentions had been paid to MI as a new filed in HMI, especially for multimedia information service. In MIT, SLS (Spoken Language Systems) group used speech interface in Galaxy project for online service in airline

T. Tan, Y. Shi, and W. Gao (Eds.): ICMI 2000, LNCS 1948, pp. 634–641, 2000.
© Springer-Verlag Berlin Heidelberg 2000

bulletin, weather forecast and map query. In Carnegie Mellon university, its ISL lab applied multiple modalities, i.e., facial feeling, lip-reader, gesture, speech, etc. to improve communications between human and machines.   In Europe, Amodeus-2 and MIAMI projects were set in ES-PRIT program to investigate the model, structure, representation and integration of HMI.

Recently, content-based retrieval (CBR) techniques have been developed as an important branch in multimedia information service. Comparing with traditional text-based systems, CBR can provide more efficient and effective solutions for different applications. CBR has intrinsic interrelationship with MI because multimedia requires multi-modal interfaces during its input / output, representation and understanding. Visual media, like image, graph or video, need corresponded optical devices and representation forms. For music or speech, acoustical instruments and methods are demanded.

There are many significant MI techniques that have been presented in CBR systems. It has been our aim to conclude these techniques in MI's opinion, which can be classified into three levels, namely traditional GUI, multimedia MI and intelligent MI. The detail discussion of MI in CBR will be given in the following sections.

2   Traditional Human-Machine Interfaces

Traditional HMI means character-based user interface (CUI), graphic user interface (GUI) and multimedia user interface. CUI and GUI are commonly used for many software systems, and multimedia UI here refers the interface interacts through multimedia.

2.1   From CUI to GUI

CUI is the first human-machine interface form, which adopts text-based language with certain rules as commands between the user and computers. CUI can be developed into two stages, namely formal language and natural language. Although command interface requires well-trained skills to remember the interaction semantics and easily goes awry, it is still appreciated by professional users for its high flexibility, efficiency and effectiveness.

GUI is the main stream of current UI techniques, which is widely used in many computer systems, such as Macintosh, Microsoft Windows, Unix and etc., where window, icon, menu, dialog and other graphic controls are the main forms for HMI. Pointing device, such as mouse, is applied in GUI to complement keyboard. Window-based message-driven mechanism is the kernel technique of GUI.

GUI needs not additional burden to master the formal commands, where menu and other graphic controls will help the users in convenient HMI. GUI is independent to different cultures or languages, which can be applied for different users of the world.

Although GUI has apparent advantages than text-based interface, it has some intrinsic limitations in I/O imbalance, low efficiency and weak interactivity. For professional users, GUI is less efficient and lower flexible than CUI. Shortcut is one kind of implementation of CUI in GUI. Moreover, it is hard for GUI to support the representation and interaction of non-spatial information.

2.2   Multimedia User Interface

Multimedia is taken as a transitional technique before intelligent user interface and natural human-machine interactions. Multimedia usually refers to text, graphics, image, audio and video (cartoon). CUI and GUI has utilized the prior two, multimedia user interface will apply the latter ones for representation of different information media.

Most of the multimedia can be classified into two categories namely visual media (VM) and acoustical media (AM). Tactual or olfactory media are not main streams in multimedia.

Three representative forms of VM are graphics, image and video. Pointing devices (i.e. mouse, digitizer) and optical devices (scanner, camera, video camera, etc.) are mainly used for input of graphics and image /video respectively. Monitor and printer are two common used output devices. In order to acquire or display high resolution VM, some accessory devices are needed for image / video capturing, 2D/3D or frame display accelerating, etc. AM includes audio, speech and music. Microphone, speaker, as well as some music mixer and MIDI synthesizer, are usually adapted for its input and output. Through cooperation of these interactive devices, it helps to improve HMI by efficient and effective information intercommunion.

Multimedia has enormous potential ability to improve the representation and interaction of information. It supports parallel, associational and symbolic information processing, which can reduce or eliminate the multi-vocal connotations and noise in the process of HMI.

3   Multimodal Interfaces for CBR of Multimedia

Although multimedia user interface has enriched UI techniques, it is only limited in media storage, transfer and representation. For content-based multimedia retrieval, it requires more work on feature extraction, understanding and abstraction, which is the most important difference between multimedia interface and multimodal interface.

To improve the effectiveness and efficiency in CBR of multimedia, MI can been used in media representation, feature representation and interactive retrieval.

3.1   Media Representation

Media representation involves media visualization or acousticalization of compressed and uncompressed media for management and retrieval. VM has spatial relationship, but for video and AM, temporal relationship should also be represented in reasonable forms.

Media representation should consider it is convenient for media processing and manipulation includes interactive editing, filtering, trans-coding and feature extraction, especially in compressed domain. String, tree and graph are commonly used for operation. Suitable representative forms and interface for interactive operation should be selected according to different media types. For example, composed music can be represented by series of structured element or metadata, the corresponded MI should consider at least two aspects: one is visualization of the structure; the other is acousticalization of the metadata.

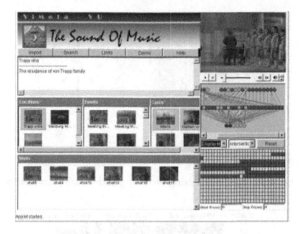

Fig. 1. Hierarchical representation of video in visual interface of Vimeta-VU system

Fig 1 is the main interface of Vimeta-VU system designed by Visual Information Processing (VIP) group of New South Wales University, Australia. Key frame series, hierarchical graph (scene, shot and frame) and corresponded mapping list (physical continuous frames) are used to represent the spat-temporal relationship between meta-data in videos.

3.2   Features Representation and Interactive Retrieval

There are three kinds of features for different media: The first one is media attribute, such as media format, length and hierarchical description, etc. Visual structure using tree /graph can be used to represent it (see Fig1).   The second is visual/acoustic features, such as color, texture, shape, motion or tone and timbre, visual or acoustic modality is required for representation (see Fig2).   The third is semantic content of media, which is directly related with perceptual understanding and can be effectively described using combination of text and MI (see Fig3).

Unlike text, multimedia retrieval requires more interactions to specify user's requirements and give satisfied response. According to text, VM and AM, different modalities should be integrated to fulfill the query task. The main problems here include:

1) How to input of user's requirements?
2) How to understand user's input?
3) How to give effective response efficiently?
4) How to demonstrate the output media to the user friendly?

Moreover, the system should also has the ability to allow the user execute some further queries based on previous results. The detail comparison of the query mechanisms is discussed in section IV.

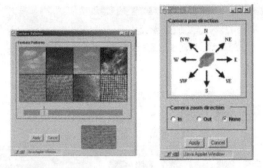

Fig. 2. Texture and motion represented using browsing and sketch-map in VideoQ system

Fig. 3. Text combined with MI for semantic query of Content-based News Management (CBNM) System by the Authors. Semantic entries, including person, place and topic/event, are represented by relevant image with text

3.3 Intelligent Interfaces for Multimedia Retrieval

To improve the retrieval effectiveness, intelligent interface techniques are specially emphasized in CBR of multimedia, which include NLU (natural language understanding), recognition of speech, character, symbol, gesture and face, etc.

Multimedia signal processing, pattern recognition and knowledge-based machine learning are the theoretic bases of the intelligent interfaces. Visual or acoustic features will be extracted automatically through multimedia signal analysis, semantic concepts can be learned by relevance feedback in MARS (University of Illinois at Urbana-Champaign) and Image Finder (Microsoft Research Center of Beijing).

Take videos for example: Symbols, includes business or traffic tags, as well as captions can be recognized and indexed for retrieval. Speeches in videos are also considered to understand the scene and index the whole video. Face, gesture, facial expression, and even for talking lips are all studied to help the understanding and indexing of the complex media.

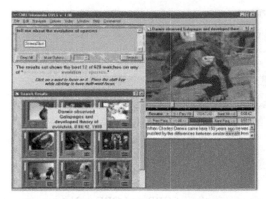

Fig. 4. Video query using NLU in Informedia system: The user inputs "tell me about the evolution of species", the system gives the video of "Darwin observed Galapagos and developed theory of evolution" for response by automatic extraction of the key words "evolution" and "species"

Fig 4 is an example of intelligent interface in video retrieval using NLU by Carnegie Mellon University namely Informedia. The system can understand user's requirements by automatic extraction of important words from inputted natural language sentence, and give relevant response by semantic matching.

4   Integrated Query Interfaces

For efficient and effective retrieval of multimedia, a great deal of MI techniques has been delivered from media representation, feature extraction and interactive query.

Following is some representative query methods:

- QBB (Query by random browsing or navigation in categories). QBB retrieves media similar with HTML and usually for semantic queries.
- QBE (Query by example).   QBE retrieves media by matching of candid object and sample object specified by users. Visual /acoustic similarities are analyzed and measured automatically according to corresponded features, such as find similar color, texture or tone.
- QBS (Query by sketch).   QBS allows the user to input his requirement by sketch submission, esp. for shape query. Here the shape can be geometric or abstract ones. For example, find images with circle, find images with specified color histogram, or find music in some amplitude curves.
- QBF (Query by feedback).   QBF retrieves media by interactive feedback. When the user gets a group of results, different scores or weight will be assigned to evaluate whether it is closed to the requirements. Improved results will be applied by the system through automatic analysis of these evaluations.
- QBT (Query by text, which can be extracted from speech, captions or NLU). QBT is expansion of traditional text-based retrieval.

In practical systems, such as Netra (see Fig 5) and VideoQ of UCSB and Colimia Univ. respectively, are always been combined with weighed features of color, shape, texture, etc., as well as inputted text (keyword).

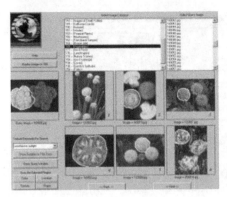

Fig. 5. Image retrieval by combination of QBB and QBE in Netra system by UCSB: Images with similar regions to the sample one can be retrieved from different image categories

5   Conclusion

Multimodal user interface and CBR techniques are both very important for information interaction between human and machines. They have same background in terms of segmentation, recognition and understanding of actual scenes. The difference between them is that the former uses online recognition of modality status, but the latter adopts online and offline processing of query process and media data respectively. While MI emphasizes more on interactive ways, CBR focuses more on information representation and retrieval, but they have same purpose for efficient and effective information intercommunion. Just as different UI techniques should be integrated, but not substitute each other, for different applications. MI and CBR will supplement each other in the development of information engineering towards general multimedia applications with higher intelligence.

Acknowledgements

This research is supported by the UGC (Polyu 1.42.37.A050) and (Polyu 119/ 96E) grants.

References

1.   Bolt, R. A.: The Human Interface. California: Lifetime Learning Press (1984)
2.   Sheiderman, B.: Direct Manipulation. A Step Beyond Programming Languages. IEEE Computer, Vol. 16. No. 8 (1983)
3.   Hartson, H. R., etc.: The UAN: A User- Oriented Representation for Direct Manipulation User Interfaces. ACM Trans. on Information Systems. Vol. 8, No. 3 (1990) 181-203
4.   Card, S. K., etc.: The Psychology of Human Computer Interaction. Hillsdale, N. J. (ed.): Lawrence Erlbaum (1983)
5.   Garve, W. W.: Auditory Icons. Using Sound in Computer Interface. Human-Computer Interface, Vol. 2 (1986)

6. Hauptmann, A. G., Mcavinney, P.: Gestures with Speech for Graphic Manipulation. Int. J. of Man-Machine Studies. Vol. 18. No. 2 (1993)
7. Burdea, G., Coiffet, P.: Virtual Reality Technology. John Wiley and Sons, Inc. New York. (1994)
8. Lin, Y., Chen M., etc.: An Architecture for Multimodal Agent Interactive System. Proc. of the 5th Int. Con. On CAD/CG'97. Beijing. Int. Academic Press, (1997)
9. Wang, J.: Integration of Eye-Gaze, Voice and Manual Response in Multimodal User Interface. In Proc. of the IEEE Int. Conf. on System, Man and Cybernetics (1995)
10. Bolognessi, T., etc.: Introduction to the ISO Specification Language LOTOS. Computer Networks and ISDN Systems, Vol. 14. (1987) 25-59
11. Idris, F., Panchanatban, S.: Review of Image and Video Indexing Techniques. Univ. of Ottawa, Canada (1996)
12. Rui, Y., Huang, T. S., Chang, S. F.: Image Retrieval: Current Techniques, Promising Directions and Open Issues. J. of Visual Communication and Image Representation, Vol. 10. (1999) 1-23
13. Chang, S. K., etc.: Reality Bites-Progressive Querying and Result Visualization in Logical and VR Spaces. http://www.unisa.it/gencos.dir/chang/365/real.htm
14. Chang, S. F., etc.: A Fully Automated Content-based Video Search Engine Supporting Spatial-Temp acoustical Queries. IEEE Trans. On Circuits and Systems for Video Technology, Vol. 8. No. 5. (1998)
15. Hauptmann, A., Witbrock, M.: Informedia: News-on-Demand Multimedia Information Acquisition and Retrieval. Intelligent Multimedia Retrieval. Mark, T. Maybury, (ed.), AAAI Press (1997) 213-223
16. Smith, J. R., Chang, S. F.: VisualSEEK: A Fully Automated Content-based Image Query System. ACM Multimedia96, Boston, MA, Nov. 20 (1996)
17. Smith, M., Kanade, T.: Video Skimming and Characterization through the Combination of Image and Language Understanding. IEEE Int. Workshop ICCV98, India (1998)
18. Deng, Y. N., Manjunath, B. S.: Content-based Search of Video Using Color, texture and Motion. Proc. of IEEE on IP, Vol. 2. CA (1997) 534-537
19. Zhang H J, etc. Video Paring, Retrieval and Browsing: An Integrated and Content-based Solution. ACM Multimedia (1995) 15-24
20. Ren, J. C., etc: A Self-Extensible Model for Content-based Video Retrieval. Int. Workshop MMWS2000, Hong Kong (2000) 259-262
21. Yeo, B. L., Yeung, M. M.: Classification, Simplification and Dynamic Visualization of Scene Transition Graphs for Video Browsing. Storage and Retrieval for Image and Video Databases VI. SPIE Vol. 3321. Jan. (1998) 60-70
22. Servetto, S., etc.: A Region-based Representation of Images in Mars. Special issue on Multimedia Signal Processing, J. on VLSI Signal Processing. Oct. (1998)
23. IBM: QBIC-IBM's Query by Image Content, http://wwwqbic.almaden.ibm.com/
24. Castagno, R., Ebrahimi, T., Kunt, M.: Video Segmentation Based on Multiple Features for Interactive Multimedia Applications, IEEE Trans. On Circuits and Systems for Video Technology, Vol.8, No.5, Sep. (1998)
25. Rodger, J. M., etc,: Towards the Digital Music Library: Tune Retrieval from Acoustic Input. In Proceedings of DL'96, (1996)
26. Wiggins, etc.: A Framework for the Evaluation of Music Representation Systems. Computer Music Journal, Vol. 17, No. 3 (1993)

Design of Retrieval Method of Map Targets in the Multimodal Interface

Jia Kebin   Shen Lansun   Zhang Hongyuan   Chang Wu

Laboratory of Signal & Information Processing,
Beijing Polytechnic University, Beijing, China 100022
E_mail: siplnet@bjpu.edu.cn

Abstract. Multimodal interface is a part of computer technologies and plays an important role in many fields. The multimodal interface, which is efficient, flexible, and convenient, supports the system of CBIR strongly. In this paper, the method, by which map targets are retrieved in multimodal interface, has been studied. According to the characteristic that the different users do not have the same interest in the map targets, a concept model, which is called multi-user conceptual implication(MUCI), has been established to reflect the difference of map targets. The MUCI not only corresponds to hierarchies of categories of map targets, but also implicates the degree of user's interest of categories. Once the map targets that users are interested in have been searched in MUCI, the targets of actual map can be quickly and accurately retrieved according to their color and shape and retrieval algorithms. The targets that are not important or less important can be ignored. The established method of this model, processes of reasoning and retrieval have been described in paper. In addition, the design and manipulation of data structure of map in multimodal interface have been illustrated. This multimodal interface has been used successfully in resource information system of a large oil field and the results are satisfying.

Keywords multimodal interface, target retrieval, multi-user conceptual implication (MUCI), data structure

1 Introduction

Multimodal interface is an important part of computer technologies. The results of its research and application dissolve the technologies of computer into every aspects of human's living. It has been accepted by different classes and has great influence on traditional manners, in which people live and work[1-2].

CBIR(Content-based Image Retrieval) has become a focal point in the field of computer vision and multimedia database. Map, as a special image, is make up of different elements of map's object. For users, these elements can be separated into significant targets and background scenery objects. For example, in the map of city, there are target elements that denote the building, playground, park, highroad, river and so on. Having used the technology of digital image processing and recognizing to segment and recognize targets, the indexing by similitude, the system of CBIR has been established. This indexing technology has been widely applied in GIS, city plan, military command, and resource management. The multimodal interface, which is efficient, flexible, and convenient, supports the systems strongly. Not only does it

T. Tan, Y. Shi, and W. Gao (Eds.): ICMI 2000, LNCS 1948, pp. 642-649, 2000.
© Springer-Verlag Berlin Heidelberg 2000

provide various visual tools of CBIR, but also organizes, assorts, and controls various complex algorithms of image processing, logic operations, and browser methods, which are used in CBIR. All operational processes that are irrelative to users are sightless, and are automatically finished by its control organizations. By this way, users can browser directly indexing results, and capacities of search and query have been greatly improved[3-4].

However, there is a problem in many applications that not all map targets are of equal importance to the users. The retrieval of map targets should reflect this difference[5]. And the system can adjust the amount of targets displayed on screen according to the degree of users' interest. The applications of this search method can improve not only the real-time capacities of the chain of command in military fight but also efficiency of resource management of ocean. But the multimodal interface, which is used for map retrieval now, does not have this capacity.

In the design of multimodal interface for map retrieval, a tree structure of multi-user conceptual implication(MUCI) is introduced in this paper. The MUCI not only corresponds to hierarchies of categories of map targets, but also implicates the degree of user's interest of categories. Once the map targets that users are interested in have been searched in MUCI, the targets of actual map can be quickly and accurately retrieved according to their color and shape and retrieval algorithms. The targets that are not important or less important can be ignored. The established method, retrieval principle, data structure and handle method of targets, with which the map targets are retrieved in the multimodal interface, will be described in following sections.

2 MUCI for Map Targets

MUCI is a tree structure that has many conceptual hierarchies. Each map target belongs to a particular category according to its conceptual level. All the map targets are divided into a number of target subcategories. When one of these subcategories is regarded as a category, it can be further divided into a number of subcategories. This process of dividing continues until each subcategory contains only one map target. The MUCI can be regarded as a three–dimension tree, in which y axis represents the category, x axis represents the different targets of same category, and z axis represent the different users. Each node j contains a weight W_{ij}, which represents the degree to which the user i interested in node target j. Here $i = 1,...,n$, denotes the n different users, and $j = 1, ..., m$, denotes m nodes.

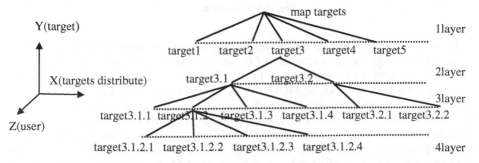

Fig. 1. Structure of MUCI tree

The weight of every target node in MUCI represents the degree to which the user is interested in it. Because its correctness will directly influence the accuracy of retrieval results of map targets, the weights need to be assigned according to the significant degree of targets. There are many methods to calculate weights[6]. Here statistical method has been used to quantitatively count the weight of every target. In order to distinguish the degree to which the user is interested in targets, a greater weight that has higher access frequency of user's application needs to be assigned. The weight can be counted according to equation (1) or equation (2):

$$W_{ij} = \frac{\ln(X_{ij})}{\sum_{i=1}^{n} \ln(X_{ij})} \tag{1}$$

$$W_{ij} = \frac{\ln(X_{ij})}{Max\{\ln(X_{ij})\}|i = 1,2,\ldots,n} \tag{2}$$

Where X_{ij} denotes the applied function number of the user i accesses the target j, W_{ij} denotes the weight of user i to node j.

3 Retrieval Operation of Map Targets

3.1 Design of Data Structure

S1 = {U_ID, SUBCATEGORY, S_WEIGHT, LEAF, POINTER}
（a） Data structure of MUCI tree
S2 = {T_ID, TM_NAME, TM_PATH, LM_NAME}
（b） Data structure of associate target with images
S3 = {U_ID, U_NAME, THRESHOLD}
（c） Data structure of user

Fig. 2. Data structure

In order to describe and handle the MUCI tree and finish different retrieval tasks, it is needed to design the data structure of retrieval system. The data structure consists of several distinctive tables (shown as Fig.2). S1 is the structure of MUCI tree. Each row corresponds to one node of MUCI that is a target or a target category. It contains five columns, in which U_ID means the user ID, SUBCATEGORY means target subcategory name, S_WEIGHT represents relative significance of subcategory, LEAF is an indicator showing whether or not the subcategory is a leaf, and POINTER is a pointer pointing the parent node of this node. S2 is the table that associates targets with target images. According to name of local map LM_NAME and target T_ID, the store path(TM_PATH) and image name(TM_NAME) of target image can be uniquely detected. S3 is the user's table, which contains three columns, in which U_ID means user's ID, U_NAME means the name of user, and THRESHOLD represents threshold of which the user accesses target.

3.2 Calculation of Priority Value of Target Retrieval

The final displayed target objects are monomial targets to which leaf nodes correspond. To determine which targets the users are interested in among the numerous leaf nodes in MUCI, the priority value of target nodes must be calculated. Here the method of reliability has been introduced.

For user i, if a selected node is node k(m) in MUCI, and the nodes on the path from tree root to it are k(1), k(2), ..., and k(m-1), then there are two methods to calculate the priority value of node k(m):

1. method of probability theory

The priority value of leaf node k(m) is the weight product of all the nodes on the path:

$$P_{ik(m)} = \prod_{l=1}^{m} W_{ik(l)} \tag{3}$$

Where i = 1,2,…,n denotes the different users, $W_{ik(l)}$ denotes the the weight with which user i accesses to node k(l).

2 method of fuzzy_set theory

The priority value of leaf node k(m) is the minimum value of all weights on path:

$$P_{ik(m)} = Min\{W_{ik(l)}\} \quad l=1,2,…,m \tag{4}$$

3.3 The Retrieval Process of Map Target Object

With computing and manipulating its concept model, map targets can be retrieved. Fig. 3 illustrates the principle of the retrieval process.

1. According to retrieval requests of different users, we obtain users' code U_ID. We also get the target item special to the user from the MUCI structure in Fig. 2.
2. Based on formula (3) or (4), we calculate the display priority of every leaf node in the table. First, we find leaf nodes (LEAF=1) from the target items which are obtained from step (1). Then according to their parent pointers, we step up to the tree tip to calculate their display priorities, and save the results in the new added data item PRIORITY. Finally we wipe off the target items which are not leaf nodes.
3. According to the user code U_ID , we get the threshold of the item visited by the user from the user table (shown as Figure 2) . We compare the threshold value with display priorities of target items gotten in the previous step, and filter the items whose values are less than the threshold.
4. According to the local area map which is displayed (LM_NAME), we find the monomial image named corresponding to target and the path under which it is

saved in the image database. Then we overlap all of these monomial targets to form the target image that the user U_ID wants.

5. We use the image display function to show the user the image processed in step (4).

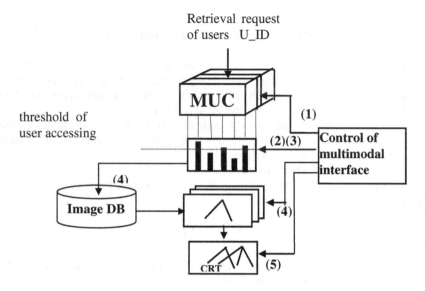

Fig. 3. The map targets retrieval

The whole process is controlled by the multimodal interface. The degree of targets that user is interested in is set by the threshold of the targets. The smaller the threshold value is, the more targets are displayed. So we can alter the threshold to adjust the degree to which different users are interested in the target dynamically.

4 Design and Manipulation of Map Data Structure[7]

4.1 Layer and Expression of Map

The retrieval of map content is a process that many kinds of target elements are retrieved in a large area. It requires not only the functions such as whole map browse, local area magnification, but also the special operations such as layer, lapping, physical zoom, relative zoom etc, and the retrieval function. In order to realize these functions, we must design and manipulate a data structure that can reflect the contents of map.

A map consists of many kinds of target and background elements, so we can make it layered and processed. An original map can be divided into many local area maps that have the same size. Each local area map can be considered as a result of overlapping the background map and the many layer maps, each of them have monomial target element. A map can be divided into target layers in two ways:

- A map is divided by different target classes in it. Each layer has a class.
- A map is divided by colors of targets. The targets that have the same color are in the same layer.

In the former way, we can use segmentation and recognition methods based on color and shape of maps to recognize target objects and then get maps of different target classes that are saved in different image files. In this way, an original local area map can be expressed as the data table shown in Fig. 4, in which LM_NAME is for the names of local area maps, T_ID is for the codes of target classes. Each layer map has an image name and a path, which constitute the saving pointers of layer maps .

Local map = {LM_NAME, T_ID, F_NAME, F_PATH}

Fig. 4. Data structure of layered map

4.2 The Manipulation of Map Data Structure

The operations of the map, such as layer, browse, filteration, and zoom, are done by manipulating the designed image data structure, as shown in fig.5. First, a pop-up window shows the contractible whole map. Users can choose the interested area by the mouse. Once the users selected the interested area, another window will show the amplified map of the selected area. Users then can do several kinds of operation in every local map. Filteration is realized by choosing the right filtering window or the menu of filtering. Users can select monomial target category and show the monolayer image every time. They also can select multiple target categories and show the multilayer image by adding them. Users can change the selection freely in the display process. Once the selection is changed, a filtering operation will be triggered. In the data structure (shown in fig. 4), the monomial target category image can be retrieved by the target categories names selected by the user (represented by the black target category codes shown in the fig.5). And then these monomial target category images are added by the adding program and shown in the display window.

Physical zoom is to simulate the different effects caused by the different distances between the user and the image. The system provides a physic zoom scrollbar on the operation interface. When the scrollbar moves forward, the map is amplified and displayed in detail. On the contrary, the map is reduced and displayed roughly. To realize the physical zoom, practical maps with different detail degrees can be displayed alternatively by changing scrollbar.

Relative zoom is a zoom operation of relative sense. It does not change the size of image. With the zoom amplification, there are more and more target categories displayed in the same local image. On the contrary, the target categories will turn fewer and fewer. In order to reflect the difference of various users' interest degree to targets and improve the retrieval efficiency, the relative zoom is calculated by the model of MUCI. When a user want to change the zoom, the system sorts out the related target categories and calculates the display priority of the target category(the priority is represented by the length of the vertical Line). The threshold with which the user accessed target is used as the default line of relative zoom. Every target category's priority is compared with the current line of relative zoom, And then all the target categories bigger than the line of relative zoom, is displayed according to its code (represented by the black code in the fig.5). It is designed to change the zoom by changing the value of relative zoom line. When the value of relative zoom line is small, it shows more target categories. Otherwise it shows fewer. Thus, we realize the aim of relative zoom.

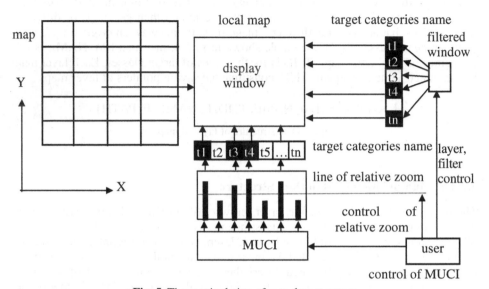

Fig. 5. The manipulation of map data structure

5 Conclusion

The multimodal interface technology plays an important role in the current information society. But there is not an all-purpose commercial system. So it needs to design and develop the multmodal interface suitable for the special application field.

The multimodal interface introduced in this paper has been successfully used in the resource map retrieval system of a large oil field and achieved the satisfactory results. The method introduced here is referential and irradiative to the general design of multimodal interface.

References

1. Burnet M M, Mcintyre D W. Visual programming. Computer, 1995, 28(3):14-16
2. Solariyappan M, Poston T, at el. Interactive visualization for rapid noninvasive cardiac assessment. IEEE Computer, 1996, 29(1):55-61
3. Kato T, Kurita T. Visual interaction with electronic art gallery. Proc. of Int. Conf on Database and Expert Systems Applications DEXA'90, 1990, Aug. 234-240
4. Kato T, Kurita T. A cognitive approach to visual interaction. Proc. of Int. Conf. On Multimedia Information Systems MIS'91, 1991, Jan. 109-120

5. Tanaka M, Ichikawa T. A visual user interface for map information retrieval based on semantic significance. IEEE Trans. Software Eng., 1988, 14(5):.666-670
6. Miller J C, .Miller J N. Statics for analytical chemistry. England: Ellis Harwood Limited, 1984
7. Jia kebin. The methodological study of query by image content using semantic index[doctoral dissertation]. University of Science and Technology of China, 1998

A First-Personness Approach
to Co-operative Multimodal Interaction

Pernilla Qvarfordt and Lena Santamarta

Natural Language Processing Laboratory, Department of Computer and Information Science,
SE-581 83 Linköpings universitet, Sweden

{perqv, lensa}@ida.liu.se

Abstract. Using natural language in addition to graphical user interfaces is often used as an argument for a better interaction. However, just adding spoken language might not lead to a better interaction. In this article we will look deeper into how the spoken language should be used in a co-operative multimodal interface. Based on empirical investigations, we have noticed that for multimodal information systems efficiency is especially important. Our results indicate that efficiency can be divided into functional and linguistic efficiency. Functional efficiency has a tight relation to solving the task fast. Linguistic efficiency concerns how to make the contributions meaningful and appropriate in the context. For linguistic efficiency user's perception of first-personness [1] is important, as well as giving users support for understanding the interface, and to adapt the responses to the user. In this article focus is on linguistic efficiency for a multimodal timetable information system.

1   Introduction

Brennan [2] argues that "human conversation is inherently cooperative" (p. 396). We agree that conversation is a co-operative activity, but using conversation in human-computer interaction will not inherently result in a good interaction. Maybury [3] argues that "fundamental to cooperative interaction are mechanisms that support media interpretation, generation, (language and representation) translation, and summarization" (p.15). In this paper we argue that having the required technology in order to offer users co-operative multimodal interaction is not enough. Instead focus on the users is equally essential when designing such systems. We agree with McGlashan that "from the user's point of view, whether the interaction is co-operative or not is judged solely on the basis of what system says" [4].

Closely related to co-operation is communication. Grice [5] talks about a co-operative effort in communication. A problem is how that effort is handled, one solution is to engage the user in a multimodal interaction, e.g. Fais, Loken-Kim and Park [6] stated that "one way to improve the performance of [natural language systems] ... is to supplement their processing capabilities with multimedia technologies designed to lessen the burden on the processing system" (p. 364). Their focus is, however, not to lessen the user's effort or engage in a dialogue on the user's term, but on the systems capabilities.

T. Tan, Y. Shi, and W. Gao (Eds.): ICMI 2000, LNCS 1948, pp. 650–657, 2000.

Another approach involves constraining the dialogue in telephone applications, in order to reduce the number of speech recognition errors.

In this paper we will discuss how to reduce the users' co-operative effort, by using Laurel's [1] concept of first-personness. Her article "Interface as Mimesis" is often used as an argument against natural language systems. Interestingly, Laurel does not argue against natural language systems. She even says that "good interfaces are ultimately multimodal" (p. 72). However, she do argue against poor design of interactive systems. She claims that the designer must let the user be in a first-person relation with the computer.

> "The first-personness is enhanced by an interface that enables inputs
> and outputs that are more nearly like in their real-world referents, in
> all relevant sensory modalities" ([1], p. 77)

If a system shall give real-world referents, in our view, the system must offer the users natural language interaction. Laurel also describes three aspects that contributes the first-personness; interactive frequency, interactive range, and interactive significance. Interactive frequency is a measure of how often the user can give input. Interactive range is a measure of the variety of choices available to the user at any moment, and finally, interactive significance is a measure of the impacts of user's choices and actions on the whole.

In what follows, we will discuss how these concepts can be used when designing co-operative multimodal interaction for a timetable information system. We will start by giving some background to co-operative interaction and communication, and then describe our investigations for gaining knowledge on how to design the interaction.

2   Background

In order to achieve a co-operative interaction the system needs to produce utterances which are perceived by the user as natural, coherent and helpful within the context of use [7]. The question is what makes utterances natural and helpful?

Clark and Shaeffer [8] argues that one function for a co-operative dialogue is to make sure that the participants have a common understanding of the dialogue, that they share a common ground. This theory has served as a basis for developing both spoken dialogue systems [9], and graphical user interfaces [10].

Brennan and Hulteen [9] believe that the response to the user should be adaptive to the system's processing. In this way the user knows where the interaction was "interrupted" and can easily correct the information given to the system. They also emphasise the importance to give both positive evidence and negative evidence of understanding in the response to the user.

The strategies proposed by Brennan and Hulteen are developed from theories of human-human communication. Other results, however, show that users do not use the same language when talking to a computer as when talking to another human (cf. [11], [12], [13]). Cheepen and Monaghan [13] talks about the transactional and interactional goals in the discourse. Transactional goals are task-oriented and interactional goals are person-oriented, i.e. to promote a relationship between the speakers. They argue that the kind of goals pursued in dialogues with a computer are primarily transactional. The user is interested in performing a particular task with the system, and not to small talk. For this reason Cheepen and Monaghan argues that what is natural in a human-human dialogue is not natural in a human-computer dialogue.

However, not only the type of spoken feedback given to the user is important, but also how the interaction between user and system is designed. Chu-Carroll and Nickerson [14] have noticed that in a telephone application mixed initiative and automatic adaptation of the dialogue strategies led to better performance and user satisfaction. Interesting to note is that the system's adaptations matched the expectations of the users.

3   Design Case

We have been working with the domain of timetable information of public transportation in the city of Linköping. The MALIN system (Multimodal Application of LINlin) is a prototype system under development, which consists of processing modules for interpretation, generation, dialogue management and knowledge manager. These in turn consult various knowledge sources such as the timetable database, a geographical information system, a domain model, dialogue models, lexicon, grammar etc. The system is describe in more detail in [15].

An evaluation of the first interface is presented in [16]. It allowed only multimodal input, and we are currently redesigning the interface to handle also multimodal output in addition to multimodal input. The system is designed as an information kiosk, and the spoken output from the system will be presented by an animated head.

The user shall be able to interact with the system using speech and either pointing or pen input. The system is based on the collaborative interface agent paradigm [17], [18]. Fill-in forms for questions, results and a map are all shown at the same time. Our system can be classified as a simple service system [19]. Such systems require in essence only that the user identifies certain entities, parameters of the service, to the system providing the service. Once they are identified the service can be provided.

3.1   Expectation on the Interaction

We believe that how the system fulfils the user's expectations is an important factor when measuring efficiency and effectiveness. For this reason, one of the first steps in the design of our multimodal co-operative system was to investigate what expectations the users had. We therefore distributed a questionnaire with 41 different properties to students at Linköping University. A total of 114 students filled in the questionnaire, 74 students had experience from searching for bus timetable information on the Internet and 32 had not. The students had to rate how important they believed each property was for a timetable information system for public transportation (buses) on the Internet, on a 1-7 scale, where 1 was equivalent to "not important at all", and 7 "very important." We will not here presents the complete result, just the parts that are interesting in this context.

The five most important properties were (means in parenthesis); *trustworthy* (6.68), *efficient* (6.58), *relevant* (6.50), *stable* (6.41) *and fast* (6.39). Interesting to note, is that students, who had no experience in searching for bus timetable information on Internet, did rate the properties *simple, responsive, productive, spontaneous, willing to learn, tempting*, and *supervising* as significantly (p<.05), see Table 1, more important than those students with experience. Non-experienced students also had a strong preference

Table 1. Compairison between students having experience from bus timetable on the Internet and those how have not for properties with a signigicant difference (p.<.05)

Property	Sig. (2-tailed)	Experienced		Unexperienced	
		Mean rating	Std. Deviation	Mean rating	Std. Deviation
Simple	.043	5.95	1.34	6.47	.80
Responsive	.021	4.58	1.63	5.34	1.23
Productive	.005	4.26	1.58	5.16	1.22
Spontaneous	.003	3.59	1.55	4.56	1.29
Willing to learn	.019	3.70	1.59	4.47	1.34
Tempting	.013	3.55	1.61	4.38	1.36
Supervising	.034	3.31	1.70	4.06	1.41

(p<.10) for the properties *comfortable*, *inviting*, *pleasant*, and *encouraging*. None of these properties are in the top ten in the average rating, simple is at place 11, however, most was rated above 4 amongst non experienced students. Worth noting is that many of these properties imply an agency from the computer, such as responsive, willing to learn.

3.2   Expert Evaluation

We have also developed a lofi-prototype of our multimodal timetable information system, MALIN. The prototype was evaluated in an expert evaluation. Three experts participated in the study, all had experience from designing interactive systems and conducting usability studies, but none had participated in the design of Malin or had any experience in designing multimodal interfaces. The experts tried the prototype using a scenario. They were encouraged to give comments about the prototype during usage, and to think-aloud. Afterwards they were given some questions. The question of interest here is "What would you like system to say to you?"

The experts came with several suggestions, both on the type of information the system should give and how it should behave. The comments can be divided into five groups:

- *Initiate a dialogue:* "It could say 'Where do you want to go from?' and things like that."
- *Give further/extended information:* "It could say: 'If you are in a hurry, you can walk somewhat longer' and things you do not think of when using the system." "Give a short introduction, 'With this system you can see where thingis are', both in text and in speech."
- *Clarify concepts used in the interface:* "I believed that 'nearest' meant 'fastest', it could clarify things like that." "It could clarify what is shown in the interface, e.g. 'The best time is shown in the grey field.'"
- *Clarification:* "Asking for clarification, when I have underspecified a question."

- *Behaviour:* "It should be supportive"

These suggestions have in common that the spoken comments from the system give some added value to the graphical interface. Interesting for this conclusion is that the subjects did not like the system to say something after every action, instead they said: "You can see what it has filled in."

4   Suggestion for Co-operative Dialogue

As shown from the results of the student questionnaires, users expect a timetable information system to be efficient. We can also see that speed is very important, but that efficiency is something more than just speed. In analogue with Cheepen and Monaghan's [13] discussion about transactional and interactional goals in conversation, and the fact that our system is a simple service system, the transactional goals should be of more importance for the users.

We would, however, like to focus on the part of efficiency that is not explained by speed. We believe that efficiency in co-operative multimodal dialogue systems can be divided into two, functional efficiency and linguistic efficiency. Functional efficiency means that the system is offering appropriate functions for solving a task efficiently. Linguistic efficiency means that responses are meaningful and are adapted to the context. This means that e.g. information given visually does not need to be repeated verbally. As shown from the expert evaluation, this is also something proposed by the usability experts. We believe that linguistic efficiency is especially important to reduce the co-operative effort of the user. Since linguistic efficiency also strive for adapting to the context, we suggest that, this also can contribute to the property *trustworthy.*

4.1   Input

As mentioned above, Laurel [1] discerns three interactional aspects that affect first-personness: Interactive frequency, Interactive range, and Interactive significance.

Interactive range is, as stated above, a measure of what choices are available for the user at any moment. In our case, all choices are available at any. The system is also designed to enable user input whenever, even while the system is speaking, which means that the user may interrupt the system or change her mind in the middle of a task. As in human communication barge-ins and changes of subject do always result in a momentary poorer interaction from which the dialogue should recover as soon as possible. However, this could require some additional turns. We believe that this can give users some of the wanted added value. Barge-ins also preserves the user's first-personness, since she is always in control of the dialogue.

One problematic feature of natural language systems and the aspect of interactive range is language understanding capabilities the systems have. We have for that reason given visual cues about what kind of questions the systems can answer. The interactive range of the system is also delimited by the "world" described in the application, i.e. geographical information and public transportation in Linköping. Thus the system cannot answer question about tourist information in the same city, or public transportation in another city.

The users' choices and actions have maximal significance in our design, as it is the user who decides (within the range of the context) what the system will do. Adaptability, both of interaction and language gives a maximal significance not only to what to user do but also to how it is done.

4.2   Output

Laurel [1] focuses her discussion about first-personness on the input to the system and on the constraints of the system. However, we believe that preserving the user's perception of first-personness is also a feature of the output of the system. In this section, we put forward a proposal on what is required of the output from a co-operative multimodal system in order to maintain first-personness and how spoken output supports it.

The major role of spoken language in our system, besides dialogue supporting, is to help the user to achieve a fast and easy understanding of the graphical information. This means that the system directs the user's attention to where important information is presented and complements the graphical information, e.g. by informing how graphical symbols are to be interpreted. To do this in an effective way, the spoken output has to be tailored towards the user's information needs and linguistic preferences. This implies that the same information can be presented in different ways, but a certain form will be more suitable as an answer to a question than to another.

Dialogue openings and closings, as well as giving feedback, is done via speech and gestures (compare with the expert evaluation above). The system may also ask for a confirmation or a repetition to reduce uncertainty. It is important that multimodal feedback is not perceived as intruding or irritating by the user. Instead the feedback should be perceived as co-operative. Therefore, the spoken feedback should be short and use the everyday words of the user.

The system has a sophisticated domain knowledge management capable to handle complicated spatial and temporal references [20]. Most of the database information is presented in the graphical interface, therefore it is very important that the spoken output supports its interpretation. There are two important points: first to direct the user's attention to critical data and to complement the graphical representation with explanations when needed. When the information given by the user have been relaxed in order to find a solution to the task; this is conveyed to the user as the relation between what the user said and the solution. For instance, if the user said she wanted to go to "Skäggetorpskyrkan" (The Church of Skäggetorp) and there is neither a suburb nor a bus stop with that name, the system realises that it is a church in the suburb "Skäggetorp" that is close to a bus stop. The system finds the nearest bus stop and indicates both on the map highlighting them. Then using speech and gestures, the system shows the geographical relation between the two.

Another important factor is for the system to use the referent expressions chosen by the user to avoid misunderstanding, i.e. to tailor its language to the use's linguistic preferences. For example, if the user refers to the railway station as "the railway station" the system should do so too, although the official name is "the central station". When the system refers to items on the screen, pronouns (and/or other appropriate referring expressions) are used. If needed they are accompanied by diectic actions.

To generate co-operative utterances in context, it is also very important to ensure coherence between the user's and the system's turns; i.e. to ensure that the same concepts are in focus. This is conveyed by means of prosody as well as lexical choice and syntactic structure, and results in the fact that the same data is presented in different ways depending on which information request they have to fulfil.

The students questionnaires showed that depending on user's experience, their expectations differ, and that non experienced users expects more agency from the interface. Results from the first interface [16] show that users' performance and preference also differ depending on their domain knowledge. This suggest that the spoken responses should be flexible and adapt to the user. For example, a user that barges in frequently can be given less spoken feedback, while long silences from the user can evoke more spoken initiatives from the system.

5   Conclusion

In this article we have argued that the user's co-operative effort needs to be reduced in co-operative multimodal interfaces. In order to do so, the efficiency of the interface plays an important role, especially in dialogues with transactional goals in simple service systems. The efficiency can be divided into functional and linguistic efficiency. Functional efficiency has a tight relation to solving the task fast. Linguistic efficiency concerns how to make the contributions meaningful and appropriate in the context.

We believe that a co-operative multimodal dialogue system working as we propose supports first-personness, as it always adheres to the user and it does not impose a way of interacting or of expressing. However, this is still up to the user do decide. The next step is therefore to conduct an experiment, in order to see if the user perceive the system as efficient, and co-operative, and if the user perceives first-personness in the interaction.

We have focused on discussing linguistic efficiency, which consist of preserving users' experience of first-personness, giving users support for interpretation the interface, and to adapt the responses to the user. In this article we have giving examples of linguistic efficiency of a multimodal timetable information system.

References

1.   Laurel, B.: Interface as Mimesis. In D. A. Norman and S. W. Draper (Eds.) *User Centered Systems Design, New Perspectives on Human-Computer Interaction*. Hillsdale, NJ: Lawrence Erlbaum (1986) 67-85

2.   Brennan, S.: Conversation as Direct Manipulation. In B. Laurel (Ed.) *The Art of Human-Computer Interface Design. Reading*, MA: Addison-Wesley Publishing Company (1990) 393-416

3.   Maybury, M.: Toward Cooperative Multimedia Interaction. In H. Bunt, R.-J. Buen, and T. Borghuis (eds.) *Multimodal Human-Computer Communication, Systems, Techniques, and Evaluation. Lecture Notes in Artificial Intelligence 1374*. Springer Verlag (1998) 13-38

4.   McGlashan, S.; Towards Multimodal Dialogue Management. In *Proceedings of Twente Workshop on Language Technology 11*, Enschede, The Netherlands (1996)

5.   Grice, H. P.: Logic and Conversation (From William James Lectures, Harvard University, 1967). In P. Cole, and J. Morgan (Eds.) *Syntax and Sematics 3: Speech Acts*. New York: Academic Press (1975)

6.   Fais, L., Loken-Kim, K.-H., and Park, Y.-D.,:. Speaker's Responses to Requests for Repetition in a Multimodal Language Processing Environment. In H. Bunt, R.-J. Buen, and T. Borghuis (eds.) *Multimodal Human-Computer Communication, Systems, Techniques, and Evaluation. Lecture Notes in Artificial Intelligence 1374*. Springer Verlag (1998) 264-278

7.   McGlashan, S., Fraser, N. M., Gilbert, G. N., Bilange, E., Heisterkamp, P., and Youd, N. J.: Dialogue Management for Telephone Information Systems. In *Proceedings of the International Conference on Applied Language Processing*, Trento, Italy (1992)

8.   Clark, H. H., and Schaefer, E. F.: Contribution to Discourse. *Cognitive Science*. Vol. 13, (1989) 259-294

9.   Brennan, S. E. and Hulteen, E. A.: Interaction and feedback in a spoken language system. In *AAAI-93 Fall Symposium on Human-Computer Collaboration: Reconciling Theory, Syntesizing Practice. AAAI Technical Report FS-93-05* (1993) 1–5

10. Pérez-Quiñones, M. A., and Sibert, J. L.: A Collaborative Model of Feedback in Human-Computer Interaction. In *Proceeding of Conference on Human Factors in Computing, (CHI 96)*, Vancouver, Canada (1996) 316-323

11. Jönsson, A. and Dahlbäck, N.: Talking to a Computer is not Like Talking to Your Best Friend. In *Proceedings of The first Scandinivian Conference on Artificial Intelligence*, Tromsø, Norway (1988)

12. Cheepen, C., and Monaghan, J.: Designing for Naturalness in Automated Dialogues. In Y. Wilks (Ed.) *Machine Conversation*. Boston: Kluwer Academic Publishers (1999) 127-142

13. William, D. and Cheepen, C.: "Just speak naturally": Designing for naturalness in automated spoken dialogues. In *Proceedings of Conference on Human Factors in Computing Systems (CHI'98)*, Los Angeles (1998) 243–244

14. Chu-Carroll, J., and Nickerson, J. S.: Evaluating Automatic Dialogue Strategy Adaptation for a Spoken Dialogue System. In *Proceedings of 1st Meeting of the North American Chapter of the Association of Computational Linguistics*, Seattle, WA (2000) 202-209

15. Dahlbäck, N., Flycht-Eriksson, A., Jönsson, A, and Qvarfordt, P.: An Architecture for Multi-Modal Natural Dialogue Systems. In *Proceedings of ESCA Tutorial and Research Workshop (ETRW) on Interactive Dialogue in Multi-Modal Systems*, Germany (1999) 53-56

16. Qvarfordt, P., and Jönsson, A.: Effects on Using Speech in Timetable Information System for the WWW. In *Proceedings of the ICSLP'98*, Sydney, Australia (1998) 1635-1638

17. Bunt, H., Ahn, R., Beun, R.-J., Borghuis, T., and van Overveld, K.: Multimodal Cooperation with the DENK system. In H. Bunt, R.-J. Buen, and T. Borghuis (Eds.) *Multimodal Human-Computer Communication, Systems, Techniques, and Evaluation. Lecture Notes in Artificial Intelligence 1374*. Springer Verlag (1998) 39-67

18. Sidner, C. L., Boettner, C., and Rich, C.: Lessons Learned in Building Spoken Language Collaborative Interface Agents. In *Proceedings of ANLP/NNAACLP 2000 Workshop on Conversational Systems*, Seattle, WA (2000) 1-6

19. Hayes, P. J. and Reddy, D. R.: Steps toward graceful interaction in spoken and written man-machine communication. *International Journal of Man-Machine Studies* 19 (1983) 231– 284

20. Flycht-Eriksson, A.: A Domain Knowledge Manager for Dialogue Systems. In *Proceedings of European Conference on Artificial Intelligence (ECAI'00)*. Berlin: Germany (2000)

A Pragmatic Semantic Reliable Multicast Architecture for Distance Learning

TAN Kun   SHI Yuanchun   XU Guangyou

Dept. Of Computer Science and Technology of Tsinghua University
Beijing 100084, P.R. of China
tank@sist.tsinghua.edu.cn

Abstract. Although IP multicast has been rapidly evolving in the last few years, the current Internet is not fully multicast-enabled because of the lack of robust inter-domain routing and controlling protocols. In this paper, we present an alternative architecture for data distribution on Internet called PSRM that realizes semantic reliable, loose synchronized, multicast data delivery. PSRM is not based on global IP multicast, instead it uses a hybrid way to combine unicast delivery with multicast delivery. In PSRM, IP multicast is used only in local area, or multicast domain, to enhance the performance, and these multicast domains are organized into a spanning tree by TCP connections. As PSRM is based upon ALF protocol architecture, we use application-defined semantics to adapt content in a heterogeneous environment. PSRM have been implemented and demonstrated by a prototype distance learning application.

1   Introduction

In recent years, distance learning has increasingly become one of the most important applications on Internet and is being discussed and studied by various universities, institutes and companies. Comparing with other forms of distance learning, one more attractive way is to provide live lectures via Internet. Teachers and students, distributed on Internet, can take part in a virtual classroom and hold a lecture. A blackboard is shared among all participants in the virtual classroom on which teaching materials and dynamic annotations can be placed. The information on the blackboard, called "*blackboard events*", shall be sent to each attendee reliably, orderly and synchronously. This requires a reliable data transmission from a sender to multiple receivers.

A number of reliable multicast protocols have been proposed up-to-now [2][3][4]. The basic idea of most of these protocols is to use the IP Multicast infrastructure [1] for routing, and adds additional functionality at end host to support reliability. However, they are fundamentally challenged by the instinct heterogeneity of the Internet. That is, the characteristics of receivers have varied in a large range. Many previous work have addressed this problem [5][6], but most of them handle with stream media such as video/audio where loss is bearable to some extent, while seldom consider the scenario in reliable multicast. [7] provides a solution for reliable multicast transcoding, but it works rather as a proxy server for end PDA users and

T. Tan, Y. Shi, and W. Gao (Eds.): ICMI 2000, LNCS 1948, pp. 658-665, 2000.

provides no architecture for reliable multicast. In addition, the deployment of IP Multicast over the wide-area is not comprehensively realized. There are two major reasons: one is the lack of robust inter-domain multicast routing and controlling protocols, and the other is the hard address allocation of limited multicast addresses.

In this paper, we introduce our comprehensive work on an alternative architecture that does not rely on a global multicast-enabled network, called pragmatic semantic reliable multicast (PSRM) architecture. In our approach, we take IP Multicast only as a technology that can be available in some local environments (e.g. LAN or small administrative scope) and exploited to improve the performance. We provide a hybrid way that combines unicast and multicast in single distribution architecture. There are three key points in the architecture. Firstly, we limit IP multicast in local area, or multicast domain, and connect each multicast domain by unicast protocols such as TCP. Thus, we can create a multicast session among different domains even if routers between domains are not multicast-enabled. Secondly, we avoid the troublesome session allocation and discovery problem of IP multicast in large scope, and provide a more simple and natural way to naming and locating the sessions (See section 3.1). Finally, transcoding can be performed before forwarding data from one domain to another domain. Thus, it provides a proxy-based way to cope with the heterogeneity problem (See section 3.3).

The remainder of this paper is organized as following. The next section discusses some issues in supporting reliable multicast and presents our design choices. Section 3 presents the PSRM architecture in details. In section 4, we describe the current implementation status on PSRM. Finally, after the discussion of some related work, we summarize this paper.

2   Issues in Supporting Reliable Multicast

In this section, we discuss two key issues in supporting reliable multicast: the semantics of reliable multicast and the application level framing protocol architecture.

2.1   Semantics of Reliable Multicast

Unlike unicast communication, there is no consensus in the semantics of reliable multicast. In different applications, the characteristics of underlying reliable multicast can be classified into three categories: 1) *Strict reliability with synchronization*. It requires the *same data* should be sent from source to all receivers and the data reception is loosely synchronized among the receivers. That is, all receivers should get the data almost in same time. 2) *Partial reliability with synchronization*. In this case, data received by each attendee can have slightly difference. In a heterogeneous session, slow receivers intend to receive smaller data in order to catch up the speed of the whole session, while fast receivers pursue for high quality. 3) *Strict reliability with no synchronization*. This suits for those applications that can bear long time waiting but are sensitive to data loss, e.g. file distribution.

For live blackboard for distance learning, we have chosen the second reliable semantics. We support only *partial reliability*, or *application-defined semantics reliability* (as introduced in [7]). That is, after transcoding, we shall keep the data information unchanged, but the data itself may be changed to other forms.

2.2 Application Level Framing

Up-to-now, researchers in multicast society have a common idea that there is no *one-fits-all* protocol that can optimally serve the needs of all types of multicast applications. Application Level Framing has been proposed to help adapt transport-level services to the needs of specific application by reflecting applications' semantics into the network protocol. Instead of using packages, ALF-based protocol takes *Application Data Units* as the fundamental communication units. This is important for PSRM to base itself upon ALF architecture, and leverage the application-defined semantics encapsulated in ADUs to support partial reliability.

3 Definition of PSRM

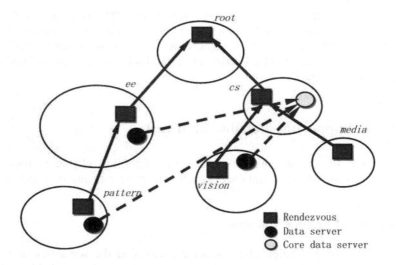

Fig. 1. The overview of PSRM

In the current Internet world, IP Multicast is available only in some local environments like LAN or small scale Intranet with carefully administrated bounders. We take these bounded networks as *multicast domains*. Our goal is to tunnel the domains with well-understood unicast protocols such as TCP so that "Multicast"[1] can be achieved in a global "dial tone". We take a multicast domain as *homogeneous*, because in a local network, all clients are likely to bear the similar network conditions. Thus, we consider only the difference between domains and the connections that "tunnel" domains up. In each domain, there is a designated server

---

[1] From now, we take the word *multicast* only as *distribution of data in one-to-many mode*.

named *rendezvous server*. A rendezvous server is a manager of the multicast session. Relying on it, hosts can create, bootstrap, and eliminate sessions.

In this section, we shall detail PSRM architecture. The following subsection describes session control scheme. In subsection 3.2, we detail the data communication model. And in the last subsection, we discuss some issues on transcoding.

3.1 Session Control

Addressing and Naming. Each session in PSRM is identified a pair of <*domain name: session name*>. The *domain name* is a human-readable name of the multicast domain which the session creator is in. The structure of the *domain name* is based on the "dotted" notation of the Domain Name System. We organize rendezvous servers into a tree, select an arbitrary server as *root*, and give each rendezvous server a name consists of several parts joined by the period. We take a multicast domain name as its rendezvous server's name. A *session name* is any string that the creator assigns to name the session. The session name does not need to be unique in global, but should be unique in the domain. For examples, *media.cs.root:mediacourse* and *security.cs.root:data-encrypt* are all legal names to PSRM sessions.

Hosts can create as many PSRM sessions as they like individually without coordination with hosts outside the domain.

Session Maintenance. To initialize a session, a host must contact its rendezvous to register a session name[2]. The rendezvous then allocates a multicast address for the session, and report the (*address, port*) to the creator. After that, the rendezvous creates a *data server*[3] and assigns the IP of *data server* to the session name. This *data server* acts as a *core data server* in the session. The creator can use the multicast address and the port to contact the *data server*.

To join a session, a host should know the *domain-name:session-name* of the session. Firstly, the host sends a request-to-join to the local rendezvous. Then, the rendezvous resolves the session identity and finds the *core server*. After that, a new *data server* is created by the rendezvous and connected to the *core server* with TCP[4]. At last, the host gets a reply from the rendezvous with a local-allocated (address, port) pair that can be used to contact local *data server*.

A session is closed and, therefore, unregistered, if the *core server* is shutdown. A *core server* will be down, if any of the following condition is matched: 1) an authorized request-to-close message is performed. 2) if the *core server* is found to be alone. That is, no *data server* or end host is connected to it. 3) if the *core server* has been idle for a specific long time.

---

[2] How a host can find its rendezvous is beyond this paper.

[3] The data server can be a process in the same machine of rendezvous or in another machine if cluster is used.

[4] If there has existed a data server associates to the session, the address and port of it are reported to the host. And if the core server has too many branches, a new core will be selected. However, this is far beyond this paper.

3.2 Data Transmission and Recovery

After contacting local data server, any host can send data to the domain and the data sever forwards the data to other multicast domains. The data delivery in PSRM is quite straightforward. When source sends out data to local multicast domain, the data server receives and forwards it along all its TCP links to other data servers. When receiving data from another linked data server, a data server multicast the data to its domain, and forward them to all linked data server other than the incoming one.

Although TCP links between data servers provide link-reliable among domains, data may get lost within a multicast domain. In order to provide local data recovery, every data server should cache all the session data, and reply for any request for lost data from its local hosts[5].

3.3 Application-Defined Transcoding

PSRM is based on ALF protocol architecture. That means content transmitted in PSRM architecture are in the units of application layer frames or ADUs. PSRM data servers can leverage the application-define semantics embedded in the ADUs to perform transcoding while keeping the information that data carry unchanged. The transcoding progress can be performed in two dimensions: in fidelity or in modality, as described in [8].

Data servers can automatically decide the proper transcoding methods according to the dynamic measurements of network links and the predefined configuration. These measurements can be RTT or the throughput of the TCP connection, or other end-to-end measurements that can find congestion or detect available bandwidth more precisely. User's interests can also be reflected into transcoding policy by assigning various priorities to different content. The data with higher priority have privilege to allocate bandwidth ahead of the less urgent data.

4 Implementing PSRM

We have instantiated the PSRM architecture in windows platform and tested it in our lab. networking environment. PSRM has been implemented in two parts: (a) rendezvous server, and (b) data server. Although rendezvous server and data server can run on different machines, in our current implementation, a rendezvous server creates data servers as a separate process in the same machine for simplicity.

The rendezvous server maintains the domain name, propagates, updates and caches the domain name information as DNS servers.

The data server can be depicted as fig 2. The multicast agent is the interface to the multicast domain, while the TCP links agent maintains the TCP connections to other data servers. The monitoring agent monitors the network condition of the TCP link and reports the critical parameters to the control center. According to the monitoring

---

[5] It seems quite costly if every data server has to cache all data, but in practice, it is necessary to do so especially the connection between data servers has narrow bandwidth.

reports and application-specific policy, the transformation agent transforms incoming data to proper form and then forwards the converted data to other data servers. All data received have been cached in each data server for local recovery.

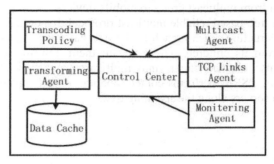

Fig. 2. Data sever structure

A demonstration based on the PSRM architecture, named ARTEMIS [9] has been shown. It implements a blackboard on which HTML documents and dynamic annotations can be placed. Hosts on two LANs (each is a multicast domain) can take part in a session. To demonstrate the transcoding feature, a HP Jornada PDA is dialed into the network and forms another domain that contains only one host and one rendezvous/data server.

The transcoding policy has been configured as following: a) text has higher priority than images; b) HTML is translated to plain text before sending to PDA domain; c) images are compressed and convert to jpeg format before sending to PDA domain.

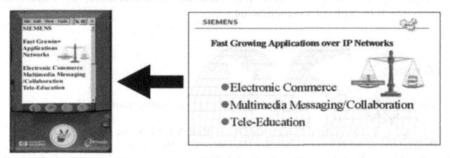

Fig. 3. ARTEMIS: Transcoding has been performed before forwarding data to PDA domain

5 Related Work

The notion of proxy is not new. The first proxy mechanism is proposed with HTTP. Proxies have been used in many applications, such as security firewall, data caching and pre-fetching, etc. Proxies are also implemented as filtering agents to improve the performance of current Web services [10]. However, beyond just filters, proxies can be designed to support much complicated work such as transcoding. IBM Co. has built a proxy-based information service, *IBM transcoding service* [11] [13], to provide proper information according to clients' capability. Similar services are also built by Intel [16] and Spyglass [17]. A. Fox [14] builds a proxy that "distillates" the images in real-time. In multicast, [5] is a proxy dealing with video/audio and [6] is a proxy that transfers data from multicast group to mobile devices.

RMTP [2] is a tree-based reliable multicast protocol that organizes members into a hierarchical tree structure. Each branch in the tree has a designated receiver (DR) to receive ACKs from its children and aggregate them upwards to the sender. Like RMTP, PSRM also clusters clients into multicast groups. However, besides just transferring ACKs, PSRM performs content adaptation and forwards the transcoded data to other data servers along the spanning tree.

6 Summary

In this paper, we present an alternative architecture for data distribution on Internet, called Pragmatic Semantic Reliable Multicast. PSRM realizes partial reliable (or semantic reliable), loose synchronized, multicast data delivery. PSRM is not based on global IP multicast, instead it uses a hybrid way to combine unicast delivery with multicast delivery. In PSRM, IP multicast is used only in local area, or multicast domain, to enhance the performance, and these multicast domains are organized into a spanning tree by TCP connections. As PSRM is based upon ALF protocol architecture, we can use application-defined semantics to transcode content before forwarding it to other domains. Thus, it provides a proxy-based way to cope with the heterogeneity problem. We have presented an implementation of PSRM and demonstrate it by a prototype application.

References

1. Stephen E. Deering, "Multicast Routing in a Datagram Internetwork", Ph.D. thesis, Stanford University, Dec. 1991.
2. John C. Lin and Sanjoy Paul, "RMTP: A Reliable Multicast Transport Protocol", in Proceedings of IEEE Infocom'96, San Francisco. CA, Mar. 1996.
3. Sally Floyd, Van Jacobson, et. al., "A Reliable Multicast Framework for Light-weight Sessions and Application Level Framing", in Proceedings of ACM SIGCOMM'95, Boston, MA, Aug. 1995.
4. Tony Speakman, Dino Farinacci, et. al., "PGM Reliable Transport Protocol Specification", Internet-draft, June 1999
5. Elan Amir, Steven McCanne and Hui Zhang, "An Application Level Video Gateway", in Proceedings of ACM Multimedia, 1995
6. Steven McCanne, Van Jacobson and M. Vetterli, "Receiver-driven Layered Multicast", in Proceedings of ACM SIGComm, 1996
7. Yatin Chawathe, Steve A. Fink, Steven McCanne, and Eric A. Brewer, "A Proxy Architecture for Reliable Multicast in Heterogeneous Environments", in Proceedings of ACM Multimedia, 1998

8. John R. Smith, Rakesh Mohan and Chung-Sheng Li, "Transcoding Internet Content for Heterogeneous Client Devices", in Proceedings of IEEE ISCAS, May, 1998
9. Tan Kun, Liou Shih-Ping, et. al., Artemis: Providing Live Course for Distant Learning, in Proceedings of Fourth International Workshop on CSCW in Design 1999
10. Edith Cohen, Balachander Krishnamurthy, et. al., Improving End-to-End Performance of the Web Using Server Volumes and Proxy Filters, ACM Comm. 98
11. IBM transcoding solution and service-Extending the reach and exploiting the value of data, Special Report, IBM Co.
12. Richard Han, Pravin Bhagwat, et. al., Dynamic Adaptation in an Image Transcoding Proxy for Mobile Web Browsing, IEEE Personal Communications Magazine, 1998
13. Internet Transcoding for Universal Access, IBM Co., available at: http://www.research .ibm. com/networked_data_systems/transocding/
14. A. Fox and E. A. Brewer, Reducing WWW Latency and Bandwidth Requirements by Real-Time Distillation, In: Fifth International World Wide Web Conference, May 6-10, 1996, Paris, France
15. R. Mohan, J. R. Smith, et al., Adapting Multimedia Internet Content for Universal Access, IEEE Trans. On Multimedia, Vol. 1, No. 1, Mar 1999
16. Intel Quick Web. available at: http://www.intel.com/quickweb
17. Spyglass-Prism. available at: http://www.spyglass/products/prism

A Framework for Supporting Multimodal Conversational Characters in a Multi-agent System

Yasmine Arafa, and Abe Mamdani

Intelligent and Interactive Systems Section, Department EEE, Imperial College of Science, Technology and Medicine, Exhibition Road, London SW7 2BT, +44 (0) 207 594 6319, y.arafa@ic.ac.uk

ABSTRACT. This paper discusses the computational framework for enabling multimodal conversational interface agents embodied in lifelike characters within a multi-agent environment. It is generally argued that one of the problems with such interface characters today is their inability to respond believably or adequately to the context of an interaction and the surrounding environment. Affective behaviour is used to better express responses to interaction context and provide more believable visual expressive responses. We describe an operational approach to enabling the computational perception required for the automated generation of affective behaviour through inter-agent communication in multi-agent real-time environments. The research is investigating the potential of extending current agent communication languages so as they not only convey the semantic content of knowledge exchange but also they can communicate affective attitudes about the shared knowledge. Providing a necessary component of the framework required for autonomous agent development with which we may bridge the gap between current research in psychological theory and practical implementation of social multi-agent systems.

Keywords. Multimodal Conversational Agents; Personal Service Assistants; Human-Computer Interface; Multi-agent Systems.

1. INTRODUCTION

Multimodal conversational agents are autonomous Interface Agents that employ intelligence and adaptive reasoning methods to provide active, autonomous and collaborative services to users of a virtual environment [9, 16]. Interface agents differ from customary interfaces in that they are expected to change behaviour and actions autonomously according to users' actions and the surrounding system environment as an interaction progresses. The Conversational Agent (CA) metaphor aims towards providing effective highly personalised services. Personifying the CA with a context generated lifelike character is a visual dimension to providing personalised services. The motivation for this type of personalisation is that an animated figure, eliciting quasi-human capabilities, may add an expressive dimension to the agent's conversational features, which can add to the effectiveness and personalisation of the interface and the virtual interactive experience on the whole. Furthermore, there is strong evidence that Affect has major influence on learning and recall [4], reasoning and decision making [7], both effecting system usability and efficiency.

Particularly important capabilities of an agent that must interact with the human user is *Believability,* in terms of how the agents behaves and expresses itself and that these expressions are appropriate to the context of the interaction [3]. *Affect* has been proven to be important in enhancing the expressive or visual believability of an agent [3]. Affective computing is "computing that relates to, arises from or deliberately influences emotions" [15]. For the agent to effectively achieve believable lifelike behaviour it must have the appropriate knowledge to handle and reason about *affect* so as to produce the believable response. Since the agent works in a multi-agent system (MAS) environment this knowledge is maintained and exchanged through agent communication. In a real-time multi-agent environment the CA inhabits a world which is dynamic and unpredictable. To be autonomous, it must be able to perceive its environment and decide its actions to reach the goals defined by its behavioural models. To

T. Tan, Y. Shi, and W. Gao (Eds.): ICMI 2000, LNCS 1948, pp. 666-673, 2000.

visually represent the behaviour, the relevant actions and behaviour must be transformed into visual motional actions. Therefore the design of an animated conversational agent system requires components to endow them with perception, behaviour processing and generation, action selection, and behaviour interpretation into believable graphical representation. In this paper we discuss issues related to the perception required to support a multimodal conversational agent.

2. PERCEPTION THROUGH AGENT COMMUNICATION

In order for the conversational agent to select the appropriate actions, the behavioural system needs to be aware and able to perceive the state of the surrounding environment. This environment includes the user, other service agents, processes and events. Perception of events are decomposed into different classes: expected or desired events, events occurring in the surrounding world, and potential anticipated events which may or may not occur depending on the progress of a situation. We consider how a communication language can provide a structure that conveys this perception.

Communication is the intentional exchange of information brought about by the production and perception of clues drawn from a shared system of conversational primitives [1]. Agent communication languages like KQML [6], and FIPA ACL [7] define a set of general primitives corresponding to speech acts (or communicative acts) [10], so known as performatives. These performatives define the protocols of dialogue between agents. They define how communication is conducted through defined protocols for a communication dialogue, rather than what is actually being communicated. Whereas, What is the core message which is included in the content part of a communicative act.

Inter-agent communication is the means by which conversation is mediated between an agent and the agent society wherein it is situated. We use this communication to acquire the information required for agent's affective perception on both the how and what dimensions. We consider the development of CA perception as a process of two separate stages:

- **inter-agent** interaction between the various entities within a MAS society. We further consider three levels of inter-agent communication at which affect may be conveyed:
  - **content level:** referring to the actual raw message or object to be communicated among entities;
  - **intentional level:** expressing the intentions of agent communicative acts (performatives of an agent communication language); and
  - **conversational level:** protocols that govern the conversations shared between agents when exchanging dialogue,
- **CA affect model:** dealing with the agents' inner behaviour (knowledge representation, reasoning, learning, etc.), the agents social and affective behaviour, and the generation of appropriate behaviour states that are transformed into scripts for visual embodiment in the interface.

Accordingly we make the distinction between the affect indicators that may be conveyed at the intentional level, which expresses semantics of communicative act performatives, and at the content level, which expresses facts about the domain. Such layering facilitates the successful integration of the language to applications while providing a conceptual framework for its understanding. Although current primitives could be extended to distinctively convey an affective message, the existing primitives capture many of our intuitions about what constitutes affect from the communicative act irrespective of application. We consider that semantic description could provide a model of affect that is useful for modelling the overall behaviour.

3. AGENT COMMUNICATIVE ACTS

We use the performatives of agent communication language in terms of their communicative functions rather than being just a communication medium, and we provide a representation of the meaning in terms of how they can be used for affect modelling. This representation forms the base upon which we define a set of inference rules describing the affect process ongoing in agent conversation, and also has implications on how a receiver agent will interpret a conveyed message. These rules take into account the conversation instigator's conception of the context at hand, including the experience, type and degree of relation between both communicants and a model of the receiver 's personality.

We consider affect in agent communication at the Intentional level to define the intentional purpose of a performative. Based on the work by Conte and Castelfranchi [5] a communicative act models a set of goals and beliefs that agents maintain when in conversation. At this level the intentional meaning is represented in terms of propositions that are declarative representations of semantic primitives conveying information about communication intention and from which we may induce some kind of attitude and/or emotions. Pelachaud and Poggi interpret the meaning of a communicative act on two dimensions [14] (in their work on analysis communicational signals to generate facial expressions): Pragmatic Interpretation as the general objective expected in terms of a physical action, and Semantic Interpretation as prepositional content. In the context of FIPA's agent communication language ACL [7], we take several communicative act primitives and consider how an agent may interpret the intended message in terms of both Pragmatic and Semantic content proposition interpretations. We take for example: primitives like Request, and Inform, and composites[1] like Agree or Refuse, to show pragmatic interpretation:

* **Inform:** A_s goal is that the recipient A_r believe the content message conveyed.
  $<A_s, inform(A_r, m)>$
     where m is the belief conveyed.

* **Agree:** A_s goal is to convey a general purpose agreement to a previously submitted request to perform some action. Agree is a composite communicative act that A_s (according to ACL protocol) informs the receiver A_r of provisional agreement to perform the action, pending some given precondition is true.
  $<A_s, agree(A_r, <A_s, act>, \phi))> \equiv$
      $< A_s, inform(A_r I_{As} Done(<A_s, act>, \phi))>$
     Where ϕ is the object that needs to be agreed.

We must now narrow down the semantic interpretation to induce some affect. For example: the *agree* primitive infers a behaviour of positive nature since both parties are agreeing. Similarly, the *refuse* primitive infers a behaviour of negative nature since something is being rejected. However, in real-time agent conversation communicative acts are not standalone and cannot alone determine behavioural inferences because they are uttered within particular context. What may have a positive impact on one individual may have neutral or yet negative impact on another. Moreover, a primitive like *inform* does not convey any affect. Nevertheless, these interpretations serve as the innate knowledge an agent may start with which can be further focused to the context at hand through the communicative act content. We use the prepositional content to provide a more context sensitive interpretation. In our system we present information that specifies context by using meta-data description defined by a Content Object Description Language.

---

[1] Primitive Communicative Acts (CA): are those whose actions are defined atomically, i.e. they are not defined in terms of other acts. Composite CAs are the converse, where one AC is the object of another [7].

4. CONTENT LANGUAGE

The content language described, here, will provide the specific more context-oriented information. We use meta-level representation to convey more specific semantics of the data being exchanged. For the implemented components to be fully converged and effective in a real-time MAS, the agents must have some knowledge about the surrounding environment which includes affective knowledge about internal and external states. The agents must have a semantic and contextual understanding of the information being exchanged. This requires a theoretical framework for representing knowledge and belief of agents interacting within an agent society. This includes frameworks for representing uncertain knowledge about the surrounding environment evolving with experience and time, awareness of the implication of time constraints, and context-based behaviour. For this purpose we use the notion of meta-level knowledge representation which are annotations of data[2] [12], being manipulated between the agents in a MAS. This provides an understanding of the content being handled and hence provides better awareness of the environment. The abstractions hold physical and conceptual meaning as well as affective states. Offering a structured way of co-ordinating content and domain data for agents without the agents having to understand the raw content; as well as an automated process for returning information, and feedback regarding the state of the environment to agents in a language the agents can understand [2].

By including meta-representation of the data in the content part of the communication act, we provide an added dimension to agent comprehension of the actual meaning within the content exchanged. The content object not only serves as data that agents manipulate but also serve as a basis for defining a social framework. Since any information can contain vast amounts of embedded knowledge, content object descriptions can therefore have potential to convey associated and inferred perception. Adding social attributes that can collectively serve as choices that influence agents' perception about the underlying semantics of the content and attributes to it's behavioural traits, provides an improvisational basis that serves as a key aspect in determining a CA's character.

These content object descriptions essentially ground five important aspects of information sharing between agents and entities [12]:

- **Affect model**: defines emotional and personality attributes;
- **Domain model**:   defines how the content is manipulated by the particular service within a given domain;
- **Task model**: defines the tasks the content object may initiate and maintain;
- **Action model**:defines the sequence of actions that should be executed when the object is interacted with;
- **Media content**:   is intended, only, for visual objects and defines the physical properties, types and values of objects to be rendered at a presentation system.

The affect model includes indicators to emotions expected when a defined content object is being handled in a normal or non-context sensitive way. These indicators are associated, using a set of selection rules, with a predefined intended behaviour. This planned behaviour is then modified to incorporate context-sensitive variables to select the appropriate affect-based behaviour. The resulting behaviour is then mapped to a visual (animated) representation. Variations of animated expression (or intonation) can be realised by altering and adjusting the intensity of the behaviour according to the immediate situation.

---

[2] Meta-representation of content objects was initiated in the EC ACTS project KIMSAC AC030.

Content object Description Model and Structure: Content objects are intended for the exchange of knowledge during agent communication as well as to facilitate for content reuse. The contextual semantic, visualisation entities, as well as the relationships among objects are what explicitly model the objects. The objects are abstracted encapsulations of this information, providing a means by which an agent can interpret the underlying semantics of the object being used and by a presentation system to understand the mechanisms and services required for rendering the multimedia objects. Hence, content objects can be objects of multimedia, services, affective feedback or any other type of data.

Object type   specifies the multimedia type, service,

Content-based   describes the object content which will contribute to its functional semantics,

Non-content-based (includes conceptual & physical properties)

Conceptual properties include:

affect indicators: indicators to an expected emotion the content may convey

domain   subset of the ontology

tasks   denoting actions or functional roles

scripts   to represent a sequence of instructions

Physical properties, attributes that aid in the physical representation of visual objects

Content object Description Structure

The content object data model addresses aspects related to data presentation, manipulation, semantic and conceptual properties, and content-based retrieval. It provides a set of requirements based on which the perception mechanism can function. The content object model is to contain many content object descriptions, which will allow the selection of content objects for potentially any purpose. The purpose of this model is to dynamically generate objects for the knowledge exchange to render or for use with other applications that require objects of similar characteristics. The importance of the flexibility of the model is not only to facilitate content object reuse within one application but moreover to allow other applications to also reuse these dynamically generated content objects. The base Content object model consists of:

- Object Description, providing an annotated view of raw data;
- Visual Description, aiding a presentation service in describing the temporal and spatial relationships among different objects for rendering objects;
- Content Semantics, giving a semantic view of the stored objects;
- Service Description, instantiated to an application domain;
- Initial Affect Indicators that a content object will convey in a non-context sensitive environment. These may change drastically according to the past and immediate states of the agent within an MAS and the state of the environment.

These initial indicators alone cannot determine behavioural reactions. The impact would depend on past experiences, the current mental state and the environment. So the actual influence of such indicators will be determined by these variable factors that are governed by the context in which the content is being conveyed. Nevertheless, these indicators serve as the innate knowledge an agent may start with, and based on which may influence its behaviour. If we are effectively able to use such variables in defining the context, we may effectively determine the attitude form for the visual realisation of a CA's character.

5. POSITIONING WITHIN THE OVERALL ARCHITECTURE

We have discussed input to the perception module from an agents' communication language point of view. We now brief on how affect is modelled within the overall architecture. The system is composed of three modules:
the Head, which deals with perception, continuous memory, reasoning, and behaviour selection and generation; the Heart, which maintains and manipulates the affect models of emotion and personality; and the body, which deals with behaviour action execution and visual representation. The system architecture is delineated in *figure 1*.

The perception system provides the state of the world to the behavioural and emotional modules through agent communication and content object descriptions. When content objects are fed into the perception system it is unwrapped to extract the initial indicators and feed into the behaviour system. The behaviour system then uses this information, along with information of past experiences and memory to select the appropriate behavioural response. The resulting behaviour is fed into the action module to generate the appropriate script for animated visual representation.

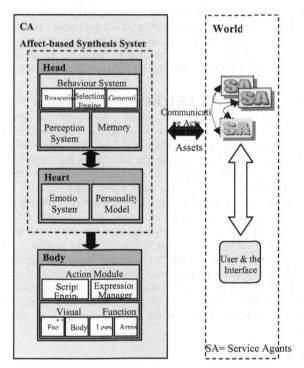

Figure 1 - System Overview

We use emotion to describe short-term variations in internal mental states, describing focused meaning pertaining to specific incidences or situations. The emotional model is based on the description of emotions made by Ortony *et. al.* [13]. We view emotion as brief short termed, and focused with respect to a particular matter. We assume that a character need not exhibit all the attributes of the emotion or personality definition to be a successful affect model. It needs to incorporate at least some of the more basic features of affect.

To realise an Emotion model, we focus on the Arousal -Valence space of emotional response dimensions [8]. Where Arousal denotes the intensity of emotion and ranges from negative, neutral, to positive; and Valence denotes the basic types of emotional response and ranges from negative (sad), neutral, to positive (happy).

We use personality to characterise long-term patterns of emotion, and behaviour associated with the synthetic character. Personality is the general behaviour characteristics that do not arise from and are not pertaining to any particular matter. We model the broad qualities that include individually distinctive, and consistently enduring yet subject to influence and change. Psychologists have characterised five basic dimensions of personality, so known as, the Five Factor model or Big Five [11] of independent traits.

To realise a Personality model, we use two of these traits which we deemed appropriate for a CA's synthetic character in that they are easily distinguishable and have crucial influence on behaviour. We use Extraversion and Agreeableness to maintain a two-dimensional personality profile of the character. Extraversion, so as the CA may openly express its emotion to the world, and a tendency to Agreeableness because being a personal assistant by nature of function must agree to autonomously perform services for, and on behalf the user. These personality factors influence the choice of communicative act performatives and can be used to infer some elements of personality from agent dialogue. Our system provides a range of personality profiles which users may choose from at first encounter: Dominant, Submissive, Friendly, and Active. Future work will provide combinations and degree variations.

6. SUMMARY AND CONCLUSIONS

For the implemented components, required to support a conversational agent, to be fully converged and effective in a multi-agent system the agents must have some knowledge about the surrounding environment which would include affective knowledge about internal and external states. The agents must have a semantic and contextual understanding of the information being exchanged. This requires a theoretical framework for representing knowledge and belief of agents interacting with other agents. This includes frameworks for representing uncertain knowledge about the surrounding environment evolving with experience and time, awareness of the implication of time constraints, and context-based behaviour. For this purpose we use the notion of meta-level knowledge representation, which are annotations of objects being manipulated between the agents in a multi-agent system. This provides an understanding of the content being handled and hence a better awareness of the environment. These abstractions hold physical and conceptual meaning and also include affective states. This offers a structured way of co-ordinating content and domain data for agents without agents having to understand the raw content; as well as an automated process for returning information, and changes regarding the state of the environment to agents in a language the agents can understand.

7. ACKNOWLEDGEMENT

The work supported by a grant from the EU ESPRIT projects funding for the MAPPA project, under grant number EP28831.

8. REFERENCES

[1]   AI A Modern Approach, R. Russell.c

[2]   Arafa, Y., & Mamdani, E. Real-time Chracters with Artificial Hearts, *IEEE Conf. On Systems, Man and Cybernetics, pp* 456-564, 1999.

[3]   Bates, J. The Role of Emotion in Believable Agents. In Communication of the ACM (37)7:122-125, '94.

[4]   Bower, G.H., & Cohen P.R. Emotional Influences in Memory & Thinking:Data & Theory. Clack. & Fiske Eds, Affect & Cognition, pp 291-233: Lawrence Erlbaum Association Publishers, 1982.

[5]   Conte, R. & Castelfranchi, C., Cognitive and Social Action. UCL Press Limited, UC L, 1995.

[6]   Finin, T.& Fritzson, R.; KQML: A Language & Protocol for Knowledge & Information Exchange, *Proc. 19th Intl. DAI Workshop*, pp 127–136,1994.

[7]   FIPA-ACL99 http://www.fipa.org/spec/fipa99spec.htm

[8]   Lang, P. The Emotion Probe: Studies of Motivation & Attention,American Psychologist 50-5:372-385 '95.

[9]   Maes, P. 1994. Agents that Reduce the Work Overload. Communications of ACM 37(7):31-40.

[10] Maybury, M. Communicative acts for generating natural language arguments. *Proc. 11th Conf. AAAI*, pp 357-364, 1993.

[11] McCrae, R. and Costa P. T. The structure of Interpersonal Traits: Wiggin's Circumplex & the Five Factor Model" Journal of Personality & Social Psychology 56(5): 586-595, 1989.

[12] Mc Guigan, R., Delorme, P., Grimson, J., Charlton, P., Arafa, Y. 1998. The Reuse of Multimedia Objects by Software Agents in the Kimsac System. OOIS'98.

[13] Ortony, A., Clore, G., & Collins A. The cognitive Structure of Emotions. Cambridge Uni. Press. 1990.

[14] Pelachaud, C. & Poggi, E., Facial Performative in a Conversational System. In WECC'98, The First Workshop on Embodied Conversational Characters.

[15] Picard, R., '97. Affective Computing. The MIT Press.

[16] Wooldridge, M. and Jennings, N.. Intelligent Agents: Theory & Practice. Knowledge Engineering Review 10(2):115-152. 1995.

Author Index